Graduate Texts in Mathematics 108

Graduate Texts in Mathematics

1 TAKEUTI/ZARING. Introduction to Axiomatic Set Theory. 2nd ed.
2 OXTOBY. Measure and Category. 2nd ed.
3 SCHAEFFER. Topological Vector Spaces.
4 HILTON/STAMMBACH. A Course in Homological Algebra.
5 MACLANE. Categories for the Working Mathematician.
6 HUGHES/PIPER. Projective Planes.
7 SERRE. A Course in Arithmetic.
8 TAKEUTI/ZARING. Axiomatic Set Theory.
9 HUMPHREYS. Introduction to Lie Algebras and Representation Theory.
10 COHEN. A Course in Simple Homotopy Theory.
11 CONWAY. Functions of One Complex Variable. 2nd ed.
12 BEALS. Advanced Mathematical Analysis.
13 ANDERSON/FULLER. Rings and Categories of Modules.
14 GOLUBITSKY/GUILLEMIN. Stable Mappings and Their Singularities.
15 BERBERIAN. Lectures in Functional Analysis and Operator Theory.
16 WINTER. The Structure of Fields.
17 ROSENBLATT. Random Processes. 2nd ed.
18 HALMOS. Measure Theory.
19 HALMOS. A Hilbert Space Problem Book. 2nd ed., revised.
20 HUSEMOLLER. Fibre Bundles. 2nd ed.
21 HUMPHREYS. Linear Algebraic Groups.
22 BARNES/MACK. An Algebraic Introduction to Mathematical Logic.
23 GREUB. Linear Algebra. 4th ed.
24 HOLMES. Geometric Functional Analysis and its Applications.
25 HEWITT/STROMBERG. Real and Abstract Analysis.
26 MANES. Algebraic Theories.
27 KELLEY. General Topology.
28 ZARISKI/SAMUEL. Commutative Algebra. Vol. I.
29 ZARISKI/SAMUEL. Commutative Algebra. Vol. II.
30 JACOBSON. Lectures in Abstract Algebra I: Basic Concepts.
31 JACOBSON. Lectures in Abstract Algebra II: Linear Algebra.
32 JACOBSON. Lectures in Abstract Algebra III: Theory of Fields and Galois Theory.
33 HIRSCH. Differential Topology.
34 SPITZER. Principles of Random Walk. 2nd ed.
35 WERMER. Banach Algebras and Several Complex Variables. 2nd ed.
36 KELLEY/NAMIOKA et al. Linear Topological Spaces.
37 MONK. Mathematical Logic.
38 GRAUERT/FRITZSCHE. Several Complex Variables.
39 ARVESON. An Invitation to C^*-Algebras.
40 KEMENY/SNELL/KNAPP. Denumerable Markov Chains. 2nd ed.
41 APOSTOL. Modular Functions and Dirichlet Series in Number Theory.
42 SERRE. Linear Representations of Finite Groups.
43 GILLMAN/JERISON. Rings of Continuous Functions.
44 KENDIG. Elementary Algebraic Geometry.
45 LOÈVE. Probability Theory I. 4th ed.
46 LOÈVE. Probability Theory II. 4th ed.
47 MOISE. Geometric Topology in Dimensions 2 and 3.

continued after Index

R. Michael Range

Holomorphic Functions and Integral Representations in Several Complex Variables

With 7 Illustrations

Springer-Verlag
New York Berlin Heidelberg Tokyo

R. Michael Range
Department of Mathematics and Statistics
State University of New York at Albany
Albany, NY 12222
U.S.A.

AMS Classifications: 32-01, 32-02

Library of Congress Cataloging in Publication Data
Range, R. Michael.
Holomorphic functions and integral representations in several complex variables.
(Graduate texts in mathematics; 108)
Bibliography: p.
Includes index.
1. Holomorphic functions. 2. Integral representations. 3. Functions of several complex variables.
I. Title. II. Series.
QA331.R355 1986 515.9′8 85-30309

Typeset by Asco Trade Typesetting Ltd., Hong Kong.
Printed and bound by R.R. Donnelley and Sons, Harrisonburg, Virginia.
Printed in the United States of America.

9 8 7 6 5 4 3 2 1

ISBN 0-387-96259-X Springer-Verlag New York Berlin Heidelberg Tokyo
ISBN 3-540-96259-X Springer-Verlag Berlin Heidelberg New York Tokyo

To my family
SANDRINA,
OFELIA, MARISA, AND ROBERTO

Preface

The subject of this book is Complex Analysis in Several Variables. This text begins at an elementary level with standard local results, followed by a thorough discussion of the various fundamental concepts of "complex convexity" related to the remarkable extension properties of holomorphic functions in more than one variable. It then continues with a comprehensive introduction to integral representations, and concludes with complete proofs of substantial global results on domains of holomorphy and on strictly pseudoconvex domains in $\mathbb{C}^n$, including, for example, C. Fefferman's famous Mapping Theorem.

The most important new feature of this book is the systematic inclusion of many of the developments of the last 20 years which centered around integral representations and estimates for the Cauchy–Riemann equations. In particular, integral representations are the principal tool used to develop the global theory, in contrast to many earlier books on the subject which involved methods from commutative algebra and sheaf theory, and/or partial differential equations. I believe that this approach offers several advantages: (1) it uses the several variable version of tools familiar to the analyst in one complex variable, and therefore helps to bridge the often perceived gap between complex analysis in one and in several variables; (2) it leads quite directly to deep global results without introducing a lot of new machinery; and (3) concrete integral representations lend themselves to estimations, therefore opening the door to applications not accessible by the earlier methods.

The Contents and the opening paragraphs of each chapter will give the reader more detailed information about the material in this book.

A few historical comments might help to put matters in perspective. Already by the middle of the 19th century, B. Riemann had recognized that the description of all complex structures on a given compact surface involved complex *multidimensional* "moduli spaces." Before the end of the century, K. Weierstrass, H. Poincaré, and P. Cousin had laid the foundation of the local theory and generalized important global results about holomorphic functions from regions in the complex plane to product domains in $\mathbb{C}^2$ or in $\mathbb{C}^n$. In 1906, F. Hartogs discovered domains in $\mathbb{C}^2$ with the property that all functions holomorphic on it necessarily extend holomorphically to a strictly larger domain, and it rapidly became clear that an understanding of this new phenomenon—which does not appear in one complex variable—would be a central problem in multidimensional function theory. But in spite of major contributions by Hartogs, E.E. Levi, K. Reinhardt, S. Bergman, H. Behnke, H. Cartan, P. Thullen, A. Weil, and others, the principal global problems were still unsolved by the mid 1930s. Then K. Oka introduced some brilliant new ideas, and from 1936 to 1942 he systematically solved these problems one after the other. However, Oka's work had much more far-reaching implications. In 1940, H. Cartan began to investigate certain algebraic notions implicit in Oka's work, and in the years thereafter, he and Oka, independently, began to widen and deepen the algebraic foundations of the theory, building upon K. Weierstrass' Preparation Theorem. By the time the ideas of Cartan and Oka became widely known in the early 1950s, they had been reformulated by Cartan and J.P. Serre in the language of sheaves. During the 1950s and early 1960s, these new methods and tools were used with great success by Cartan, Serre, H. Grauert, R. Remmert, and many others in building the foundation for the general theory of "complex spaces," i.e., the appropriate higher dimensional analogues of Riemann surfaces. The phenomenal progress made in those years simply overshadowed the more constructive methods present in Oka's work up to 1942, and to the outsider, Several Complex Variables seemed to have become a new abstract theory which had little in common with classical complex analysis.

The solution of the ∂-Neumann problem by J.J. Kohn in 1963 and the publication in 1966 of L. Hörmander's book in which Several Complex Variables was presented from the point of view of the theory of partial differential equations, signaled the beginning of a reapproachment between Several Complex Variables and Analysis. Around 1968–69, G.M. Henkin and E. Ramirez—in his dissertation written under H. Grauert—introduced Cauchy-type integral formulas on strictly pseudoconvex domains. These formulas, and their application shortly thereafter by Grauert/Lieb and Henkin to solving the Cauchy–Riemann equations with supremum norm estimates, set the stage for the solution of "hard analysis" problems during the 1970s. At the same time, these developments led to a renewed and rapidly increasing interest in Several Complex Variables by analysts with widely differing backgrounds.

First plans to write a book on Several Complex Variables reflecting these latest developments originated in the late 1970s, but they took concrete form only in 1982 after it was discovered how to carry out relevant global constructions directly by means of integral representations, thus avoiding the need to introduce other tools at an early stage in the development of the theory. This emphasis on integral representations, however, does not at all mean that coherent analytic sheaves and methods from partial differential equations are no longer needed in Several Complex Variables. On the contrary, these methods are and will remain indispensable. Therefore, this book contains a long motivational discussion of the theory of coherent analytic sheaves as well as numerous references to other topics, including the theory of the $\bar{\partial}$-Neumann problem, in order to encourage the reader to deepen his or her knowledge of Several Complex Variables. On the other hand, the methods presented here allow a rather direct approach to substantial global results in $\mathbb{C}^n$ and to applications and problems at the present frontier of knowledge, which should be made accessible to the interested reader without requiring much additional technical baggage. Furthermore, the fact that integral representations have led to the solution of major problems which were previously inaccessible would suggest that these methods, too, have earned a lasting place in complex analysis in several variables.

In order to limit the size of this book, many important topics—for which fortunately excellent references are available—had to be omitted. In particular, the systematic development of global results is limited to regions in $\mathbb{C}^n$. Of course, Stein manifolds are introduced and mentioned in several places, but even though it is possible to extend the approach via integral representations to that level of generality, not much would be gained to compensate for the additional technical complications this would entail. Moreover, it is my view that the reader who has reached a level at which Stein manifolds (or Stein spaces) become important should in any case systematically learn the relevant methods from partial differential equations and coherent analytic sheaves by studying the appropriate references.

I have tried to trace the original sources of the major ideas and results presented in this book in extensive Notes at the end of each chapter and, occasionally, in comments within the text. But it is almost impossible to do the same for many Lemmas and Theorems of more special type and for the numerous variants of classical arguments which have evolved over the years thanks to the contributions of many mathematicians. Under no circumstances does the lack of a specific attribution of a result imply that the result is due to the author. Still, the expert in the field will perhaps notice here and there some simplifications in known proofs, and novelties in the organization of the material. The Bibliography reflects a similar philosophy: it is not intended to provide a complete encyclopedic listing of all articles and books written on topics related to this book. I believe, however, that it does adequately document the material discussed here, and I offer my sincerest apologies for any omissions or errors of judgment in this regard. In addition, I have included a perhaps somewhat random selection of quite recent articles for the sole purpose of guiding the reader to places in the literature from where he or she may begin to explore specific topics in more detail, and also find the way back to other (earlier) contributions on such topics. Altogether, the references in the Bibliography, along with all the references quoted in them, should give a fairly complete picture of the literature on the topics in Several Complex Variables which are discussed in this book.

We all know that one learns best by doing. Consequently, I have included numerous exercises. Rather than writing "another book" hidden in the exercises, I have mainly included problems which test and reinforce the understanding of the material discussed in the text. Occasionally the reader is asked to provide missing steps of proofs; these are always of a routine nature. A few of the exercises are quite a bit more challenging. I have not identified them in any special way, since part of the learning process involves being able to distinguish the easy problems from the more difficult ones.

The prerequisites for reading this book are: (1) A solid knowledge of calculus in several (real) variables, including Taylor's Theorem, Implicit Function Theorem, substitution formula for integrals, etc. The calculus of differential forms, which should really be part of such a preparation, but too often is missing, is discussed systematically, though somewhat compactly, in Chapter III. (2) Basic complex analysis in one variable. (3) Lebesgue measure in $\mathbb{R}^n$, and the elementary theory of Hilbert and Banach spaces as it is needed for an understanding of L^p spaces and of the orthogonal projection onto a closed subspace of L^2. (4) The elements of point set topology and algebra. Beyond this, we also make crucial use of the Fredholm alternative for perturbations of the identity by compact operators in Banach spaces. This result is usually covered in a first course in Functional Analysis, and precise references are given.

Before beginning the study of this book, the reader should consult the Suggestions for the Reader and the chart showing the interdependence of the chapters, on pp. xvii–xix.

It gives me great pleasure to express my gratitude to the three persons who have had the most significant and lasting impact on my training as a mathematician. First, I want to mention H. Grauert. His lectures on Several Complex Variables, which I was privileged to hear while a student at the University of Göttingen, introduced me to the subject and provided the stimulus to study it further. His early support and his continued interest in my mathematical development, even after I left Göttingen in 1968, is deeply appreciated. I discussed my plans for this book with him in 1982, and his encouragement contributed to getting the project started. Once I came to the United States, I was fortunate to study under T.W. Gamelin at UCLA. He introduced me to the Theory of Function Algebras, a fertile ground for applying the new tools of integral representations which were becoming known around that time, and he took interest in my work and supervised my dissertation. Finally, I want to mention Y.T. Siu. It was a great experience for me—while a "green" Gibbs Instructor at Yale University—to have been able to continue learning from him and to collaborate with him.

Regarding this book, I am greatly indebted to my friend and collaborator on recent research projects, Ingo Lieb. He read drafts of virtually the whole book, discussed many aspects of it with me, and made numerous helpful suggestions. W. Rudin expressed early interest and support, and he carefully read drafts of some chapters, making useful suggestions and catching a number of typos. S. Bell, J. Ryczaj, and J. Wermer also read portions of the manuscript and provided valuable feedback. Students at SUNY at Albany patiently listened to preliminary versions of parts of this book; their interest and reactions have been a positive stimulus. My colleague R. O'Neil showed me how to prove the real analysis result in Appendix C.

I thank JoAnna Aveyard, Marilyn Bisgrove, and Ellen Harrington for typing portions of the manuscript. Special thanks are due to Mary Blanchard, who typed the remaining parts and completed the difficult job of incorporating all the final revisions and corrections. B. Tomaszewski helped with the proof-reading. The Department of Mathematics and Statistics of the State University of New York at Albany partially supported the preparation of the manuscript.

I would also like to acknowledge the National Science Foundation for supporting my research over many years. Several of the results incorporated in this book are by-products of projects supported by the N.S.F.

Finally, I want to express my deepest appreciation to my family, who, for the past few years, had to share me with this project. Without the constant encouragement and understanding of my wife Sandrina, it would have been difficult to bring this work to completion. My children's repeated questioning if I would ever finish this book, and the fact that early this past summer my 6-year-old son Roberto started his own "book" and proudly finished it in one month, gave me the necessary final push.

R. Michael Range

Contents

CHAPTER IV

CHAPTER V

CHAPTER VI

CHAPTER VII

Suggestions for the Reader

This book may be used in many ways as a text for courses and seminars, or for independent study, depending on interest, background, and time limitations. The following are just intended as a few suggestions. The reader should refer to the chart on page xix showing the interdependence of chapters in order to visualize matters more clearly.

(1) The obvious suggestion is to cover the entire book. Typically this will require more than two semesters. If time is a factor, certain sections may be omitted: natural candidates are §3 in Chapter I, §4, §5 in II, §2 in IV, §2, §3, §6 in VI, and, if necessary, parts of VII.
(2) Another possibility is a first course in Several Complex Variables, to be followed by a course which will emphasize the general theory, i.e., complex spaces, sheaves, etc. Such an introductory course could include I, §2.1, §2.7–§2.10, and §3 in II, III as needed, §1, §3 in IV, §1, §2 in V, and VI.
(3) A first course in Several Complex Variables which emphasizes recent developments on analytic questions, in preparation for studying the relevant research literature on weakly (or strictly) pseudoconvex domains, could be based on the following selection: §1, §2 in I, §1–3 in II, III as needed, §1, §3, §4 in IV, V, and VII. This could be done comfortably in a year course.
(4) The more advanced reader who is familiar with the elements of Several Complex Variables, and who primarily wants to learn about integral representations and some of their applications, may concentrate on Chapters IV (I advise reading §3 in III beforehand!), V, and VII.
(5) Finally, I have found the following selection of topics quite effective for a one-semester introduction to Several Complex Variables for students with limited technical background in several (real) variables: §1 and §2.1–§2.5 in I, §1–§3 in II, §1 (without 1.8), §4, §5, and §6 (if time) in VI. In order to handle

Chapter VI, one simply states without proof the vanishing theorem $H^1_{\bar{\partial}}(K) = 0$, i.e., the solvability of the Cauchy–Riemann equations in neighborhoods of K, for a compact pseudoconvex compactum K in $\mathbb{C}^n$. In case $n = 1$, this result is easily proved by reducing it to the case where the given (0, 1)-form $f d\bar{z}$ has compact support. This procedure, of course, does not work in general because multiplication by a cutoff function destroys the necessary integrability condition in case $n > 1$. Assuming $H^1_{\bar{\partial}}(K) = 0$, it is easy to solve the Levi problem (cf. §1.4 in V), and one can then proceed directly with Chapter VI. Notice that only the vanishing of $H^1_{\bar{\partial}}$ is required in Chapter VI, so all discussions involving $(0, q)$ forms for $q > 1$ can be omitted! In such a course it is also natural to present a proof of the Hartogs Extension Theorem based on the (elementary) solution of $\bar{\partial}$ with *compact supports* (see Exercise E.2.4 and E.2.5 in IV for an outline, or consult Hörmander's book [Hör 2]).

Within each chapter Theorems, Lemmas, Remarks, etc., are numbered in one sequence by double numbers; for example, Lemma 2.1 refers to the first such statement in §2 in that same chapter. A parallel sequence identifies formulas which are referred to sometime later on; e.g., (4.3) refers to the third numbered formula in §4. References to Theorems, formulas, etc., in a *different* chapter are augmented by the Roman numeral identifying that chapter.

Interdependence of the Chapters

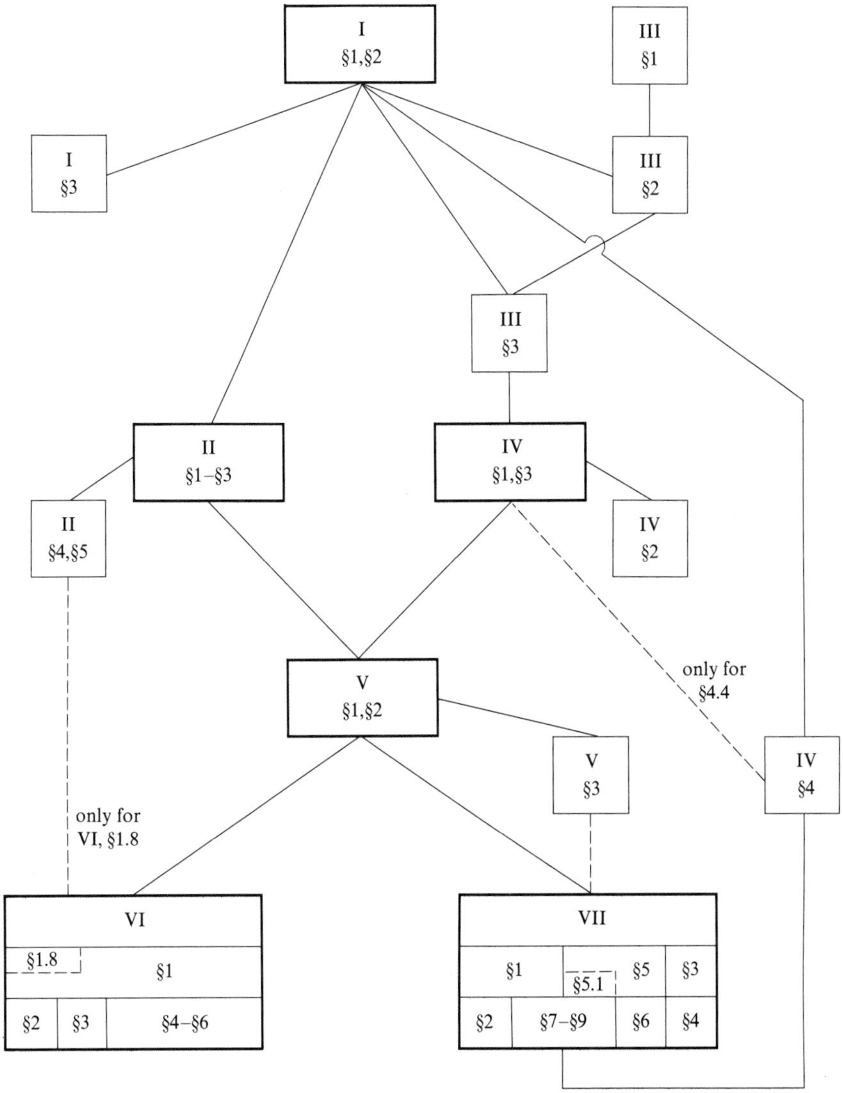

CHAPTER I

Elementary Local Properties of Holomorphic Functions

In §1 and §2 of this chapter we present the standard local properties of holomorphic functions and maps which are obtained by combining basic one complex variable theory with the calculus of several (real) variables. The reader should go through this material rapidly, with the goal of familiarizing himself with the results, notation, and terminology, and return to the appropriate sections later on, as needed. The inclusion at this stage of holomorphic maps and of complex submanifolds, i.e., the level sets of nonsingular holomorphic maps, is quite natural in several variables. In particular, it allows us to present elementary proofs of two results which distinguish *complex* analysis from *real* analysis, namely: (i) *the only compact complex submanifolds of* $\mathbb{C}^n$ *are finite sets*, and (ii) *the Jacobian determinant of an injective holomorphic map from an open set in* $\mathbb{C}^n$ *into* $\mathbb{C}^n$ *is nowhere zero.* Section 3, which gives an introduction to analytic sets, may be omitted without loss of continuity. We have included it mainly to familiarize the reader with a topic which is fundamental for many aspects of the general theory of several complex variables, and in order to show, by means of the Weierstrass Preparation Theorem, how algebraic methods become indispensable for a thorough understanding of the deeper local properties of holomorphic functions and their zero sets.

§1. Holomorphic Functions

1.1. Complex Euclidean Space

We collect some basic facts, notations, and terminology, which will be used throughout this book.

$\mathbb{R}$ and $\mathbb{C}$ denote the field of real, respectively complex numbers; $\mathbb{Z}$ and $\mathbb{N}$

denote the integers, respectively nonnegative integers, while we use $\mathbb{N}^+$ for the positive integers.

For $n \in \mathbb{N}^+$, the ***n*-dimensional complex number space**

$$\mathbb{C}^n = \{z: z = (z_1, \ldots, z_n), \quad z_j \in \mathbb{C} \quad \text{for} \quad 1 \le j \le n\}$$

is the Cartesian product of n copies of $\mathbb{C}$. $\mathbb{C}^n$ carries the structure of an n-dimensional complex vector space. The standard **Hermitian inner product** on $\mathbb{C}^n$ is defined by

$$(1.1) \qquad (a, b) = \sum_{j=1}^{n} a_j \bar{b}_j, \qquad a, b \in \mathbb{C}^n.$$

The associated **norm** $|a| = (a, a)^{1/2}$ induces the Euclidean metric in the usual way: for $a, b \in \mathbb{C}^n$, $\operatorname{dist}(a, b) = |a - b|$.

The (open) **ball of radius** $r > 0$ **and center** $a \in \mathbb{C}^n$ is defined by

$$(1.2) \qquad B(a, r) = \{z \in \mathbb{C}^n: |z - a| < r\}.$$

The collection of balls $\{B(a, r): r > 0 \text{ and rational}\}$ forms a countable neighborhood basis at the point a for the topology of $\mathbb{C}^n$.

The topology of $\mathbb{C}^n$ is identical with the one arising from the following identification of $\mathbb{C}^n$ with $\mathbb{R}^{2n}$. Given $z = (z_1, \ldots, z_n) \in \mathbb{C}^n$, each coordinate z_j can be written as $z_j = x_j + iy_j$, with $x_j, y_j \in \mathbb{R}$ (i is the imaginary unit $\sqrt{-1}$). The mapping

$$(1.3) \qquad z \to (x_1, y_1, \ldots, x_n, y_n) \in \mathbb{R}^{2n}$$

establishes an $\mathbb{R}$-linear isomorphism between $\mathbb{C}^n$ and $\mathbb{R}^{2n}$, which is compatible with the metric structures: a ball $B(a, r)$ in $\mathbb{C}^n$ is identified with a Euclidean ball in $\mathbb{R}^{2n}$ of equal radius r. Because of this identification, all the usual concepts from topology and analysis on real Euclidean spaces $\mathbb{R}^{2n}$ carry over immediately to $\mathbb{C}^n$. In the following, we shall freely use such standard results and terminology.

In particular, we recall that $D \subset \mathbb{C}^n$ is **open** if for every $a \in D$ there is a ball $B(a, r) \subset D$ with $r > 0$, and that an open set $D \subset \mathbb{C}^n$ is *connected* if and only if D is *pathwise connected*. Unless specified otherwise, D will usually denote an open set in $\mathbb{C}^n$; such a D will also be called a **domain**, or **region**. Notice that we do not require a domain to be connected. We shall say that a subset Ω of D is **relatively compact in** D, and denote this by $\Omega \subset\subset D$, if the closure $\bar{\Omega}$ of Ω is a compact subset of D.

The *topological boundary* of a set $A \subset \mathbb{C}^n$ will be denoted by bA (rather than the more commonly used ∂A, as the symbol ∂ generally has different meaning in complex analysis (see §1.2)).

Given a domain D, $\delta_D(z) = \sup\{r: B(z, r) \subset D\}$ denotes the (Euclidean) distance from $z \in D$ to the boundary of D. If $D \neq \mathbb{C}^n$, then $0 < \delta_D(z) < \infty$ for all $z \in D$, and δ_D extends to a continuous function on $\bar{D}$ by setting $\delta_D(z) = 0$ for $z \in bD$. One has $\delta_D(z) = \inf\{|z - \zeta|: \zeta \in bD\}$. The distance between two sets A, B is given by $\operatorname{dist}(A, B) = \inf\{|a - b|: a \in A, b \in B\}$. Notice that if $\Omega \subset\subset D$,

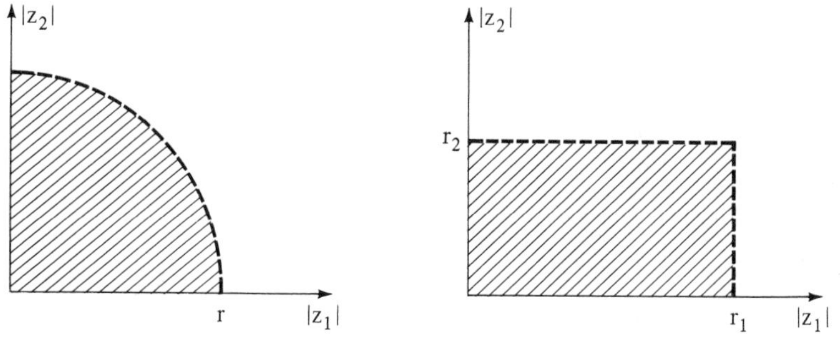

$\tau(B(0, r))$ $\tau(P(0, (r_1, r_2)))$

Figure 1. Representations of ball and polydisc in absolute space.

then $\operatorname{dist}(\Omega, bD) > 0$, or, equivalently, there is $\gamma > 0$ such that $\delta_D(z) \geq \gamma$ for all $z \in \Omega$; conversely, if D is bounded (i.e., $D \subset B(0, r)$ for some $r < \infty$) and $\gamma > 0$, then $\{z \in D: \delta_D(z) > \gamma\} \subset\subset D$.

Often it is convenient to use another system of neighborhoods: **the (open) polydisc** $P(a, r)$ **of multiradius** $r = (r_1, \ldots, r_n)$, $r_j > 0$, **and center** $a \in \mathbb{C}^n$ is the product of n open discs in $\mathbb{C}$:

$$P(a, r) = \{z \in \mathbb{C}^n: |z_j - a_j| < r_j, 1 \leq j \leq n\}. \tag{1.4}$$

More generally, a **polydomain** is the product of n planar domains.

Notice that

$$P(a, (r_1, \ldots, r_n)) \subset B(a, R)$$

whenever $\Sigma r_j^2 < R^2$, *and that*

$$B(a, \rho) \subset P(a, (r_1, \ldots, r_n))$$

for $\rho \leq \min\{r_j: 1 \leq j \leq n\}$.

In order to represent certain sets in $\mathbb{C}^n$ geometrically, it is convenient to consider the image $\tau(D)$ of D in **absolute space** $\{(r_1, \ldots, r_n) \in \mathbb{R}^n: r_j \geq 0$ for $j = 1, \ldots, n\}$, under the map $\tau: a \to (|a_1|, \ldots, |a_n|)$. For example, $B(0, r)$ and $P(0, (r_1, r_2))$ in $\mathbb{C}^2$ have the representations shown in Figure 1.

If $n > 2$, we sometimes write $z = (z', z_n)$, where $z' = (z_1, \ldots, z_{n-1}) \in \mathbb{C}^{n-1}$. For example, if $0 < r_j < 1$, $1 \leq j \leq n$, the domain

$$H(r) = \{z \in \mathbb{C}^n: z' \in P'(0, r'), |z_n| < 1\} \cup \{z \in \mathbb{C}^n, z' \in P'(0, 1), r_n < |z_n| < 1\}$$

can be represented schematically by Figure 2.

The pair $(H(r), P(0, 1))$ is called a (Euclidean) **Hartogs figure**; its significance will become clear in Chapter II.

Notice that

$$\tau^{-1}(r) = \{(r_1 e^{i\theta_1}, \ldots, r_n e^{i\theta_n}): 0 \leq \theta_j \leq 2\pi \qquad \text{for } 1 \leq j \leq n\}$$

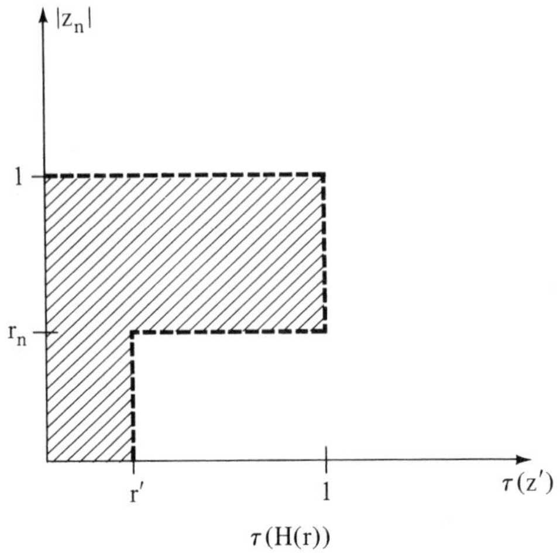

Figure 2. Representation of $H(r)$ in absolute space.

is an n-dimensional real torus. So the representation in absolute space is reasonable only for sets which are circled in the following sense.

Definition. A set $\Omega \subset \mathbb{C}^n$ is **circled** (around 0) if for every $a \in \Omega$ the torus

$$\tau^{-1}(\tau(a)) = \{z \in \mathbb{C}^n: z = (a_1 e^{i\theta_1}, \ldots, a_n e^{i\theta_n}), 0 \le \theta_j \le 2\pi\}$$

lies in Ω as well. A **Reinhardt domain** (centered at 0) is an open circled (around 0) set in $\mathbb{C}^n$. A Reinhardt domain D is **complete** if for every $a \in D$ one has $P(0, \tau(a)) \subset D$.

It is clear how to define the corresponding concepts for arbitrary centers.

$B(0, r)$ and $P(0, r)$ are complete Reinhardt domains, while the domain $H(r)$ in Figure 2 is a Reinhardt domain which is not complete.

Reinhardt domains appear naturally when one considers power series or Laurent expansions of holomorphic functions (see §1.5 and Chapter II, §1). Observe that a complete Reinhardt domain in $\mathbb{C}$ (centered at 0) is an open disc with center 0; what are the Reinhardt domains in $\mathbb{C}$?

1.2. The Cauchy–Riemann Equations

For $D \subset \mathbb{R}^n$, open, and $k \in \mathbb{N} \cup \{\infty\}$, $C^k(D)$ denotes the space of k times continuously differentiable complex valued functions on D; we also write $C(D)$ instead of $C^0(D)$. We shall use the standard multi-index notation: if $\alpha = (\alpha_1, \ldots, \alpha_n) \in \mathbb{N}^n$ and $x = (x_1, \ldots, x_n) \in \mathbb{R}^n$, one sets

$$|\alpha| = \alpha_1 + \cdots + \alpha_n, \quad \alpha! = \alpha_1! \cdot \cdots \cdot \alpha_n!,$$

$$x^\alpha = x_1^{\alpha_1} \cdot \ldots \cdot x_n^{\alpha_n}, \quad \alpha \ge 0 (>0) \quad \text{if } \alpha_j \ge 0 (>0) \qquad \text{for } 1 \le j \le n,$$

(1.5) $$D^\alpha = \frac{\partial^{|\alpha|}}{\partial x_1^{\alpha_1} \ldots \partial x_n^{\alpha_n}}.$$

For $f \in C^k(D)$, $k < \infty$, we define the C^k **norm of** f **over** D by

(1.6) $$|f|_{k,D} = \sum_{\substack{\alpha \in \mathbb{N}^n \\ |\alpha| \le k}} \sup_{x \in D} |D^\alpha f(x)|;$$

we write $|f|_D$ instead of $|f|_{0,D}$, and if D is clear from the context, we may write $|f|_k$ instead of $|f|_{k,D}$. The space $B^k(D) = \{f \in C^k(D): |f|_k < \infty\}$ is complete in the C^k norm $|\cdot|_k$, and hence $B^k(D)$ is a Banach space. Similarly, the space $C^k(\bar{D}) := \{f \in C^k(D): D^\alpha f$ extends continuously to $\bar{D}$ for all $\alpha \in \mathbb{N}^n$ with $|\alpha| \le k\}$, with the norm $|\cdot|_{k,D}$, is also a Banach space.

Turning to $\mathbb{C}^n = \mathbb{R}^{2n}$ with coordinates $z_j = x_j + \sqrt{-1}\, y_j$, one introduces the partial differential operators

(1.7) $$\frac{\partial}{\partial z_j} = \frac{1}{2}\left(\frac{\partial}{\partial x_j} + \frac{1}{i}\frac{\partial}{\partial y_j}\right), \qquad \frac{\partial}{\partial \bar{z}_j} = \frac{1}{2}\left(\frac{\partial}{\partial x_j} - \frac{1}{i}\frac{\partial}{\partial y_j}\right).$$

The following rules are easily verified:

$$\overline{\partial f/\partial z_j} = \partial \bar{f}/\partial \bar{z}_j, \qquad \overline{\partial f/\partial \bar{z}_j} = \partial \bar{f}/\partial z_j.$$

The multi-index notation (1.5) is extended to the operators (1.7) as follows: for $\alpha, \beta \in \mathbb{N}^n$,

(1.8) $$D^{\alpha\bar{\beta}} = \frac{\partial^{|\alpha|+|\beta|}}{\partial z_1^{\alpha_1} \ldots \partial z_n^{\alpha_n} \partial \bar{z}_1^{\beta_1} \ldots \partial \bar{z}_n^{\beta_n}}.$$

We write D^α for $D^{\alpha 0}$ and $D^{\bar{\beta}}$ for $D^{0\bar{\beta}}$; this should cause no confusion with (1.5).

Notice that $f \in C^k(D)$ if and only if $D^{\alpha\bar{\beta}} f \in C(D)$ for all α, β with $|\alpha| + |\beta| \le k$.

We now introduce the class of functions which is the principal object of study in this book.

Definition. Let $D \subset \mathbb{C}^n$ be open. A function $f: D \to \mathbb{C}$ is called **holomorphic** (on D) if $f \in C^1(D)$ and f satisfies the system of partial differential equations

(1.9) $$\frac{\partial f}{\partial \bar{z}_j}(z) = 0 \qquad \text{for } 1 \le j \le n \text{ and } z \in D.$$

The space of holomorphic functions on D is denoted by $\mathcal{O}(D)$. More generally, if Ω is an arbitrary subset of $\mathbb{C}^n$, we denote by $\mathcal{O}(\Omega)$ the collection of those functions which are defined and holomorphic on some open neighborhood of Ω, with the understanding that two such functions define the same element in $\mathcal{O}(\Omega)$ if they agree on a neighborhood of Ω.[1] A function f is said to be **holomorphic at the point** $a \in \mathbb{C}^n$ if $f \in \mathcal{O}(\{a\})$.

The following result is an immediate consequence of the definitions and standard calculus.

[1] This identification can be formalized by introducing the language of *germs of functions* (see Chapter VI, §4).

Theorem 1.1. *For any subset Ω of $\mathbb{C}^n$, $\mathcal{O}(\Omega)$ is closed under pointwise addition and multiplication. Any polynomial in $z_1, \ldots, z_n$ with complex coefficients is holomorphic on $\mathbb{C}^n$, and hence, by restriction, is in $\mathcal{O}(\Omega)$. If $f, g \in \mathcal{O}(\Omega)$ and $g(z) \neq 0$ for all $z \in \Omega$, then $f/g \in \mathcal{O}(\Omega)$.*

Equation (1.9) is called the system of **(homogeneous) Cauchy–Riemann equations**. Notice that any function f which satisfies (1.9) satisfies the Cauchy–Riemann equations in the z_j-coordinate for any j, and hence is holomorphic in each variable separately. It is a remarkable phenomenon of complex analysis —discovered by F. Hartogs in 1906 [Har 2]—that conversely, *any* function $f: D \to \mathbb{C}$ which is holomorphic in each variable separately is holomorphic, as defined above. This shows that the requirement that $f \in C^1(D)$ can be dropped in the definition of holomorphic function. The main difficulty in Hartogs' Theorem is to show that a function f which satisfies (1.9) is locally bounded. Assuming that f is bounded, it is quite elementary to show that (1.9) implies $f \in C^\infty(D)$ (see Exercise E.1.3 and Corollary 1.5 below).

In order to appreciate the strength of Hartogs' Theorem, the reader should notice that the function $f: \mathbb{R}^2 \to \mathbb{R}$ defined by $f(0) = 0$ and $f(x, y) = xy/(x^4 + y^4)$ for $(x, y) \neq 0$ is C^∞ (even real analytic) in each variable separately, but is not bounded at 0.

The **inhomogeneous system of Cauchy–Riemann equations**

$$\partial f/\partial \bar{z}_j = u_j, \qquad 1 \le j \le n, \tag{1.10}$$

where $u_1, \ldots, u_n$ are given C^1 functions on D, will also be very important for the study of holomorphic functions. For $n = 1$, the system (1.10) is determined (i.e., one equation for one unknown function, or two real equations for the two real functions $\operatorname{Re} f$, $\operatorname{Im} f$), while for $n > 1$ (1.10) is overdetermined (more equations than unknowns). This fact makes life in several variables harder, and it accounts for many of the differences between the cases $n = 1$ and $n > 1$. Notice that if there is a solution $f \in C^2(D)$ of (1.10), then the functions $u_1, \ldots, u_n$ must satisfy the **necessary integrability conditions**

$$\partial u_j/\partial \bar{z}_k = \partial u_k/\partial \bar{z}_j, \qquad 1 \le j, k \le n; \tag{1.11}$$

(1.11) always holds in case $n = 1$, while it is quite restrictive in case $n > 1$.

We now give another interpretation for the solutions of the homogeneous Cauchy–Riemann equations. Let $f \in C^1(D)$; its **differential** df_a **at** $a \in D$ is the unique $\mathbb{R}$-linear map $\mathbb{R}^{2n} \to \mathbb{R}^2$ which approximates f near a in the sense that $f(z) = f(a) + df_a(z - a) + o(|z - a|)$.[1] In terms of the real coordinates $(x_1, y_1, \ldots, x_n, y_n)$ of $\mathbb{C}^n$, one has

$$df_a = \sum_{j=1}^{n} \left[\frac{\partial f}{\partial x_j}(a)(dx_j)_a + \frac{\partial f}{\partial y_j}(a)(dy_j)_a \right], \tag{1.12}$$

[1] We use a standard notation from analysis: if $A = A(x)$ is an expression which depends on $x \in \mathbb{R}^n$, the statements $A = O(|x|)$, and $A = o(|x|)$ mean, respectively, that $|A(x)| \le C|x|$ as $|x| \to 0$ for some constant C, and $\lim_{|x| \to 0} |A(x)|/|x| = 0$.

where dx_j, dy_j are the differentials of the coordinate functions, i.e.,

$$dx_j(w) = u_j,\ dy_j(w) = v_j \qquad \text{for } w = (u_1, v_1, \ldots, u_n, v_n).$$

Via the identification $\mathbb{R}^{2n} = \mathbb{C}^n$ and $\mathbb{R}^2 = \mathbb{C}$, the differential df_a can be viewed as a map $\mathbb{C}^n \to \mathbb{C}$ which is $\mathbb{R}$-linear, though not necessarily $\mathbb{C}$-linear. In particular, the differentials dx_j and dy_j are *not* linear over $\mathbb{C}$; for example, if $\zeta = (1, 0, \ldots, 0) \in \mathbb{R}^{2n}$, then $i\zeta = (0, 1, 0, \ldots, 0)$, so that $dx_1(i\zeta) = 0$, while $idx_1(\zeta_1) = i$. In complex analysis one therefore considers the differentials $dz_j = dx_j + idy_j$ (this is $\mathbb{C}$-linear) and $d\bar{z}_j = dx_j - idy_j$ (this is conjugate $\mathbb{C}$-linear[1]) of the complex coordinate functions z_j, $1 \le j \le n$. A simple computation shows that

$$df_a = \sum_{j=1}^{n} \frac{\partial f}{\partial z_j}(a)(dz_j)_a + \sum_{j=1}^{n} \frac{\partial f}{\partial \bar{z}_j}(a)(d\bar{z}_j)_a. \tag{1.13}$$

The first sum in (1.13) is denoted by ∂f_a, or $\partial f(a)$, the second sum by $\bar{\partial} f_a$, or $\bar{\partial} f(a)$. So one can say that

$$f \in C^1(D) \textit{ is holomorphic} \Leftrightarrow \bar{\partial} f = 0 \Leftrightarrow df = \partial f. \tag{1.14}$$

Theorem 1.2. *A function $f \in C^1(D)$ satisfies the Cauchy–Riemann equations at the point $a \in D$ if and only if its differential df_a at a is $\mathbb{C}$-linear. In particular, $f \in \mathcal{O}(D)$ if and only if df is $\mathbb{C}$-linear at every point.*

Proof. Since ∂f_a is obviously $\mathbb{C}$-linear for any $a \in D$, one implication is trivial. For the other implication, suppose $\beta_k = \partial f/\partial \bar{z}_k(a) \neq 0$ for some k. Let $\alpha_k = \partial f/\partial z_k(a)$, and $w = (0, \ldots, 1, 0, \ldots, 0) \in \mathbb{C}^n$, with the 1 in the kth place. Then $df_a(w) = \alpha_k + \beta_k$, and $df_a(iw) = \alpha_k i - \beta_k i = i(\alpha_k - \beta_k) \neq idf_a(w)$, so that df_a is not linear over $\mathbb{C}$. ■

We shall discuss these matters more systematically and in coordinate-free form in Chapter III, §2.2; for the present, let us mention though that it is Equation (1.13) for the differential of a C^1 function which motivates the definition of the operators $\partial/\partial z_j$ and $\partial/\partial \bar{z}_j$ in (1.7).

1.3. The Cauchy Integral Formula on Polydiscs

As in the case of one complex variable, the basic local properties of holomorphic functions follow from an integral representation formula, which is most easily established on polydiscs. Later we will consider an analogous formula on the ball and on more general domains (see Chapter IV, §3.2 and Chapter VII, §1).

[1] A map $l\colon V \to W$ between two complex vector spaces V and W is conjugate $\mathbb{C}$-linear if l is linear over $\mathbb{R}$ and if $l(\lambda v) = \bar{\lambda} l(v)$ for all $\lambda \in \mathbb{C}$, $v \in V$.

Theorem 1.3. *Let $P = P(a, r)$ be a polydisc in $\mathbb{C}^n$ with multiradius $r = (r_1, \ldots, r_n)$. Suppose $f \in C(\bar{P})$, and f is holomorphic in each variable separately, i.e., for each $z \in \bar{P}$ and $1 \le j \le n$, the function $f_{z_j}(\lambda) = f(z_1, \ldots, z_{j-1}, \lambda, z_{j+1}, \ldots, z_n)$ is holomorphic on $\{\lambda \in \mathbb{C}: |\lambda - a_j| < r_j\}$. Then*

$$(1.15) \qquad f(z) = (2\pi i)^{-n} \int_{b_o P} \frac{f(\zeta)\, d\zeta_1 \ldots d\zeta_n}{(\zeta_1 - z_1)\ldots(\zeta_n - z_n)} \qquad \text{for } z \in P,$$

where $b_o P = \{\zeta \in \mathbb{C}^n: |\zeta_j - a_j| = r_j, 1 \le j \le n\}$.

Notice that the region of integration $b_o P$ in (1.15) is strictly smaller than the topological boundary bP of P in case $n > 1$. $b_o P$ is called the **distinguished boundary of** P, and in many situations it plays the same role as the unit circle in one complex variable (see Theorem 1.8 below for an example).

The integral in (1.15) is an example of an n-form integrated over the real n-dimensional manifold $b_0 P$ (see Chapter III, §1). In terms of the standard parametrization

$$\zeta_j = a_j + r_j e^{i\theta_j}, \qquad 0 \le \theta_j \le 2\pi,\ 1 \le j \le n$$

of $b_o P(a, r)$, one has

$$(1.16) \quad \int_{b_o P(a,r)} g(\zeta)\, d\zeta_1 \ldots d\zeta_n = i^n r_1 \ldots r_n \int_{[0,2\pi]^n} g(\zeta(\theta)) e^{i\theta_1} \ldots e^{i\theta_n} d\theta_1 \ldots d\theta_n$$

for any $g \in C(b_o P)$. For the time being, the reader may simply view the left side in (1.16) as a shorthand notation for the right side.

PROOF. We use induction over the number of variables n. For $n = 1$ one has the classical Cauchy integral formula, which we assume as known. Suppose $n > 1$, and that the theorem has been proved for $n - 1$ variables. For $z \in P$ fixed, apply the inductive hypothesis with respect to $(z_2, \ldots, z_n)$, obtaining

$$(1.17) \quad f(z_1, z_2, \ldots, z_n) = (2\pi i)^{-n+1} \int_{b_o P'(a', r')} \frac{f(z_1, \zeta_2, \ldots, \zeta_n)\, d\zeta_2 \ldots d\zeta_n}{(\zeta_2 - z_2)\ldots(\zeta_n - z_n)},$$

where $a' = (a_2, \ldots, a_n)$, $r' = (r_2, \ldots, r_n)$. For $\zeta_2, \ldots, \zeta_n$ fixed, the case $n = 1$ gives

$$(1.18) \qquad f(z_1, \zeta_2, \ldots, \zeta_n) = (2\pi i)^{-1} \int_{|\zeta_1 - a_1| = r_1} \frac{f(\zeta_1, \ldots, \zeta_n)\, d\zeta_1}{\zeta_1 - z_1}.$$

Now substitute (1.18) into (1.17) and transform the iterated integral over $\{|\zeta_1 - a_1| = r_1\} \times b_o P'(a', r')$ into an integral over $b_o P$—use the parametrization (1.16). ∎

Remark 1.4. The continuity of f was used only at the end of the proof. Weaker conditions on f, for example f bounded and measurable, would work just as well by basic results in integration theory. On the other hand, it is important for the applications given below that (1.15) is an integral over $b_o P$, and not just an iterated integral.

Corollary 1.5. *Suppose $f \in C(D)$ (or just bounded on D) is holomorphic in each variable separately. Then $f \in C^\infty(D)$ and, in particular, $f \in \mathcal{O}(D)$. For any $\alpha \in \mathbb{N}^n$, $D^\alpha f \in \mathcal{O}(D)$.*

Proof. Apply Theorem 1.3 to a polydisc $P(a, r) \subset\subset D$; in (1.15) it is legitimate to differentiate under the integral sign as often as needed. ■

Theorem 1.6 (Cauchy estimates). *Let $f \in \mathcal{O}(P(a, r))$. Then, for all $\alpha \in \mathbb{N}^n$,*

$$|D^\alpha f(a)| \leq \frac{\alpha!}{r^\alpha} |f|_{P(a,r)}; \tag{1.19}$$

$$|D^\alpha f(a)| \leq \frac{\alpha!(\alpha_1 + 2)\dots(\alpha_n + 2)}{(2\pi)^n r^{\alpha+2}} \|f\|_{L^1(P(a,r))}. \tag{1.20}$$

Note that $r^\alpha = r_1^{\alpha_1}\dots r_n^{\alpha_n}$, and for $m \in \mathbb{Z}$, $\alpha + m = (\alpha_1 + m, \dots, \alpha_n + m)$; for $1 \leq p \leq \infty$, $L^p(D)$ denotes the space of functions on D with $|f|^p$ Lebesgue integrable over D (with respect to Lebesgue measure on $\mathbb{R}^{2n}$), and $\|f\|_{L^p(D)} = (\int_D |f|^p)^{1/p}$.

Proof. Fix $0 < \rho < r$. Apply Theorem 1.3 to $P(a, \rho) \subset\subset P(a, r)$ and differentiate under the integral sign, obtaining

$$D^\alpha f(a) = \frac{\alpha!}{(2\pi i)^n} \int_{b_o P(a,\rho)} \frac{f(\zeta)\, d\zeta_1 \dots d\zeta_n}{(\zeta - a)^{\alpha+1}} \tag{1.21}$$

(see (1.5) for the multi-index notation used). After an obvious estimation of (1.21) and taking the limit $\rho \to r$, (1.19) follows. For (1.20), use (1.16) in (1.21), multiply by $\rho^{\alpha+1}$ and estimate, obtaining

$$|D^\alpha f(a)| \rho^{\alpha+1} \leq \frac{\alpha!}{(2\pi)^n} \int_{[0,2\pi]^n} |f(\zeta(\theta))| \rho_1 \dots \rho_n \, d\theta_1 \dots d\theta_n. \tag{1.22}$$

The desired inequality follows after integrating (1.22) over $0 \leq \rho_j \leq r_j$, $1 \leq j \leq n$, and transforming the n-fold integral in polar coordinates into a volume integral. ■

The estimate (1.20) is often used in the following form.

Corollary 1.7. *For each $\alpha \in \mathbb{N}^n$, $1 \leq p \leq \infty$, and $\Omega \subset\subset D$ there is a constant $C = C(\alpha, p, \Omega, D)$ such that*

$$|D^\alpha f|_\Omega \leq C \|f\|_{L^p(D)} \quad \textit{for all } f \in \mathcal{O}(D) \cap L^p(D). \tag{1.23}$$

The space $\mathcal{O}(D) \cap L^p(D)$ of **holomorphic L^p functions on D** will be denoted by $\mathcal{O}L^p(D)$.

PROOF. Fix $0 < \delta < \operatorname{dist}(\Omega, bD)$ and let $r = \delta/\sqrt{n}$. Then (1.20) holds for each $a \in \Omega$, and since $\|f\|_{L^1(P(a,r))} \leq \text{constant} \cdot \|f\|_{L^p(P(a,r))}$, (1.23) follows. ■

Another consequence of the Cauchy integral formula is the following version of the maximum principle. A different form of the maximum principle is discussed in §1.6., Corollaries 1.22 and 1.23.

Theorem 1.8. *For $P = P(a, r)$ and $z \in \bar{P}$ one has*

$$|f(z)| \leq |f|_{b_o P} \text{ for all } f \in C(\bar{P}) \cap \mathcal{O}(P). \tag{1.24}$$

The space of functions $C(\bar{P}) \cap \mathcal{O}(P)$ is known as the **polydisc algebra**, and is denoted by $A(P)$. It is a subalgebra of $C(\bar{P})$ which is closed in the norm $|\cdot|_P$ (this follows from Theorem 1.9. below). We re-emphasize that $b_o P$ is strictly smaller than the topological boundary if $n > 1$. In the language of Uniform Algebras, (1.24) says that $b_o P$ is a **boundary** for the polydisc algebra $A(P)$; in fact, $b_o P$ is the smallest closed boundary, the so-called **Shilov boundary**, of $A(P)$.

PROOF. It is enough to prove (1.24) for $z \in P$. From (1.15) it follows by an obvious estimate that there is a constant C_z such that $|f(z)| \leq C_z |f|_{b_o P}$ for all $f \in A(P)$. Hence, for $k = 1, 2, \ldots$, since $f^k \in A(P)$ for $f \in A(P)$, one obtains

$$|f(z)|^k = |f^k(z)| \leq C_z |f^k|_{b_o P} \leq C_z (|f|_{b_o P})^k;$$

this implies $|f(z)| \leq C_z^{1/k} |f|_{b_o P}$, and (1.24) follows by letting $k \to \infty$. ■

1.4. Sequences and Compactness in Spaces of Holomorphic Functions

As in the case of functions of one complex variable, the Cauchy integral formula implies strong convergence theorems. We say that a sequence $\{f_j: j = 1, 2, \ldots,\} \subset C(D)$ **converges compactly in** D if $\{f_j\}$ converges uniformly on each compact subset of D. It is well known that $C(D)$ is closed under compact convergence.

Theorem 1.9. *Suppose $\{f_j: j = 1, 2, \ldots,\} \subset \mathcal{O}(D)$ converges compactly in D to the function $f: D \to \mathbb{C}$. Then $f \in \mathcal{O}(D)$, and for each $\alpha \in \mathbb{N}^n$,*

$$\lim_{j \to \infty} D^\alpha f_j = D^\alpha f$$

compactly in D.

The proof of Theorem 1.9 is the same as in the classical case $n = 1$ and will be omitted. Combined with Corollary 1.7, Theorem 1.9 implies the following result.

Corollary 1.10. *For any $1 \leq p \leq \infty$ the space $\mathcal{O}L^p(D)$ is a closed subspace of $L^p(D)$, and hence $\mathcal{O}L^p(D)$ is a Banach space.*

Unless stated otherwise, we will always consider $\mathcal{O}(D)$ equipped with the natural topology in which convergent sequences are precisely those which converge compactly. This topology is, in fact, metrizable, as follows. Fix an increasing sequence of compact sets $\{K_\nu\}$, such that

$$(1.25)\qquad \begin{aligned} &\text{(i)}\ K_1 \subset\subset \operatorname{int} K_2 \subset\subset \ldots K_\nu \subset\subset \operatorname{int} K_{\nu+1} \subset\subset \ldots \subset D \\ &\text{(ii)}\ \bigcup_{\nu=1}^{\infty} K_\nu = D. \end{aligned}$$

A sequence $\{K_\nu\}$ which satisfies (1.25) is called a **normal exhaustion** of D. It is obvious that $\lim_{j\to\infty} f_j = f$ compactly in D if and only if $\lim f_j = f$ uniformly on each K_ν. For $f, g \in C(D)$, one then defines

$$(1.26)\qquad \delta(f, g) = \sum_{\nu=1}^{\infty} 2^{-\nu} \frac{|f - g|_{K_\nu}}{1 + |f - g|_{K_\nu}}.$$

Lemma 1.11. *The function δ defined by (1.26) is a metric on $C(D)$. A sequence $\{f_j\} \subset C(D)$ converges compactly to f if and only if $\lim_{j\to\infty} \delta(f_j, f) = 0$. The topology on $C(D)$ defined by δ is independent of the choice of the normal exhaustion $\{K_\nu\}$.*

The proof is left to the reader.

Theorem 1.9 can now be restated: *$\mathcal{O}(D)$ is a closed subspace of $C(D)$, and every partial differentiation D^α: $\mathcal{O}(D) \to \mathcal{O}(D)$, $\alpha \in \mathbb{N}^n$, is continuous.*

The spaces $C(D)$ and $\mathcal{O}(D)$ are important examples of so-called **Fréchét spaces**. These are vector spaces V which are complete metrizable topological spaces, so that the vector space operations in V are continuous.

A subset S in a Fréchét space, or more generally, in a topological vector space V, is called **bounded** if for every neighborhood U of 0 in V there is $\lambda > 0$ such that $S \subset \lambda U$. The reader should convince himself that this definition of a bounded set is equivalent to the familiar one in normed linear spaces.

Lemma 1.12. *A subset $S \subset C(D)$ (or $\subset \mathcal{O}(D)$) is bounded if and only if for every compact $K \subset D$ one has*

$$\sup\{|f|_K, f \in S\} < \infty.$$

The proof is left to the reader.

The following characterization of compact sets in $\mathcal{O}(D)$ is of fundamental importance; it should be compared to the analogous characterization in finite dimensional vector spaces (i.e., for $\mathbb{R}^n$ or $\mathbb{C}^n$).

Theorem 1.13. *A subset $S \subset \mathcal{O}(D)$ is compact if and only if S is closed and bounded.*

Proof. As the classical proof for $n = 1$ generalizes to $n > 1$, we only give an outline. Since $\mathcal{O}(D)$ is complete metrizable, a closed set $S \subset \mathcal{O}(D)$ is compact if and only if every sequence $\{f_j\} \subset S$ has a convergent subsequence. The essential part of the theorem thus involves showing that every bounded sequence $\{f_j\} \subset \mathcal{O}(D)$ has a convergent subsequence (i.e., $\mathcal{O}(D)$ has the Bolzano–Weierstrass property).

Fix a normal exhaustion $\{K_\nu\}$ of D. If $\{f_j\} \subset \mathcal{O}(D)$ is bounded, Lemma 1.12 and Corollary 1.7 imply that $\{f_j\}$ has uniformly bounded first order derivatives on each K_ν, and hence, via the Mean Value Theorem, one sees that $\{f_j|_{K_\nu}, j = 1, 2, \ldots,\}$ is uniformly equicontinuous for each ν. The theorem of Ascoli–Arzela, combined with a Cantor diagonal sequence argument, then gives a subsequence $\{f_{j_l}, l = 1, 2, \ldots,\}$ which converges uniformly on each K_ν, $\nu = 1, 2, \ldots$; thus $\{f_{j_l}, l = 1, 2, \ldots,\}$ converges compactly in D. ∎

By Corollary 1.7, any subset $S \subset \mathcal{O}L^p(D)$, $1 \leq p \leq \infty$, which is bounded in L^p-norm, is also bounded in $\mathcal{O}(D)$. By Theorem 1.13, S has compact closure in $\mathcal{O}(D)$, but not necessarily in $L^p(D)$. In order to obtain a relatively compact subset of L^p we must restrict to some $\Omega \subset\subset D$, as compact convergence in D implies convergence in $L^p(\Omega)$ for any $\Omega \subset\subset D$ and $1 \leq p \leq \infty$. One thus obtains the following result.

Theorem 1.14. *Let $\Omega \subset\subset D$, and suppose $1 \leq p, q \leq \infty$. Then the restriction of $f \in \mathcal{O}(D)$ to Ω defines a compact linear map*

$$\mathcal{O}L^p(D) \to \mathcal{O}L^q(\Omega).$$

Recall that a linear map $B_1 \to B_2$ between two Banach spaces is called compact if the image of a bounded set in B_1 is relatively compact in B_2. See also Exercise E.1.7 for a related statement.

1.5. Power Series

We briefly recall first the basic facts about multiple series; that is, formal expressions

$$(1.27) \qquad \sum_{\nu \in \mathbb{N}^n} b_\nu, \qquad b_\nu = b_{\nu_1 \ldots \nu_n} \in \mathbb{C}.$$

If $n > 1$, the index set $\mathbb{N}^n$ does not carry any natural ordering, so that there is no canonical way to consider $\sum b_\nu$ as a sequence of (finite) partial sums as in case $n = 1$. The ambiguity is avoided if one considers (absolutely) convergent series, defined as follows.

Definition. The multiple series $\sum_{\nu \in \mathbb{N}^n} b_\nu$ is called convergent[1] if

$$\sum_{\nu \in \mathbb{N}^n} |b_\nu| = \sup\left\{ \sum_{\nu \in \Lambda} |b_\nu| : \Lambda \text{ finite} \right\} < \infty.$$

It is well known that the convergence of $\sum b_\nu$, as defined above, is necessary and sufficient for the following to hold.

Given any bijection $\sigma: \mathbb{N} \to \mathbb{N}^n$, the ordinary series

$$\sum_{j=0}^{\infty} b_{\sigma(j)}$$

converges in the usual sense to a limit $L \in \mathbb{C}$ which is independent of σ. This number L is called the limit (or sum) of the multiple series, and one writes

$$L = \sum_{\nu \in \mathbb{N}^n} b_\nu.$$

In particular, if $\sum b_\nu$ converges, its limit can be computed from the *homogeneous expansion*

$$L = \sum_{k=0}^{\infty} \left(\sum_{|\nu|=k} b_\nu \right). \tag{1.28}$$

Furthermore, for any permutation τ of $\{1, \ldots, n\}$, the iterated series

$$\sum_{\nu_{\tau(n)}=0}^{\infty} \left(\cdots \left(\sum_{\nu_{\tau(1)}=0}^{\infty} b_{\nu_1 \ldots \nu_n} \right) \cdots \right) \tag{1.29}$$

converges to L as well. Conversely, if $b_\nu \geq 0$, the convergence of any one of the iterated series (1.29) implies the convergence of $\sum b_\nu$.[2]

A **power series in n complex variables** $z_1, \ldots, z_n$ centered at the point $a \in \mathbb{C}^n$ is a multiple series $\sum_{\nu \in \mathbb{N}^n} b_\nu$ with terms

$$b_\nu = c_\nu (z - a)^\nu = c_{\nu_1 \ldots \nu_n} (z_1 - a_1)^{\nu_1} \ldots (z_n - a_n)^{\nu_n},$$

where $c_\nu \in \mathbb{C}$ for $\nu \in \mathbb{N}^n$. To simplify notation we will only consider power series centered at $a = 0$ in this section.

Definition. The domain of convergence $\Omega = \Omega(\{c_\nu\})$ of the power series

$$\sum_{\nu \in \mathbb{N}^n} c_\nu z^\nu \tag{1.30}$$

is the interior of the set of points $z \in \mathbb{C}^n$ for which (1.30) converges.

[1] Since functions f are defined to be (Lebesgue) integrable if $|f|$ is integrable, we take the liberty to drop the word "absolutely". In fact, convergent series are precisely the elements in $L^1(\mathbb{N}^n, \mu)$, where μ is counting measure.

[2] These results can be viewed as special cases of the Fubini–Tonelli theorem in integration theory.

Notice that (1.30) always converges for $z = 0$, but if $n > 1$, $\Omega(\{c_\nu\})$ may be empty even if (1.30) converges at some point $z \neq 0$. For example, the power series

$$\sum_{\substack{\nu_1 \geq 0 \\ \nu_2 > 0}} \nu_1! z_1^{\nu_1} z_2^{\nu_2}$$

converges for $z = (z_1, 0)$, but not for $z = (z_1, z_2)$ if $z_1, z_2 \neq 0$; hence its domain of convergence is empty.

The following result, known as Abel's Lemma, gives the basic general result about convergence of power series.

Lemma 1.15. *Suppose $c_\nu \in \mathbb{C}$ for $\nu \in \mathbb{N}^n$ and that for some $w \in \mathbb{C}^n$*

$$\sup_{\nu \in \mathbb{N}^n} |c_\nu w^\nu| = M < \infty. \tag{1.31}$$

Let $r = \tau(w) = (|w_1|, \ldots, |w_n|)$. Then the power series $\sum c_\nu z^\nu$ converges on the polydisc $P(0, r)$. Moreover, the convergence is **normal** *in the following sense: if $K \subset P(0, r)$ is compact and $\varepsilon > 0$ is arbitrary, there is a finite set $\Lambda = \Lambda(K, \varepsilon) \subset \mathbb{N}^n$, such that*

$$\sum_{\nu \notin \Lambda} |c_\nu z^\nu| < \varepsilon \qquad \text{for all } z \in K.$$

Proof. Given $K \subset\subset P(0, r)$, choose $0 < \lambda < 1$, such that $K \subset P(0, \lambda r)$. For $z \in P(0, \lambda r)$ one obtains from (1.31) that

$$|c_\nu z^\nu| \leq |c_\nu w^\nu| \lambda^{|\nu|} \leq M \lambda^{|\nu|} \qquad \text{for } \nu \in \mathbb{N}^n.$$

Since $\sum_{\nu \in \mathbb{N}^n} \lambda^{|\nu|} = (\sum_{j=0}^{\infty} \lambda^j)^n < \infty$, the result follows. ∎

Corollary 1.16. *The domain of convergence Ω of the power series $\sum c_\nu z^\nu$ is a (possibly empty) complete Reinhardt domain, and Ω is the interior of the set of points $w \in \mathbb{C}^n$ which satisfy* (1.31). *The convergence is* **normal** *in Ω.*

Theorem 1.17. *A power series $f(z) = \sum c_\nu z^\nu$ with nonempty domain of convergence Ω defines a holomorphic function $f \in \mathcal{O}(\Omega)$. Moreover, for $\alpha \in \mathbb{N}^n$, the series of derivatives $\sum c_\nu (D^\alpha z^\nu)$ converges compactly to $D^\alpha f$ on Ω, and*

$$D^\alpha f(0) = \alpha! \, c_\alpha. \tag{1.32}$$

Proof. We fix a bijection $\sigma \colon \mathbb{N} \to \mathbb{N}^n$. Then

$$f(z) = \lim_{k \to 0} \sum_{j=0}^{k} c_{\sigma(j)} z^{\sigma(j)}$$

compactly on Ω. Since the partial sums are holomorphic, Theorem 1.9 implies $f \in \mathcal{O}(D)$ and

$$D^\alpha f(z) = \lim_{k \to \infty} \sum_{j=0}^{k} c_{\sigma(j)} (D^\alpha z^{\sigma(j)})$$

on Ω. Thus, for fixed $\alpha \in \mathbb{N}^n$ and $w \in \Omega$,

$$\sup_{\nu \in \mathbb{N}^n} \left| c_\nu (D^\alpha z^\nu) \Big|_{z=w} \right| < \infty.$$

Corollary 1.16 implies that Ω is contained in the domain of convergence of the power series $\sum c_\nu (D^\alpha z^\nu)$. Equation (1.32) follows by evaluating $D^\alpha f(z) = \sum c_\nu (D^\alpha z^\nu)$ at $z = 0$. ∎

In fact, $\sum c_\nu z^\nu$ and $\sum c_\nu (D^\alpha z^\nu)$ have the same domain of convergence (see Exercise E.1.8).

The domains of convergence of power series in several variables exhibit a much greater variety than in one variable. We give some examples in Figure 3 (the reader should verify the statements made).

Clearly a Hartogs domain $H(r)$ (see Figure 2) is not the domain of convergence of a power series; every power series which converges on $H(r)$ must necessarily converge on the polydisc $P(0, 1)$ (use Lemma 1.15). We will show in Chapter II, §1, that every $f \in \mathcal{O}(H(r))$ can be represented on $H(r)$ by a convergent power series, which therefore defines a holomorphic extension of f to $P(0, 1)$!

Not every complete Reinhardt domain is the precise domain of convergence of some power series (except in case $n = 1$, of course). We will discuss the characterization of domains of convergence of power series in Chapter II, §3.8.

$\tau(\Omega)$

(1.33) $\quad \sum_{\nu_1, \nu_2 \geq 0} z_1^{\nu_1} z_2^{\nu_2},$

(1.34) $\quad \sum_{\nu=0}^{\infty} (z_1 z_2)^\nu,$

(1.35) $\quad \sum_{\nu_1, \nu_2 \geq 0} \frac{\nu_1}{\nu_2!} z_1^{\nu_1} z_2^{\nu_2},$

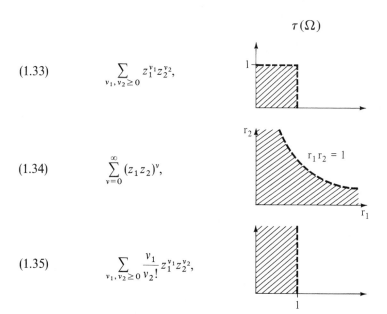

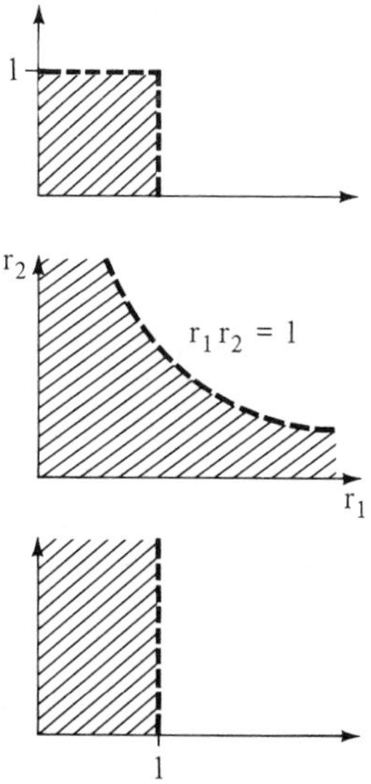

Figure 3. Domains of convergence of some power series.

1.6. Taylor Expansion and Identity Theorems

We now show that every holomorphic function can be represented locally by a convergent power series. Together with Theorem 1.17, this shows that the space $\mathcal{O}(D)$ can also be defined in terms of power series. This is the approach taken, for example, in [GuRo] or [Nar 3].

Theorem 1.18. *Let $f \in \mathcal{O}(P(a, r))$. Then the Taylor series of f at a converges to f on $P(a, r)$, that is,*

$$f(z) = \sum_{\nu \in \mathbb{N}^n} \frac{D^\nu f(a)}{\nu!}(z - a)^\nu \qquad \text{for } z \in P(a, r).$$

Proof. In the Cauchy integral formula (1.15), applied to $z \in P(a, \rho) \subset\subset P(a, r)$, one expands $(\zeta - z)^{-1} = (\zeta_1 - z_1)^{-1} \ldots (\zeta_n - z_n)^{-1}$ into a multiple geometric series

$$(\zeta - z)^{-1} = \sum_{\nu \in \mathbb{N}^n} \frac{(z - a)^\nu}{(\zeta - a)^{\nu+1}}, \tag{1.36}$$

which converges uniformly for $\zeta \in b_o P(a, \rho)$, since $|z_j - a_j|/|\zeta_j - a_j| \leq |z_j - a_j|/\rho_j < 1$ for such ζ and all $1 \leq j \leq n$. It is therefore legitimate to substitute (1.36) into (1.15) and to interchange summation and integration, leading to

$$f(z) = \sum_{\nu \in \mathbb{N}^n} \left[(2\pi i)^{-n} \int_{b_o P(a, \rho)} \frac{f(\zeta)\, d\zeta_1 \ldots d\zeta_n}{(\zeta - a)^{\nu+1}} \right] (z - a)^\nu \tag{1.37}$$

for $z \in P(a, \rho)$. By (1.21), or by Theorem 1.17, the coefficient of $(z - a)^\nu$ in (1.37) equals $D^\nu f(a)/\nu!$. ∎

Theorem 1.19. *Let $D \subset \mathbb{C}^n$ be connected. If $f \in \mathcal{O}(D)$ and there is $a \in D$, such that $D^\alpha f(a) = 0$ for all $\alpha \in \mathbb{N}^n$, then $f(z) = 0$ for $z \in D$. In particular, if there is a nonempty open set $U \subset D$, such that $f(z) = 0$ for $z \in U$, then $f \equiv 0$ on D.*

Proof. Theorem 1.18 implies that the set $\Omega = \{z \in D: D^\alpha f(z) = 0 \text{ for all } \alpha \in \mathbb{N}^n\}$ is open. By continuity of $D^\alpha f$, Ω is also closed, and since the hypothesis says that $\Omega \neq \varnothing$, the connectedness of D implies $\Omega = D$. ∎

Remark 1.20. The hypothesis in Theorem 1.19 will hold if f vanishes on a set E which is "thick" enough. For example, in $\mathbb{C}^1$ it suffices that E has an accumulation point in D, but this is clearly not enough if $n > 1$. The function $f(z_1, z_2) = z_1$ is zero on $\{(0, z_2): z_2 \in \mathbb{C}\}$, but $f \not\equiv 0$. An obvious necessary and sufficient condition is that E not be contained in the zero set of a nontrivial holomorphic function; but this is really a tautology, unless one has more precise geometric information about such zero sets. We will consider this question in §3. Here we mention one case which shows that more than topological or measure theoretic properties are involved: *Suppose $f \in \mathcal{O}(D)$,*

$a \in D$, and $f(a + x) = 0$ for all x in a neighborhood of 0 in $\mathbb{R}^n$; then $D^\alpha f(a) = 0$ for all α, and *hence $f \equiv 0$ on D.*

Theorem 1.20. *Let D be connected. Then $\mathcal{O}(D)$ is an integral domain.*

PROOF. Suppose $f, g \in \mathcal{O}(D)$ and $f(z) \cdot g(z) = 0$ for $z \in D$. If $f \not\equiv 0$, there is $a \in D$ with $f(a) \neq 0$, and hence $f(z) \neq 0$ in a neighborhood U of a. But then $g(z) = 0$ for $z \in U$, which implies $g(z) = 0$ for all $z \in D$ by Theorem 1.19. ■

The following result is an easy generalization of the corresponding classical one variable result.

Theorem 1.21. *Let D be connected and suppose $f \in \mathcal{O}(D)$ is not constant. Then $f(\Omega)$ is open for any open set $\Omega \subset D$.*

PROOF. It is enough to show that for any ball $B(a, r) \subset D$, $f(B(a, r))$ is a neighborhood of $f(a)$. Theorem 1.19 implies that $f|_{B(a,r)}$ is not constant, otherwise f would have to be constant on D. Choose $p \in B(a, r)$ such that $f(p) \neq f(a)$, and define $h(\lambda) = f(a + \lambda p)$ for $\lambda \in \Delta = \{\lambda \in \mathbb{C}: |\lambda| \leq 1\}$. Then h is nonconstant on $\bar{\Delta}$ and holomorphic—just compute $\partial h/\partial \bar{\lambda} = 0$, or see Theorem 2.3. By the known one variable result (cf. [Ah1], p. 132), $h(\bar{\Delta}) \subset f(B(a, r))$ is a neighborhood of $h(0) = f(a)$. ■

Corollary 1.22. *Suppose $f \in \mathcal{O}(D)$ and that $|f|$ has a local maximum at the point $a \in D$. Then f is constant on the connected component of D containing a.*

Corollary 1.23. *Suppose $D \subset\subset \mathbb{C}^n$ and $f \in A(D) = C(\bar{D}) \cap \mathcal{O}(D)$. Then*

$$|f(z)| \leq |f|_{bD} \qquad \text{for all } z \in \bar{D}.$$

EXERCISES

E.1.1. Show that an open set D in $\mathbb{C}^n$ is connected if and only if D is *pathwise connected.* (i.e., if $P, Q \in D$, there is a continuous map $\varphi: [0, 1] \to D$ with $\varphi(0) = P$, $\varphi(1) = Q$.)

E.1.2. Let D be open in $\mathbb{C}^n$. For $j = 1, 2, \ldots$ define

$$K_j = \{z \in D: \delta_D(z) \geq 1/j \text{ and } |z| \leq j\}.$$

Show that K_j is compact, $K_j \subset$ interior K_{j+1} and $D = \bigcup_{j=1}^\infty K_j$. (This shows that every open set D in $\mathbb{C}^n$ has a normal exhaustion.)

E.1.3. Show that if D is open in $\mathbb{C}^n$ and $f: D \to \mathbb{C}$ is holomorphic in each variable separately and locally bounded (i.e., for all $a \in D$ there is a neighborhood $U_a \subset D$ of a such that $f|_{U_a}$ is bounded), then f is continuous on D.

E.1.4. Show that $\mathcal{O}L^p(\mathbb{C}^n) = \{0\}$ for $1 \leq p < \infty$ and that $\mathcal{O}L^\infty(\mathbb{C}^n) = \mathbb{C}$.

E.1.5. Let $C(D)$ be the space of continuous functions on $D \subset \mathbb{C}^n$, with the topology of compact convergence.

(i) For $K \subset D$, compact, $\varepsilon > 0$, and $g \in C(D)$, set $U(g; K, \varepsilon) = \{f \in C(D): |f - g|_K < \varepsilon\}$. Show that if $\{K_j\}$ is a normal exhaustion of D, then $\{U(g; K_j, 1/l): j, l = 1, 2, \ldots\}$ is a neighborhood basis for g.
(ii) Prove in detail that $C(D)$ is metrizable.

E.1.6. Prove Lemma 1.12.

E.1.7. Show that if $\Omega \subset\subset D \subset \mathbb{C}^n$ are open, then the restriction of $f \in \mathcal{O}(D)$ to $f|_\Omega \in \mathcal{O}(\Omega)$ defines a *compact* map $\mathcal{O}(D) \to \mathcal{O}(\Omega)$. (This means that there is a neighborhood $V \subset \mathcal{O}(D)$ of 0, such that its image in $\mathcal{O}(\Omega)$ has compact closure.)

E.1.8. Prove that a power series $\sum c_\nu z^\nu$ and the derived series $\sum c_\nu(D^\alpha z^\nu)$ have equal domain of convergence for every multi-index $\alpha \in \mathbb{N}^n$.

E.1.9. Let D be open in $\mathbb{C}^n$ and let $\Delta = \{z \in \mathbb{C}: |z| < 1\}$. Show that for $N \in \mathbb{N}^+$, every $f \in \mathcal{O}(D \times \Delta^N)$ has a power series representation

$$f(z, w) = \sum_{\nu \in \mathbb{N}^N} a_\nu(z) w^\nu$$

with coefficients $a_\nu \in \mathcal{O}(D)$, which converges compactly on $D \times \Delta^N$.

E.1.10. A domain D in $\mathbb{C}^n$ is called a **Hartogs domain** if $z = (z', z_n) \in D$ implies that $(z', e^{i\theta} z_n) \in D$ for all $0 \le \theta \le 2\pi$. Show that every function f holomorphic on a Hartogs domain D has a Laurent series expansion with respect to z_n,

$$f(z) = \sum_{-\infty}^{\infty} a_j(z') z_n^j,$$

which converges compactly on D, and whose coefficients are holomorphic in z'.

E.1.11. Let $P = P(0, (r_1, \ldots, r_n)) \subset \mathbb{C}^n$ be a polydisc. If $\zeta \in bP$ satisfies $|\zeta_l| = r_l$ for some l, the set

$$P'(l, \zeta) = \{z \in \bar{P}: z_l = \zeta_l, |z_j| < r_j \text{ for } j \ne l\}$$

can be viewed as a polydisc in $\mathbb{C}^{n-1}$. Show that if $f \in A(P)$, then f restricts to a holomorphic function on $P'(l, \zeta)$ in $n - 1$ variables.

E.1.12. Let P be a bounded polydisc in $\mathbb{C}^n$. Show that if $S \subset bP$ satisfies $|f(z)| \le |f|_S$ for all $z \in \bar{P}$ and $f \in A(P)$, then S contains the distinguished boundary $b_0 P$ of P. (Together with Theorem 1.8, this shows that $b_0 P$ is the Shilov boundary of $A(P)$.)

E.1.13. Let $D \subset \mathbb{C}^n$ be connected and suppose $f: D \times D \to \mathbb{C}$ is holomorphic in the $2n$ complex variables $(z, w) \in D \times D$. Show that if there is a point $p \in D$ with $\bar{p} \in D$, such that $f(z, \bar{z}) = 0$ for all z in a neighborhood of p, then $f(z, w) = 0$ for all $(z, w) \in D \times D$. (Hint: Introduce new coordinates $u = z + w$, $v = z - w$.)

§2. Holomorphic Maps

2.1. The Derivative of a Holomorphic Map

Let $D \subset \mathbb{C}^n$ be open and consider a map $F: D \to \mathbb{C}^m$. By writing $F = (f_1, \ldots, f_m)$ and $f_k = u_k + \sqrt{-1} v_k$, where u_k, v_k are real valued functions on D, we can view $F = (u_1, v_1, \ldots, u_m, v_m)$ as a map from $D \subset \mathbb{R}^{2n}$ into $\mathbb{R}^{2m}$. If F is differentiable at $a \in D$, its differential $dF(a)\colon \mathbb{R}^{2n} \to \mathbb{R}^{2m}$ is a linear transformation with matrix representation given by the (real) Jacobian matrix

$$J_{\mathbb{R}}(F) = \begin{bmatrix} \dfrac{\partial u_1}{\partial x_1} & \dfrac{\partial u_1}{\partial y_1} \cdots \dfrac{\partial u_1}{\partial y_n} \\ \dfrac{\partial v_1}{\partial x_1} & \cdots\cdots\cdots\cdots \\ \vdots & \\ \dfrac{\partial v_m}{\partial x_1} & \cdots\cdots \dfrac{\partial v_m}{\partial y_n} \end{bmatrix}$$

evaluated at a.

The map $F\colon D \to \mathbb{C}^m$ is called *holomorphic* if its (complex) components $f_1, \ldots, f_m$ are holomorphic functions on D. If F is holomorphic, its differential $dF(a)$ at $a \in D$ is a complex linear map $\mathbb{C}^n \to \mathbb{C}^m$ (this follows from Theorem 1.2), with complex matrix representation

$$F'(a) = \begin{bmatrix} \dfrac{\partial f_1}{\partial z_1}(a) \cdots \dfrac{\partial f_1}{\partial z_n}(a) \\ \vdots \\ \dfrac{\partial f_m}{\partial z_1}(a) \cdots \dfrac{\partial f_m}{\partial z_n}(a) \end{bmatrix}.$$

We call $F'(a)$ **the derivative (or complex Jacobian matrix) of the holomorphic map F at** a.

Lemma 2.1. *If $D \subset \mathbb{C}^n$ and $F\colon D \to \mathbb{C}^n$ is holomorphic, then*

$$\det J_{\mathbb{R}}F(z) = |\det F'(z)|^2 \geq 0$$

for $z \in D$.

Proof. After a permutation of the rows and columns one can write

$$\det J_{\mathbb{R}}F = \det \begin{bmatrix} \left(\dfrac{\partial u_k}{\partial x_j}\right) & \left(\dfrac{\partial u_k}{\partial y_j}\right) \\ \left(\dfrac{\partial v_k}{\partial x_j}\right) & \left(\dfrac{\partial v_k}{\partial y_j}\right) \end{bmatrix},$$

where the four blocks on the right are real $n \times n$ matrices. Adding $i = \sqrt{-1}$ times the bottom blocks to the top and using the Cauchy–Riemann equations $\partial f_k/\partial \bar{z}_j = 0$, i.e., $\partial u_k/\partial x_j = \partial v_k/\partial y_j$ and $\partial u_k/\partial y_j = -\partial v_k/\partial x_j$, one obtains

$$\det J_{\mathbb{R}}F = \det \begin{bmatrix} \left(\dfrac{\partial u_k}{\partial x_j} + i\dfrac{\partial v_k}{\partial x_j}\right) & \left(i\dfrac{\partial u_k}{\partial x_j} - \dfrac{\partial v_k}{\partial x_j}\right) \\ \left(\dfrac{\partial v_k}{\partial x_j}\right) & \left(\dfrac{\partial u_k}{\partial x_j}\right) \end{bmatrix}.$$

Now subtract $i = \sqrt{-1}$ times the left blocks from the right side; it follows that

$$\det J_{\mathbb{R}} F = \det \begin{bmatrix} \left(\dfrac{\partial f_k}{\partial x_j}\right) & \vdots & 0 \\ \cdots\cdots & \vdots & \cdots\cdots \\ * & \vdots & \left(\dfrac{\partial \bar{f}_k}{\partial x_j}\right) \end{bmatrix} = \det F' \cdot \overline{\det F'},$$

where we have used that $\partial f/\partial z_j = \partial f/\partial x_j$ for holomorphic f. ■

2.2. Composition and the Chain Rule

We now discuss the important result that the composition of holomorphic maps is again holomorphic; in particular, this implies that the definition of holomorphic functions is independent of the Euclidean coordinates of $\mathbb{C}^n$.

Lemma 2.2. *Let $D \subset \mathbb{C}^n$ and $\Omega \subset \mathbb{C}^m$ be open sets. If $F = (f_1, \ldots, f_m): D \to \Omega$ is holomorphic and $g \in \mathcal{O}(\Omega)$, then $g \circ F \in \mathcal{O}(D)$; moreover, for $a \in D$ and $1 \leq j \leq n$,*

$$\frac{\partial(g \circ F)}{\partial z_j}(a) = \sum_{k=1}^{m} \frac{\partial g}{\partial w_k}(F(a)) \frac{\partial f_k}{\partial z_j}(a). \tag{2.1}$$

Proof. We give two proofs of this result. The first one is based on power series, while the second uses a complex version of the real chain rule, which is useful in other contexts as well.

Suppose $a \in D$, $F(a) = b \in \Omega$. Choose a polydisc $P(b, \varepsilon) \subset\subset \Omega$, such that

$$g(w) = g(b) + \sum_{|\nu| \geq 1} \frac{D^\nu g}{\nu!}(b)(w - b)^\nu,$$

with normal convergence on $P(b, \varepsilon)$. By continuity of F, there is a polydisc $P(a, \delta) \subset D$, such that $F(z) \in P(b, \varepsilon)$ for $z \in P(a, \delta)$. Hence, for $z \in P(a, \delta)$,

$$g(F(z)) = g(F(a)) + \sum_{|\nu| \geq 1} \frac{D^\nu g}{\nu!}(b)(F(z) - b)^\nu,$$

with normal convergence on $P(a, \delta)$. Since the terms of the series are holomorphic, it follows that $g \circ F \in \mathcal{O}(P(a, \delta))$; moreover, for any $1 \leq j \leq n$,

$$\frac{\partial(g \circ F)}{\partial z_j}(a) = \sum_{|\nu| \geq 1} \frac{D^\nu g}{\nu!}(b) \left[\frac{\partial}{\partial z_j}(F(z) - b)^\nu\right](a).$$

In the latter series, the terms with $|\nu| > 1$ are 0, and (2.1) follows.

For the second proof, if F and g are only differentiable, then $g \circ F$ is differentiable on D, and the (real) chain rule implies (use (1.13)!) that

$$\frac{\partial(g \circ F)}{\partial z_j} = \sum_{k=1}^{m} \left[\left(\frac{\partial g}{\partial w_k} \circ F\right) \frac{\partial f_k}{\partial z_j} + \left(\frac{\partial g}{\partial \bar{w}_k} \circ F\right) \frac{\partial \bar{f}_k}{\partial z_j}\right], \tag{2.2}$$

$$(2.3) \qquad \frac{\partial(g \circ F)}{\partial \bar{z}_j} = \sum_{k=1}^{m} \left[\left(\frac{\partial g}{\partial w_k} \circ F \right) \frac{\partial f_k}{\partial \bar{z}_j} + \left(\frac{\partial g}{\partial \bar{w}_k} \circ F \right) \frac{\partial \bar{f}_k}{\partial \bar{z}_j} \right],$$

for any $1 \le j \le n$. If, in addition, F and g are holomorphic, then (2.3) implies $\bar{\partial}(g \circ F) = 0$, i.e., $g \circ F \in \mathcal{O}(D)$, and (2.2) implies (2.1). ∎

By applying Lemma 2.2 to each component of a holomorphic map G, one immediately obtains the following result.

Theorem 2.3. *Suppose $F\colon D \to \Omega \subset \mathbb{C}^m$ and $G\colon \Omega \to \mathbb{C}^l$ are holomorphic maps. Then $G \circ F\colon D \to \mathbb{C}^l$ is holomorphic and*

$$(G \circ F)'(z) = G'(F(z)) \cdot F'(z) \qquad \text{for } z \in D.$$

2.3. The Implicit Mapping Theorem

The study of solution sets of analytic equations, that is, the common zero set of one or several holomorphic functions, is of fundamental importance in the theory of several complex variables. A brief introduction into the more elementary properties of such sets, called *analytic sets*, will be presented in §3. Here we first deal with the easier case of *nonsingular* equations.

Theorem 2.4. *Let $D \subset \mathbb{C}^n$ and let $F = (f_1, \ldots, f_m)\colon D \to \mathbb{C}^m$ be holomorphic. Suppose $m \le n$, $F(a) = 0$ for some $a \in D$, and*

$$(2.4) \qquad \det \left[\frac{\partial f_k}{\partial z_j}(a) \right]_{\substack{k=1,\ldots,m \\ j=n-m+1,\ldots,n}} \neq 0.$$

Then there are $\varepsilon' > 0$, $\varepsilon'' > 0$, and a holomorphic map $h = (h_1, \ldots, h_m)\colon B'(a', \varepsilon') \to B''(a'', \varepsilon'')$, where $a' = (a_1, \ldots, a_{n-m})$, $a'' = (a_{n-m+1}, \ldots, a_n)$, with the following property:

$$(2.5) \qquad \begin{aligned} &\text{if } z = (z', z'') \in B'(a', \varepsilon') \times B''(a'', \varepsilon''), \text{ then} \\ &F(z', z'') = 0 \qquad \text{if and only if } z'' = h(z'). \end{aligned}$$

In case $m = n$, the theorem means that h is constant, and hence $z = a$ is the only solution of $F(z) = 0$ in a neighborhood of a. If $m < n$, the theorem means geometrically that the set $\{z \in D\colon F(z) = 0\}$ is, near a, the graph of a holomorphic map h in $n - m$ variables (see Figure 4).

Proof. Lemma 2.1, applied to the map $\tilde{F}$, defined by $\tilde{F}(z'') = F(a', z'')$ in a neighborhood of a'', shows that $\det J_{\mathbb{R}}\tilde{F}(a'') \neq 0$. Hence the implicit mapping theorem from real calculus (see [Nar 4], §1.3) can be applied, yielding $\varepsilon', \varepsilon'' > 0$ and a C^1 map $h = (h_1, \ldots, h_m)\colon B'(a', \varepsilon') \to B''(a'', \varepsilon'')$, such that (2.5) holds. To complete the proof we must show that h is holomorphic near a.

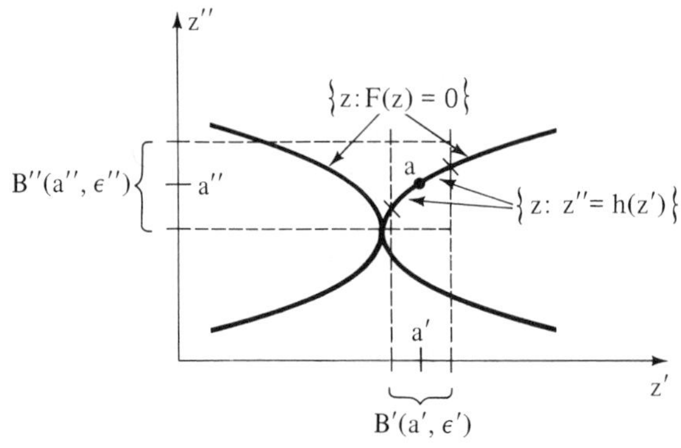

Figure 4. Local representation of the solution set of $F(z) = 0$ as the graph of the holomorphic function h.

Since f_k is holomorphic and $f_k(z', h(z')) \equiv 0$ for $z' \in B'(a', \varepsilon')$ and $1 \le k \le m$, one obtains, by applying $\partial/\partial\bar{z}_l$, $1 \le l \le n - m$, and using (2.3), that

$$\sum_{j=1}^{m} \frac{\partial f_k}{\partial z_{n-m+j}}(z', h(z')) \cdot \frac{\partial h_j}{\partial \bar{z}_l}(z') = 0, \qquad 1 \le k \le m. \tag{2.6}$$

By (2.4), the matrix of the system of linear equations (2.6) is nonsingular at $z' = a'$, and hence on $B'(a', \varepsilon')$ for sufficiently small ε'. Therefore $(\partial h_j/\partial \bar{z}_l)(z') = 0$ for $1 \le j \le m$ and $1 \le l \le n - m$, so that h is holomorphic on $B'(a', \varepsilon')$. ■

The hypotheses (2.4) in the theorem is equivalent, except for a renumbering of the components of F, to the statement that the derivative F' has maximal rank $= \min(n, m)$ at the point a. We say that F **is nonsingular at** a if $F'(a)$ has maximal rank; F is **nonsingular (on** D**)**, if F is nonsingular at every $a \in D$.

It is easy to see that in case $F: D \to \mathbb{C}^m$ is nonsingular at $a \in D$ and $m > n$, the conclusion is the same as in Theorem 2.4 for $m = n$, namely $z = a$ is an isolated zero of F. In fact, even more is true: F is injective on a neighborhood of a (see Corollary 2.6 below).

2.4. Biholomorphic Maps

We now consider in more detail the equidimensional case $m = n$.

Theorem 2.5. *Suppose $D \subset \mathbb{C}^n$ and the holomorphic map $F: D \to \mathbb{C}^n$ is nonsingular at a (i.e.,* det $F'(a) \neq 0$*). Then there are open neighborhoods U of a and W of $b = F(a)$, such that $F|_U: U \to W$ is a homeomorphism with holomorphic inverse $H: W \to U$.*

Proof. We introduce the map $G(w, z) = F(z) - w$ from $\mathbb{C}^n \times D$ into $\mathbb{C}^n$. By hypothesis, $G(b, a) = 0$ and

$$\det\left[\frac{\partial g_k}{\partial z_j}\right]_{\substack{k=1,\ldots,n\\ j=1,\ldots,n}}(b, a) = \det F'(a) \neq 0.$$

Therefore Theorem 2.4 gives a holomorphic map H from a neighborhood W of b into a ball $B(a, \varepsilon) \subset D$, such that for $(w, z) \in W \times B$ one has $G(w, z) = 0$, i.e. $w = F(z)$, if and only if $z = H(w)$. It follows that $H: W \to U = F^{-1}(W)$ is the desired holomorphic inverse of $F|_U$. ■

Corollary 2.6. *Suppose $D \subset \mathbb{C}^n$ and $F: D \to \mathbb{C}^m$ is holomorphic and nonsingular at $a \in D$. If $m \geq n$, then there is a neighborhood U of a, such that $F|_U$ is injective.*

Proof. Since rank $F'(a) = n$, after renumbering the components of $F = (f_1, \ldots, f_m)$, one can assume that $\tilde{F} = (f_1, \ldots, f_n)$ is nonsingular at a. Theorem 2.5 now implies that $\tilde{F}$, and hence also F, is injective on some neighborhood U of a. ■

Let D_1, D_2 be open sets in $\mathbb{C}^n$, resp. $\mathbb{C}^m$; we say that the map $F: D_1 \to D_2$ is **biholomorphic** if F is a holomorphic homeomorphism with holomorphic inverse $F^{-1}: D_2 \to D_1$. If F is biholomorphic, it follows from the chain rule that $(F^{-1})'(F(z))$ is the inverse matrix of $F'(z)$; in particular, F is nonsingular, and $m = n$. The open sets D_1 and D_2 are called **biholomorphically equivalent** if there is a biholomorphic map $F: D_1 \to D_2$. $F: D_1 \to D_2$ is called **biholomorphic at** $a \in D_1$ if there is a neighborhood U of a, such that $F|_U: U \to F(U)$ is biholomorphic. Theorem 2.5 can now be reformulated: *If $D \subset \mathbb{C}^n$ and $F: D \to \mathbb{C}^n$ is holomorphic and nonsingular at $a \in D$, then F is biholomorphic at a.*

If $F: U \to W$ is biholomorphic, with $F(z) = w = (w_1, \ldots, w_n)$, we also say that $(w_1, \ldots, w_n)$ is a **holomorphic**, or **complex coordinate system** on U. A function $h(z)$ on U can then be expressed in terms of the w-coordinates, i.e., by considering $h \circ (F^{-1})(w)$, and the analytic properties of h do not depend on the choice of coordinates. It will often be useful to introduce special holomorphic coordinates in order to simplify the geometry. We will see this technique at work in the following sections.

Remark. The results discussed here and in the preceding section are the obvious analogues of well known theorems in real calculus. More surprising is the fact that an injective holomorphic map F from $D \subset \mathbb{C}^n$ into $\mathbb{C}^n$ is necessarily nonsingular, and hence biholomorphic from D onto $F(D)$. No comparable result exists in real calculus: consider the map $f: \mathbb{R} \to \mathbb{R}$ given by $f(x) = x^3$! In case $n = 1$, this result is an easy consequence of the residue theorem, but for $n > 1$ the proof is more subtle. We will discuss it in §2.8, after we have introduced the concept of complex submanifold in §2.6.

Example. It is well known that every biholomorphic map $F: \mathbb{C} \to \mathbb{C}$ is necessarily linear, i.e., of the form $F(z) = az + b$ for some constants $a, b \in \mathbb{C}$. In contrast, the group of automorphisms $\mathrm{Aut}(\mathbb{C}^2) = \{F: \mathbb{C}^2 \to \mathbb{C}^2$, biholomorphic} is much larger: every entire function $h: \mathbb{C} \to \mathbb{C}$ defines a biholomorphic map $F_h: \mathbb{C}^2 \to \mathbb{C}^2$ by setting $F_h(z, w) = (z + h(w), w)$.

2.5. The Biholomorphic Inequivalence of Ball and Polydisc

The Riemann Mapping Theorem states that a connected, simply connected domain in the complex plane is either $\mathbb{C}$ itself or else it is biholomorphic to the open unit disc. The following result shows that it is impossible to find a higher dimensional analog of Riemann's Theorem which involves only topological conditions.

Theorem 2.7. *There exists no biholomorphic map*

$$F: P(0, 1) \to B(0, 1)$$

between polydisc and ball in $\mathbb{C}^n$ *if* $n > 1$.

This fact was discovered by H. Poincaré in 1907 ("Les fonctions analytiques de deux variables et la représentation conforme," Rend. Circ. Mat. Palermo **23**(1907), 185–220). Poincaré's original proof was based on a computation and comparison of the groups of holomorphic automorphisms of ball and bidisc which fix the origin. The proof given below is more direct and elementary, and its basic idea is applicable in much more general settings (see Exercise E.II.2.12).

PROOF. For simplicity, we consider the case $n = 2$; the argument easily generalizes to arbitrary $n \geq 2$. Let $\Delta = \{\zeta \in \mathbb{C}: |\zeta| < 1\}$ be the open unit disc in $\mathbb{C}$. Suppose $F = (f_1, f_2): \Delta \times \Delta \to B = B(0, 1) \subset \mathbb{C}^2$ is biholomorphic. We will show that for each fixed $w \in \Delta$ the holomorphic map $F_w: \Delta \to B$ defined by

$$F_w(z) = \left(\frac{\partial f_1}{\partial w}(z, w), \frac{\partial f_2}{\partial w}(z, w)\right)$$

satisfies

$$(*) \qquad \lim_{z \to b\Delta} F_w(z) = 0.$$

This immediately gives a contradiction, as follows: $(*)$ implies that F_w extends continuously to $\bar{\Delta}$, with boundary values 0. Since F_w is holomorphic on Δ, it follows that $F_w \equiv 0$ on Δ, i.e., $F(z, w)$ is independent of w, and F could not be one-to-one.

To prove $(*)$ it is enough to show that every sequence $\{z_\nu\} \subset \Delta$ with $|z_\nu| \to 1$ has a subsequence $\{z_{\nu_j}\}$ with $\lim_{j\to\infty} F_w(z_{\nu_j}) = 0$. Given such a sequence $\{z_\nu\}$, an application of Montel's Theorem to the bounded sequence $\{F(z_\nu, \cdot)$, $\nu = 1, 2, \dots\}$ of holomorphic maps $F(z_\nu, \cdot): \Delta \to B$ in the second variable gives a subsequence $\{z_{\nu_j}\}$, such that $\{F(z_{\nu_j}, \cdot)\}$ converges compactly in Δ to a holomorphic map $\varphi: \Delta \to \bar{B}$. Since F is biholomorphic, we must have $F(z_\nu, w) \to$

bB for every $w \in \Delta$ as $z_\nu \to b\Delta$; hence $\varphi(\Delta) \subset bB$, i.e., if $\varphi = (\varphi_1, \varphi_2)$, then $|\varphi_1(w)|^2 + |\varphi_2(w)|^2 = 1$ for all $w \in \Delta$. By applying $\partial^2/\partial\bar{w}\partial w$ to this equation one obtains $|\varphi_1'(w)|^2 + |\varphi_2'(w)|^2 = 0$, so $\varphi' \equiv 0$ on Δ. Since

$$F_w(z_{\nu_j}, w) \to \varphi'(w) \qquad \text{as } j \to \infty,$$

the desired conclusion follows. ∎

Theorem 2.7 shows that simply connected domains in dimension two or higher are much more "analytically rigid" than in the plane. Related to this theme, it has long been known that in $\mathbb{C}^2$ there exist simply connected domains whose only holomorphic automorphism is the identity (cf. [BeTh], p. 169). In 1935 W. Rothstein ("Zur Theorie der analytischen Abbildungen im Raum zweier komplexer Veränderlichen," Diss. Univ. Münster, 1935) found the first *domain of holomorphy* with these properties. (See Chapter II, §2.1, for the definition of this concept.) More recently, D. Burns and S. Shnider [BuSh] showed that "almost every" sufficiently small C^∞ perturbation of the unit ball in $\mathbb{C}^2$ has no holomorphic automorphism besides the identity, and hence, in particular, is not biholomorphically equivalent to the ball.[1] The situation in higher dimensions is thus considerably more complicated than in the plane, and a great deal of progress has been made in this area during the last decade. The result of Burns and Shnider mentioned above makes use of *Fefferman's Mapping Theorem*, a fundamental result dealing with the C^∞ extension to the boundary of biholomorphic maps, which was proved in 1974 by C. Fefferman [Fef]. This theorem made it possible to apply some classical results of E. Cartan on biholomorphic invariants of hypersurfaces in $\mathbb{C}^2$. We will prove Fefferman's Theorem in Chapter VII, §8.

2.6. Complex Submanifolds

We now introduce a (local) generalization of the concept of complex linear subspace of $\mathbb{C}^n$ which is invariant under complex coordinate changes. Usually there is no need to introduce this concept in function theory in one complex variable, as the relevant sets are either open subsets of $\mathbb{C}$ or discrete, but in several variables these so-called complex submanifolds appear naturally as the solution sets of nonsingular systems of holomorphic equations, and, together with the more general concept of analytic set, they are a very important tool for proofs by induction over the number of variables.

Definition. A set $M \subset \mathbb{C}^n$ is called a **complex submanifold** (of $\mathbb{C}^n$), if for every point $P \in M$ there are a holomorphic coordinate system $(w_1, \ldots, w_n)$ on a

[1] It is known that—just as in case $n = 1$—all holomorphic automorphisms of the unit ball B in $\mathbb{C}^n$ are rational, and that the group $\mathrm{Aut}(B)$ is transitive. This was proved by K. Reinhardt [Rei] in case $n = 2$ (see also [BeTh], p. 162); the reader may find a proof for arbitrary n in [Rud 3].

neighborhood U of P, and an integer k, $0 \le k \le n$, such that

$$(2.7) \qquad M \cap U = \{z \in U: w_j(z) = 0 \quad \text{for } j > k\}.$$

The integer k appearing in (2.7) is called the **(complex) dimension of** M **at** P, and it is denoted by $k = \dim_{\mathbb{C}} M_P$, or simply $\dim M_P$. $\dim M_P$ is independent of the holomorphic coordinate system appearing in the definition. In fact, if $w^{\#}$ is another such coordinate system on U, with $M \cap U = \{z \in U: w_j^{\#}(z) = 0 \text{ for } j > k^{\#}\}$, then $w^{\#} \circ w^{-1}: w(U) \to w^{\#}(U)$ is biholomorphic. The sets $\Omega = w(U) \cap \{w_j = 0 \text{ for } j > k\}$ and $\Omega^{\#} = w^{\#}(U) \cap \{w_j^{\#} = 0 \text{ for } j > k^{\#}\}$ can be viewed as open subsets of $\mathbb{C}^k$, respectively $\mathbb{C}^{k\#}$, and the restriction $w^{\#} \circ w^{-1}|_{\Omega}$: $\Omega \to \Omega^{\#}$ is biholomorphic; this implies $k = k^{\#}$.

Notice that $\dim M_P$ is locally constant on M, and hence is constant on each connected component of M. The **dimension of** M is defined by

$$\dim M = \sup_{P \in M} \dim_{\mathbb{C}} M_P.$$

Every open set $D \subset \mathbb{C}^n$ is a complex submanifold of dimension n at every point. Conversely, if $\dim M_P = n$, there is a neighborhood U of P in $\mathbb{C}^n$, such that $M \cap U = U$. If E is a k-dimensional complex affine subset of $\mathbb{C}^n$, then E and $E \cap D$ are complex submanifolds of dimension k at every point. Also, a set $S \subset \mathbb{C}^n$ is discrete (i.e., every point is isolated) if and only if S is a complex submanifold of dimension 0. The details are left to the reader.

The following characterization gives more interesting examples.

Theorem 2.8. *A subset M of $\mathbb{C}^n$ is a complex submanifold if and only if for every $P \in M$ there are a neighborhood U of P, an open ball $B^{(k)}(a, \varepsilon) \subset \mathbb{C}^k$, and a nonsingular holomorphic map $H: B^{(k)}(a, \varepsilon) \to \mathbb{C}^n$, such that*

$$(2.8) \qquad H(B^k(a, \varepsilon)) = M \cap U.$$

A map H which satisfies all the conditions in Theorem 2.8 is called a **local parametrization of** M **at** P.

PROOF. Suppose first that M is a complex submanifold, and let $w = (w_1, \ldots, w_n): U \to W$ be a coordinate system on the neighborhood U of P which satisfies (2.7), with $k = \dim M_P$. By shrinking W and U we may assume that $W = B(a, \varepsilon)$, where $a = w(P)$ and $\varepsilon > 0$. Let $a' = (a_1, \ldots, a_k)$ and set $\tilde{H} = w^{-1}$; the map $H: B^{(k)}(a', \varepsilon) \to \mathbb{C}^n$ defined by $H(w_1, \ldots, w_k) = \tilde{H}(w_1, \ldots, w_k, 0, \ldots, 0)$ has all the required properties.

Conversely, if $H: B^{(k)}(a, \varepsilon) \to \mathbb{C}^n$ is a local parametrization of M at P, we may assume that $H(a) = P$. Since H is nonsingular, there are vectors $u_{k+1}, \ldots, u_n \in \mathbb{C}^n$ which, together with the vectors $\partial H/\partial w_1(a), \ldots, \partial H/\partial w_k(a)$, form a basis for $\mathbb{C}^n$. Define the map $\tilde{H}$ for $w \in B(\tilde{a}, \varepsilon) \subset \mathbb{C}^n$, $\tilde{a} = (a_1, \ldots, a_k, 0, \ldots, 0)$, by

$$\tilde{H}(w) = H(w_1, \ldots, w_k) + w_{k+1}u_{k+1} + \cdots + w_n u_n.$$

Then $\tilde{H}$ is biholomorphic at $\tilde{a}$, and $w = \tilde{H}^{-1}$ is a coordinate system on a neighborhood U of P which satisfies (2.7). ■

Theorem 2.9. *Let* $D \subset \mathbb{C}^n$ *and suppose* $F: D \to \mathbb{C}^m$ *is nonsingular. Then for every* $a \in D$ *the level set*

$$L_a(F) = \{z \in D: F(z) = F(a)\}$$

is a complex submanifold of dimension $\max(0, n - m)$ *at every point.*

Proof. Let $P \in L_a(F)$. If $m \geq n$, then P is isolated in $L_a(F)$ by Corollary 2.6, and the theorem is proved. We now assume $m < n$. After replacing F by $F^{\#} = F - F(a)$ and renumbering the coordinates, the hypotheses of Theorem 2.4 are satisfied for $F^{\#}$. It follows that in a neighborhood U of P, $\{z \in U: F^{\#}(z) = 0\}$ is the graph of a holomorphic map $h: B^{(k)}(P', \varepsilon) \to \mathbb{C}^m$, with $k = n - m$. The graph map $H(w) = (w, h(w)): B^{(k)}(P', \varepsilon) \to \mathbb{C}^n$ then defines a local parametrization of $\{z \in U: F^{\#}(z) = 0\} = L_a(F) \cap U$. The result now follows from Theorem 2.8. ■

Example. In the theory of one complex variable the Riemann surface S of $f(z) = \sqrt{z}$ is typically described as a "branched" covering of $\mathbb{C}$ or of the closed Riemann sphere. Only after a deeper investigation of the "branching point" 0 does one realize that in a more abstract sense the complex structure of S near 0 is the same as near every other point on S. Now consider $M = \{(z, w) \in \mathbb{C}^2: w^2 - z = 0\}$; M is a complex submanifold of $\mathbb{C}^2$ of dimension 1 (use Theorem 2.9!), and M gives a concrete representation of S (without the point at ∞). The projection $\pi_1: M \to \mathbb{C}$ onto the first coordinate exhibits the familiar branched covering of $M = S$ over $\mathbb{C}$, and the projection $\pi_2: M \to \mathbb{C}$ onto the second coordinate represents the function "$\sqrt{z}$" on M. There is an obvious way, made precise in the following section, in which these functions are "holomorphic on M".

2.7. Function Theory on Complex Submanifolds

The local parametrizations of a complex submanifold $M \subset \mathbb{C}^n$ can be used to define the concept of a holomorphic function on M.

Definition. The function $f: M \to \mathbb{C}$ is holomorphic at $P \in M$ if $f \circ H^{-1}$ is holomorphic at $H^{-1}(P)$ for a local parametrization H of M at P. f is holomorphic on M if f is holomorphic at P for every $P \in M$.

The definition is independent of the particular local parametrization H. (see Exercise E.2.10). Holomorphic maps $F: M \to \mathbb{C}^m$ are now defined in the obvious way. We leave it to the reader to verify the following results.

Theorem 2.10. *A function* $f: M \to \mathbb{C}$ *is holomorphic at* $P \in M$ *if and only if there are a neighborhood* U *of* P *(in* $\mathbb{C}^n$*) and* $\tilde{f} \in \mathcal{O}(U)$*, such that* $\tilde{f}|_{M \cap U} = f|_{M \cap U}$.

Theorem 2.11. *Let* H *be a local parametrization of* M *at* P*, with image* $M \cap U$*. Then* $H^{-1}: M \cap U \to \mathbb{C}^{\dim M_P}$ *is holomorphic.*

The reader should convince himself that all the results in §1 which do not involve explicitly the Euclidean coordinates will remain true if the open set $D \subset \mathbb{C}^n$ is replaced by a complex submanifold of $\mathbb{C}^n$. All the other results still apply locally once a local parametrization has been fixed.

Remark. The reader familiar with differentiable submanifolds of $\mathbb{R}^n$, i.e., curves, surfaces, etc., will have recognized the obvious formal similarities between those concepts and the theory of complex submanifolds. But there are some surprising differences as well, as evidenced by the following result, which has no counterpart for differentiable or even real analytic submanifolds of $\mathbb{R}^n$.

Theorem 2.12. *Let $M \subset \mathbb{C}^n$ be a complex submanifold and suppose that M is compact. Then M consists of finitely many points.*

PROOF. It is enough to show: if the given M is also connected, then M is a single point. For each $j = 1, \ldots, n$, the restriction to M of the coordinate function z_j is a holomorphic function on M. Since $z_j(M) \subset \mathbb{C}$ is compact, the open mapping theorem (Theorem 1.21) implies that $z_j|M$ is a constant p_j. Hence $M = \{(p_1, \ldots, p_n)\}$. ■

2.8. Injective Holomorphic Maps

We now prove the result announced in the Remark at the end of §2.4. For this, we will need the following information about the zero set $Z(f, U) = \{z \in U : f(z) = 0\}$ of a holomorphic function f defined on U.

Lemma 2.13. *Let f be a holomorphic function on the connected region D in $\mathbb{C}^n$. Suppose $Z(f, D) \neq \varnothing$ and $f \not\equiv 0$. Then there exists an open set $U \subset D$ such that $Z(f, U)$ is a nonempty complex submanifold of U of dimension $n - 1$.*

PROOF. In case there is a point $P \in Z(f, D)$ with $df(P) \neq 0$, the statement follows immediately from Theorem 2.9. In order to apply this result in the general case, we consider higher order derivatives of f as follows. Let

$$\Lambda = \{\lambda \in \mathbb{N} : D^\alpha f(z) = 0 \qquad \text{for all } z \in Z(f, D) \text{ and } |\alpha| \leq \lambda\}.$$

Since $f \not\equiv 0$, the Identity Theorem 1.19 implies that Λ is finite. So there is $\beta \in \mathbb{N}^n$ with $|\beta| = \max \Lambda$, such that the differential $d(D^\beta f)(P) \neq 0$ for some $P \in Z(f, D)$, and such that

$$Z(f, D) \subset Z(D^\beta f, D). \tag{2.9}$$

By Theorem 2.9, for every sufficiently small neighborhood U of P, the set $Z(D^\beta f, U)$ is an $(n - 1)$ dimensional complex submanifold of U. We now complete the proof by showing that U can be chosen so that $Z(f, U) = Z(D^\beta f, U)$.

After a holomorphic change of coordinates near P, we may assume that $P = 0$ and that

$$(2.10) \qquad Z(D^\beta f, W) = \{(w', w_n) \in W : w_n = 0\}$$

for some neighborhood W of 0. Choose $\delta_n > 0$ sufficiently small so that $f(0', w_n)$ has a zero of some positive order k at $w_n = 0$, and no other zero on $\Delta = \{|w_n| \leq \delta_n\}$. By continuity of f and Rouché's Theorem ([Ahl], p. 153), there is $\delta' > 0$ such that the number of zeroes of $f(w', \cdot)$ in Δ is constant for $w' \in P(0', \delta')$, i.e., equals $k > 0$. We clearly may assume that $U = P(0, (\delta', \delta_n)) \subset W$. Thus, for each $w' \in P(0', \delta')$ there is at least one $w_n \in \Delta$ with $(w', w_n) \in Z(f, U)$. Moreover, by (2.9) and (2.10), if $(w', w_n) \in Z(f, U)$, then $w_n = 0$. Hence we have shown that $Z(f, U) = \{w \in U : w_n = 0\} = Z(D^\beta f, U)$. ∎

Remark. In §3 we will refine some of the arguments used in the preceding proof in order to obtain more precise local information about the zero set of a holomorphic function.

Theorem 2.14. *Let $D \subset \mathbb{C}^n$ and suppose that the holomorphic map $F: D \to \mathbb{C}^n$ is injective. Then* $\det F'(z) \neq 0$ *for all $z \in D$, and hence F is biholomorphic from D onto $F(D)$.*

The proof of Theorem 2.14 will involve induction over the number of variables n. We assume as known the classical case $n = 1$ (see [Ahl], Theorem 4.11). Given the induction hypotheses that the theorem has been proved for $n - 1 > 0$ variables, we first prove the following technical lemma.

Lemma 2.15. *Under the above assumption, if F is as in Theorem* 2.14, *then $F'(a) \neq 0$ at a point $a \in D$ implies* $\det F'(a) \neq 0$.

Proof of Lemma 2.15. After renumbering we may assume that $F = (f_1, \ldots, f_n)$ and $\partial f_n/\partial z_n(a) \neq 0$. If $w(z) = (z_1, \ldots, z_{n-1}, f_n(z))$, then $\det(\partial w_k/\partial z_j)(a) \neq 0$, so that $w = (w_1, \ldots, w_n)$ defines holomorphic coordinates in a neighborhood of a. In these coordinates, $\tilde{F} = F \circ w^{-1}$ is given by

$$(2.11) \qquad \tilde{F}(w) = (g_1(w), \ldots, g_{n-1}(w), w_n)$$

with $g_1, \ldots, g_{n-1}$ holomorphic at $b = w(a)$. We write $w = (w', w_n)$, where $w' = (w_1, \ldots, w_{n-1})$, and define $G(w') = (g_1(w', b_n), \ldots, g_{n-1}(w', b_n))$. Then G is an injective holomorphic map in $(n - 1)$ variables in a neighborhood of $b' = (b_1, \ldots, b_{n-1})$ so that, by inductive assumption, $\det G'(b') \neq 0$. But this and (2.11) imply that $\det \tilde{F}'(b) \neq 0$, and hence $\det F'(a) \neq 0$, as well. ∎

Returning to the proof of the theorem, notice that $h = \det F' \in \mathcal{O}(D)$. Suppose $Z(h) = \{z \in D : h(z) = 0\} \neq \varnothing$. It then follows from Lemma 2.13 that $Z(h)$ contains a complex submanifold $M \neq \varnothing$ of dimension $= n - 1 > 0$. By Lemma 2.15, $F'(z) = 0$ for all $z \in Z(h)$, and hence $F' \equiv 0$ on M. But this implies that F is locally constant on M—just express F in terms of local parametrizations of M—and since $\dim M > 0$, F could not be injective. This contradiction shows that $Z(h)$ must be empty.

Remark 2.16. It is crucial in Theorem 2.14 that we are in the equidimensional case. For example, the map $f(z) = (z^2, z^3)$ from $\mathbb{C}$ into $\mathbb{C}^2$ is injective, but f is singular at 0.

Exercises

E.2.1. Let $D \subset \mathbb{C}^n$ be open and suppose $F = (f_1, \ldots, f_n)\colon D \to \mathbb{C}^n$ has components in $C^1(D)$. Show that

$$\det J_{\mathbb{R}} F = \det \begin{bmatrix} \left(\dfrac{\partial f_k}{\partial z_j}\right) & \left(\dfrac{\partial f_k}{\partial \bar{z}_j}\right) \\ \left(\dfrac{\partial \bar{f}_k}{\partial z_j}\right) & \left(\dfrac{\partial \bar{f}_k}{\partial \bar{z}_j}\right) \end{bmatrix}.$$

E.2.2. Suppose $D_1 \subset \mathbb{C}^n$ is open and $F = D_1 \to D_2 \subset \mathbb{C}^m$ is biholomorphic. Show that $n = m$.

E.2.3. Let $S \subset \mathbb{C}^n$ be a subset. Show that S is discrete if and only if S is a complex submanifold of dimension 0.

E.2.4. Let Δ^2 be the unit bidisc in $\mathbb{C}^2$. Show that every $f \in \operatorname{Aut}(\Delta^2)$ is of the form $f = (f_1, f_2)$, where f_1 and f_2 depend each on only one variable and $f_1, f_2 \in \operatorname{Aut}(\Delta)$. (Hint: By using an automorphism of the above simple type, reduce the general case to the case where $f(0) = 0$.)

E.2.5. Carry out the proof of Theorem 2.7 for arbitrary n.

E.2.6. Prove **Cartan's Uniqueness Theorem**: If $D \subset \mathbb{C}^n$ is a bounded connected region in $\mathbb{C}^n$ with $0 \in D$, and $F\colon D \to D$ is a holomorphic map with $F(0) = 0$ and $F'(0) =$ identity matrix, then F is the identity map. (Hint: Let $F(z) = z + P_k(z) + O(|z|^{k+1})$ be the beginning of the Taylor series of F, where P_k is homogeneous of degree $k \geq 2$, and apply Cauchy estimates to the iterates $F^j = F \circ \cdots \circ F$ (j times)).

E.2.7. Let $B_n \subset \mathbb{C}^n$ be the unit ball and set $G_n = \{(w', w_n) \in \mathbb{C}^n\colon \operatorname{Im} w_n > |w'|^2\}$. G_n is called the **Siegel upper half-space**. Define $\varphi(z) = (w_1, \ldots, w_n)$ by $w_j = z_j/(1 + z_n)$ for $1 \leq j \leq n - 1$ and $w_n = i(1 - z_n)/(1 + z_n)$.

(i) Show that $\varphi\colon B_n \to G_n$ is biholomorphic. (φ is called the **Cayley transform.**)
(ii) The boundary $bG_n = \{(z', t + i|z'|^2)\colon z' \in \mathbb{C}^{n-1}, t \in \mathbb{R}\}$ is naturally identified with $\mathbb{C}^{n-1} \times \mathbb{R} = \{(z', t)\}$. Show that the multiplication

$$(z', t) \cdot (\zeta', \tau) = (z' + \zeta', t + \tau + 2 \operatorname{Im}(z', \zeta'))$$

turns bG_n into a group which is non-abelian if $n > 1$. (This group is called the **Heisenberg group** of order $n - 1$.)

E.2.8. Generalize the example after Theorem 2.9 to find a concrete realization of the Riemann surface of the inverse of a polynomial $p(z) = \sum_{j=0}^{n} a_j z^j$ with $a_n \neq 0$.

E.2.9. (i) Show that $M = \{(z, w) \in \mathbb{C}^2 : z^2 = w^3\}$ is not a complex submanifold of $\mathbb{C}^2$. (Hint: Consider a local "parametrization" $H = (h_1, h_2)$ of M near 0 and show that $H'(0) = 0$.)
(ii) Show that M defined in (i) is homeomorphic to $\mathbb{C}$.

E.2.10. Let $M \subset D \subset \mathbb{C}^n$ be a complex submanifold near the point $P \in M$. Suppose H_1 and H_2 are local parametrizations of M near P. Show that for a function $f: M \to \mathbb{C}$ the function $f \circ H_1^{-1}$ is holomorphic at $H_1^{-1}(P)$ if and only if $f \circ H_2^{-1}$ is holomorphic at $H_2^{-1}(P)$.

E.2.11. Prove Theorem 2.10.

E.2.12. Prove Theorem 2.11.

E.2.13. Let M_1 and M_2 be closed, connected complex submanifolds of the region $D \subset \mathbb{C}^n$. Suppose there is an open neighborhood U of $P \in M_1 \cap M_2$ such that $U \cap M_1 = U \cap M_2$. Show that $M_1 = M_2$.

E.2.14. Let $f \in \mathcal{O}(D)$. Prove that the set of regular points in the zero set $Z(f)$ of f in D is dense in $Z(f)$.

§3. Zero Sets of Holomorphic Functions

3.1. The Riemann Removable Singularity Theorem

We first discuss an elementary result—Lemma 3.2 below—about the structure of the zero set of a holomorphic function which, nevertheless, has several interesting applications, including the generalization of the classical one variable theorem of Riemann on removable singularities. For this purpose, it is necessary to single out one of the coordinates.

Definition. A function f holomorphic at $a = (a', a_n) \in \mathbb{C}^n$, with $f(a) = 0$, is said to be **z_n-regular of order** $k \in \mathbb{N}^+$ at a, if $g(z_n) = f(a', z_n)$, has a zero of order k at $z_n = a_n$, i.e., if

$$g(a_n) = g'(a_n) = \cdots = g^{(k-1)}(a_n) = 0,\ g^{(k)}(a_n) \neq 0. \tag{3.1}$$

Lemma 3.1. *Suppose $f \in \mathcal{O}(B(a, \varepsilon))$, $f(a) = 0$, but f is not identically zero. Then, after a suitable complex linear coordinate change, f is z_n-regular of some order $k \geq 1$ at a.*

Proof. By hypothesis there is $p \in B(a, \varepsilon)$, $p \neq a$, such that $f(p) \neq 0$. After applying an affine complex linear coordinate change, one may assume that $p - a$ lies in the z_n-axis, i.e., $p = (a', p_n)$, $p_n - a_n \neq 0$. Then $g(z_n) = f(a', z_n)$ is holomorphic and nonconstant on $\{|z_n - a_n| < \varepsilon\}$. By the Identity Theorem 1.19 there is $k \in \mathbb{N}$, such that (3.1) holds, and $k \geq 1$, since $g(a_n) = 0$. ∎

Example. The function $f(z_1, z_2) = z_1 \cdot z_2 + z_1^6$ satisfies $f(0, z_2) \equiv 0$, so it is not z_2-regular at 0. Introducing the coordinates $w_1 = z_2$, $w_2 = z_1$, f becomes w_2-regular of order 6. On the other hand, in the coordinates $u_1 = z_1 - z_2$, $u_2 = z_2$, f becomes u_2-regular of order 2. See Exercise E.3.1 for a precise statement involving the choice of a minimal order of regularity.

Without loss of generality we will limit ourselves to the case $a = 0$.

Lemma 3.2. *Suppose f is holomorphic at 0, $f(0) = 0$, and f is z_n-regular of order $k \geq 1$ at 0. Then for each sufficiently small $\delta_n > 0$ there is $\delta' > 0$, such that for each fixed $z' \in P(0', \delta')$ the equation $f(z', z_n) = 0$ has precisely k solutions (counted with multiplicities) in the disc $\{|z_n| \leq \delta_n\}$.*

Proof. By hypothesis, for each sufficiently small $\delta_n > 0$, $g(z_n) = f(0', z_n)$ is holomorphic on $|z_n| \leq \delta_n$, g has a zero of order k at 0, and $g(z_n) \neq 0$ for $0 < |z_n| \leq \delta_n$. By continuity of f and Rouché's Theorem, there is $\delta' > 0$ such that the conclusion of the Lemma holds for all $z' \in P(0', \delta')$. ■

We see that, locally near 0, the zero set of f consists of a "branched covering" over $P(0', \delta')$ with at most k sheets which are glued together at some points. We will see later, as a consequence of the Weierstrass Preparation Theorem, that under suitable hypothesis there will be k *distinct* sheets over "most" $z' \in P(0', \delta')$, each of which will be an $(n-1)$-dimensional complex submanifold at the k points lying over z'.

Corollary 3.3. *The zeroes of a function holomorphic in 2 or more variables are never isolated.*

In order to deal with a somewhat more general situation we say that a subset E of $D \subset \mathbb{C}^n$ is **thin**, if for every point $p \in D$ there are a ball $B(p, \varepsilon)$ and a function $f \in \mathcal{O}(B(p, \varepsilon))$, f not constant, such that $f(z) = 0$ for $z \in E \cap B$. Notice that if $E \subset D$ is thin, its closure (in D) is thin, and, by the Identity Theorem, E is nowhere dense.

Theorem 3.4. *Let E be a thin subset of $D \subset \mathbb{C}^n$. Let $h \in \mathcal{O}(D - E)$ and suppose h is locally bounded on D (i.e., for all $\Omega \subset\subset D$, h is bounded on $\Omega - E$). Then there is $H \in \mathcal{O}(D)$ such that $H = h$ on $D - E$.*

Proof. Since E is nowhere dense, the extension H—if it exists—is determined uniquely by h. Therefore it is enough to construct a holomorphic extension of h to a neighborhood of an arbitrary point $p \in E$. Without loss of generality we may assume $p = 0$ and $E = \{z : f(z) = 0\}$, where f is holomorphic at 0, and—in view of Lemma 3.1—f is z_n-regular of some order k. With $\delta = (\delta', \delta_n)$ as in Lemma 3.2, so that $P(0, \delta) \subset\subset D$, the function

$$H(z', z_n) = (2\pi i)^{-1} \int_{|\zeta| = \delta_n} \frac{h(z', \zeta)\, d\zeta}{z_n - \zeta} \tag{3.2}$$

clearly is defined and holomorphic on $P(0, \delta)$. Now, for z' fixed, the function $h(z', \cdot)$ is holomorphic on $|z_n| \le \delta_n$, with the possible exception of finitely many points, namely the k zeroes of f lying over z'. By the classical one variable theorem of Riemann (see [Ahl], p. 124), $h(z', \cdot)$ extends holomorphically to $|z_n| \le \delta_n$, and therefore the integral in (3.2) equals $h(z', z_n)$ if $(z', z_n) \notin E$. Thus $H = h$ on $P(0, \delta) - E$. ■

This proof is based on the following general principle: in order to find an extension of h, assume the existence of the required extension and apply an appropriate integral representation formula—the Cauchy integral formula (3.2) in the case just discussed. The main difficulty then involves proving that the integral indeed defines an *extension* of the given function. We will see much more striking applications of this principle in Chapter II, §1, and in Chapter IV, Theorem 2.1.

Remark 3.5. As in case $n = 1$, weaker growth conditions for h are sufficient for the existence of a holomorphic extension across thin sets. For example, the conclusion of Theorem 3.4 holds if $h \in \mathcal{O}(D - E)$ is only assumed to be locally in L^2 (see Exercise E.3.2). On the other hand, the function $h(z) = z_n^{-1}$ is holomorphic on $\mathbb{C}^n - E$, $E = \{z: z_n = 0\}$, h is locally in L^p for any $p < 2$ (but not in L^2!), and clearly h has no holomorphic extension across E.

Corollary 3.6. *Let E be a thin subset of $D \subset \mathbb{C}^n$. If D is connected, so is $D - E$.*

Proof. Since $D - E$ will be connected if $D - \bar{E}$ is, we may assume that E is closed. Suppose $U \neq \varnothing$ is an open and closed subset of $D - E$. We must show that $U = D - E$. Define the function h by setting $h(z) = 0$ for $z \in U$ and $h(z) = 1$ for $z \in (D - E) - U$. Then h is bounded and holomorphic on $D - E$, so by Theorem 3.4 there is $H \in \mathcal{O}(D)$ with $H = h$ on $D - E$. As D is connected, the Identity Theorem implies $H \equiv 0$ on D, and hence $(D - E) - U = \varnothing$. ■

Finally, we state another property of thin sets which follows from the geometric information contained in Lemma 3.2.

Theorem 3.7. *Let E be a thin subset of $D \subset \mathbb{C}^n$. Then the $2n$-dimensional Lebesgue measure of E is zero. In particular, if D is connected and $f \in \mathcal{O}(D)$ vanishes on a set of positive measure, then $f \equiv 0$ on D.*

The proof is left to the reader.

3.2. Analytic Sets

In several complex variables it is important to study not just the zero set of one holomorphic function, but also of several functions, i.e., of holomorphic maps. In §2 we already studied the case of regular maps, which led us to the concept of a complex submanifold. The general case is quite a bit more complicated. Here we briefly discuss some of the relevant concepts and some simple examples, mainly in order to familiarize the reader with the basic terminology. For further studies the interested reader should consult some of the specialized literature, for example R. Gunning [Gun] and R. Narasimhan [Nar 2].

Definition. A subset A of the region $D \subset \mathbb{C}^n$ is called **analytic** in D if A is closed in D and if for every $p \in A$ there are an open neighborhood U_p of p in D and a holomorphic map $H_p: U_p \to \mathbb{C}^{l_p}$, such that

$$U_p \cap A = \{z \in U_p: H_p(z) = 0\}. \tag{3.3}$$

Stated differently, (3.3) means that $U_p \cap A$ is the common zero set of the components $h_1^{(p)}, \ldots, h_{l_p}^{(p)}$ of H_p.

It readily follows from the definition that $A_1 \cup A_2$ and $A_1 \cap A_2$ are analytic sets in D whenever A_1 and A_2 are analytic in D. An analytic set $A \subset D$ is said to be **reducible** if A can be written as $A = A_1 \cup A_2$, where A_1, A_2 are analytic in D and $A_1 \neq A$, $A_2 \neq A$. A is said to be **irreducible** if A is not reducible.

A point $p \in A$ of an analytic set A is called a **regular point** of A—or A is said to be regular at p—if there is a neighborhood U of p, such that $A \cap U$ is a complex submanifold of U, and a **singular point** otherwise. The set of regular points is denoted by $\mathscr{R}(A)$; it is the maximal complex submanifold contained in A. The set $\mathscr{S}(A) = A - \mathscr{R}(A)$ is called the **singular set** of A.

We discuss some examples: (1) Every closed (in D) submanifold M of $D \subset \mathbb{C}^n$ is analytic in D, with $\mathscr{R}(M) = M$; in particular, D itself is analytic in D. (2) The set $A_1 = \{(z, w) \in \mathbb{C}^2: z^2 - w^3 = 0\}$ is analytic in $\mathbb{C}^2$, $(0, 0)$ is a singular point of A_1, and $\mathscr{R}(A_1) = A_1 - \{0\}$ (see Exercise E.2.9). The map $t \to (t^3, t^2)$ establishes a homeomorphism between $\mathbb{C}$ and A_1, so the topological structure of A_1 is still very simple, even near the singular set. (3) The analytic set $A_2 = \{(z_1, z_2): z_1 \cdot z_2 = 0\}$ in $\mathbb{C}^2$ also has $(0, 0)$ as its only singular point, but in contrast to the previous example, no neighborhood of $(0, 0)$ in A_2 is homeomorphic to an open set in $\mathbb{C}$; in fact $U \cap A_2 - \{0\}$ is disconnected for any such neighborhood U. Still, the singularity of A_2 arises in a simple way: A_2 is reducible. In fact $A_2 = C_1 \cup C_2$, where $C_i = \{(z_1, z_2): z_i = 0\}$, $i = 1, 2$ are complex submanifolds, and $\{0\} = C_1 \cap C_2$. (4) Let $B = \{z \in \mathbb{C}^3: z_3^2 - z_1 z_2 = 0\}$; we leave it to the reader to check that $B - \{0\}$ is a *connected* complex submanifold of dimension 2, and that B is irreducible (see Exercise E.3.5). We claim that B is not locally Euclidean at 0, that is, no neighborhood U of 0 in B has the topological structure of an open ball (in $\mathbb{R}^4$).[1] In fact, if $U \subset B$ is such a locally Euclidean neighborhood, then $U - \{0\}$ would have to be simply connected. But this is not possible, as the map $\pi: \mathbb{C}^2 - \{0\} \to B - \{0\}$, given by $\pi(t_1, t_2) = (t_1^2, t_2^2, t_1 t_2)$ is a two-sheeted covering map. So $\mathscr{S}(B) = \{0\}$.

[1] The reader unfamiliar with covering spaces may omit the argument which follows.

Theorem 3.8. *Let A be an analytic set in the connected region D in $\mathbb{C}^n$. If $A \neq D$, then A is thin, and hence $D - A$ is connected.*

PROOF. Since the second statement in the conclusion follows from the first, by Corollary 3.6, it is enough to show that if A is not thin, then $A = D$. For each $p \in A$ we choose a connected neighborhood U_p and a holomorphic map $H_p: U_p \to \mathbb{C}^{l_p}$ such that (3.3) holds. If A is not thin, there must be at least one $p \in A$, such that all the components of H_p are identically zero on U_p. Hence, $A \cap U_p = U_p$, and the interior $\mathring{A}$ of A is not empty. If we can show that $\mathring{A}$ is closed in D, the connectedness of D will imply that $\mathring{A} = D$, and we are done. So, take $q \in b\mathring{A} \cap D$. Then $\mathring{A} \cap U_q$ is open and nonempty, and the components of H_q are zero on $\mathring{A} \cap U_q$. By the Identity Theorem, they must be zero on all of U_q. This implies $U_q \subset A$, so $q \in \mathring{A}$ and $\mathring{A}$ is closed in D. ∎

Remark. It is natural to ask whether one obtains a more general notion of analytic set by considering solution sets of an infinite (rather than finite) collection of holomorphic equations. The following theorem, whose proof requires more detailed information about local properties of rings of holomorphic functions, shows that this is not the case (see [GuRo], Theorem II.E.3).

Theorem 3.9. *If $\mathscr{F} \subset \mathcal{O}(D)$ and $A = \{z \in D: f(z) = 0$ for all $f \in \mathscr{F}\}$, then A is analytic in D.*

3.3. The Weierstrass Preparation Theorem

If f is holomorphic at $0 \in \mathbb{C}^n$, $f(0) = 0$, and f is z_n-regular of order $k \geq 1$, it follows from Lemma 3.2 that for each $z' \in P(0', \delta')$ there is a unique normalized polynomial in z_n of degree k,

$$\omega(z', z_n) = z_n^k + a_{k-1}(z')z_n^{k-1} + \cdots + a_0(z'), \tag{3.4}$$

such that $f(z', \cdot)$ and $\omega(z', \cdot)$ have the same zeroes (counting multiplicities) in $|z_n| < \delta_n$: $\omega(z', z_n)$ is simply the product $\prod_{j=1}^k (z_n - \varphi_j(z'))$, where $\varphi_1(z'), \ldots, \varphi_k(z')$ are the zeroes of $f(z', \cdot)$ in $|z_n| < \delta_n$. Therefore $f = \omega \cdot u$ for some nonvanishing function u on $P(0, \delta)$. It is a remarkable fact, first proved by K. Weierstrass, that both ω and u are holomorphic (Theorem 3.10 below). This result, for which several different proofs are now known (see [Nar 2]), is one of the cornerstones of local function theory in several variables.

A very interesting account of the history of the **Weierstrass Preparation Theorem** and of its far-reaching consequences was given by H. Cartan ([Car], 875–888). The proof given here follows classical arguments of Simart, first presented in É. Picard's 1893 Traité d'Analyse, Vol. II, and later in [Osg], 83–89.

We first introduce some terminology. A function ω as in (3.4) is called a **pseudopolynomial** (in z_n) **at** 0 if the coefficients $a_0, \ldots, a_{k-1}$ are holomorphic functions in z' at $0'$; ω is called a **distinguished** pseudopolynomial, or **Weierstrass polynomial at** 0, if, in addition, $a_0(0') = \cdots = a_{k-1}(0') = 0$.

Theorem 3.10. *Let f be holomorphic at 0, $f(0) = 0$, and suppose f is z_n-regular of order $k \geq 1$. Then there is a unique factorization*

(3.5) $$f = \omega \cdot u$$

on some polydisc $P(0, \delta)$, where $\omega \in \mathcal{O}(P(0', \delta'))[z_n]$[1] *is a distinguished pseudopolynomial of degree k at 0, $u \in \mathcal{O}(P(0, \delta))$, and $u \neq 0$ on $P(0, \delta)$.*

Proof. The uniqueness of the factorization (3.5) is obvious in view of the preceding remarks. In order to prove that the coefficients of ω are holomorphic, we choose $P(0, \delta)$ as in Lemma 3.2. Notice that $a_0(z'), \ldots, a_{k-1}(z')$ are the elementary symmetric functions of the zeroes $\varphi_1(z'), \ldots, \varphi_k(z')$ of $f(z', \cdot)$ in $|z_n| < \delta_n$. The φ_j's will, in general, not be holomorphic. However, it is a well known consequence of the residue theorem that for any $m \in \mathbb{N}$,

(3.6) $$S_m(z') = \sum_{j=1}^{k} \varphi_j^m(z') = \frac{1}{2\pi i} \int_{|\zeta| = \delta_n} \frac{\zeta^m (\partial f/\partial \zeta)(z', \zeta)\, d\zeta}{f(z', \zeta)}, \quad z' \in P(0', \delta')$$

(see [Ahl], p. 153–154). Since $f(z', \zeta) \neq 0$ for $|\zeta| = \delta_n$, (3.6) implies that $S_m \in \mathcal{O}(P(0', \delta'))$. Finally, it is known from algebra that any symmetric function of $\varphi_1, \ldots, \varphi_k$, and therefore also a_j, $0 \leq j \leq k - 1$, is a polynomial in $S_0, S_1, \ldots$. It follows that $a_j \in \mathcal{O}(P(0', \delta'))$, and since $f(0', z_n) = z_n^k \cdot g(z_n)$, with $g(0) \neq 0$, one must have $a_j(0) = 0$ for $j = 0, \ldots, k - 1$. Thus $\omega(z', z_n)$ is indeed a distinguished pseudopolynomial at 0.

It remains to be shown that $u = f/\omega$ is holomorphic. From the construction of ω it is clear that $u(z', \cdot)$ is holomorphic on $|z_n| \leq \delta_n$ for each fixed z'. Therefore

(3.7) $$u(z', z_n) = \frac{1}{2\pi i} \int_{|\zeta| = \delta_n} \frac{(f/\omega)(z', \zeta)\, d\zeta}{\zeta - z_n}.$$

The function $(f/\omega)(z', \zeta)$ is holomorphic in z' for $|\zeta| = \delta_n$, since $\omega(z', \zeta) \neq 0$ and ω is holomorphic. It then readily follows from (3.7) that $u \in \mathcal{O}(P(0, \delta))$. ■

Theorem 3.10 can be viewed as a generalization of the Implicit Function Theorem 2.4 in the case of one equation ($m = 1$). In fact, if $\partial f/\partial z_n(0) \neq 0$, f is z_n-regular of order 1, and the theorem implies that for $z \in P(0, \delta)$, $f(z', z_n) = 0$ if and only if $\omega(z', z_n) = z_n - a_0(z') = 0$, i.e., $z_n = a_0(z')$, where a_0 is holomorphic. The reader should consult [GuRo], Chapter I.B, for a "complex variable proof" of the general Implicit Function Theorem.

[1] If R is a ring, $R[z_n]$ denotes the polynomial ring with coefficients in R.

3.4. The Zero Set of a Single Function

The Weierstrass Preparation Theorem reduces the local study of analytic sets to certain pseudopolynomial equations which can be handled by a sophisticated combination of algebraic and analytic techniques. As an introduction we discuss some of the simpler results in the case of a single equation, a case which is considerably easier than, but still quite representative of, the general case.

We consider a pseudopolynomial $\omega \in R[X]$, where $R = \mathcal{O}(P(0, \delta))$ for some polydisc $P(0, \delta) \subset \mathbb{C}^n$, and we denote its zero set by

$$Z(\omega) = \{(z, w) \in P(0, \delta) \times \mathbb{C}: \omega(z, w) = 0\}.$$

Notice that R is an integral domain (Theorem 1.20), and therefore R has a well defined quotient field which we denote by Q.

By Theorem 2.9, every point p in $Z(\omega)$ at which $(\partial\omega/\partial X)(p) \neq 0$ will be a regular point of $Z(\omega)$. We therefore analyze the common zero set of ω and $\partial\omega/\partial X$.

Lemma 3.11. *Suppose $\omega \in R[X]$ is irreducible in $Q[X]$. Let $E \subset P(0, \delta)$ be the set of points z, such that $\omega(z, \cdot)$ has at least one zero of multiplicity greater than one. Then E is thin.*

Proof. The Euclidean algorithm being valid in $Q[X]$, the polynomials ω and $\partial\omega/\partial X$ have a greatest common divisor, which must be 1, since ω is irreducible. Hence there are $\varphi, \psi \in Q[X]$, such that

$$\varphi\omega + \psi\partial\omega/\partial X = 1. \tag{3.8}$$

Let $h \in R$, $h \neq 0$ be a common denominator for all the coefficients of φ and ψ. Equation (3.8) implies

$$(h\varphi)\omega + (h \cdot \psi)\partial\omega/\partial X = h, \tag{3.9}$$

where $h \cdot \varphi$ and $h \cdot \psi \in R[X]$.

We now interpret (3.9) as an equation for functions on $P(0, \delta) \times \mathbb{C}$. Notice that if $z \in E$, i.e., there is $w \in \mathbb{C}$ such that $\omega(z, w) = \partial\omega/\partial X(z, w) = 0$, then (3.9) implies that $h(z) = 0$. Since $h \neq 0$ in R, h is not identically zero on $P(0, \delta)$ and the lemma is proved. ■

We summarize the main consequences.

Theorem 3.12. *Let $\omega \in R[X]$ be a Weierstrass polynomial of degree k which is irreducible in $Q[X]$, and let $\pi: \mathbb{C}^n \times \mathbb{C} \to \mathbb{C}^n$ be the projection. Then there is a thin subset $E \subset P(0, \delta)$, such that the following statements hold.*

(i) *$Z(\omega) - \pi^{-1}(E)$ is an n-dimensional complex submanifold of $(P(0, \delta) - E) \times \mathbb{C}$;*

(ii) $Z(\omega) - \pi^{-1}(E)$ *is dense in* $Z(\omega)$;
(iii) $\pi|_{Z(\omega)-\pi^{-1}(E)}: Z(\omega) - \pi^{-1}(E) \to P(0, \delta) - E$ *is a k-sheeted covering*;
(iv) $Z(\omega) - \pi^{-1}(E)$ *is connected.*

PROOF. Let $E \subset P(0, \delta)$ be the thin set given by Lemma 3.11. Part (i) is a direct consequence of Lemma 3.11 and Theorem 2.9. For (ii), let $p \in Z(\omega)$ with $a = \pi(p) \in E$. Then ω is z_{n+1}-regular at p. Let U be an arbitrary neighborhood of p. By Lemma 3.2 there is $\gamma > 0$, such that for each $z \in P(a, \gamma)$, $\omega(z, \cdot)$ will have at least one zero λ with $(z, \lambda) \in U$. Any such z which is not in E gives a point $q = (z, \lambda) \in (Z(\omega) - \pi^{-1}(E)) \cap U$. (iii) and (iv) require some familiarity with covering spaces; the details are left to the interested reader (see Exercise E.3.7). ■

Remark 3.13. The reader may be tempted to conclude that $Z(\omega) \cap \pi^{-1}(E)$ is the singular set of $Z(\omega)$. Unfortunately, the situation is more complicated. Even when this set is the exact "branch locus" of the covering exhibited by Theorem 3.12, it may still contain regular points. For example, consider $\omega = z_3^2 - z_1 z_2$, so that $Z(\omega)$ is the analytic set B discussed in §3.2; notice that $\partial\omega/\partial z_3 = 2z_3$, and hence

$$Z(\omega) \cap \pi^{-1}(E) = \{z \in Z(\omega): z_1 \cdot z_2 = 0\}.$$

But we had seen that $Z(\omega)$ is regular at every point $p \neq 0$.

In order to apply Theorem 3.12 in case of arbitrary Weierstrass polynomials, one needs a factorization into irreducible pseudopolynomials.

Lemma 3.14. *Suppose* $\omega_1, \omega_2 \in Q[X]$ *are monic polynomials such that* $\omega_1 \cdot \omega_2 \in R[X]$. *Then* ω_1 *and* ω_2 *are in* $R[X]$.

PROOF. Write $\omega = \omega_1 \cdot \omega_2 = X^k + a_{k-1}X^{k-1} + \cdots + a_0$, where $a_j \in R = \mathcal{O}(P(0, \delta))$, $0 \le j \le k-1$. Since the coefficients of $\omega(z, X)$ are locally bounded on $P(0, \delta)$, so are its roots $\varphi_1(z), \ldots, \varphi_k(z)$. If $\omega_i = X^{k_i} + b_{k_i-1}^{(i)} X^{k_i-1} + \cdots + b_0^{(i)}$, $i = 1, 2$, with $b_j^{(i)} \in Q$, $0 \le j \le k_i - 1$, let $h \in R$, $h \neq 0$, be a common denominator for all the coefficients $b_j^{(i)}$. Then $E = \{z \in P(0, \delta): h(z) = 0\}$ is thin, and $b_j^{(i)} \in \mathcal{O}(P(0, \delta) - E)$, since the quotient of holomorphic functions is holomorphic wherever the denominator is $\neq 0$.

For $z \notin E$, the zeroes of $\omega_1(z, \cdot)$ respectively $\omega_2(z, \cdot)$ are among the zeroes of $\omega(z, \cdot)$; hence the coefficients $b_j^{(i)}(z)$ are elementary symmetric functions of certain subsets of $\{\varphi_1(z), \ldots, \varphi_k(z)\}$. In view of the remark at the beginning of the proof, it follows that all coefficients $b_j^{(i)}$ are locally bounded on D. Therefore, by Theorem 3.4, each $b_j^{(i)}$ has a holomorphic extension across the thin set E, i.e., $b_j^{(i)} \in R$. ■

Theorem 3.15. *Let* $\omega \in R[X]$ *be a pseudopolynomial. Then*

$$\omega = \omega_1 \cdot \omega_2 \cdot \cdots \cdot \omega_r,$$

where each $\omega_i \in R[X]$ is a pseudopolynomial which is irreducible in $Q[X]$. If ω is distinguished, so are $\omega_1, \ldots, \omega_r$.

PROOF. Let $\omega = \omega_1 \cdot \cdots \cdot \omega_r$ be the factorization of ω into irreducible monic polynomials in $Q[X]$ of degree ≥ 1. Repeated application of Lemma 3.14 gives $\omega_i \in R[X]$, $1 \leq i \leq r$. Finally, if $\omega(0, X) = X^k$, one must have $\omega_i(0, X) = X^{k_i}$, $k_1 + \cdots + k_r = k$, i.e., each ω_i is distinguished as well. ∎

By combining Theorem 3.12 and Theorem 3.15 one can now show that Theorem 3.12 remains true for arbitrary Weierstrass polynomials $\omega \in R[X]$, except for part (iv). Unless ω is irreducible, $Z(\omega) - \pi^{-1}(E)$ need not be connected. Also, the number of sheets of the covering may now be smaller than the degree of ω. The reader may find more details in Exercise E.3.7.

To conclude this brief introduction into analytic sets, let us mention that the local description of the zero set $Z(\omega)$ of an irreducible Weierstrass polynomial given in Theorem 3.12 remains true for arbitrary analytic sets in the following form. Suppose A is an analytic set in $\mathbb{C}^n$ with $0 \in A$, which is **irreducible at 0** (this means that $A \cap P(0, \varepsilon)$ is irreducible for all small $\varepsilon > 0$). Then, after a suitable complex linear change of coordinates, there are an integer k with $0 \leq k \leq n$, a neighborhood U of 0, and a thin set $E \subset \pi(U) \subset \mathbb{C}^k$, where $\pi = \mathbb{C}^n \to \mathbb{C}^k$ is the projection onto the first k coordinates, such that the following holds:

(i) *$A \cap U - \pi^{-1}(E)$ is a k-dimensional complex submanifold of U, which is dense in $A \cap U$;*
(ii) *π: $A \cap U - \pi^{-1}(E) \to \pi(U) - E$ is a finitely sheeted covering map;*
(iii) *$A \cap U - \pi^{-1}(E)$ is connected.*

The integer k is called the **dimension of A at** 0; it clearly depends only on A and not on the choice of coordinates, and it is known to agree with the *topological dimension* of the set A at 0. For more details about this so-called **Local Parametrization Theorem** the reader may consult [Gun] and [Nar 2].

EXERCISES

E.3.1. Let f be holomorphic at $0 \in \mathbb{C}^n$.

(i) Show that f has a unique homogeneous expansion

$$f(z) = \sum_{k=0}^{\infty} p_k(z),$$

where p_k is a homogeneous polynomial of order k (i.e., $p_k(\lambda z) = \lambda^k p_k(z)$ for $\lambda \in \mathbb{C}$, $z \in \mathbb{C}^n$).

(ii) The minimal k in (i) such that $p_k \not\equiv 0$ is called the **order of f at** 0. Show that if f has order k at zero, then after a suitable linear change of coordinates, f is z_n-regular of order k.

E.3.2. Let D be open in $\mathbb{C}^n$ and let $E \subset D$ be thin. Show that every $f \in \mathcal{O}(D - E)$ which is locally in L^2 (i.e., every $a \in D$ has a neighborhood $U_a \subset D$, such that $f|_{U_a - E} \in L^2(U_a - E)$) has a holomorphic extension across E.

E.3.3. Show that a thin set in $\mathbb{C}^n$ has zero $2n$-dimensional Lebesgue measure.

E.3.4. Let M be a complex submanifold of $D \subset \mathbb{C}^n$. Show that M is irreducible (as an analytic set) if and only if M is connected.

E.3.5. Let A be an analytic set and let $\mathscr{R}(A)$ be the set of regular points of A.

(i) Show that if $\mathscr{R}(A)$ is dense in A (this is true for every analytic set), then A is irreducible if $\mathscr{R}(A)$ is connected. (The converse is true also, but is much deeper.)

(ii) Show that the analytic set

$$B = \{z \in \mathbb{C}^3 : z_1 z_2 - z_3^2 = 0\}$$

is irreducible.

E.3.6. *Continuity of roots.* Let f be holomorphic at 0 and z_n-regular of some finite order. Show that there is $\delta > 0$, such that if $\varphi: U' \to \mathbb{C}$ defined in a neighborhood U' of $O' \in \mathbb{C}^{n-1}$ satisfies $|\varphi(z')| < \delta$ and $f(z', \varphi(z')) = 0$ for $z' \in U'$, then φ is continuous at O'.

E.3.7. Consider the setup in Theorem 3.12.

(i) Show that if ω is irreducible, then $Z(\omega) - \pi^{-1}(E)$ is connected.

(ii) Show by an example that $Z(\omega) - \pi^{-1}(E)$ need not be connected for an arbitrary Weierstrass polynomial ω.

E.3.8. Suppose A_1 and A_2 are analytic sets and let $P \in A_1 \cap A_2$. Suppose that for every neighborhood U of P one has $U \cap A_1 \neq U \cap A_2$. Show that P is a singular point of $A = A_1 \cup A_2$. (Hint: Use E.2.13.)

E.3.9. Let f be holomorphic at 0. Show that there is a polydisc $P(0, \delta)$ such that $A = \{z \in P(0, \delta): f(z) = 0\}$ is a finite union $A = \bigcup_{i=1}^{l} A_i$ of irreducible analytic sets $A_1, \ldots, A_l$ in $P(0, \delta)$.

Notes for Chapter I

The origins of much of the material in this chapter are "lost in antiquity"; certainly most of it was known to K. Weierstrass. One of the earliest systematic presentations is in the 1924 edition of the book by W.F. Osgood [Osg]. The first investigations of holomorphic maps between domains invariant under rotations in the coordinate axis (i.e., Reinhardt domains in the present terminology) are due to K. Reinhardt [Rei]; major progress in the theory of holomorphic maps was made in the early 1930s by H. Cartan (see, for example, [Car], 141–254, 255–275, and 336–369). A very readable account of Cartan's fundamental results is given by R. Narasimhan [Nar 3]. As noted in the text, the nonequivalence of ball and polydisc in more than one variable was discovered by H. Poincaré. Many other proofs of this result are now known; in particular, it is a special case of the general results of H. Cartan. The proof given here is based on ideas of R. Remmert and K. Stein [ReSt 2], as presented in [Nar 3]. The regularity of injective holomorphic maps in the equidimensional case (Theorem 2.14) is due to Clements (Bull. Amer. Math. Soc. **18**(1912), 451–456) (cf. [Osg], p. 149). Osgood's presentation ([Osg], 141–149) is rather difficult to follow; later proofs (for example, [Nar 3], 86–89) were still quite involved. The simple proof given here is due to J.P. Rosay [Ros]; the completely elementary proof of Lemma 2.13, which usually is obtained as a consequence of the Weierstrass Preparation Theorem (see Theorem 3.12), simplifies matters even further. The systematic investigation of analytic sets was begun by R. Remmert and K. Stein [ReSt 1]. Their deeper properties are now incorporated in the theory of coherent analytic sheaves [GrRe 2].

CHAPTER II

Domains of Holomorphy and Pseudoconvexity

In 1906 F. Hartogs discovered the first example exhibiting the remarkable extension properties of holomorphic functions in more than one variable. It is this phenomenon, more than anything else, which distinguishes function theory in several variables from the classical one-variable theory. Hartogs' discovery marks the beginning of a genuine several-variable theory, in which fundamental new concepts like *domains of holomorphy* and the various notions of convexity used to characterize them have become indispensable. In particular, the property now generally referred to as "*pseudoconvexity*" originates with Hartogs, and even today it still is one of the richest sources of intriguing phenomena and deep questions in complex analysis. (See, for example, the remarks at the end of §2.8.) We will say more about this in Chapter VII.

In this chapter, after an introduction to some of the elementary extension phenomena in §1, we give a rather detailed discussion of pseudoconvexity by first following—if not in detail, at least in spirit—the early work of F. Hartogs and E.E. Levi, then by introducing the fundamental concept of a *strictly pseudoconvex domain*, and finally ending up with a general definition of pseudoconvexity involving the existence of a C^2 strictly plurisubharmonic exhaustion function. This latter version of pseudoconvexity, even though it lacks the intuitive geometric appeal of Hartogs' original version, is very convenient for extending function theoretic results from relatively compact subsets of a domain D to the domain D itself, as we shall see in Chapter VI. In §3 we discuss *holomorphic convexity*, an intrinsic global characterization of domains of holomorphy which was introduced in 1932 by H. Cartan and P. Thullen, and which has since taken a central place in the modern theory of several complex variables. By constructing a real analytic strictly plurisubharmonic exhaustion function on a holomorphically convex domain, one easily sees directly that such a domain is pseudoconvex. The converse of this—the so-called Levi problem—is much harder. We will prove it for strictly pseudoconvex domains in Chapter V, and for the general case in Chapter VI.

The discussion of pseudoconvexity is completed in §5 by showing the equivalence of the various notions of pseudoconvexity introduced earlier, and of several other notions which are useful in various contexts. The principal tool here is the class of general plurisubharmonic functions introduced, independently, by K. Oka and P. Lelong in the early 1940s, whose basic properties we collect in §4. In fact, one may well say that the study of pseudoconvexity is equivalent to the study of plurisubharmonic functions. Let us mention though that this book has been arranged in such a way that the later chapters (except for §1.8 in Chapter VI) are independent of §4 and §5 of this chapter.

§1. Elementary Extension Phenomena

We begin by presenting some of the basic elementary techniques which are used to construct holomorphic extensions of *all* holomorphic functions on certain domains to a larger domain.

1.1. Extensions by Means of the Cauchy Integral Formula

The starting point is the following fundamental fact discovered by F. Hartogs in 1906 [Har 1].

Theorem 1.1. *Let n be ≥ 2 and suppose that $0 < r_j < 1$ for $1 \leq j \leq n$. Then every function f holomorphic on the domain*

$$H(r) = \{z \in \mathbb{C}^n : |z_j| < 1 \text{ for } j < n, \quad r_n < |z_n| < 1\}$$
$$\cup \{z \in \mathbb{C}^n : |z_j| < r_j \text{ for } j < n, \quad |z_n| < 1\}$$

(see Figure 2.) has a unique holomorphic extension $\hat{f}$ to the polydisc $P(0, 1)$.

Proof. The uniqueness of the extension is an immediate consequence of the Identity Theorem. In order to obtain the desired extension $\hat{f}$ we will write down an integral formula for $\hat{f}$. Fix $r_n < \delta < 1$; then

$$\hat{f}(z', z_n) = (2\pi i)^{-1} \int_{|\zeta|=\delta} \frac{f(z', \zeta)}{\zeta - z_n}\, d\zeta \tag{1.1}$$

defines a function on $P(0, (1', \delta))$. It is easy to see that $\hat{f}$ is continuous on $P(0, (1', \delta))$ and holomorphic in each variable separately. By Corollary I.1.5, $\hat{f}$ is holomorphic. Since for $z' \in P(0, r')$ the function $f(z', \cdot)$ is holomorphic on $|z_n| < 1$, (1.1) implies $\hat{f}(z', z_n) = f(z', z_n)$ for $(z', z_n) \in P(0, (r', \delta))$. The Identity Theorem now implies $\hat{f} = f$ on $H(\mathrm{r}) \cap P(0, (1', \delta))$, so that $\hat{f}$ does indeed extend f to the polydisc $P(0, 1)$. ∎

The kind of argument used in the preceding proof can be adapted to a variety of situations; for example, the reader should prove the following result.

Theorem 1.2. *Let $n \geq 2$ and suppose U is a neighborhood of the boundary bP of a polydisc $P \subset \mathbb{C}^n$, such that $U \cap P$ is connected. Then every $f \in \mathcal{O}(U)$ has a holomorphic extension to P.*

Corollary 1.3. *Let U be open in $\mathbb{C}^n$ and $a \in U$. If $n \geq 2$, then every $f \in \mathcal{O}(U - \{a\})$ extends holomorphically across a.*

We see that holomorphic functions in two or more variables, in contrast to the situation in one variable, cannot have isolated singularities. The reader should check Exercise E.1.3 for a result which generalizes Corollary 1.3 to extension across complex submanifolds of dimension $\leq n - 2$.

1.2. Laurent Series

The extension of holomorphic functions can often be obtained by first expanding into a Laurent series and then showing that, under suitable geometric hypothesis, the Laurent series is actually a power series which converges on a larger domain.

Proposition 1.4. *Suppose $0 < R_j < \infty$ and $0 \leq r_j \leq R_j$ for $1 \leq j \leq n$, and let $K(r, R) = \{z \in \mathbb{C}^n : r_j \leq |z_j| \leq R_j, 1 \leq j \leq n\}$. Then every $f \in \mathcal{O}(K)$, has a unique representation*

$$f(z) = \sum_{\nu \in \mathbb{Z}^n} c_\nu z^\nu \qquad \text{for } z \in K(r, R). \tag{1.2}$$

The series (1.2) converges uniformly and absolutely on $K(r, R)$. Moreover, for any $\rho = (\rho_1, \ldots, \rho_n) > 0$ with $r \leq \rho \leq R$,

$$c_\nu = (2\pi i)^{-n} \int_{b_0 P(0,\rho)} f(\zeta)\zeta^{-\nu-1}\, d\zeta_1 \ldots d\zeta_n \qquad \text{for } \nu \in \mathbb{Z}^n, \tag{1.3}$$

and

$$\text{if } r_l = 0 \text{ for some } l, \text{ then } c_\nu = 0 \text{ whenever } \nu_l < 0. \tag{1.4}$$

Remarks. The series (1.2) is called the **Laurent series** of f on $K(r, R)$. It is a particular type of multiple series, with index set $\mathbb{Z}^n$ rather than $\mathbb{N}^n$; the remarks made in I.§1.5 apply to the present situation as well. The integral in (1.3) is defined as in I. (1.16).

PROOF. We first prove that uniform convergence in (1.2) implies (1.3) and (1.4); in particular, this implies the uniqueness of the Laurent expansion. Given ρ as in the Theorem, the uniform convergence of (1.2) implies that

$$(1.5)\qquad \int_{b_0P(0,\rho)} f(\zeta)\zeta^\alpha\, d\zeta = \sum_{\nu\in\mathbb{Z}^n} c_\nu \int_{b_0P(0,\rho)} \zeta^\nu\zeta^\alpha\, d\zeta = \begin{cases}(2\pi i)^n c_\nu & \text{if } \nu+\alpha = (-1,\ldots,-1)\\ 0 & \text{otherwise.}\end{cases}$$

Thus (1.3) follows. Furthermore, if $r_l = 0$ for some $1 \le l \le n$ and if $\nu_l < 0$, the integrand in (1.3) is holomorphic in ζ_l for $0 \le |\zeta_l| \le R_l$; so integration in ζ_l gives zero by Cauchy's Theorem, and (1.4) holds.

In order to prove the existence part we use induction on the number of variables n. The classical case $n = 1$ is known (see [Ahl], p. 184), so we assume that $n \ge 2$ and that the theorem holds for $n - 1 > 0$ variables. Given $f \in \mathcal{O}(K(r, R))$ we choose $\lambda \in \mathbb{R}$ with $0 < \lambda < 1$, so that f is defined and holomorphic on the larger set $K_\lambda = K(\lambda r, R/\lambda)$. Define $K'_\lambda = \{z' \in \mathbb{C}^{n-1}: \lambda r_j \le |z_j| \le R_j/\lambda \text{ for } 1 \le j \le n-1\}$ and $K_{n,\lambda} = \{z_n \in \mathbb{C}: \lambda r_n \le |z_n| \le R_n/\lambda\}$, and set $M = |f|_{K_\lambda} < \infty$. For each $z_n \in K_{n,\lambda}$ one has $f(\cdot, z_n) \in \mathcal{O}(K'_\lambda)$; so, by inductive hypothesis,

$$(1.6)\qquad f(z', z_n) = \sum_{\nu'\in\mathbb{Z}^{n-1}} c_{\nu'}(z_n)(z')^{\nu'} \qquad \text{for } (z', z_n) \in K'_\lambda \times K_{n,\lambda},$$

and, according to (1.3) in case of $n - 1$ variables,

$$(1.7)\qquad c_{\nu'}(z_n) = (2\pi i)^{n-1} \int_{b_0P(0,\rho')} f(\zeta', z_n)(\zeta')^{-\nu'-1}\, d\zeta_1 \ldots d\zeta_{n-1}$$

for any $\rho' \in \mathbb{R}^{n-1}$ with $\rho' > 0$ and $\lambda r' \le \rho' \le R'/\lambda$. Equation (1.7) implies that $c_{\nu'} \in \mathcal{O}(K_{n,\lambda})$; so the case $n = 1$ gives

$$(1.8)\qquad c_{\nu'}(z_n) = \sum_{\nu_n\in\mathbb{Z}} c_{\nu',\nu_n} z_n^{\nu_n} \qquad \text{for } z_n \in K_{n,\lambda},$$

and

$$(1.9)\qquad c_{\nu',\nu_n} = (2\pi i)^{-1} \int_{|\zeta|=\rho_n} c_{\nu'}(\zeta)\zeta^{-\nu_n-1}\, d\zeta$$

for $\rho_n > 0$ with $\lambda r_n \le \rho_n \le R_n/\lambda$. Substituting (1.8) into (1.6) and writing $(\nu', \nu_n) = \nu \in \mathbb{Z}^n$, one obtains

$$(1.10)\qquad f(z) = \sum_{\nu'\in\mathbb{Z}^{n-1}} \left(\sum_{\nu_n\in\mathbb{Z}} c_\nu z^\nu\right) \text{ for } z \in K_\lambda.$$

We now claim that

$$(1.11)\qquad |c_\nu z^\nu| \le M\lambda^{|\nu_1|+\cdots+|\nu_n|} \qquad \text{for } z \in K(r, R).$$

Since $\lambda < 1$, (1.11) readily implies that the series $\sum c_\nu z^\nu$ converges uniformly and absolutely on $K(r, R)$, and hence its limit agrees with the iterated series (1.10), i.e., with $f(z)$. This will complete the proof of the theorem.

In order to prove (1.11), notice that by estimating (1.7) and (1.9) it follows that

$$(1.12) \qquad |c_\nu| \le M/\rho^\nu$$

for all $\rho \in \mathbb{R}^n$, $\rho > 0$, with $\lambda r \le \rho \le R/\lambda$. Given $z \in K(r, R)$ with $z_j \ne 0$ for $1 \le j \le n$ and $\nu \in \mathbb{Z}^n$, we choose $\rho_1, \ldots, \rho_n$ as follows: for j with $\nu_j \ge 0$, set $\rho_j = |z_j|/\lambda$, and if $\nu_j < 0$, set $\rho_j = \lambda |z_j|$. Then (1.12) implies

$$|c_\nu z^\nu| \le M \frac{|z_1|^{\nu_1} \cdots |z_n|^{\nu_n}}{\rho^\nu} = M \lambda^{|\nu_1| + \cdots + |\nu_n|}.$$

Finally, if $z \in K(r, R)$ satisfies $z_l = 0$ for some l, then one must have $r_l = 0$; so, by letting $\rho_l \to 0$ in (1.12), it follows that $c_\nu = 0$ if $\nu_l < 0$, and if $\nu_l \ge 0$, (1.11) is obvious. So (1.11) holds for all $z \in K(r, R)$. ∎

Theorem 1.5. *Let D be a connected Reinhardt domain with center 0. Then every $f \in \mathcal{O}(D)$ has a Laurent series representation*

$$(1.13) \qquad f(z) = \sum_{\nu \in \mathbb{Z}^n} c_\nu z^\nu \qquad \textit{for } z \in D,$$

which converges normally on D. Moreover, if $D \cap \{z \in \mathbb{C}^n : z_l = 0\} \ne \varnothing$ for some $1 \le l \le n$, then $c_\nu = 0$ for $\nu_l < 0$.

Proof. Suppose $f \in \mathcal{O}(D)$. For $a \in D$, we choose $K(r, R) \subset D$ so that a lies in the interior of $K(r, R)$. By Proposition 1.4, $f(z) = \sum c_\nu(a) z^\nu$ uniformly and absolutely on $K(r, R)$, and the coefficients $c_\nu(a)$, $\nu \in \mathbb{Z}^n$, are independent of the particular choice of r and R. Moreover, if a and a^* are two points in the interior of $K(r, R)$, then $c_\nu(a) = c_\nu(a^*)$. So, for each $\nu \in \mathbb{Z}^n$, $a \to c_\nu(a)$ is locally constant on D, and hence globally constant, since D is connected. This proves the existence of the representation (1.13) valid on D. The uniform absolute convergence on each $K(r, R)$ and a standard compactness argument show that (1.13) converges normally in D. Finally, if for some l, $1 \le l \le n$, there is $a \in D$ with $a_l = 0$, then we can choose r and R with $r_l = 0$, so that $a \in K(r, R) \subset D$; hence, by Proposition 1.4, $c_\nu = c_\nu(a) = 0$ for $\nu_l < 0$. ∎

1.3. Extension by Power Series

Theorem 1.5 has some very striking consequences.

Theorem 1.6. *Let D be a connected Reinhardt domain with center 0, and suppose that $D \cap \{z \in \mathbb{C}^n : z_l = 0\} \ne \varnothing$ for all $l = 1, 2, \ldots, n$ (this holds, in particular, if $0 \in D$). Then every $f \in \mathcal{O}(D)$ has a convergent power series expansion on D. Moreover, this power series defines a holomorphic extension of f to the smallest complete Reinhardt domain $\bigcup_{w \in D} P(0, \tau(w))$ which contains D.*

Proof. Let $f \in \mathcal{O}(D)$. Theorem 1.5 gives a Laurent expansion $f(z) = \sum_{\nu \in \mathbb{Z}^n} c_\nu z^\nu$ on D, and the hypotheses on D imply that $c_\nu = 0$ unless $\nu \ge 0$. Hence $f(z) = \sum_{\nu \in \mathbb{N}^n} c_\nu z^\nu$ for $z \in D$, and the Theorem now follows from the results in Chapter I, §1.5. ∎

Corollary 1.7. *Let D be a complete Reinhardt domain. Then the Taylor series of $f \in \mathcal{O}(D)$ at the center of D converges on D.*

In case $n = 1$, the hypothesis in Theorem 1.6 implies that D is a disc, and hence the theorem does not give any new information. But as soon as $n \geq 2$, there obvioulsy exist many noncomplete Reinhardt domains for which the Theorem gives a nontrivial conclusion. For example, both Theorems 1.1 and 1.2 are immediate consequences of Theorem 1.6. Another example is given by a spherical shell $G(r, R) = \{z \in \mathbb{C}^n : r < |z| < R\}$, where $0 \leq r < R$: if $n > 1$, Theorem 1.6 applies, and every holomorphic function on $G(r, R)$ extends to a holomorphic function on $B(0, R)$! Thus, if $n > 1$, every function holomorphic on the boundary of a ball (or polydisc, by Theorem 1.2) extends holomorphically to the interior. It is remarkable that this sort of result remains true for *arbitrary* domains with connected boundary! This was already known to Hartogs, who used the techniques discussed in this paragraph combined with some rather complicated and obscure geometric arguments. Simpler proofs are now available which make use of more powerful global methods. We shall return to these matters in Chapter IV. (In fact, none of the material in the rest of this chapter is needed for Chapter IV.)

Exercises

E.1.1. Let $P \subset \mathbb{C}^n$ be a polydisc and suppose $n \geq 2$. Show that every $f \in \mathcal{O}(bP)$ has a holomorphic extension to $\bar{P}$.

E.1.2. Let $B = B(0, R)$ be a ball of radius R in $\mathbb{C}^n$, $n \geq 2$, and suppose U is a neighborhood of a point $P \in bB$. Show that if $\Omega = \{z \in U : |z| > R\}$, then there is a neighborhood $W \subset U$ of P, such that every $f \in \mathcal{O}(\Omega)$ has a holomorphic extension to $W \cup \Omega$.

E.1.3. Let M be a closed complex submanifold of the region D in $\mathbb{C}^n$. Suppose $n \geq 2$ and that $\dim_P M \leq n - 2$ for all $P \in M$. Show that every $f \in \mathcal{O}(D - M)$ extends holomorphically across M.

E.1.4. Let $B = B(0, R)$ be a ball in $\mathbb{C}^n$ and $n \geq 2$. Let $A(B) = C(\bar{B}) \cap \mathcal{O}(B)$.

(i) Show that if $f \in A(B)$ satisfies $f(a) = 0$ for some $a \in B$, then there is $p \in bB$, such that $f(p) = 0$.
(ii) If $f \in A(B)$ satisfies $|f| = 1$ on bB, then f is constant.

Remark. It has recently been shown by A.B. Alexsandrov [Ale] and E. Løw [Løw] that there are many nonconstant **inner functions** on B, that is, functions $f \in H^\infty(B) = \mathcal{O}L^\infty(B)$, whose boundary values $f^* \in L^\infty(bB)$ satisfy $|f^*| = 1$ almost everywhere on bB – $f^*(\zeta) := \lim_{r \to 1} f(r\zeta)$ exists for almost all $\zeta \in bB$ by a generalization of Fatou's Theorem (see [Rud 3], Chapter 5). This solved the long outstanding *Inner Function Problem.*

E.1.5. Let $n \geq 2$ and suppose $K \subset \mathbb{C}^n$ is compact and $\mathbb{C}^n - K$ is connected. Show that every $f \in \mathcal{O}(\mathbb{C}^n - K)$ with

$$\limsup_{|z| \to \infty} |f(z)| < \infty$$

is constant.

§2. Natural Boundaries and Pseudoconvexity

En 1906, F. Hartogs a découvert une restriction très curieuse, à la quelle sont soumis les domaines d'holomorphie, et par cette decouverte même, je pense, a commencé le développement récent de la théorie des fonctions analytiques de plusieurs variables.

K. Oka, 1941 (from the introduction to "Domaines Pseudoconvexes"; English transl.: [Oka], p. 48)

The phenomenon of simultaneous extension of all holomorphic functions from one domain to a strictly larger one raises the question of characterizing those domains for which this phenomenon does not occur: these are the so-called domains of holomorphy. Since Hartogs' pioneering work in 1906, this question has been the motivation for many of the most important developments in the theory of several complex variables. In this paragraph we introduce the reader to the "curious restriction" referred to by Oka, and which is known as pseudoconvexity, by first following the early work of Hartogs and Levi; we shall then present those definitions of pseudoconvexity which will be used in later chapters. A more complete discussion of pseudoconvexity, based on plurisubharmonic functions, will be given in §5.

2.1. Domains of Holomorphy

In order to formulate precisely the statement that a holomorphic function $f \in \mathcal{O}(D)$ has no holomorphic extension across a boundary point $p \in bD$, one must take into account the fact familiar in one complex variable that the process of analytic continuation may lead to different function elements at p, depending on the particular approach to p chosen.

Definition. A holomorphic function f on D is **completely singular** at $p \in bD$ if for every connected neighborhood U of p there is no $h \in \mathcal{O}(U)$ which agrees with f on some connected component of $U \cap D$. D is called a **weak domain of holomorphy** if for every $p \in bD$ there is $f_p \in \mathcal{O}(D)$ which is completely singular at p. D is called a **domain of holomorphy** if there exists $f \in \mathcal{O}(D)$ which is completely singular at every boundary point $p \in bD$.

The concept of *weak* domain of holomorphy is not standard; it is, in fact, equivalent to the concept of domain of holomorphy (see §3.6), but this result is not elementary. On the other hand, it is usually much easier to verify that certain domains are *weak* domains of holomorphy, and this property is obviously sufficient to prevent the simultaneous extension of all holomorphic functions. This (formally) weaker concept of domain of holomorphy thus provides a convenient terminology at the introductory level. Notice, for example, that every open set D in the complex plane is obviously a *weak* domain of holomorphy (for $p \in bD$, take $f_p = (z - p)^{-1}$), but that such D is a domain of holomorphy is a rather deep result in classical complex analysis.

This observation, combined with the results in §1, shows that there is no several-variable analog of the function $(z - p)^{-1}$. However, in special cases, there is a reasonable substitute. For example, if p is a boundary point of the ball $B(0, R)$, the function $f_p(z) = [R^2 - (z, p)]^{-1}$ is holomorphic on $B(0, R)$ and completely singular at p. More generally, the following holds.

Lemma 2.1. *Every convex domain in $\mathbb{C}^n$ is a weak domain of holomorphy.*

The notion of convexity used here is the usual one in (real) linear spaces: $\Omega \subset \mathbb{R}^n$ *is* **convex** *if Ω contains every line segment whose endpoints lie in Ω.* This condition is equivalent to saying that *for every $p \notin \Omega$ there is a hyperplane H through p, so that Ω lies on one side of H* (see Exercise E.2.2.).

Proof. Let $p \in bD$. Since D is convex, one can find an $\mathbb{R}$-linear function $l = l_p: \mathbb{C}^n \to \mathbb{R}$, such that the hyperplane $\{z: l(z) = l(p)\}$ separates D and p, i.e., we may assume $l(z) < l(p)$ for $z \in D$. We can write $l(z) = \sum_{j=1}^n \alpha_j z_j + \sum_{j=1}^n \beta_j \bar{z}_j$, with $\alpha_j, \beta_j \in \mathbb{C}$, and since l is real valued, one must have $\beta_j = \bar{\alpha}_j$ for $j = 1, \ldots, n$. Hence $l(z) = \operatorname{Re} h(z)$, where $h(z) = 2\sum_{j=1}^n \alpha_j z_j$ is complex linear. It follows that $f_p = [h - h(p)]^{-1}$ is holomorphic on D and completely singular at p. ∎

2.2. Hartogs Pseudoconvexity

The extension property described in Theorem 1.1 leads directly to the formulation of a geometric condition which must be satisfied by weak domains of holomorphy. We first remove the dependence on the Euclidean coordinates in Theorem 1.1, and, for convenience, we specialize to compact sets. We set

$$\Gamma = \{z \in \mathbb{C}^n: z_j = 0 \text{ for } j < n, |z_n| \le 1\}$$
$$\cup \{z \in \mathbb{C}^n: z_j = 0 \text{ for } j < n - 1, |z_{n-1}| \le 1, |z_n| = 1\},$$

and

$$\hat{\Gamma} = \{z \in \mathbb{C}^n: z_j = 0 \text{ for } j < n - 1, |z_{n-1}| \le 1, |z_n| \le 1\};$$

we call the pair $(\Gamma, \hat{\Gamma})$ the **(standard) Hartogs frame** in $\mathbb{C}^n$. Note that $\Gamma = \hat{\Gamma}$ for $n = 1$. A pair $(\Gamma^*, \hat{\Gamma}^*)$ of compact sets in $\mathbb{C}^n$ is called a **Hartogs figure** if there is a biholomorphic map $F: \hat{\Gamma} \to \hat{\Gamma}^*$, such that $F(\Gamma) = \Gamma^*$.

Lemma 2.2. *Let $(\Gamma^*, \hat{\Gamma}^*)$ be a Hartogs figure. Then every $f \in \mathcal{O}(\Gamma^*)$ has a holomorphic extension $\hat{f} \in \mathcal{O}(\hat{\Gamma}^*)$.*

Proof. Let $F: \hat{\Gamma} \to \hat{\Gamma}^*$ be the biholomorphic map given by the hypothesis. If $f \in \mathcal{O}(\Gamma^*)$, then $g = f \circ F \in \mathcal{O}(\Gamma)$, and just as in the proof of Theorem 1.1, one

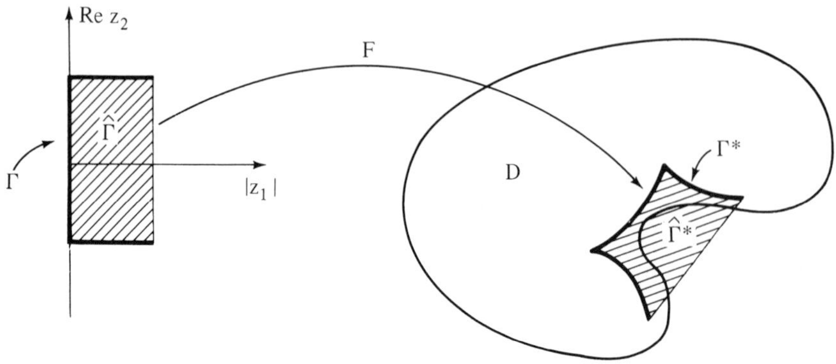

Figure 5. D is not Hartogs pseudoconvex.

sees that for $\varepsilon > 0$ sufficiently small,

$$\hat{g}(z', z_n) = \frac{1}{2\pi i} \int_{|\zeta|=1+\varepsilon} \frac{g(z', \zeta)}{\zeta - z_n} \, d\zeta$$

defines a holomorphic extension of g to a neighborhood of $\hat{\Gamma}$, i.e., $\hat{g} \in \mathcal{O}(\hat{\Gamma})$ and $g = \hat{g}$ in a neighborhood of Γ. It follows that $\hat{f} = \hat{g} \circ (F^{-1}) \in \mathcal{O}(\hat{\Gamma}^*)$ is the desired extension of f. ■

Definition. A domain $D \subset \mathbb{C}^n$ is called **Hartogs pseudoconvex** ($=$ ***H*-pseudoconvex**) if for every Hartogs figure $(\Gamma^*, \hat{\Gamma}^*)$ with $\Gamma^* \subset D$ one has $\hat{\Gamma}^* \subset D$ as well.

Figure 5 illustrates a region $D \subset \mathbb{C}^2$ which is not Hartogs pseudoconvex. Every function $f \in \mathcal{O}(D)$ extends holomorphically to a neighborhood of $\hat{\Gamma}^*$.

Notice that every domain $D \subset \mathbb{C}$ is trivially H-pseudoconvex. The following statements are also immediate consequences of the definition.

(2.1) *The intersection of finitely many H-pseudoconvex domains is H-pseudoconvex.*

(2.2) *If $D_1 \subset D_2 \subset \ldots$ is an increasing sequence of H-pseudoconvex domains, then $\bigcup_{j=1}^{\infty} D_j$ is H-pseudoconvex.*

Theorem 2.3. *A weak domain of holomorphy is Hartogs pseudoconvex.*

PROOF. Suppose that $D \subset \mathbb{C}^n$ is not H-pseudoconvex. We will show that there is $p \in bD$ such that *no* $f \in \mathcal{O}(D)$ is completely singular at p, and therefore D cannot be a weak domain of holomorphy. Let $(\Gamma^*, \hat{\Gamma}^*)$ be a Hartogs figure with $\Gamma^* \subset D$ but $\hat{\Gamma}^* \not\subset D$. If $f \in \mathcal{O}(D)$, then $f \in \mathcal{O}(\Gamma^*)$, and by Lemma 2.2, there is a holomorphic extension $\hat{f} \in \mathcal{O}(\hat{\Gamma}^*)$ of f. Let U be an open, connected neighborhood of $\hat{\Gamma}^*$, such that $f \in \mathcal{O}(U)$. Since $U \cap D \neq \emptyset$ and $U \not\subset D$, it follows that $U \cap bD \neq \emptyset$. Since $\hat{f} = f$ in a neighborhood of $\Gamma^* \subset U \cap D$, the Identity Theorem implies that $\hat{f} \equiv f$ on the connected component of $U \cap D$ which contains Γ^*. Hence f is not completely singular at every point $p \in U \cap bD$. ∎

Corollary 2.4. *Every convex domain in $\mathbb{C}^n$ is Hartogs pseudoconvex.*

2.3. Differentiable Boundaries

In 1910, E.E. Levi discovered another pseudoconvexity condition for domains of holomorphy with differentiable boundary. We shall discuss it in §2.6 after we have reviewed the relevant concepts from real calculus and some of their complex analogues.

Definition. The open set $D \subset \mathbb{R}^n$ is said to have **differentiable boundary** bD **of class** C^k, $1 \leq k \leq \infty$, **at the point** $p \in bD$ if there are an open neighborhood U of p and a real valued function $r \in C^k(U)$ with the following properties:

$$U \cap D = \{x \in U : r(x) < 0\}; \tag{2.3}$$

$$dr(x) \neq 0 \qquad \text{for } x \in U. \tag{2.4}$$

bD is of class C^k if it is of class C^k at every $p \in bD$.

Notice that (2.3) and (2.4) imply

$$U \cap bD = \{x \in U : r(x) = 0\} \text{ and } U - \bar{D} = \{x \in U : r(x) > 0\}. \tag{2.5}$$

Any function $r \in C^k(U)$ which satisfies (2.3) and (2.4) is called a **(local) defining function for D at p.** If U is a neighborhood of bD, a function $r \in C^k(U)$ with (2.3) and (2.4) is called a **global defining function**, or simply a **defining function**, for D.

The relationship between different defining functions is clarified by the following lemma.

Lemma 2.5. *Suppose r_1 and r_2 are two local defining functions for D of class C^k in a neighborhood U of $p \in bD$. Then there exists a positive function $h \in C^{k-1}(U)$ such that*

$$r_1 = h \cdot r_2 \text{ on } U; \tag{2.6}$$

$$dr_1(x) = h(x) \cdot dr_2(x) \qquad \text{for } x \in U \cap bD. \tag{2.7}$$

Proof. Clearly the conditions imposed on h determine h uniquely, and $h = r_1/r_2$ is C^k and positive on $U - bD$. Now fix $q \in U \cap bD$. After a local change of coordinates of class C^k near q, one may assume that $q = 0$, $U \cap bD = \{x \in U:$

$x_n = 0\}$ and $r_2(x) = x_n$. For $x' = (x_1, \dots, x_{n-1})$ near 0, we have $r_1(x', 0) = 0$. By the fundamental theorem of calculus,

$$r_1(x', x_n) = r_1(x', x_n) - r_1(x', 0) = x_n \int_0^1 \frac{\partial r_1}{\partial x_n}(x', tx_n)\, dt,$$

and the integral on the right is clearly a function of class C^{k-1} near $x = 0$. As the latter statement is independent of the particular choice of C^k-coordinates, (2.6) holds with $h \in C^{k-1}$ near q. Equation (2.7) is now obvious if $k - 1 \geq 1$, and if $k = 1$, (2.7) follows from (2.8) below, which is an easy consequence of the definitions.

(2.8) *If f is differentiable at $0 \in \mathbb{R}^n$, $f(0) = 0$, and h is continuous at 0, then $h \cdot f$ is differentiable at 0 and $d(hf)_0 = h(0)\, df_0$.*

Finally, Equation (2.7) combined with (2.4), implies that $h(x) \neq 0$ for $x \in U \cap bD$; since $h > 0$ on $U - bD$, continuity implies that h is positive on U. ■

Remark. To say that $D \subset \mathbb{R}^n$ has C^k boundary at $p \in bD$ according to the definition above, is equivalent to the statement that for some neighborhood U of p, $bD \cap U$ is a closed, real C^k submanifold of U of dimension $n - 1$.

Lemma 2.5 implies that the space

$$T_p(bD) = \left\{ \xi \in \mathbb{R}^n : dr_p(\xi) = \sum_{j=1}^n \frac{\partial r}{\partial x_j}(p)\xi_j = 0 \right\}$$

does not depend on the choice of the defining function r; $T_p(bD)$ is called the **tangent space to** bD **at** p. After a translation and rotation, one can always achieve that $p = 0$ and that $(\operatorname{grad} r)(p) = (0, \dots, 0, (\partial r/\partial x_n)(p))$, with $(\partial r/\partial x_n)(p) > 0$, so that $T_p(bD) = \{x \in \mathbb{R}^n : x_n = 0\}$. In this situation we say that $x' = (x_1, \dots, x_{n-1})$ are **tangential coordinates** at $p \in bD$, while x_n is the (outer) **normal coordinate**. It follows from the Implicit Function Theorem that if bD is of class C^k near $p = 0$ and $x_1, \dots, x_{n-1}$ are tangential coordinates, one can choose a local defining function r of the form

$$r(x', x_n) = x_n - \varphi(x_1, \dots, x_{n-1}), \tag{2.9}$$

where φ is of class C^k on a neighborhood of 0 in $T_0(bD)$, and $d\varphi_0 = 0$.

Lemma 2.6. *If $D \subset\subset \mathbb{R}^n$ has C^k boundary, then there is a global C^k defining function r for D.*

Proof. By compactness of bD there are finitely many open sets $U_1, \dots, U_l$ and local defining functions $r_\nu \in C^k(U_\nu)$, $\nu = 1, \dots, l$, so that $bD \subset \bigcup_{\nu=1}^l U_\nu$ and (2.3) and (2.4) hold for each r_ν. Choose functions $\varphi_\nu \in C_0^\infty(U_\nu)$ so that $0 \leq \varphi_\nu \leq 1$ and $\sum_{\nu=1}^l \varphi_\nu(x) = 1$ in a neighborhood $U \subset \bigcup_{\nu=1}^l U_\nu$ of bD, and define

$r = \sum \varphi_\nu r_\nu$. Then $r \in C^k(\mathbb{R}^n)$, and for $x \in bD$ one has $r(x) = 0$ and $dr_x \neq 0$. (This follows from $(\partial r_\nu/\partial x_n)(p) > 0$ for all ν if x_n is the normal coordinate at p.) After shrinking U, we may assume that $dr_x \neq 0$ for all $x \in U$. Finally, we verify that $U \cap D = \{x \in U : r(x) < 0\}$. In fact, if $x \in U \cap D$, then $x \in U_\nu \cap D$ for some ν with $\varphi_\nu(x) > 0$, and hence $r(x) \leq \varphi_\nu(x) r_\nu(x) < 0$. Conversely, if $r(x) < 0$ for some $x \in U$, then there is ν with $r_\nu(x) < 0$, and hence $x \in U_\nu \cap D \subset U \cap D$. ■

Remark 2.7. We leave it to the reader to verify that one can modify the defining function $r \in C^k(\mathbb{R}^n)$ for D given by Lemma 2.6, so that

$$D = \{x \in \mathbb{R}^n : r(x) < 0\}.$$

Of course, it is not possible to also achieve $dr_x \neq 0$ for all $x \in \mathbb{R}^n$.

2.4. The Complex Tangent Space

If $D \subset \mathbb{C}^n$ has C^k boundary at $p \in bD$ (for this we of course think of D as a subset of $\mathbb{R}^{2n}$), the complex structure of $\mathbb{C}^n$ induces an additional structure on the (real) tangent space $T_p(bD)$ as follows. Under the usual identification of $\mathbb{R}^{2n}$ and $\mathbb{C}^n$, $T_p(bD)$ is a subset of $\mathbb{C}^n$. Multiplication by $i = \sqrt{-1}$ defines a linear isomorphism $\mathbb{C}^n \to \mathbb{C}^n$; $T_p(bD)$ and its image $iT_p(bD)$ under this map are real subspaces of $\mathbb{C}^n$ of dimension $2n - 1$, and hence

$$T_p^{\mathbb{C}}(bD) = T_p(bD) \cap iT_p(bD) \tag{2.10}$$

is a real $(2n - 2)$-dimensional subspace of $T_p(bD)$ which is closed under multiplication by i. Thus $T_p^{\mathbb{C}}(bD)$ is a complex subspace of $\mathbb{C}^n$, of complex dimension $n - 1$. $T_p^{\mathbb{C}}(bD)$ is called the **complex tangent space to bD at p.** Notice that $T_p^{\mathbb{C}}(bD)$ is nontrivial only if $n \geq 2$.

The following Lemma gives an algebraic characterization of $T_p^{\mathbb{C}}(bD)$.

Lemma 2.8. *If r is a local defining function for $D \subset \mathbb{C}^n$ at $p \in bD$, then*

$$T_p^{\mathbb{C}}(bD) = \left\{ t \in \mathbb{C}^n : \partial r_p(t) = \sum_{j=1}^n \frac{\partial r}{\partial z_j}(p) t_j = 0 \right\}.$$

PROOF. Since r is real valued, we have $dr_p = \partial r_p + \bar{\partial}_p r = 2 \operatorname{Re} \partial r_p$. According to (2.10),

$$T_p^{\mathbb{C}}(bD) = \{t \in \mathbb{C}^n : dr_p(t) = dr_p(it) = 0\}.$$

Since ∂r_p is $\mathbb{C}$-linear,

$$\operatorname{Re}[\partial r_p(it)] = \operatorname{Re}[i \partial r_p(t)] = -\operatorname{Im} \partial r_p(t).$$

So

$$T_p^{\mathbb{C}}(bD) = \{t \in \mathbb{C}^n : \operatorname{Re} \partial r_p(t) = \operatorname{Im} \partial r_p(t) = 0\}. \quad ■$$

2.5. Convexity with Respect to Holomorphic Curves

If $D \subset \mathbb{R}^n$ is convex and $p \in bD$, then clearly there is no line segment L through p such that $L - \{p\} \subset D$. We now show that Hartogs pseudoconvex domains satisfy an analogous condition obtained by replacing line segments with complex 1-dimensional submanifolds. This is the first of a number of results which exhibit pseudoconvexity as a complex version of (linear) convexity.

Theorem 2.9. *Suppose $D \subset \mathbb{C}^n$ is Hartogs pseudoconvex and has C^1 boundary near $p \in bD$. Then there is no complex one-dimensional submanifold M with p in its interior, and with $M - \{p\} \subset D$.*

PROOF. The result is clearly true if $n = 1$, so we will assume $n \geq 2$. We will show that the existence of M with the properties stated in the theorem implies that there is a Hartogs figure $(\Gamma^*, \hat{\Gamma}^*)$ with $\Gamma^* \subset D$ but $\hat{\Gamma}^* \not\subset D$. Given such an M, after applying a holomorphic coordinate change in a neighborhood U of p we may assume that $p = 0$, $U \cap M = \{(z_1, 0, \ldots, 0): |z_1| < 2\delta\}$, and that the local defining function for D is of the form $r = x_n - \varphi(z', y_n)$, with $d\varphi_0 = 0$ (see (2.9)). Since we assume that $M \cap U - \{0\} \subset D$, we have $r(z_1, 0, \ldots, 0) < 0$ for $0 < |z_1| < 2\delta$, and therefore $\varphi(z_1, 0, \ldots, 0) \geq 0$ for $|z_1| < 2\delta$. By continuity of r it follows that for sufficiently small $\eta > 0$, $r(z)$ remains negative for $z \in K_1 = \{z = (z_1, 0, \ldots, z_n): |z_1| = \delta, |z_n + \eta| \leq \eta\}$, i.e., $K_1 \subset D \cap U$. Furthermore $r(z_1, 0, \ldots, 0, -\eta) = -\eta - \varphi(z_1, 0, \ldots) < 0$ for $|z_1| \leq \delta$, i.e., $K_2 = \{(z_1, 0, \ldots, -\eta): |z_1| \leq \delta\} \subset D \cap U$ as well. We set $\Gamma^* = K_1 \cup K_2$ and $\hat{\Gamma}^* = \{(z_1, 0, \ldots, z_n): |z_1| \leq \delta, |z_n + \eta| \leq \eta\}$. Then $\Gamma^* \subset D \cap U$, while $\hat{\Gamma}^* \not\subset D$, since $0 \in \hat{\Gamma}^*$. The situation is illustrated in Figure 6.

Finally, notice that $(\Gamma^*, \hat{\Gamma}^*)$ is a Hartogs figure in U (this property is independent of the particular holomorphic coordinates on U!). In fact, the map $z = F(t)$ given by $z_1 = \delta t_n$, $z_j = t_{j-1}$ for $2 \leq j \leq n - 1$, and $z_n = \eta t_{n-1} - \eta$

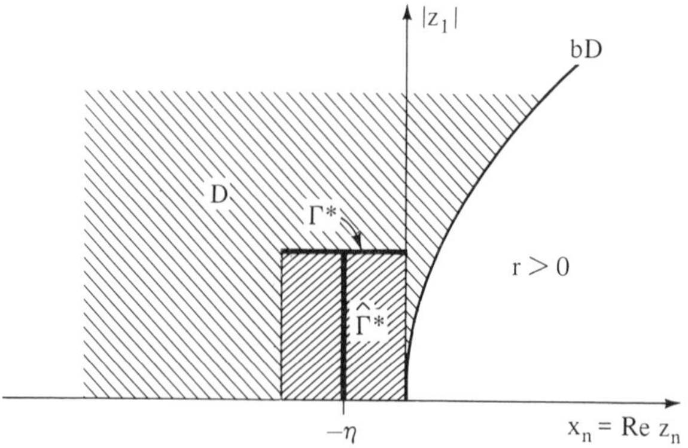

Figure 6. The Hartogs figure $(\Gamma^*, \hat{\Gamma}^*)$.

establishes the required biholomorphic correspondence between the standard Hartogs frame and $(\Gamma^*, \hat{\Gamma}^*)$. ■

2.6. Levi Pseudoconvexity

For a domain $D \subset \mathbb{R}^n$ with C^2 boundary, convexity is characterized by a differential condition in terms of a defining function r, as follows (see Exercise E.2.5):

(2.11) *If D is convex near $p \in bD$, then*

$$\sum_{j,k=1}^{n} \frac{\partial^2 r}{\partial x_j \partial x_k}(p)\xi_j\xi_k \geq 0 \qquad \textit{for all } \xi \in T_p(bD).$$

(2.12) *If $p \in bD$ and*

$$\sum_{j,k=1}^{n} \frac{\partial^2 r}{\partial x_j \partial x_k}(p)\xi_j\xi_k > 0 \qquad \textit{for all } \xi \in T_p(bD),\ \xi \neq 0, \tag{2.13}$$

then there is a (convex) neighborhood U of p, such that $D \cap U$ is convex.

One says that D is **strictly convex at** $p \in bD$ if (2.13) holds. The statements (2.11) and (2.12) are independent of the choice of defining function.

In 1910, E.E. Levi [Lev 1] discovered that domains of holomorphy with C^2 boundary satisfy a complex analogue of (2.11). This condition turns out to be extremely useful; not only is it easily computable, but its "strict" version—correponding to (2.13)—is, at least locally, also sufficient for (weak) domains of holomorphy. In order to gain a better geometric understanding of Levi's condition, we shall first examine what part of (2.11) is invariant under biholomorphic coordinate changes. After all, even though (linear) convexity is *not* preserved under biholomorphic maps, complex analytic properties like domain of holomorphy or Hartogs pseudoconvexity certainly should be invariant under such maps, and this must then also hold for any differential characterization of pseudoconvexity.

Suppose $r \in C^2(U)$ is a defining function for $D \subset \mathbb{C}^n$ near $p \in bD$. Let us write the Taylor expansion of r at p in complex form:

$$r(p+t) = r(p) + 2\,\mathrm{Re}(\partial r_p(t) + Q_p(r; t)) + L_p(r, t) + o(|t|^2), \tag{2.14}$$

where $t = (t_1, \ldots, t_n) \in \mathbb{C}^n$,

$$Q_p(r; t) = \frac{1}{2}\sum_{j,k=1}^{n} \frac{\partial^2 r}{\partial z_j \partial z_k}(p)t_j t_k, \tag{2.15}$$

and

$$L_p(r; t) = \sum_{j,k=1}^{n} \frac{\partial^2 r}{\partial z_j \partial \bar{z}_k}(p)t_j \bar{t}_k. \tag{2.16}$$

The real Hessian of r at $\xi \in \mathbb{R}^{2n}$ is given by

$$\frac{1}{2} \sum_{j,k=1}^{2n} \frac{\partial^2 r}{\partial \xi_j \partial \xi_k}(p)\xi_j \xi_k = 2 \operatorname{Re} Q_p(r; t) + L_p(r; t), \tag{2.17}$$

where t is related to ξ by $t_j = \xi_{2j-1} + \sqrt{-1}\xi_{2j}$, $1 \leq j \leq n$. If r satisfies (2.11), (2.17) implies

$$2 \operatorname{Re} Q_p(r; t) + L_p(r; t) \geq 0 \qquad \text{for all } t \in T_p(bD). \tag{2.18}$$

Since $Q_p(r; \sqrt{-1}t) = -Q_p(r; t)$ and $L_p(r; \sqrt{-1}t) = L_p(r; t)$, it follows from (2.18), applied to t and $\sqrt{-1}t$, that

$$L_p(r; t) \geq 0 \qquad \text{for all } t \in T_p^{\mathbb{C}}(bD). \tag{2.19}$$

We shall see shortly (Lemma 2.12, below), that condition (2.19) is invariant under biholomorphic coordinate changes in a neighborhood of p. It is precisely this condition which Levi found for domains of holomorphy. Equation (2.19) is usually referred to as the **Levi condition**. Note that (2.19) is independent of the particular defining function r for D. In fact, if r_1 is another C^2 defining function for D near p then, by Lemma 2.5, $r_1 = h \cdot r$, where $h > 0$ is of class C^1 near p. A straightforward computation (use (2.8) for the second order derivatives!) gives

$$L_p(r_1; t) = 2 \operatorname{Re}[\partial h_p(t) \cdot \overline{\partial r_p(t)}] + h(p) L_p(r; t). \tag{2.20}$$

Thus $L_p(r_1; t) = h(p)L_p(r; t)$ for $t \in T_p^{\mathbb{C}}(bD)$, and since $h(p) > 0$, (2.19) holds for r_1 if and only if it holds for r.

Definition. A domain D in $\mathbb{C}^n$ with C^2 boundary is said to be **Levi pseudoconvex at** $p \in bD$ if the Levi condition (2.19) holds for some (and hence all) C^2 defining function r for D near p. D is said to be **Levi pseudoconvex** if the Levi condition holds at all points $p \in bD$. D is said to be **strictly Levi pseudoconvex at** p, if $L_p(r, t) > 0$ **for all** $t \in T_p^{\mathbb{C}}(bD)$, $t \neq 0$. The Hermitian form $L_p(r; t)$ defined by (2.16) is called the **Levi form**, or **complex Hessian** of the C^2 function r at the point p.

The results of the computations we just carried out can now be summarized as follows.

Lemma 2.10. *A convex domain in* $\mathbb{C}^n$ *with* C^2 *boundary is Levi pseudoconvex.*

The converse is, of course, false. Notice that every domain in $\mathbb{C}^1$ with C^2 boundary is (strictly) Levi pseudoconvex! The remarkable fact discovered by Levi is that every domain of holomorphy with C^2 boundary is Levi pseudoconvex. This is a consequence of Theorem 2.3 and the following result.

Theorem 2.11. *If* $D \subset \mathbb{C}^n$ *is Hartogs pseudoconvex and* D *has* C^2 *boundary, then* D *is Levi pseudoconvex.*

In the proof we will make use of the invariance of Levi pseudoconvexity under holomorphic coordinate changes.

Lemma 2.12. *Suppose $D \subset \mathbb{C}^n$ has C^2 boundary near $p \in bD$ and $w = F(z)$ is a biholomorphic map on a neighborhood U of p. Let $\Omega = F(U \cap D)$. Then Ω is Levi pseudoconvex at $q = F(p)$ if and only if D is Levi pseudoconvex at p.*

Proof. Let r be a C^2 defining function for D near p; then $\rho = r \circ F^{-1}$ is a C^2 defining function for Ω near q. From $r(z) = \rho \circ F(z)$ one obtains by a straightforward calculation that for all $t \in \mathbb{C}^n$

$$\partial r_p(t) = \partial \rho_q(F'(p)t) \tag{2.21}$$

and

$$L_p(r; t) = L_q(\rho; F'(p)t). \tag{2.22}$$

Since $F'(p)$ is a nonsingular matrix, (2.21) shows that multiplication by $F'(p)$ defines an isomorphism from $T_p^{\mathbb{C}}(bD)$ onto $T_{F(p)}^{\mathbb{C}}(b\Omega)$. Equation (2.22) then implies the desired conclusion. ∎

Proof of Theorem 2.11. We will show that if the Levi condition does not hold at some point $p \in bD$, then one can find a one-dimensional complex submanifold M through p with $M \cap B(p, \varepsilon) - \{p\} \subset D$ for ε sufficiently small. According to Theorem 2.9, this would contradict the H-pseudoconvexity of D, and Theorem 2.11 will be proved.

So suppose there are $p \in bD$ and $t \in T_p^{\mathbb{C}}(bD)$ such that $L_p(r; t) < 0$, where r is a defining function near p. After a holomorphic change of coordinates we can assume that $p = 0$, $t = (1, 0, \ldots, 0)$ and that $\partial r_0 = dz_n$. Then $(\partial^2 r/\partial z_1 \partial \bar{z}_1)(0) = L_0(r; (1, 0, \ldots, 0)) < 0$, so $\eta = -(\partial^2 r/\partial z_1 \partial \bar{z}_1)(0) > 0$. From the Taylor expansion of r at 0 (see (2.14)) we obtain

$$\begin{aligned} r(z_1, 0, \ldots, 0, z_n) &= 2 \operatorname{Re} h(z) - \eta |z_1|^2 + O(|z_1||z_n|) + O(|z_n|^2) \\ &\quad + o(|z_1|^2 + |z_n|^2), \end{aligned} \tag{2.23}$$

where $h(z) = z_n + Q_0(r; (z_1, 0, \ldots, 0, z_n))$ is holomorphic. Since $\partial h/\partial z_n(0) \neq 0$, the set $M = \{z \in \mathbb{C}^n\colon h(z) = 0, z_2 = \cdots = z_{n-1} = 0\}$ is a complex one-dimensional submanifold at $0 \in M$. For $z \in M$ one has $|Q_0(r; z)| = O(|z_1|^2 + |z_n|^2)$; it follows that if $\varepsilon > 0$ is sufficiently small, then $|z_n| = O(|z_1|^2)$ for $z \in M \cap B(0, \varepsilon)$. This implies that 0 is the only point in $M \cap B(0, \varepsilon)$ with $z_1 = 0$, and from (2.23) one then obtains

$$r(z) = -\eta |z_1|^2 + o(|z_1|^2) \qquad \text{for } z \in M \cap B(0, \varepsilon). \tag{2.24}$$

After shrinking ε further, (2.24) implies that $r(z) < 0$ for $z \in M \cap B(0, \varepsilon)$ and $z \neq 0$, i.e., $M \cap B(0, \varepsilon) - \{0\} \subset D$. ∎

Even though the Levi condition at a point $p \in bD$ is independent of the defining function, the behavior of the Levi form off the boundary bD depends very much on the chosen defining function. For example, if $D = \{z \in \mathbb{C}^n : x_n < 0\}$ with defining function $r(z) = x_n$, then clearly $L_p(r; t) \equiv 0$ for all $p, t \in \mathbb{C}^n$; however, the function $\rho = x_n(1 - x_n|z|^2)$ is also a defining function for D near 0, and $L_z(\rho; t) < 0$ for $t \neq 0$ and all points z close to 0 with $\operatorname{Re} z_n \neq 0$ and $z_j = 0$ for $1 \leq j \leq n - 1$. The following result is therefore of some interest.

Lemma 2.13. *Suppose D has C^2 boundary and is Levi pseudoconvex near $p \in bD$. Then there is a C^2 defining function r for D on a neighborhood U of p such that at all points $z \in U$ one has*

$$L_z(r; t) \geq 0 \qquad \text{for all } t \text{ with } \partial r_z(t) = 0. \tag{2.25}$$

Proof. Without loss of generality we may assume that $p = 0$ and that $\operatorname{Re} z_1$, $\operatorname{Im} z_1, \ldots, \operatorname{Re} z_{n-1}, \operatorname{Im} z_{n-1}$, and $y_n = \operatorname{Im} z_n$ are tangential coordinates for bD at 0. Thus there is a local C^2 defining function r for D on a product neighborhood $U = U' \times (-\gamma, \gamma) \times (-\gamma, \gamma)$ of the form $r(z) = x_n - \varphi(z', y_n)$ (see (2.9)). By hypothesis, if U is sufficiently small, the Levi condition (2.25) for this defining function r holds for $z \in U$ with $x_n = \varphi(z', y_n)$. But (2.25) is now obviously independent of x_n, hence (2.25) holds at all points $z \in U$. ■

Remark. We must emphasize that Lemma 2.13 is a *local* result. If $D \subset\subset \mathbb{C}^n$ is Levi pseudoconvex, it is in general not possible to find a global defining function r on some neighborhood U of bD, so that (2.25) holds at all points $z \in U$. This question had remained unresolved for quite some time; an interesting and quite nontrivial counterexample was discovered only recently by K. Diederich and J.E. Fornaess [DiFo 2].

2.7. Strictly Plurisubharmonic Functions

The Levi condition involves the complex Hessian (or Levi form) of a defining function for D restricted to the complex tangent space of bD. We now single out those functions r whose complex Hessian $L_p(r; \cdot)$ is positive definite on all of $\mathbb{C}^n$; these functions turn out to be of fundamental importance in several complex variables.

Definition. A real valued C^2 function r defined on an open set $D \subset \mathbb{C}^n$ is said to be **strictly plurisubharmonic at** $p \in D$, if

$$L_p(r; t) > 0 \qquad \text{for } t \in \mathbb{C}^n, t \neq 0.$$

r is said to be **strictly plurisubharmonic** on D if r is strictly plurisubharmonic at every point $p \in D$.

Example. If $r(z) = |z|^2 = \sum_{j=1}^n z_j \bar{z}_j$, then $\partial^2 r / \partial z_j \partial \bar{z}_k = \delta_{jk}$, so $L_z(r; t) = |t|^2$ for all $z \in \mathbb{C}^n$, and we see that r is strictly plurisubharmonic on $\mathbb{C}^n$.

The reader may wonder at this point about the terminology. The connection with subharmonic and harmonic functions will become clear in §4, where the concept of plurisubharmonic function will be discussed systematically.

If r is strictly plurisubharmonic at p, then $\gamma = \min\{L_p(r, t): |t| = 1\}$ is positive, and hence $L_p(r; t) \geq \gamma|t|^2$ for $t \in \mathbb{C}^n$, by bilinearity. If $0 < c < \gamma$, the continuity of the second order derivatives of r implies

$$(2.26) \qquad L_z(r; t) \geq c|t|^2 \qquad \textit{for } t \in \mathbb{C}^n \textit{ and all } z \in U,$$

where U is some neighborhood of p, i.e., r is strictly plurisubharmonic at all points near p as well.

It easily follows from the definition that if r_1 and r_2 are strictly plurisubharmonic, so are $r_1 + r_2$ and cr_1 if $c > 0$. Also, the proof of Lemma 2.12 (in particular, formula (2.22)) shows that the definition of strictly plurisubharmonic function is invariant under holomorphic coordinate changes. This allows to define strictly plurisubharmonic functions on complex submanifolds (see Exercise E.2.9).

Proposition 2.14. *Suppose $D \subset\subset \mathbb{C}^n$ has C^2 boundary and is strictly Levi pseudoconvex. Then one can find a defining function r for D on a neighborhood U of bD which is strictly plurisubharmonic on U.*

Remark. It is possible to take for U a neighborhood of $\bar{D}$ (see Exercise E.2.8).

PROOF. By Lemma 2.6 there is a global C^2 defining function φ for D. We will show that for $A > 0$ sufficiently large, the function $r_A = \exp(A\varphi) - 1$ will do. In fact, it is obvious that r_A is a defining function for D. A straightforward computation gives

$$(2.27) \qquad L_z(r_A; t) = Ae^{A\varphi}[A|\partial\varphi_z(t)|^2 + L_z(\varphi; t)].$$

Let $S = bD \times \{t \in \mathbb{C}^n: |t| = 1\}$. Since $L_z(\varphi; t)$ is continuous on the compact set S, the subset $K = \{(z, t) \in S: L_z(\varphi; t) \leq 0\}$ is compact as well. The strict Levi pseudoconvexity of D implies that $|\partial\varphi_z(t)|^2 > 0$ whenever $(z, t) \in K$. Let $C = \min\{|\partial\varphi_z(t)|^2: (z, t) \in K\} > 0$ and $M = \min\{L_z(\varphi; t): (z, t) \in S\}$. If A is so large that $A \cdot C + M > 0$, it then follows from (2.27) that $L_z(r_A; t) > 0$ on S. Thus r_A is strictly plurisubharmonic on bD, and hence also on a neighborhood U of bD. (See (2.26) and the statements preceding it.) ∎

2.8. Strictly Pseudoconvex Domains

We now use strictly plurisubharmonic functions to define another important class of domains.

Definition. A bounded domain D in $\mathbb{C}^n$ is called **strictly pseudoconvex** if there are a neighborhood U of bD and a strictly plurisubharmonic function

$r \in C^2(U)$ such that

$$(2.28) \qquad D \cap U = \{z \in U : r(z) < 0\}.$$

Notice that we do not require that $dr_z \neq 0$ for $z \in bD$, so that a strictly pseudoconvex domain does not necessarily have a C^2 boundary. Proposition 2.14 shows that a bounded strictly Levi pseudoconvex domain is strictly pseudoconvex. Conversely, it is easy to see that every strictly pseudoconvex domain with C^2 boundary is strictly Levi pseudoconvex (see Exercise E.2.8). The simplest example of a strictly pseudoconvex domain is a ball $B(p, \varepsilon)$. The function $r(z) = |z - p|^2 - \varepsilon^2$ is strictly plurisubharmonic, and $B(p, \varepsilon) = \{z \in \mathbb{C}^n : r(z) < 0\}$. Other examples are found in the Exercises.

We now prove a partial converse to the fact that domains of holomorphy with C^2 boundary are Levi pseudoconvex (Theorem 2.11 combined with Theorem 2.3). This result is, essentially, due to E.E. Levi [Lev 2].

Theorem 2.15. *Let D be strictly pseudoconvex. Then every point $p \in bD$ has a neighborhood Ω such that $\Omega \cap D$ is a (weak) domain of holomorphy.*

For the proof of Theorem 2.15 we will use the following technical result, which is of fundamental importance for many constructions on strictly pseudoconvex domains.

Proposition 2.16. *Let U be open in $\mathbb{C}^n$ and suppose $r \in C^2(U)$ is strictly plurisubharmonic on U. Given $W \subset\subset U$, there are positive constants $c > 0$ and $\varepsilon > 0$, such that the function $F^{(r)}(\zeta, z)$ defined on $U \times \mathbb{C}^n$ by*

$$(2.29) \quad F^{(r)}(\zeta, z) = \sum_{j=1}^{n} \frac{\partial r}{\partial \zeta_j}(\zeta)(\zeta_j - z_j) - \frac{1}{2} \sum_{j,k=1}^{n} \frac{\partial^2 r}{\partial \zeta_j \partial \zeta_k}(\zeta)(\zeta_j - z_j)(\zeta_k - z_k)$$

satisfies the estimate

$$(2.30) \quad 2 \operatorname{Re} F^{(r)}(\zeta, z) \geq r(\zeta) - r(z) + c|z - \zeta|^2 \qquad \text{for } \zeta \in W \text{ and } |z - \zeta| < \varepsilon.$$

Proof. From formula (2.14), with $p = \zeta \in U$ and $t = z - \zeta$, we see that the Taylor expansion of $r(z)$ at ζ is given by

$$(2.31) \qquad r(z) = r(\zeta) - 2 \operatorname{Re} F^{(r)}(\zeta, z) + L_\zeta(r; z - \zeta) + o(|\zeta - z|^2).$$

By (2.26), if $\overline{W} \subset U$ is compact, there is $c > 0$, such that $L_\zeta(r; z - \zeta) \geq 2c|\zeta - z|^2$ for $\zeta \in W$ and $z \in \mathbb{C}^n$. Taylor's Theorem and the uniform continuity on W of the derivatives of r up to order 2 imply that the error term $o(|\zeta - z|^2)$ in (2.31) is uniform in $\zeta \in W$, that is, there is $\varepsilon > 0$, so that $|o(|\zeta - z|^2)| \leq c|\zeta - z|^2$ for $\zeta \in W$ and $|z - \zeta| < \varepsilon$. Equation (2.30) now follows by using these estimates in (2.31) and rearranging. ∎

The function $F^{(r)}(\zeta, z)$ is called the **Levi polynomial of r at ζ**; it is a quadratic holomorphic polynomial in z. Notice that it may happen that $F^{(r)}(\zeta, \cdot) \equiv 0$ for certain points ζ; from (2.30) it then follows that $r(z) \geq r(\zeta) + c|z - \zeta|^2$ for z near ζ, i.e., $r(z)$ has a strict local minimum at such a point ζ.

Proof of Theorem 2.15. Suppose $r \in C^2(U)$ is strictly plurisubharmonic in a neighborhood U of bD, so that (2.28) holds. Choose c, ε as in Proposition 2.16, so that (2.30) holds for $\zeta \in bD$. For $\zeta \in bD$ one has $r(\zeta) = 0$, and $r(z)$ cannnot have a local minimum at ζ, so $F^{(r)}(\zeta, \cdot) \not\equiv 0$, and (2.30) implies that $\operatorname{Re} F^{(r)}(\zeta, z) > 0$ for $z \in D$ with $|z - \zeta| < \varepsilon$ (choose ε so small that $B(\zeta, \varepsilon) \subset U$ for $\zeta \in bD$). If $p \in bD$ is fixed, let $\Omega = B(p, \varepsilon/2)$. We claim that $\Omega \cap D$ is a weak domain of holomorphy. In fact, for $\zeta \in \Omega \cap bD$, $f_\zeta = [F^{(r)}(\zeta, \cdot)]^{-1}$ is holomorphic on $\Omega \cap D$ and completely singular at ζ; for any of the remaining boundary points $\zeta \in b\Omega \cap \bar{D}$ of $\Omega \cap D$, the convexity of Ω implies that there is $f_\zeta \in \mathcal{O}(\Omega)$ which is completely singular at ζ. (See Lemma 2.1 or the remarks preceding it.) ■

Another application of the Levi polynomial leads to the following characterization of strict Levi pseudoconvexity.

Theorem 2.17. *Let $D \subset \mathbb{C}^n$ have C^2 boundary near $p \in bD$. Then D is strictly Levi pseudoconvex at p if and only if there is a holomorphic coordinate system $w = w(z)$ in a neighborhood of p, so that D is strictly convex with respect to the w-coordinates.*

In other words, strict Levi pseudoconvexity is precisely the locally biholomorphically invariant formulation of strict (Euclidean) convexity!

Proof. If there are holomorphic coordinates $w = w(z)$ near $p \in bD$, so that D is strictly convex with respect to w at $w(p)$, then the same computations which proved Lemma 2.10 imply that D is strictly Levi pseudoconvex in the w-coordinates, and hence also in the original z-coordinates (see Lemma 2.12). For the other implication, let r be a C^2 strictly plurisubharmonic defining function for D near p (use Proposition 2.14, which clearly holds locally as well). Let $h(z) = F^{(r)}(p, z)$ be the Levi polynomial of r at p. From (2.29) it follows that $dh_p = -\partial r_p \neq 0$ (D has a C^2 boundary!), and hence we can find a holomorphic coordinate system $w = (w', w_n)$ near p with $w(p) = 0$ and $w_n(z) = -h(z)$. With respect to w, D has a defining function $\rho(w)$ which is related to $r(z)$ by $\rho(w(z)) = r(z)$. From (2.31) and (2.22) one obtains

$$\rho(w) = 2 \operatorname{Re} w_n + L_0(\rho; w) + o(|w|^2). \tag{2.32}$$

Clearly (2.32) is the Taylor expansion of ρ up to order 2; this implies that the (real) Hessian of ρ in the w-coordinates agrees precisely with the Levi form of ρ, which is positive definite. Thus $\{w: \rho(w) < 0\}$ is strictly convex at $w = 0$. ■

Theorem 2.17 does not hold without the assumption that bD is of class C^2 (see Exercise E.2.10). More significant is the fact that, even in case of C^2

boundary, the analogous characterization of Levi pseudoconvexity (without the strictness condition) as a biholomorphically invariant version of convexity is false. The discovery of the relevant counterexample by J.J. Kohn and L. Nirenberg in 1972 ([KoNi]; see also Exercise E.2.6) was quite a suprise; it made it clear that pseudoconvexity was understood very little up to that time. Since then, a great deal of progress has been made in the study of smoothly bounded (Levi) pseudoconvex domains, especially in the case of real analytic boundaries (see, for example, [DiFo 3], [BeFo]). But a complete understanding of the geometric and analytic properties of Levi pseudoconvexity has not yet been achieved. For example, it is not known how to characterize (by a differential condition, or some other verifiable condition) those Levi pseudoconvex domains which are *locally* biholomorphic to a (linearly) convex domain. (See [Ran 3] and [Blo] for some partial results.)

2.9. The Levi Problem

The fact that strictly Levi pseudoconvex domains are *locally* (weak) domains of holomorphy (Theorem 2.15) gives strong evidence that the Levi condition comes very close to characterizing smoothly bounded domains of holomorphy. Besides the mainly technical matter of bridging the gap between Levi pseudoconvexity and strict Levi pseudoconvexity—something that can be handled by "approximation and exhaustion" techniques (see, for example, Lemma 2.19 below)—the major problem is of course to show that a strictly pseudoconvex domain D is *globally* a (weak) domain of holomorphy. That is, one must find a function *holomorphic on* D which is completely singular at $\zeta \in bD$, not just one which is holomorphic only on $\Omega \cap D$ for some neighborhood Ω of ζ. This problem, which has become known as the Levi problem, involves the construction of global holomorphic functions with some specific local properties. Questions of this sort are among the most difficult ones in complex analysis, as one cannot just construct global *holomorphic* functions by gluing together local functions with smooth cutoff functions or partitions of unity. The Levi problem was first solved by K. Oka [Oka, VI] in 1942 in $\mathbb{C}^2$, and, in arbitrary dimension, it was solved independently by Oka [Oka, IX], H. Bremermann [Bre 1], and F. Norguet [Nor 1] in the early 1950s. The solution of the Levi problem in $\mathbb{C}^n$, and later in more abstract settings, has involved the development of many powerful methods such as plurisubharmonic functions, the algebraic "coherence theory" of local rings of holomorphic functions, the cohomology theory of coherent analytic sheaves, and global solutions and estimates for the inhomogeneous Cauchy–Riemann equations. In this book, the Levi problem will be solved in Chapter V by means of integral representations. The reader interested in getting quickly to that result may skip the remainder of this chapter and proceed directly to Chapter III (as needed) and Chapter IV.

2.10. Pseudoconvex Domains

The special properties enjoyed by strictly pseudoconvex domains are very important for establishing many fundamental global results, but these domains are not sufficiently general for dealing with even rather simple domains of holomorphy. For example, a polydisc is a (weak) domain of holomorphy, and hence Hartogs pseudoconvex, but it is not strictly pseudoconvex (Exercise E.2.11). However, it turns out that every Hartogs pseudoconvex domain is the increasing union of strictly pseudoconvex domains (see §5). The following definition of pseudoconvexity, even though formally stronger than the latter property, is, in fact, equivalent to it. It is particularly suitable for extending function theoretic results from the strictly pseudoconvex case to the general case.

A function $\varphi: D \to \mathbb{R}$ on the open set D is said to be an **exhaustion function for** D if for every $c \in \mathbb{R}$ the set $D_c = \{z \in D: \varphi(z) < c\}$ is relatively compact in D. An exhaustion function φ satisfies $\varphi(z) \to \infty$ as $z \to bD$; this is also sufficient if D is bounded.

Definition. An open set D in $\mathbb{C}^n$ is called **pseudoconvex** if there is an exhaustion function $\varphi \in C^2(D)$ for D which is strictly plurisubharmonic on D.

In §5 we will show that an open set D is pseudoconvex if and only if it is Hartogs pseudoconvex, but we shall not use this result in later chapters. Instead, in §3.7, we will prove directly that every domain of holomorphy is pseudoconvex as defined above. We now present some simple sufficient conditions for pseudoconvexity.

Lemma 2.18. *A bounded strictly pseudoconvex domain D in $\mathbb{C}^n$ is pseudoconvex.*

Proof. Let r be strictly plurisubharmonic in a neighborhood U of bD, so that $D \cap U = \{z \in U: r(z) < 0\}$. Without loss of generality we may assume that $r \in C^2(\bar{D})$ and $r < 0$ on D. A straightforward computation shows that $\rho = -\log(-r)$ is strictly plurisubharmonic on $D \cap U$. Let $m = \inf\{L_z(\rho; t): z \in D - U \text{ and } |t| = 1\} > -\infty$. Then $\varphi = \rho + (m + 1)|z|^2$ is a strictly plurisubharmonic exhaustion function for D. ■

Lemma 2.19. *A bounded Levi pseudoconvex domain D with C^3 boundary is pseudoconvex.*

Proof. Given a defining function r for D on a neighborhood of $\bar{D}$, we try to imitate the proof of Lemma 2.18. Unfortunately, $\rho = -\log(-r)$ will no longer be strictly plurisubharmonic near bD, but if $r \in C^3$, we can use the Levi condition on bD to control the Levi form of r and ρ near bD, as follows. Let U be a neighborhood of bD, so that $\partial r_z \neq 0$ for $z \in U$. If $T_z' = \{t \in \mathbb{C}^n, \partial r_z(t) = 0\}$, we let T_z'' be the orthogonal complement of T_z' in $\mathbb{C}^n$. So, for each $z \in U$, a vector $t \in \mathbb{C}^n$ has a corresponding decomposition $t = t_z' + t_z''$ which depends differentiably on z. The Levi pseudoconvexity of D says that $L_z(r; t_z') \geq 0$ for $z \in bD$. As this expression is C^1 in z (this is where we use that $r \in C^3$), it follows that, after shrinking U, there is a constant $C > 0$, so that

$$L_z(r; t_z') \geq -C|r(z)|\,|t_z'|^2 \qquad \text{for } z \in D \cap U. \tag{2.33}$$

The bilinearity of the Levi form implies

$$L_z(r; t) = L_z(r; t_z') + O(|t_z'|\,|t_z''|) + O(|t_z''|^2),$$

and since $|t_z''| = O(|\partial r_z(t)|$ for $z \in U$, one obtains from (2.33) that

$$L_z(r; t) \geq -C|r(z)|\,|t|^2 - C_1|\partial r_z(t)|\,|t| \tag{2.34}$$

for $z \in U \cap D$, where $C_1 > 0$ is another constant. For $\rho = -\log(-r)$ one now computes

$$L_z(\rho; t) = |r(z)|^{-1} L_z(r; t) + |r(z)|^{-2}|\partial r_z(t)|^2 \qquad \text{for } z \in D;$$

hence, by (2.34),

$$\begin{aligned} L_z(\rho; t) &\geq -C|t|^2 - \left(\frac{|\partial r_z(t)|}{|r(z)|}\right)(C_1|t|) + \frac{|\partial r_z(t)|^2}{|r(z)|^2} \\ &\geq -(C + C_1^2)|t|^2 \qquad \text{for } z \in U \cap D, \end{aligned} \tag{2.35}$$

where we have used the inequality $ab \leq (a/2)^2 + b^2$ in order to estimate the middle term on the right side of (2.35). It now follows as in the proof of Lemma 2.18 that for sufficiently large A, the function $\varphi = \rho + A|z|^2$ is strictly plurisubharmonic on D, and φ is obviously an exhaustion function. ∎

Remark 2.20. The hypothesis that bD is of class C^3 was used only in the proof of (2.33). Whenever one has (2.33) for some C^2 defining function r (notice that Lemma 2.13 gives such a function *locally*), the rest of the proof of Lemma 2.19 goes through unchanged. In §5.6 we shall combine these observations with some deeper properties of pseudoconvex domains in order to prove Lemma 2.19 in the case of C^2 boundaries.

The following technical result will allow us to approximate strictly pseudoconvex domains by domains with C^2 boundary. This will be important in order to consider integration of functions or differential forms over the boundary.

Proposition 2.21. *If $D \subset \mathbb{C}^n$ is pseudoconvex, then there is a strictly plurisubharmonic exhaustion function r for D such that the set of critical points $\{z \in D: dr_z = 0\}$ of r is discrete in D.*

The proof is based on the following real variable lemma which is a special case of general results in Morse theory. For the reader's convenience we have included a self-contained proof in Appendix A.

Lemma 2.22. *Let D be an open set in $\mathbb{R}^n$ and suppose $g \in C^2(D)$ is real valued. Then there is a set $E \subset \mathbb{R}^n$ of measure 0, such that for all $u \in \mathbb{R}^n - E$ the set of critical points of $g_u(x) = g(x) + (u, x)$ is discrete in D.*

PROOF OF PROPOSITION 2.21. Let $\varphi \in C^2(D)$ be a strictly plurisubharmonic exhaustion function for D and set $g(z) = \varphi(z) + |z|^2$. By Lemma 2.22 there is an $\mathbb{R}$-linear function $l(z)$, so that $r(z) = g(z) + l(z)$ has only isolated critical points. It is obvious that r is strictly plurisubharmonic on D, and one easily checks that r is an exhaustion function. ■

We say that a **compact set** $K \subset \mathbb{C}^n$ **is pseudoconvex** if K has a neighborhood basis of open pseudoconvex sets.

Corollary 2.23. *Let K be a compact pseudoconvex set in $\mathbb{C}^n$. Then K has a neighborhood basis of strictly pseudoconvex domains with C^2 boundary.*

PROOF. Let W be a neighborhood of K. By hypothesis, there is an open pseudoconvex neighborhood U of K with $U \subset W$. Let r be a C^2 strictly plurisubharmonic exhaustion function for U whose set S of critical points is discrete in U, and hence countable. If $c \in \mathbb{R} - r(S)$ is larger than $M = |r|_K$, then $K \subset D_c = \{z \in U: r(z) < c\} \subset\subset W$, D_c is strictly pseudoconvex, and $dr_z \neq 0$ for $z \in bD_c$. ■

EXERCISES

E.2.1. Show that an open set D in $\mathbb{C}^n$ is a domain of holomorphy if and only if each connected component of D is a domain of holomorphy.

E.2.2. Show that an open set Ω in $\mathbb{R}^n$ is (Euclidean) convex if and only if for each $p \notin \Omega$ there is a linear functional $l: \mathbb{R}^n \to \mathbb{R}$, such that $l(x) < l(p)$ for all $x \in \Omega$. (Geometrically speaking, Ω lies on one side of the hyperplane $\{x: l(x) = l(p)\}$; see also E.3.1.)

E.2.3. Prove in detail the statements made in (2.1) and (2.2).

E.2.4. Suppose $D \subset\subset \mathbb{R}^n$ has differentiable boundary of class C^k. Show that there is a function $r \in C^k(\mathbb{R}^n)$ such that $D = \{x \in \mathbb{R}^n: r(x) < 0\}$ and $dr(x) \neq 0$ for $x \in bD$.

E.2.5. Prove the characterizations of convexity stated in (2.11) and (2.12).

E.2.6. Define $r(z)$ for $z \in \mathbb{C}^2$ by

$$r(z) = \operatorname{Re} z_2 + |z_1|^8 + 15/7|z_1|^2 \operatorname{Re}(z_1^6)$$

and set $D = \{z \in \mathbb{C}^2: r(z) < 0\}$.

(i) Show that D is Levi pseudoconvex.
(ii) Find all points $p \in bD$ at which D is strictly Levi pseudoconvex.
(iii) J.J. Kohn and L. Nirenberg [KoNi] proved that if f is any holomorphic function on a neighborhood U of 0 with $f(0) = 0$, then the zero set of f

meets both $D \cap U$ and $U - \bar{D}$. Show that this implies that there is no holomorphic change of coordinates in some neighborhood of 0 which makes D linearly convex near 0 in the new coordinates.

(iv) Try to prove the result of Kohn and Nirenberg stated in (iii). (This is quite nontrivial!)

E.2.7. Let m_1, m_2 be positive integers. Discuss the Levi pseudoconvexity and strictly Levi pseudoconvexity at the boundary points of

$$D_{(m_1,m_2)} = \{z \in \mathbb{C}^2: |z_1|^{2m_1} + |z_2|^{2m_2} < 1\}.$$

E.2.8. Let D be a bounded strictly pseudoconvex domain in $\mathbb{C}^n$ (see the Definition in §2.8).

(i) Show that if D has C^2 boundary at a point $p \in bD$, then D is strictly Levi pseudoconvex at p.

(ii) Show that there are a neighborhood U of $\bar{D}$ and a strictly plurisubharmonic function r on U such that

$$D = \{z \in U: r(z) < 0\}.$$

If D has C^2 boundary, then r can be found so that $dr \neq 0$ on bD. (*Hint*: If D is defined by the strictly plurisubharmonic function φ on a neighborhood W of bD, there are $\varepsilon_1 < \varepsilon_2 < 0$ such that $\{z \in W: \varepsilon_1 < \varphi(z) < \varepsilon_2\} \subset\subset W \cap D$. Choose $\chi \in C^2(\mathbb{R})$ such that $\chi(t)$ is constant for $t \leq \varepsilon_1$, $\chi(t) = t$ for $t \geq \varepsilon_2$, and $\chi'(t)$, $\chi''(t) > 0$ for $\varepsilon_1 < t < \varepsilon_2$. Then consider $r = \chi \circ \varphi + \lambda |z|^2$ for a suitable cutoff function λ.)

E.2.9. Let M be a complex submanifold of $\mathbb{C}^n$. A function $\varphi: M \to \mathbb{R}$ is said to be strictly plurisubharmonic at $P \in M$ if there is a local parametrization H for M near P such that $\varphi \circ H$ is strictly plurisubharmonic near $H^{-1}(P)$.

(i) Show that the above definition does not depend on the particular choice of H.

(ii) Show that φ is strictly plurisubharmonic at P if and only if there are a neighborhood U of P in $\mathbb{C}^n$ and a strictly plurisubharmonic function $\tilde{\varphi}$ on U, such that $\varphi = \tilde{\varphi}$ on $U \cap M$.

E.2.10. Show that if $D = \{x \in \Omega: r(x) < 0\}$ for some C^2 function r on a neighborhood Ω of $\bar{D}$, which is strictly convex near bD (i.e, the (real) Hessian of r is positive definite at all points in bD), then $dr \neq 0$ on bD; that is, D has necessarily a C^2 boundary.

E.2.11. Show that a polydisc in $\mathbb{C}^n$ is strictly pseudoconvex if and only if $n = 1$.

E.2.12. Let P be a polydisc in $\mathbb{C}^n$. Show that if $n \geq 2$, then P is not biholomorphic to any strictly pseudoconvex domain in $\mathbb{C}^n$. (Hint: Generalize the proof of Theorem I.2.7.)

E.2.13. Let $\varphi: \Omega \to \mathbb{R}$ be strictly plurisubharmonic and set $D = \{z \in \Omega: \varphi(z) < 0\}$. Suppose $p \in \Omega$ and $\varphi(p) = 0$. Show that there are a neighborhood U of p and a holomorphic function $h \in \mathcal{O}(U)$, with $h(p) = 1$ and $|h(z)| < 1$ for all $z \in U \cap \bar{D} - \{p\}$. (Note that it is *not* assumed that $d\varphi_p \neq 0$!) *Remark*: It is not necessary for D to be strictly pseudoconvex at p. Show that the above conclusion holds at every boundary point of $D_{(m_1,m_2)}$ (see E.2.7).

E.2.14. (i) Let $D \subset\subset \mathbb{C}^n$ have C^2 boundary. Show that the set of strictly Levi pseudoconvex boundary points of D is not empty. (Hint: Fix $a \in D$ and consider a point $P \in bD$ at maximal distance from a.)

(ii) More generally, let D be as in (i), $P \in bD$ and let U be a neighborhood of P. Suppose $v \in C^2(U)$ is strictly plurisubharmonic, $v(P) = 0$ and $v(z) < 0$ for $z \in U \cap D$. Show that D is strictly Levi pseudoconvex at P.

§3. The Convexity Theory of Cartan and Thullen

The statement that a region D in $\mathbb{C}^n$ is a domain of holomorphy (see §2.1) involves the surrounding space $\mathbb{C}^n$, and thus it is not clear whether it describes an intrinsic property of D. It was a major achievement when in 1932, H. Cartan and P. Thullen [CaTh] discovered an intrinsic characterization of domains of holomorphy in terms of convexity conditions with respect to the algebra of holomorphic functions $\mathcal{O}(D)$. This "holomorphic convexity" is one of the most fundamental concepts in several complex variables. We will now present a detailed discussion of holomorphic convexity.

3.1. Convexity with Respect to Linear Functions

In order to exhibit the formal analogies between holomorphic convexity and linear convexity, we reformulate the latter condition so as to emphasize the role of the space of linear functions on D.

If $K \subset \mathbb{R}^n$ is compact, its **convex hull** $\hat{K}^C$ is usually defined to be the smallest closed convex set which contains K, i.e., $\hat{K}^C = \bigcap \{\Omega \subset \mathbb{R}^n: \Omega \text{ closed convex and } K \subset \Omega\}$. An alternate description of the convex hull involves the **hull** $\hat{K}_{\mathscr{L}(\mathbb{R}^n)}$ **of** K **with respect to the family** $\mathscr{L}(\mathbb{R}^n)$ **of real valued linear functions on** $\mathbb{R}^n$, defined as follows:

$$\hat{K}_{\mathscr{L}(\mathbb{R}^n)} = \left\{x \in \mathbb{R}^n: \varphi(x) \leq \sup_K \varphi \qquad \text{for all } \varphi \in \mathscr{L}\right\}.$$

It is easy to check that $\hat{K}^C = \hat{K}_{\mathscr{L}(\mathbb{R}^n)}$, and that an open set $D \subset \mathbb{R}^n$ is convex if and only if $\hat{K}_{\mathscr{L}(\mathbb{R}^n)} \subset D$ for every compact set $K \subset D$ (see Exercise E.3.1). For $K \subset D$ we now consider

$$\hat{K}_{\mathscr{L}(D)} = D \cap \hat{K}_{\mathscr{L}(\mathbb{R}^n)} = \left\{x \in D: \varphi(x) \leq \sup_K \varphi \qquad \text{for all } \varphi \in \mathscr{L}\right\}.$$

By definition, $\hat{K}_{\mathscr{L}(D)} \subset D$; but if D is convex, then $\hat{K}_{\mathscr{L}(D)}$ is a *compact* subset of D, or, equivalently, $\hat{K}_{\mathscr{L}(D)} \subset\subset D$. Conversely, if $\hat{K}_{\mathscr{L}(D)} \subset\subset D$ for every $K \subset D$, then $\hat{K}_{\mathscr{L}(\mathbb{R}^n)} = \hat{K}_{\mathscr{L}(D)} \cup (\hat{K}_{\mathscr{L}(\mathbb{R}^n)} - D)$ is a decomposition of the connected set $\hat{K}_{\mathscr{L}(\mathbb{R}^n)}$ into two disjoint closed sets, and hence $\hat{K}_{\mathscr{L}(\mathbb{R}^n)} = \hat{K}_{\mathscr{L}(D)}$, i.e., it follows that D is convex. We thus see that $D \subset \mathbb{R}^n$ is (Euclidean convex) if and only if D is convex with respect to the family $\mathscr{L}(D)$ of linear functions on D in the following general sense.

Definition. Let $\mathscr{F}$ be a family of real valued functions on the topological space X. The space X is called $\mathscr{F}$-convex if for every compact subset $K \subset X$ the $\mathscr{F}$-hull of K,

$$\hat{K}_{\mathscr{F}} = \left\{x \in X: \varphi(x) \leq \sup_K \varphi \qquad \text{for all } \varphi \in \mathscr{F}\right\}$$

is relatively compact in X.

We shall see that many important function theoretic properties of domains in $\mathbb{C}^n$ are characterized by convexity with respect to an appropriate family of functions.

3.2. Holomorphic Convexity

For a subset K of $D \subset \mathbb{C}^n$, its **holomorphically convex hull $\hat{K}_{\mathcal{O}(D)}$ in** D is defined by

$$\hat{K}_{\mathcal{O}(D)} = \{z \in D: |f(z)| \leq |f|_K \qquad \text{for all } f \in \mathcal{O}(D)\}. \tag{3.1}$$

$\hat{K}_{\mathcal{O}(D)}$ is also called the **$\mathcal{O}(D)$-hull of** K, and we will simply write $\hat{K}$ instead of $\hat{K}_{\mathcal{O}(D)}$ when the region D is clear from the context. $K \subset D$ is called **$\mathcal{O}(D)$-convex** if $\hat{K}_{\mathcal{O}(D)} = K$.

We now state some simple properties of $\mathcal{O}(D)$-hulls.

Lemma 3.1. *For $K \subset D$ the following hold:*

(i) *$K \subset \hat{K}$ and $\hat{\hat{K}} = \hat{K}$;*
(ii) *if $K_1 \subset K$, then $\hat{K}_1 \subset \hat{K}$;*
(iii) *if Ω is open and $D \subset \Omega$, then $\hat{K}_{\mathcal{O}(D)} \subset \hat{K}_{\mathcal{O}(\Omega)}$;*
(iv) *$\hat{K}$ is closed in D, and if K is bounded, so is $\hat{K}$;*
(v) *given M, $\varepsilon > 0$ and $p \in D - \hat{K}$, there is $f \in \mathcal{O}(D)$ with $|f|_K < \varepsilon$ and $|f(p)| > M$.*

Proof. (i), (ii), and (iii) are straightforward. For (iv), notice that if $f \in \mathcal{O}(D)$, the set $A_f = \{z \in D: |f(z)| \leq |f|_K\}$ is closed in D ($A_f = D$ if $|f|_K = \infty$), and $\hat{K} = \bigcap \{A_f: f \in \mathcal{O}(D)\}$. Furthermore, since the coordinate functions z_j, $1 \leq j \leq n$, are holomorphic on D, if K is bounded, it follows that $\hat{K} \subset \bar{P}(0, (r_1, \ldots, r_n))$, where $r_j = |z_j|_K < \infty$; so $\hat{K}$ is bounded as well. To prove (v), observe that $p \notin \hat{K}$ implies that there is $h \in \mathcal{O}(D)$ with $|h|_K < |h(p)|$. After multiplying with a suitable constant, one may assume that $|h|_K < 1 < |h(p)|$; now take $f = h^l$ with l sufficiently large. ∎

By considering $\mathscr{F}$-convexity with respect to the family $\mathscr{F}$ of moduli of holomorphic functions, one now obtains a very important class of domains.

Definition. The region $D \subset \mathbb{C}^n$ is called **holomorphically convex** if $\hat{K}_{\mathcal{O}(D)}$ is relatively compact in D for every compact set $K \subset D$.

We will see shortly that every open set in $\mathbb{C}$ is holomorphically convex, but the situation is different in higher dimensions. For example, if $\Omega = \{z \in \mathbb{C}^n: \frac{1}{2} < |z| < 2\}$ and $K = \{z: |z| = 1\}$, then one easily checks that $\hat{K}_{\mathcal{O}(\Omega)} = K$ if $n = 1$; but if $n > 1$, Theorem 1.6 implies that every $f \in \mathcal{O}(\Omega)$ extends to a holomorphic function $\tilde{f}$ on $B(0, 2)$. It then follows from the maximum principle applied to $\tilde{f}$ that for z with $\frac{1}{2} < |z| \leq 1$, one has $|f(z)| = |\tilde{f}(z)| \leq |\tilde{f}|_K = |f|_K$, and therefore $\{z \in \Omega: |z| \leq 1\} \subset \hat{K}_{\mathcal{O}(\Omega)}$. So clearly $\hat{K}_{\mathcal{O}(\Omega)}$ is not compact, and hence Ω is not holomorphically convex if $n > 1$.

Lemma 3.2. *Let $D \subset \mathbb{C}^n$ be holomorphically convex. Then there is a normal exhaustion $\{K_j\}$ of D by $\mathcal{O}(D)$-convex sets K_j.*

Proof. The sequence $\{K_j\}$ will be constructed inductively. Let $\{Q_\nu\}$ be some normal exhaustion. The hypothesis on D implies that $\hat{Q}_\nu$ is compact for all ν. We set $K_1 = \hat{Q}_1$; then K_1 is compact and $\hat{K}_1 = K_1$. Suppose the compact sets $K_1, \ldots, K_j$ with the desired properties have already been constructed. Choose $\nu_j \geq j$ such that $K_j \subset \operatorname{int} Q_{\nu_j}$, and set $K_{j+1} = \hat{Q}_{\nu_j}$. The sequence $\{K_j, j = 1, 2, \ldots\}$ gives the desired exhaustion. ■

The following Lemma shows how holomorphic convexity is used to construct unbounded holomorphic functions.

Lemma 3.3. *Let $\{K_j\}$ be a normal exhaustion of D by $\mathcal{O}(D)$-convex sets. Suppose $p_j \in K_{j+1} - K_j$ for $j = 1, 2, \ldots$. Then there is $f \in \mathcal{O}(D)$ such that $\lim_{j \to \infty} |f(p_j)| = \infty$.*

Proof. We construct f as the limit of a series $\sum_{\nu=1}^{\infty} f_\nu$, where $\{f_\nu\} \subset \mathcal{O}(D)$ is chosen to satisfy

$$|f_\nu|_{K_\nu} < 2^{-\nu}, \qquad \nu = 1, 2, \ldots \tag{3.2}$$

and

$$|f_j(p_j)| > j + 1 + \sum_{\nu=1}^{j-1} |f_\nu(p_j)|, \qquad j = 2, 3, \ldots. \tag{3.3}$$

Assuming the existence of such a sequence $\{f_\nu\}$, (3.2) implies that $f = \sum f_\nu$ converges compactly in D, hence $f \in \mathcal{O}(D)$, and (3.3) implies

$$|f(p_j)| \geq |f_j(p_j)| - \sum_{\nu \neq j} |f_\nu(p_j)| > j + 1 - \sum_{\nu > j} |f_\nu(p_j)| \qquad \text{for } j \geq 2.$$

It then follows from (3.2) that $\sum_{v>j}|f_v(p_j)| < \sum 2^{-v} \leq 1$, and hence $|f(p_j)| > j$. This shows that $\lim_{j\to\infty}|f(p_j)| = \infty$. Finally, the required sequence $\{f_v\}$ is constructed inductively as follows. Set $f_1 = 0$, and, if $l \geq 2$, suppose that $f_1, \ldots, f_{l-1}$ have already been found so that (3.2) and (3.3) hold. By Lemma 3.1(v), since $p_l \notin \hat{K}_l$, there is $f_l \in \mathcal{O}(D)$ with $|f_l|_{K_l} < 2^{-l}$ and such that (3.3) holds for $j = l$. ∎

We now present a simple characterization of holomorphic convexity which is very useful for finding examples.

Proposition 3.4. *The region $D \subset \mathbb{C}^n$ is holomorphically convex if and only if for every sequence $\{p_v: v = 1, 2, 3, \ldots\} \subset D$ without accumulation point in D there is $f \in \mathcal{O}(D)$ with* $\sup_v|f(p_v)| = \infty$.

Proof. We first show that the given condition is sufficient. Let $K \subset D$ be compact. We prove that $\hat{K}_{\mathcal{O}(D)}$ is compact by showing that every sequence $\{p_v\} \subset \hat{K}$ has an accumulation point $p \in \hat{K}$. In fact, if $\{p_v\} \subset \hat{K}$ and $f \in \mathcal{O}(D)$, (3.1) implies $\sup_v|f(p_v)| \leq |f|_K < \infty$; therefore the condition stated in the Proposition implies that $\{p_v\}$ must have an accumulation point $p \in D$. Since $\hat{K}$ is closed in D, we must have $p \in \hat{K}$.

For the converse, we choose a normal exhaustion $\{K_v\}$ of D with $\hat{K}_v = K_v$ (use Lemma 3.2). If $\{p_v\} \subset D$ has no accumulation point in D, it is straightforward to find sequences $\{v_j\}$ and $\{\mu_j\} \subset \mathbb{N}$, such that $p_{v_j} \in K_{\mu_{j+1}} - K_{\mu_j}$ for $j = 1, 2, \ldots$. The existence of the function $f \in \mathcal{O}(D)$ with the required property now follows from Lemma 3.3. ∎

3.3. Examples

We first settle the case of regions in the complex plane.

Lemma 3.5. *Every $D \subset \mathbb{C}^1$ is holomorphically convex.*

Proof. Suppose $\{p_v\} \subset D$ has no accumulation point in D. If $\{p_v\}$ is unbounded, the function $f(z) = z$ will be unbounded on $\{p_v\}$. Otherwise $\{p_v\}$ has an accumulation point $p \in bD$, and the function $f(z) = (z - p)^{-1} \in \mathcal{O}(D)$ is unbounded on $\{p_v\}$. ∎

The same kind of argument shows that $D \subset \mathbb{C}^n$ is holomorphically convex if for every $p \in bD$ there is $f_p \in \mathcal{O}(D)$ with $\lim_{z\in D, z\to p}|f_p(z)| = \infty$. This latter condition holds, in particular, for convex domains in $\mathbb{C}^n$ (see the proof of Lemma 2.1). Therefore we have:

Lemma 3.6. *Every convex domain in $\mathbb{C}^n$ is holomorphically convex.*

Lemma 3.7. *The intersection of finitely many holomorphically convex open sets is holomorphically convex.*

The proof is obvious. We shall see below that the finiteness condition can be removed. It is also true that an *increasing union* of holomorphically convex domains is holomorphically convex, but this result is far from elementary (H. Behnke and K. Stein [BeSt]). We will obtain it in Chapter VI as a consequence of the solution of the Levi problem.

Proposition 3.8. *The product $D = D_1 \times D_2$ of two holomorphically convex regions $D_i \subset \mathbb{C}^{n_i}$, $i = 1, 2$, is holomorphically convex. In particular, every polydomain $\Omega = \Omega_1 \times \cdots \times \Omega_n$, $\Omega_j \subset \mathbb{C}$, is holomorphically convex.*

PROOF. For the first statement it is clearly enough to show that $Q = (\widehat{K_1 \times K_2})_{\mathcal{O}(D_1 \times D_2)} \subset\subset D_1 \times D_2$ for every pair of compact sets $K_i \subset D_i$, $i = 1, 2$. Since every function $f \in \mathcal{O}(D_i)$ defines a function in $\mathcal{O}(D_1 \times D_2)$, one obtains $Q \subset (\hat{K}_1)_{\mathcal{O}(D_1)} \times D_2$ and $Q \subset D_1 \times (\hat{K}_2)_{\mathcal{O}(D_2)}$. Thus $Q \subset (\hat{K}_1)_{\mathcal{O}(D_1)} \times (\hat{K}_2)_{\mathcal{O}(D_2)} \subset\subset D_1 \times D_2$. The second statement now follows by induction and Lemma 3.5. ∎

Next we introduce a class of domains which generalize the polydiscs.

Definition. An open set $\Omega \subset\subset \mathbb{C}^n$ is called an **analytic polyhedron** if there are a neighborhood U of $\bar{\Omega}$ and finitely many functions $f_1, \ldots, f_l \in \mathcal{O}(U)$, such that

$$\Omega = \{z \in U : |f_1(z)| < 1, \ldots, |f_l(z)| < 1\}. \tag{3.4}$$

The collection of functions $f_1, \ldots, f_l$ is called a **frame** for Ω.

Proposition 3.9. *Every analytic polyhedron is holomorphically convex.*

PROOF. Suppose $\{f_1, \ldots, f_l\} \subset \mathcal{O}(U)$ is a frame for the analytic polyhedron Ω. If $K \subset \Omega$ is compact, then $r_j = |f_j|_K < 1$ for $j = 1, \ldots, l$. Clearly

$$\hat{K}_{\mathcal{O}(\Omega)} \subset \{z \in U : |f_1(z)| \leq r_1, \ldots, |f_l(z)| \leq r_l\},$$

and the latter set is relatively compact in Ω. ∎

The following two results show that analytic polyhedra are sufficiently general to approximate arbitrary holomorphically convex sets.

Proposition 3.10. *If $K \subset D$ is compact and $\mathcal{O}(D)$-convex, then K has a neighborhood basis consisting of analytic polyhedra defined by frames of functions holomorphic on D.*

PROOF. Let $U \subset\subset D$ be an open neighborhood of $K = \hat{K}_{\mathcal{O}(D)}$. Since $\hat{K}_{\mathcal{O}(D)}$ and bU are disjoint, by using (3.1) and a compactness argument, one can find finitely many open sets $W_1, \dots, W_l$ and functions $f_1, \dots, f_l \in \mathcal{O}(D)$, so that $bU \subset \bigcup_{j=1}^{l} W_j$, $|f_j|_K < 1$ and $|f_j(z)| > 1$ for $z \in W_j$, $j = 1, \dots, l$. Then $\Omega = \{z \in U: |f_j(z)| < 1 \text{ for } j = 1, \dots, l\}$ is an analytic polyhedron with $K \subset \Omega \subset\subset U$. ∎

Corollary 3.11. *Let $D \subset \mathbb{C}^n$ be holomorphically convex. Then there is a normal exhaustion $\{\bar{\Omega}_j\}$ of D, where each Ω_j is an analytic polyhedron defined by a frame of functions in $\mathcal{O}(D)$.*

PROOF. Combine Lemma 3.2 and Proposition 3.10. ∎

3.4. The Construction of Singular Functions

We now show that a holomorphically convex domain D is a domain of holomorphy by constructing a holomorphic function on D which is unbounded near every point $p \in bD$. The reader may already guess how to obtain such a function: We must apply Lemma 3.3 to a carefully chosen sequence $\{p_j\}$ which accumulates at every boundary point. In fact, in order to preclude analytic extension from all possible "sides" of bD, one needs a little bit more, as follows.

Lemma 3.12. *Let $\{K_\nu\}$ be a normal exhaustion of the region D. Then there are a subsequence $\{\nu_j\}$ of $\mathbb{N}$ and a sequence $\{p_j\}$ of points in D such that*

$$p_j \in K_{\nu_{j+1}} - K_{\nu_j} \qquad \text{for } j = 1, 2, \dots, \tag{3.5}$$

and

(3.6) *for every $p \in bD$ and every connected neighborhood U of p, each component Ω of $U \cap D$ contains infinitely many points from $\{p_j\}$.*

In the proof we will require the following elementary topological fact.

Lemma 3.13. *Suppose U is a connected neighborhood of $p \in bD$ and let $\Omega \subset U \cap D$ be a nonempty connected component of $U \cap D$. Then $b\Omega \cap (U \cap bD) \neq \varnothing$.*

PROOF OF 3.13. Since Ω is a component of the open set $U \cap D$, Ω is open (in $\mathbb{C}^n$) and closed in $U \cap D$. Since U is connected and clearly $\Omega \neq U$, Ω cannot be closed in U. Hence there is $q \in (b\Omega \cap U) - \Omega$. Since $\Omega \subset D$ and Ω is closed in $U \cap D$, we must have $q \in bD$, i.e., $q \in b\Omega \cap U \cap bD$. ∎

PROOF OF LEMMA 3.12. Let $\{a_\nu: \nu = 1, 2, \dots\}$ be an enumeration of the points of D with rational coordinates. Let $r_\nu = \operatorname{dist}(a_\nu, bD) < \infty$; then $B_\nu = B(a_\nu, r_\nu)$ is contained in D. Let $\{Q_j: j = 1, 2, \dots\}$ be a sequence of such balls B_ν which contains each B_ν infinitely many times; for example, we may choose the sequence $B_1, B_1, B_2, B_1, B_2, B_3, B_1, \dots$. Now take $K_{\nu_1} = K_1$. Proceeding inductively, assume that $l > 1$ and $p_1, \dots, p_{l-1}$ and $K_{\nu_1}, \dots, K_{\nu_l}$ have been found so that (3.5) holds for $j = 1, \ldots, l-1$; since Q_l is not contained in any compact subset of D, we may choose $p_l \in Q_l - K_{\nu_l}$ and then ν_{l+1} so that $p_l \in K_{\nu_{l+1}}$. Then (3.5) holds for all $j = 1, 2, \ldots$. We now verify that $\{p_j\}$ satisfies (3.6). Given Ω as stated in (3.6), by Lemma 3.13 there is a point $q \in b\Omega \cap U \cap bD$; thus there is $a_\mu \in \Omega$ with rational coordinates sufficiently close to q, so that $B_\mu \subset \Omega$. Since B_μ occurs infinitely many times in the sequence $\{Q_j\}$, and $p_j \in Q_j$ for $j = 1, 2, \ldots$, B_μ contains infinitely many points of $\{p_j\}$, and we are done. ∎

It is now very easy to prove the principal result of this section.

Theorem 3.14. *A holomorphically convex domain D in $\mathbb{C}^n$ is a domain of holomorphy.*

PROOF. By Lemma 3.2 we can choose a normal exhaustion $\{K_\nu\}$ of D with $\hat{K}_\nu = K_\nu$. We then apply Lemma 3.3 to the sequences $\{p_j\}$ and $\{K_{\nu_j}\}$ given by Lemma 3.12 to obtain $f \in \mathcal{O}(D)$ with $\lim_{j\to\infty} |f(p_j)| = \infty$. The proof is completed by showing that f is completely singular at every point $p \in bD$. In fact, if Ω is a component of $U \cap D$, where U is a connected neighborhood of p, suppose there is $h \in \mathcal{O}(U)$ with $f|_\Omega = h|_\Omega$. After replacing U by $U'(p) \subset\subset U$ and Ω by a component Ω' of $U' \cap D$ which meets Ω, we may assume that $|h|_{\Omega'} \leq |h|_{U'} < \infty$. Hence f would have to be bounded on Ω', and this contradicts (3.6) and $\lim |f(p_j)| = \infty$. ∎

Remark. The reader should notice that Theorem 3.14 is a rather elementary consequence of the definition of $\mathcal{O}(D)$-convexity. The only property of holomorphic functions which was used is that $\mathcal{O}(D)$ is an algebra of functions closed under compact convergence; thus these arguments could be used for many other classes of functions as well. When combined with Corollary 3.4, Theorem 3.14 gives an elementary proof of the classical result that every domain in the complex plane is a domain of holomorphy. The proof given here produces a function which is unbounded everywhere near the boundary; however, analytic continuation may already fail for functions which are smooth up to the boundary. More precisely, the following result was recently proved by D. Catlin [Cat]. *Let $D \subset\subset \mathbb{C}^n$ be holomorphically convex with boundary of class C^∞. Then there is a function $f \in \mathcal{O}(D)$ which is the restriction to D of a C^∞ function on a neighborhood of $\bar{D}$, and which, nevertheless, is completely singular at every point $p \in bD$.* The proof requires some deep results about boundary regularity of solutions to the Cauchy–Riemann equations.

3.5. A Geometric Property of $\mathcal{O}(D)$-Hulls

We shall now discuss the converse of Theorem 3.14, i.e., we show that domains of holomorphy are holomorphically convex. This fact is much deeper than the results of the preceding section, as its proof makes crucial use of the power series expansion of holomorphic functions. The heart of the matter is the following important geometric property of the holomorphically convex hull.

Theorem 3.15. *Let D be a weak domain of holomorphy in $\mathbb{C}^n$. Then*

$$\operatorname{dist}(\hat{K}_{\mathcal{O}(D)}, bD) = \operatorname{dist}(K, bD) \tag{3.7}$$

for every compact set $K \subset D$. In particular, D is holomorphically convex.

Here, $\operatorname{dist}(K, bD) = \inf\{\delta_D(a): a \in K\}$ is the Euclidean distance from K to bD. As the proof of Theorem 3.15 involves the Cauchy estimates for derivatives on polydiscs, we shall express $\delta_D(a)$ in terms of "polydisc distance functions," as follows.

For a given multiradius $r = (r_1, \ldots, r_n) > 0$ we define, for $a \in D \subset \mathbb{C}^n$,

$$\delta_D^{(r)}(a) = \sup\{\lambda > 0: a + \lambda P(0, r) \subset D\}.$$

Then $\delta_D^{(r)}(a) > 0$, and, if $D \neq \mathbb{C}^n$, $\delta_D^{(r)}(a) < \infty$ for all $a \in D$. Clearly a number $\lambda > 0$ satisfies $\lambda \leq \delta_D^{(r)}(a)$ if and only if $P(a, \lambda r) \subset D$.

Lemma 3.16. *If $D \neq \mathbb{C}^n$, then $\delta_D^{(r)}$ is continuous on D. Moreover, for $a \in D$,*

$$\delta_D(a) = \inf\{\delta_D^{(r)}(a): r > 0 \text{ and } |r|^2 = \textstyle\sum r_j^2 = 1\}. \tag{3.8}$$

PROOF. Fix $a \in D$. One easily checks that for $0 < \varepsilon < \delta_D^{(r)}(a)$ one has $\delta_D^{(r)}(a) - \varepsilon \leq \delta_D^{(r)}(z) \leq \delta_D^{(r)}(a) + \varepsilon$ for all $z \in P(a, \varepsilon r)$; this implies the continuity of $\delta_D^{(r)}$ at a. In order to prove (3.8), let η denote the infimum on the right side of (3.8). For any multiradius $r > 0$ with $|r| = 1$ one has $P(a, \lambda r) \subset B(a, \lambda)$, and hence $\delta_D^{(r)} \geq \delta_D(a)$; thus $\eta \geq \delta_D(a)$. For the reverse inequality, let $\varepsilon > 0$ be arbitrary and choose λ with $\delta_D(a) < \lambda < \delta_D(a) + \varepsilon$. Then $B(a, \lambda) \subset D$, and hence there is $r = (r_1, \ldots, r_n) > 0$ with $|r| = 1$, so that $P(a, \lambda r) \not\subset D$, i.e., $\delta_D^{(r)}(a) \leq \lambda$. This implies $\eta < \delta_D(a) + \varepsilon$ for every $\varepsilon > 0$, and we are done. ∎

We can now formulate and prove the key step in the proof of Theorem 3.15.

Proposition 3.17. *Let $K \subset D$ be a compact subset of the open set $D \subset \mathbb{C}^n$, and fix a positive multiradius $r > 0$. Suppose $\eta > 0$ satisfies $\delta_D^{(r)}(z) \geq \eta$ for all $z \in K$. Then for every $a \in \hat{K}_{\mathcal{O}(D)}$ and $f \in \mathcal{O}(D)$, the Taylor series of f at a converges on the polydisc $P(a, \eta r)$.*

PROOF. We fix a function $f \in \mathcal{O}(D)$. We shall estimate the coefficients of the Taylor series of f by means of Cauchy estimates for derivatives of f. For this one needs a uniform bound for $|f|$. We therefore fix $0 < \eta' < \eta$; then $Q = \bigcup_{a \in K} P(a, \eta' r)$ is a compact subset of D, and $M = |f|_Q < \infty$. The Cauchy estimates (Theorem I.1.6) applied to $P(a, \eta' r)$ for $a \in K$ now imply that $|D^\alpha f|_K \leq \alpha! M(\eta' r)^{-\alpha}$ for all $\alpha \in \mathbb{N}^n$. Since $D^\alpha f \in \mathcal{O}(D)$, it follows that

$$|D^\alpha f(a)| \leq \frac{\alpha! M}{(\eta' r)^\alpha} \qquad \text{for } a \in \hat{K}_{\mathcal{O}(D)} \text{ and } \alpha \in \mathbb{N}^n. \tag{3.9}$$

Equation (3.9) and Abel's Lemma imply the convergence of the Taylor series of f at $a \in \hat{K}_{\mathcal{O}(D)}$ on the polydisc $P(a, \eta' r)$, and, since $\eta' < \eta$ is arbitrary, on $P(a, \eta r)$ as well. ∎

Proof of Theorem 3.15. Clearly one has $\operatorname{dist}(\hat{K}_{\mathcal{O}(D)}, bD) \leq \operatorname{dist}(K, bD)$. We will show that if this inequality is strict, then D is not a weak domain of holomorphy; this obviously will prove the Theorem.

So suppose $\eta = \operatorname{dist}(K, bD) > \operatorname{dist}(\hat{K}_{\mathcal{O}(D)}, bD)$, and choose $a \in \hat{K}_{\mathcal{O}(D)}$ with $\delta_D(a) < \eta$. By Lemma 3.16, Equation (3.8), there is a positive multiradius r with $|r| = 1$, so that $\delta_D^{(r)}(a) < \eta \leq \delta_D^{(r)}(z)$ for all $z \in K$. We now use Proposition 3.17 to find a holomorphic extension of $f \in \mathcal{O}(D)$ across bD. In fact, since $\delta_D^{(r)}(a) < \eta$, $P(a, \eta r) \not\subset D$. Given $f \in \mathcal{O}(D)$, Proposition 3.17 shows that the Taylor series u_f of f at the point a defines a holomorphic function on $P(a, \eta r)$, which agrees with f in a neighborhood of a and hence on the connected component of $P(a, \eta r) \cap D$ which contains a. Thus every $f \in \mathcal{O}(D)$ fails to be completely singular at every boundary point $\zeta \in bD \cap P(a, \eta r)$. ∎

3.6. Characterizations of Domains of Holomorphy

We now summarize the principal results obtained in the preceding sections.

Theorem 3.18. *The following are equivalent for an open set D in $\mathbb{C}^n$.*

(i) *D is a weak domain of holomorphy.*
(ii) $\operatorname{dis}(\hat{K}_{\mathcal{O}(D)}, bD) = \operatorname{dist}(K, bD)$ *for every compact set $K \subset D$.*
(iii) *D is holomorphically convex.*
(iv) *D is a domain of holomorphy.*

Proof. Theorem 3.15 proves (i) ⇒ (ii); (ii) ⇒ (iii) is obvious; (iii) ⇒ (iv) is proved in Theorem 3.14, and (iv) ⇒ (i) is trivial. ∎

We can now generalize Lemma 3.7.

Corollary 3.19. *The interior Ω of the intersection of an arbitrary collection $\{D_\alpha : \alpha \in I\}$ of holomorphically convex domains is holomorphically convex.*

Proof. We assume that $\Omega \neq \varnothing$. If $K \subset \Omega$ is compact, $0 < d = \operatorname{dist}(K, b\Omega) \leq \operatorname{dist}(K, bD_\alpha)$ for each α, and by Theorem 3.18, $d \leq \operatorname{dist}(\hat{K}_{\mathcal{O}(D_\alpha)}, bD_\alpha)$ as well. Since $\hat{K}_{\mathcal{O}(\Omega)} \subset \hat{K}_{\mathcal{O}(D_\alpha)}$, we obtain $\operatorname{dist}(\hat{K}_{\mathcal{O}(\Omega)}, bD_\alpha) \geq d$ for all α, which implies $\operatorname{dist}(\hat{K}_{\mathcal{O}(\Omega)}, b\Omega) \geq d$. ∎

Notice that if D is not a domain of holomorphy, the best one could say—based on just the definition in §1.1—is that every $f \in \mathcal{O}(D)$ extends holomorphically to some larger set, which will depend on f. The following consequence of Theorem 3.18 and the proof of Theorem 3.15 is thus of some interest.

Corollary 3.20. *If $D \subset \mathbb{C}^n$ is not a domain of holomorphy, then there are an open connected set Ω with $\Omega \cap D \neq \emptyset$ and $\Omega \not\subset D$, and a nonempty component W of $\Omega \cap D$, such that every $f \in \mathcal{O}(D)$ has a holomorphic extension from W to Ω.*

It is natural to ask at this point if there is a "maximal" region $E(D)$ to which every $f \in \mathcal{O}(D)$ extends holomorphically. For example, the polydisc $P(0, 1)$ clearly is such a maximal region for the Hartogs domain $H(r)$. If one limits oneself to regions in $\mathbb{C}^n$, the answer is, in general, negative (see Exercise E.3.13). The situation is analogous to the problem of finding a "maximal" domain of definition for a single function like $f(z) = \sqrt{z}$, $z \in \mathbb{C} - \{0\}$, which can only be handled adequately by introducing more abstract spaces (i.e., Riemann surfaces). Similarly, in several variables, one is led to consider domains which have different layers spread over $\mathbb{C}^n$—these are called Riemann domains. One can then show that for every domain $D \subset \mathbb{C}^n$, there is a Riemann domain $E(D)$, called the **envelope of holomorphy** of D, so that every $f \in \mathcal{O}(D)$ has a holomorphic extension to $E(D)$, and $E(D)$ is "maximal" with respect to this property. The interested reader may find more details in [Nar 3] or [GrFr].

3.7. Stein Domains and Stein Compacta

The remarks made at the end of the preceding section suggest that in order to deal with certain global questions it is necessary to extend function theory from domains in $\mathbb{C}^n$ to more abstract spaces. In a pioneering paper published in 1951, Karl Stein [SteK 2] discovered a class of abstract complex manifolds that enjoy complex analytic properties similar to those of domains of holomorphy. The fundamental importance of these manifolds for global complex analysis was soon recognized, and already in 1952 H. Cartan [Car 1] referred to them as **Stein manifolds** (variétés de Stein). Among the axioms which define a Stein manifold X is the requirement that X be holomorphically convex (this is defined as in §3.2, once one knows what is meant by the algebra $\mathcal{O}(X)$ of holomorphic functions on X. The other axioms are more technical and they are trivially satisfied for any open set $D \subset \mathbb{C}^n$, or even for any (not necessarily closed) complex submanifold of $\mathbb{C}^n$. Consequently, a domain $D \subset \mathbb{C}^n$ is a Stein manifold precisely when D is holomorphically convex. We shall therefore refer to holomorphically convex open sets (in $\mathbb{C}^n$) as **Stein domains**, and we shall say that a compact set $K \subset \mathbb{C}^n$ is Stein, or that K is a **Stein compactum**, if K has a neighborhood basis of Stein domains.

Let us give some examples of Stein compacta. If Ω is an analytic polyhedron with frame $\{f_1, \ldots, f_l\} \subset \mathcal{O}(U)$, where U is a neighborhood of $\bar{\Omega}$, then $\bar{\Omega} = \{z \in U : |f_j(z)| \leq 1 \text{ for } 1 \leq j \leq l\}$ is a Stein compactum. We shall call $\bar{\Omega}$ a compact analytic polyhedron. In general, the closure $\bar{D}$ of a bounded Stein domain D is not necessarily a Stein compactum (see Exercise E.3.9). If K is compact and $\hat{K}_{\mathcal{O}(U)} = K$ for some open neighborhood U of K, then K is Stein; this follows from Proposition 3.10. The converse is, in general, false (see Exercise E.3.6).

For future reference it is convenient to reformulate some of the earlier results for compact instead of open sets.

Proposition 3.21. *A Stein compactum K has a neighborhood basis of analytic polyhedra.*

Proof. If U is an open neighborhood of K, there is a Stein domain D with $K \subset D \subset U$. Then $\hat{K}_{\mathcal{O}(D)}$ is compact, and $\mathcal{O}(D)$-convex, so, by Proposition 3.10, there is an analytic polyhedron Ω with $\hat{K}_{\mathcal{O}(D)} \subset \Omega \subset D$. Thus $K \subset \Omega \subset U$. ∎

Lemma 3.22. *If K is a Stein compactum and $h_1, \ldots, h_l \in \mathcal{O}(K)$, then*

$$K_l = \{z \in K: |h_j(z)| \leq 1 \quad \text{for } 1 \leq j \leq l\}$$

is Stein.

The proof involves a straightforward application of Proposition 3.9 and Exercise E.3.3, and is left to the reader.

Given a compact set K, the $\mathcal{O}(K)$-hull of a set $L \subset K$ is defined by

$$\hat{L}_{\mathcal{O}(K)} = \{z \in K: |f(z)| \leq |f|_L \text{ for all } f \in \mathcal{O}(K)\}. \tag{3.10}$$

In analogy to Proposition 3.10 we have:

Proposition 3.23. *Suppose $L \subset K$ and $\hat{L}_{\mathcal{O}(K)} = L$. If U is a neighborhood of L, then there are finitely many functions $h_1, \ldots, h_l \in \mathcal{O}(K)$ such that*

$$L \subset \{z \in K: |h_j(z)| \leq 1 \text{ for } 1 \leq j \leq l\} \subset U. \tag{3.11}$$

If K is Stein, so is L.

Proof. Since $K - U$ is compact and disjoint from $\hat{L}_{\mathcal{O}(K)}$, a simple compactness argument, as in the proof of Proposition 3.10, gives functions $h_1, \ldots, h_l \in \mathcal{O}(K)$ which satisfy (3.11). The other statement then follows easily from Lemma 3.22. ∎

Since a Stein domain D is a domain of holomorphy, D is Hartogs pseudoconvex (Theorem 2.3). According to results discussed later in §5, this implies that D is pseudoconvex, as defined in §2.10. We shall now prove this fact directly in the following strengthened form.

Theorem 3.24. *Let D be a Stein domain. Then there is a real analytic strictly plurisubharmonic exhaustion function for D. In particular, D is pseudoconvex.*

Proof. Let $\{K_j\}$ be a normal exhaustion of D by $\mathcal{O}(D)$-convex sets. By arguments as those used in the proof of Proposition 3.10, for each $j = 1, 2, \ldots$ we can find an open set Ω_j with $K_j \subset \Omega_j \subset K_{j+1}$ and functions $f_{j1}, \ldots, f_{jl_j} \in \mathcal{O}(D)$, such that $|f_{jl}|_{K_j} < 1$ for $l = 1, \ldots, l_j$ and $\max_l |f_{jl}(z)| > 1$ for $z \in K_{j+1} - \Omega_j$. By replacing f_{jl} with f_{jl} raised to a sufficiently high power, we can achieve that

$h_j := \sum_{l=1}^{l_j} f_{jl}$ satisfies

(3.12) $$|h_j|_{K_j} < 2^{-j} \qquad \text{and } h_j(z) > j \text{ for } z \in K_{j+1} - \Omega_j.$$

Equation (3.12) then implies that $\psi = \sum_{j=1}^{\infty} h_j$ is a continuous exhaustion function for D. We now show that ψ is, in fact, real analytic on D. Notice that $f_{jl}(z)\overline{f_{jl}(\bar{w})}$ is holomorphic in (z, w) on $D \times \bar{D} = \{(z, \bar{w}): z, w \in D\} \subset \mathbb{C}^{2n}$. Equation (3.12) implies that

$$H(z, w) = \sum_{j=1}^{\infty} \left(\sum_{l=1}^{l_j} f_{jl}(z)\overline{f_{jl}(\bar{w})} \right)$$

converges compactly on $D \times \bar{D}$, thus H is holomorphic on $D \times \bar{D}$ and $\psi(z) = H(z, \bar{z})$ is real analytic. Moreover, the series $\psi = \sum h_j$ may be differentiated term by term; therefore the Levi form of ψ satisfies

$$L_z(\psi; t) = \sum_{j=1}^{\infty} L_z(h_j; t) = \sum_{j,l} |\partial f_{jl}(z)(t)|^2 \geq 0.$$

So $\varphi = \psi + |z|^2$ is the desired strictly plurisubharmonic exhaustion function for D. ■

Corollary 3.25. *Every Stein compactum in* $\mathbb{C}^n$ *is pseudoconvex.*

3.8. Complete Reinhardt Domains

We now discuss a characterization of regions of convergence of power series in terms of a simple geometric condition which, for complete Reinhardt domains, is equivalent to holomorphic convexity. If $D \subset \mathbb{C}^n$ is a Reinhardt domain with center 0, we define $\log \tau(D) = \{\xi \in \mathbb{R}^n: \xi = (\log|z_1|, \ldots, \log|z_n|)$ for some $z = (z_1, \ldots, z_n) \in D\}$. Obviously, if D is complete, then

(3.13) $\xi \in \log \tau(D)$ *implies* $\{\eta \in \mathbb{R}^n: \eta_j \leq \xi_j \text{ for } 1 \leq j \leq n\} \subset \log \tau(D)$.

Lemma 3.26. *If* $D \subset \mathbb{C}^n$ *is the region of convergence of a power series* $\sum_{\nu \in \mathbb{N}^n} c_\nu z^\nu$, *then* $\log \tau(D)$ *is a convex subset of* $\mathbb{R}^n$.

Proof. Suppose ξ and η are points in $\log \tau(D)$; we must show that $t\xi + (1-t)\eta \in \log \tau(D)$ for $0 \leq t \leq 1$. Choose $p, q \in D$, and $\lambda > 1$, so that $\xi_j = \log|p_j|$ and $\eta_j = \log|q_j|$ for $1 \leq j \leq n$, and $\lambda p, \lambda q \in D$. Since $\sum c_\nu z^\nu$ converges at λp and λq, there is $M < \infty$ such that

$$|c_\nu|\lambda^{|\nu|}|p^\nu| \leq M \quad \text{and} \quad |c_\nu|\lambda^{|\nu|}|q^\nu| \leq M$$

for all $\nu \in \mathbb{N}^n$. It follows that

$$|c_\nu|\lambda^{|\nu|}|p^\nu|^t|q^\nu|^{1-t} \leq M \qquad \text{for } 0 \leq t \leq 1.$$

According to Lemma I.1.15, this implies that $\sum c_\nu z^\nu$ converges in a neighborhood of

$$a_t = (|p_1|^t|q_1|^{1-t}, \ldots, |p_n|^t|q_n|^{1-t}),$$

i.e., $a_t \in D$. So $t\xi + (1-t)\eta \in \log \tau(D)$ for all $0 \le t \le 1$. ∎

Lemma 3.27. *Let D be a complete Reinhardt domain. If $\log \tau(D)$ is convex, then D is convex with respect to the family $\mathcal{M} = \{z^\nu, \nu \in \mathbb{N}^n\}$ of holomorphic monomials on $\mathbb{C}^n$.*

Proof. We must show that $\hat{K}_{\mathcal{M}} \cap D \subset\subset D$ for every compact set $K \subset D$, where $\hat{K}_{\mathcal{M}} = \{z \in \mathbb{C}^n : |m(z)| \le |m|_K \text{ for all } m \in \mathcal{M}\}$. Since $\hat{K}_{\mathcal{M}}$ is obviously compact, this will follow once we show that $\hat{K}_{\mathcal{M}} \cap bD = \varnothing$. Given a compact set $K \subset D$, we cover K by finitely many polydiscs $P(0, \tau(q^{(l)})) \subset D$, $l = 1, \ldots, k$, where $q^{(l)} \in D$ and $q_j^{(l)} \ne 0$ for $1 \le j \le n$. Let $Q = \{q^{(1)}, \ldots, q^{(k)}\}$. Then $\hat{K}_{\mathcal{M}} \subset \hat{Q}_{\mathcal{M}}$, and we shall show that $\hat{Q}_{\mathcal{M}} \cap bD = \varnothing$, completing the proof of the Lemma.

Let $p \in bD$. Assume first that $p_j \ne 0$ for $1 \le j \le n$. Then $p^* = (\log|p_1|, \ldots, \log|p_n|) \in b[\log \tau(D)]$, and since $\log \tau(D)$ is assumed convex, there is a linear function $L(\xi) = \sum_{j=1}^n \mu_j \xi_j$, $\mu_j \in \mathbb{R}$, so that $L(\xi) < L(p^*)$ for all $\xi \in \log \tau(D)$. Since $\log \tau(D)$ satisfies (3.13), we must have $\mu_j \ge 0$. Let Q^* be the finite set of points in $\log \tau(D)$ which corresponds to Q. One can find rational numbers $\alpha_j > \mu_j \ge 0$ sufficiently close to μ_j, $1 \le j \le n$, so that for $\tilde{L}(\xi) = \sum \alpha_j \xi_j$ one has

$$\tilde{L}(\xi) < \tilde{L}(p^*) \qquad \text{for all } \xi \in Q^*. \tag{3.14}$$

Equation (3.14) remains true after multiplication with the (positive) common denominator of $\alpha_1, \ldots, \alpha_n$, so we may assume that $\alpha_j \in \mathbb{N}^+$. The monomial $m_\alpha(z) = z_1^{\alpha_1} \ldots z_n^{\alpha_n}$ then satisfies $|m_\alpha|_Q < |m_\alpha(p)|$, i.e., we have shown that $p \notin \hat{Q}_{\mathcal{M}}$. For any of the remaining points $p \in bD$, we may renumber the coordinates so that for some $1 \le l < n$ one has $p_1 \cdot \ldots \cdot p_l \ne 0$, while $p_{l+1} = \cdots = p_n = 0$. If $\pi_l: \mathbb{C}^n \to \mathbb{C}^l$ is the projection $z \mapsto (z_1, \ldots, z_l)$, then $\log \tau(\pi_l(D)) \subset \mathbb{R}^l$ is convex and satisfies (3.13). The preceding argument applied to $\pi_l(p)$ now gives a monomial m in $z_1, \ldots, z_l$ with $|m|_Q < |m(p)|$, so $p \notin \hat{Q}_{\mathcal{M}}$ in this case also. ∎

We now have all the necessary ingredients for the principal result of this section.

Theorem 3.28. *The following are equivalent for a complete Reinhardt domain D in $\mathbb{C}^n$ with center 0.*

(i) *D is the region of convergence of a power series.*
(ii) *$\log \tau(D) \subset \mathbb{R}^n$ is convex.*
(iii) *D is $\mathcal{M}$-convex.*
(iv) *D is holomorphically convex.*
(v) *D is a domain of holomorphy.*

Proof. The implications (i) ⇒ (ii) ⇒ (iii) are proved in Lemma 3.26 and Lemma 3.27. (iii) ⇒ (iv) is trivial, since $\mathcal{M} \subset \mathcal{O}(D)$, and (iv) ⇒ (v) was proved in §3.4

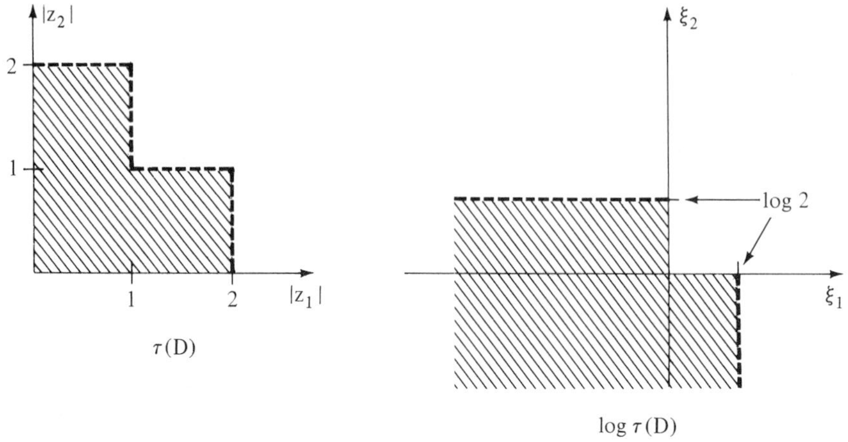

Figure 7. A complete Reinhardt domain which is not logarithmically convex.

(Theorem 3.14). To see that (v) implies (i), let $f \in \mathcal{O}(D)$ be completely singular at every boundary point of D. By Corollary 1.7, the Taylor series of f converges on D, and it obviously cannot converge on any region strictly larger than D. ■

We leave it to the reader to verify that an open set $D \subset \mathbb{C}^n$ is the region of convergence of a power series $\sum c_\nu z^\nu$ if and only if D is $\mathcal{M}$-convex.

Remark. The equivalence between (ii) and (v) solves the Levi problem for complete Reinhardt domains, in the sense that it gives a local geometric characterization for the boundary of domains of holomorphy. A Reinhardt domain D such that $\log \tau(D)$ is convex is said to be **logarithmically convex**.

Figure 7 below shows a simple example of a complete Reinhardt domain D which is not logarithmically convex.

Every function holomorphic on D extends holomorphically to a region containing the torus $\tau^{-1}(1)$; the envelope of holomorphy of D, i.e., the "largest" region with that property can be described easily (see Exercise E.3.10).

Exercises

E.3.1. Let $K \subset \mathbb{R}^n$ be compact. Show that the convex hull $\hat{K}^c$ is equal to the intersection of all closed half spaces H which contain K.

E.3.2. In §3.1 we defined the concept of the $\mathcal{F}$-hull $\hat{K}_{\mathcal{F}}$ of a set $K \subset X$. Show:

(i) $K \subset \hat{K}_{\mathcal{F}}$, and if $L = \hat{K}_{\mathcal{F}}$, then $\hat{L}_{\mathcal{F}} = L$;
(ii) if $Q \subset K$, then $\hat{Q}_{\mathcal{F}} \subset \hat{K}_{\mathcal{F}}$;
(iii) if $\mathcal{F}_1 \subset \mathcal{F}$, then $\hat{K}_{\mathcal{F}} \subset \hat{K}_{\mathcal{F}_1}$. Hence, if X is $\mathcal{F}_1$-convex, then it is also $\mathcal{F}$-convex.

E.3.3. Suppose D is holomorphically convex and $f_1, \ldots, f_l \in \mathcal{O}(D)$.

(i) Show that $\{z \in D: |f_j(z)| < 1 \text{ for } 1 \leq j \leq l\}$ is holomorphically convex.
(ii) Show that (i) is not necessarily true if D is not holomorphically convex.

E.3.4. Show that if $F: D_1 \to D_2$ is holomorphic and proper (i.e., $F^{-1}(K)$ is compact for every compact set $K \subset D_2$), and if D_2 is holomorphically convex, then so is D_1.

E.3.5. Let $D = \{z \in \mathbb{C}^n: 1 < |z| < 3\}$ and $K = \{z \in D: |z| = 2\}$. Find $\hat{K}_{\mathcal{O}(D)}$.

E.3.6. Let $\{\Delta_j: j = 1, 2, \ldots\}$ be a sequence of disjoint open discs in $\Delta = \{z \in \mathbb{C}: |z| < 1\}$, whose centers converge to 0, and set $K = \Delta - \bigcup_{j=1}^{\infty} \Delta_j$. Show that K is a Stein compactum with $K \neq \hat{K}_{\mathcal{O}(U)}$ for every open neighborhood U of K.

E.3.7. For a set $A \subset \mathbb{C}^n$ and $r > 0$ define

$$A(r) = \bigcup_{a \in A} P(a, r).$$

Show that if D is holomorphically convex and $K \subset D$ is compact with $\overline{K(r)} \subset D$, then

$$(\hat{K}_{\mathcal{O}(D)})(r) \subset \widehat{[K(r)]}_{\mathcal{O}(D)}.$$

E.3.8. Let $D \subset \mathbb{C}^n$ be a connected Stein domain in $\mathbb{C}^n$ and suppose $\Omega \subset D$ is an open subset such that the restriction map $R: \mathcal{O}(D) \to \mathcal{O}(\Omega)$ is onto.

(i) Show that $f(\Omega) = f(D)$ for all $f \in \mathcal{O}(D)$.
(ii) Show that for every compact set $K \subset D$ there is a compact set $L \subset \Omega$ such that $K \subset \hat{L}_{\mathcal{O}(D)}$.
(iii) Show that a sequence $\{f_j\} \subset \mathcal{O}(D)$ converges compactly in D if and only if it converges compactly in Ω. (This is also a consequence of the open mapping theorem for Fréchet spaces.)

E.3.9. Let $D = \{(z, w) \in \mathbb{C}^2: 0 < |z| < |w| < 1\}$.

(i) Show that D is Stein.
(ii) Show that the closure $\bar{D}$ of D is not a Stein compactum.

E.3.10. For $1 \leq j \leq n$ set $D_j = \{z \in \mathbb{C}^n: |z_\nu| < 2 \text{ for } \nu \neq j \text{ and } |z_j| < 1\}$ and $D = \bigcup_{j=1}^{n} D_j$.

(i) Prove that if $n > 1$, then every $f \in \mathcal{O}(D)$ has a holomorphic extension to the torus $T = \{z \in \mathbb{C}^n: |z_j| = 1, 1 \leq j \leq n\}$. (Hint: Consider the region of convergence of the Taylor series of $f \in \mathcal{O}(D)$.)
(ii) Show that there is a maximal region $E(D) \subset \mathbb{C}^n$ (the *envelope of holomorphy of D*), such that every $f \in \mathcal{O}(D)$ extends holomorphically to $E(D)$, and describe $E(D)$.
(iii) Use (i) to show: If U is a neighborhood of 0 in $\mathbb{C}^n$, $n \geq 2$, and $f \in \mathcal{O}(U - \mathbb{R}^n)$, then f extends holomorphically to 0.

E.3.11. Show that every closed complex submanifold M of $\mathbb{C}^n$ has the following three properties. (These properties are the defining axioms for a **Stein manifold**; so this shows that every closed complex submanifold of $\mathbb{C}^n$ is Stein.)

(i) M is $\mathcal{O}(M)$-convex.
(ii) Given two distinct points $P, Q \in M$, there is $f \in \mathcal{O}(M)$ with $f(P) \neq f(Q)$.

(iii) For every $P \in M$ there is a holomorphic coordinate system in a neighborhood of P which is given by global holomorphic functions in $\mathcal{O}(M)$.

E.3.12. Suppose D is open in $\mathbb{C}^n$ and let $\mathcal{M}$ be the set of holomorphic monomials. Show that if D is $\mathcal{M}$-convex, then D is a complete Reinhardt domain with center 0 (and hence is the region of convergence of some power series, by Theorem 3.28).

E.3.13. Let $r = (\frac{1}{2}, \frac{1}{2})$ and consider the domain $H(r) \subset \mathbb{C}^2$ (see I, Figure 2). Let W_1 be the polydisc with center $(1, 0)$ and radius $\frac{1}{4}$ in $\mathbb{C}^2$, and $W_2 = \{z \in \mathbb{C}^2\colon \operatorname{dist}(z, \gamma) < \frac{1}{4}\}$, where γ is the curve in the z_1 plane given by

$$\gamma = \{x\colon 1 \le x \le 2\} \bigcup \{2e^{i\theta}\colon 0 \le \theta \le \pi\} \bigcup \{x\colon -2 \le x \le 0\}.$$

Set $D = H(r) \cup W_1 \cup W_2$. Show that D is an open connected set which has no envelope of holomorphy in $\mathbb{C}^2$. (Hint: Consider the function $f \in \mathcal{O}(D)$ defined by a branch of $\sqrt{z_1 - i}$ on $H(r)$ and use Theorem 1.1!)

§4. Plurisubharmonic Functions

In this paragraph we discuss the generalization to several complex variables of the classical concept of subharmonic function in the complex plane. These so-called *plurisubharmonic functions* play a fundamental role in many areas of complex analysis. Strictly plurisubharmonic functions, introduced in §2.7, are a particularly useful class of plurisubharmonic functions. In §5 we shall see how plurisubharmonic functions unify the various notions of pseudoconvexity introduced in §2. For the convenience of the reader we include a review of harmonic functions in §4.1 and a detailed discussion of the elementary properties of subharmonic functions in the plane in §4.2–§4.4.

4.1. Harmonic Functions in the Complex Plane

Recall that the **Laplace operator** Δ in $\mathbb{C}$ is defined by

$$\Delta = \frac{\partial^2}{\partial x^2} + \frac{\partial^2}{\partial y^2} = 4\frac{\partial^2}{\partial z \partial \bar{z}},$$

where $z = x + iy$. A C^2 function u on a region $D \subset \mathbb{C}$ is called **harmonic** if $\Delta u = 0$ on D. We state some of the well-known elementary properties of harmonic functions (see [Ahl], for example).

(4.1) *A real valued function u is harmonic if and only if u is locally the real part of a holomorphic function. In particular, harmonic functions are C^∞, and even real analytic.*

(4.2) **The Mean Value Property.** *If u is harmonic on $D \subset \mathbb{C}$, then*

$$u(a) = \frac{1}{2\pi} \int_0^{2\pi} u(a + re^{i\theta})\, d\theta$$

whenever $\{z: |z - a| \leq r\} \subset D$.

(4.3) **The Maximum Principle.** *If u is real valued and harmonic on $D \subset \mathbb{C}$, then:*

(Strong version) *If u has a local maximum at the point $a \in D$, then u is constant in a neighborhood of a (and hence on the connected component of D which contains a).*

(Weak version) *If $D \subset\subset \mathbb{C}$ and u extends continuously to $\bar{D}$, then $u(z) \leq \max_{bD} u$ for $z \in D$.*

Notice that the strong version of the maximum principle implies the weak version.

(4.4) **The Dirichlet Problem.** *If $\Delta = \{z: |z - a| < r\}$ and $g \in C(b\Delta)$, then there is a unique continuous function u on $\bar{\Delta}$ which is harmonic in Δ, such that $u(z) = g(z)$ for $z \in b\Delta$. This harmonic extension u is given explicitly by the Poisson integral of g,* i.e.,

$$u(a + \zeta) = \frac{1}{2\pi} \int_0^{2\pi} \frac{r^2 - |\zeta|^2}{|re^{i\theta} - \zeta|^2} g(a + re^{i\theta})\, d\theta \qquad \text{for } |\zeta| < r.$$

4.2. Subharmonic Functions

The analog in one real variable of the solutions of the Laplace equation are the linear functions $l(x) = ax + b$. A function $y = u(x)$ is said to be **convex** if on any interval $[\alpha, \beta]$ in its domain $u(x)$ is less than or equal to the unique linear function l with $u(\alpha) = l(\alpha)$ and $u(\beta) = l(\beta)$. Substituting harmonic functions for linear functions in the definition above leads to the idea of **subharmonic functions**: A continuous function u is subharmonic on $D \subset \mathbb{C}$ if on every disc $\Delta \subset\subset D$ one has $u \leq h$, where $h \in C(\bar{\Delta})$ is the unique function harmonic on Δ with $h = u$ on $b\Delta$. (The function h exists by the solution of the Dirichlet problem for discs.)

For technical reasons it is convenient to include upper semicontinuous functions and to admit the value $-\infty$ in the definition of subharmonic functions. Moreover, one usually replaces discs by more general sets (although this doesn't really matter—see Theorem 4.4 below). As the Dirichlet problem cannot generally be solved in this setting, one is led to the following formulation.

Definition. A function $u\colon D \to \mathbb{R} \cup \{-\infty\}$ is called subharmonic if u is upper semicontinuous and if for every compact set $K \subset D$ and every function $h \in C(K)$ which is harmonic on the interior of K and satisfies $u \le h$ on bK it follows that $u \le h$ on K.

Recall that u is said to be **upper semicontinuous on** D if

$$\limsup_{z \to a} u(z) \le u(a) \qquad \text{for } a \in D, \tag{4.5}$$

or, equivalently,

$$\{z \in D\colon u(z) < c\} \text{ is open for every } c \in \mathbb{R}. \tag{4.6}$$

An upper semicontinuous function takes on a maximum on every compact set (though not necessarily a minimum). A function $u\colon D \to \mathbb{R}$ is continuous if and only if u and $-u$ are upper semicontinuous.

From the (weak) maximum principle one sees immediately that harmonic functions are subharmonic. We will see other examples in §4.3, after we have discussed some characterizations of subharmonic functions.

Lemma 4.1. *Let $D \subset \mathbb{C}$ be open.*

(i) *If u is subharmonic on D, so is cu for $c > 0$.*
(ii) *If $\{u_\alpha\colon \alpha \in A\}$ is a family of subharmonic functions on D such that $u = \sup u_\alpha$ is finite and upper semicontinuous, then u is subharmonic.*
(iii) *If $\{u_j, j = 1, 2, \ldots\}$ is a decreasing sequence of subharmonic functions on D, then $u = \lim_{j\to\infty} u_j$ is subharmonic.*

Proof. (i) and (ii) are obvious consequences of the definitions. In order to prove (iii), suppose $K \subset D$ is compact and $h \in C(K)$ is harmonic on $\operatorname{int}(K)$ with $h \ge u = \lim u_j$ on bK. Given $\varepsilon > 0$, $E_j = \{z \in bK\colon u_j(z) \ge h(z) + \varepsilon\}$ is a closed subset of bK for $j = 1, 2, \ldots$, $E_{j+1} \subset E_j$, and $\bigcap_{j=1}^\infty E_j = \varnothing$. By compactness of bK, there is $l \in \mathbb{N}$ with $E_l = \varnothing$. So $u_l \le h + \varepsilon$ on bK, and hence on K as well, as u_l is subharmonic. This implies $u \le h + \varepsilon$ on K for all $\varepsilon > 0$, i.e., $u \le h$ on K. ∎

As an application we present a curious property of arbitrary domains in $\mathbb{C}$.

Corollary 4.2. *For every open set D in $\mathbb{C}$ the function $u(z) = -\log \delta_D(z)$ is subharmonic on D.*

Proof. If $D = \mathbb{C}$, then $u \equiv -\infty$, and there is nothing to prove. If $D \ne \mathbb{C}$, then $u(z)$ is continuous, and for $z \in D$ one has $u(z) = \sup\{-\log|z - \zeta|\colon \zeta \in bD\}$; since $-\log|z - \zeta|$ is harmonic, and hence subharmonic on D (it is, locally, the real part of a holomorphic branch of $-\log(z - \zeta)$), the conclusion follows by Lemma 4.1. (ii). ∎

4.3. The Submean Value Property

We shall now discuss some other characterizations of subharmonic functions which are useful in various situations. In particular, it will follow that subharmonicity is a local property.

Recall from integration theory that for a Borel measure μ on a compact set K and an upper semicontinuous function $u: K \to \mathbb{R} \cup \{-\infty\}$, the integral $\int u\, d\mu$ is well defined (possibly $= -\infty$). Moreover

$$\int_K u\, d\mu = \inf\left\{\int_K \varphi\, d\mu : \varphi \in C(K) \text{ and } \varphi \geq u\right\}, \tag{4.7}$$

and $u \in L^1(K, \mu)$ if and only if $\int u\, d\mu > -\infty$.

Theorem 4.3. *Let D be open in $\mathbb{C}$. The following statements are equivalent for an upper semicontinuous function $u: D \to \mathbb{R} \cup \{-\infty\}$:*

(i) *u is subharmonic.*
(ii) *For every disc $\Delta \subset\subset D$ and holomorphic polynomial f with $u \leq \operatorname{Re} f$ on $b\Delta$, one has $u \leq \operatorname{Re} f$ on Δ.*
(iii) *For every $a \in D$ there exists a positive number $r_a < \delta_D(a)$ such that*

$$u(a) \leq \frac{1}{2\pi}\int_0^{2\pi} u(a + re^{i\theta})\, d\theta \qquad \textit{for all } 0 < r \leq r_a.$$

Remark. (iii) is called the **submean value property**. It is clearly a local property, and it is additive. Therefore we have:

Corollary 4.4. *If u_1 and u_2 are subharmonic on D, so is $u_1 + u_2$.*

Before proving the Theorem we single out an important ingredient of the proof.

Lemma 4.5. *An upper semicontinuous function u which satisfies the submean value property satisfies the strong maximum principle* (4.3.)

Proof. The argument is identical to the one which is often used to prove the maximum principle for harmonic functions. Suppose u satisfies the submean value property and u has a local maximum at $a \in D$, i.e., there is $\rho > 0$ such that $u(z) \leq u(a)$ for all z with $|z - a| \leq \rho$. We may assume that $\rho \leq r_a$. If there were a point z_0 with $r = |z_0 - a| \leq \rho$ and $u(z_0) < u(a)$, then

$$\{\theta \in [0, 2\pi]: u(a + re^{i\theta}) < u(a)\}$$

would have nonempty interior, by the upper semicontinuity of u; thus

$$\int_0^{2\pi} u(a + re^{i\theta})\, d\theta < \int_0^{2\pi} u(a)\, d\theta = 2\pi u(a),$$

in contradiction to the hypothesis. So u must be constant in a neighborhood of a. ∎

Proof of Theorem 4.3. It is obvious that (i) implies (ii). In order to show (ii) ⇒ (iii), suppose $\Delta = \{z: |z - a| < r\} \subset\subset D$ and let $\varphi \in C(b\Delta)$ with $\varphi \geq u$ on $b\Delta$. After replacing φ by its Poisson integral, we may assume that φ is continuous on $\bar{\Delta}$ and harmonic on Δ. For $\tau < 1$, the function

$$\varphi_\tau(z) = \varphi(a + \tau(z - a))$$

is harmonic in a neighborhood of $\bar{\Delta}$, and $\varphi_\tau \to \varphi$ uniformly on $\bar{\Delta}$ as $\tau \to 1$. Now $\varphi_\tau = \operatorname{Re} f_\tau$, where f_τ is holomorphic on $\bar{\Delta}$, and by considering the partial sums of the Taylor series of f_τ, it follows that for $\varepsilon > 0$, there is a holomorphic polynomial f with $u \leq \varphi \leq \operatorname{Re} f \leq \varphi + \varepsilon$ on $b\Delta$. By (ii) and the mean value property for the harmonic function $\operatorname{Re} f$, one obtains

$$u(a) \leq \operatorname{Re} f(a) = \frac{1}{2\pi}\int_0^{2\pi} \operatorname{Re} f(a + re^{i\theta})\, d\theta \leq \frac{1}{2\pi}\int_0^{2\pi} \varphi(a + re^{i\theta})\, d\theta + \varepsilon.$$

As ε is arbitrary, we have shown that

$$u(a) \leq \frac{1}{2\pi}\int_0^{2\pi} \varphi(a + re^{i\theta})\, d\theta$$

for every continuous function $\varphi \geq u$ on $b\Delta$, and thus (iii) follows from (4.7).

Finally, to show (iii) ⇒ (i), let $K \subset D$ be compact and suppose $h \in C(K)$ is harmonic on int K and $u \leq h$ on bK; we must show $u \leq h$ on K. Notice that (iii) and the mean value property for h imply the submean value property for $u - h$ on int K. Therefore, by Lemma 4.5, $(u - h)(z) \leq \max_{bK}(u - h) \leq 0$ for $z \in K$, i.e., $u \leq h$ on K. ∎

Proposition 4.6. *If f is holomorphic on D, then $|f|^\alpha$ for $\alpha > 0$ and $\log|f|$ are subharmonic on D.*

The proof is left to the reader.

The following property of the mean values of subharmonic functions is very useful.

Lemma 4.7. *If u is subharmonic on the disc $\{|z - a| < \rho\}$, then*

$$A(u; r) = \frac{1}{2\pi}\int_0^{2\pi} u(a + re^{i\theta})\, d\theta$$

is a nondecreasing function for $0 < r < \rho$.

Proof. Let $\Delta(r) = \{|z - a| < r\}$ and suppose $0 < r_1 < r_2 < \rho$. Let $\varphi \in C(b\Delta(r_2))$ satisfy $\varphi \geq u$ on $b\Delta(r_2)$. By taking the Poisson integral of φ, we may assume that $\varphi \in C(\bar{\Delta}(r_2))$ and φ is harmonic on $\Delta(r_2)$. By the mean value property, $A(\varphi; r) = \varphi(a)$ for $r \leq r_2$, and the subharmonicity of u implies $u \leq \varphi$

on $\Delta(r_2)$. Hence $A(u; r_1) \leq A(\varphi; r_1) = A(\varphi, r_2)$ for all such φ, and it follows that $A(u; r_1) \leq \inf\{A(\varphi; r_2)\colon \varphi$ continuous and $\geq u$ on $b\Delta(r_2)\} = A(u; r_2)$. ∎

4.4. The Differential Characterization

It is well known that a C^2 function $u(x)$ on an interval $I \subset \mathbb{R}$ is convex if and only if $u''(x) \geq 0$ on I. An analogous characterization holds for smooth subharmonic functions, giving a simple computational test for subharmonicity.

Proposition 4.8. *A real valued function $u \in C^2(D)$ is subharmonic on D if and only if $\Delta u \geq 0$ on D.*

Proof. We first show that $\Delta u > 0$ implies that u is subharmonic. Let $K \subset D$ be compact, $h \in C(K)$ harmonic on int K, and suppose $v = u - h \leq 0$ on bK. If $v(z) > 0$ for some $z \in K$, then v would take on its maximum at a point $a \in \operatorname{int} K$, and it would follow that $\Delta v(a) \leq 0$. Since $\Delta h = 0$, this contradicts $\Delta u(a) > 0$, so we must have $v \leq 0$, i.e., $u \leq h$, on K. Next, if $\Delta u \geq 0$, the preceding argument applied to $u_j = u + (1/j)|z|^2$ for $j = 1, 2, \ldots$ shows that u_j is subharmonic. As $u_j(z)$ decreases to $u(z)$ as $j \to \infty$, Lemma 4.1 implies that u is subharmonic as well.

To prove the converse, let u be subharmonic and suppose there is $a \in D$ such that $\Delta u(a) < 0$. By continuity, $\Delta u < 0$ on a neighborhood U of a, and hence, by the first part of the proof, $-u$ is subharmonic on U. Thus u and $-u$ are subharmonic on U, and hence u is harmonic on U (Exercise E.4.2); but this would imply $\Delta u = 0$ on U, contradicting $\Delta u(a) < 0$. So we must have $\Delta u \geq 0$ on D. ∎

Remark. By considering "weak derivatives" or "distributional derivatives," one may omit the hypothesis that $u \in C^2$ in Proposition 4.7.

4.5. Plurisubharmonic Functions

In the preceding sections we discussed harmonic and subharmonic functions of two real variables. There is an obvious extension of these concepts to several real variables, but this generalization is not very useful in complex analysis in more than one variable, primarily because the (local) equivalence between harmonic functions and real parts of holomorphic functions does not hold in more than *one* complex variable (see Exercise E.4.5). Moreover, the class of subharmonic functions in $2n$ real variables on $D \subset \mathbb{C}^n$ is not invariant under biholomorphic maps except for $n = 1$. A much more useful generalization for *complex* analysis are the *plurisubharmonic* functions, which are those functions whose restrictions to complex lines are subharmonic. We now give the formal definition.

Definition. Let D be open in $\mathbb{C}^n$. A function $u: D \to \mathbb{R} \cup \{-\infty\}$ is said to be **plurisubharmonic on** D if u is upper semicontinuous, and if for every $a \in D$ and $w \in \mathbb{C}^n$ the function $\lambda \mapsto u(a + \lambda w)$ is subharmonic on the region $\{\lambda \in \mathbb{C}: a + \lambda w \in D\}$. The class of plurisubharmonic functions on D is denoted by $PS(D)$.

Remarks. Certain properties of subharmonic functions are inherited by plurisubharmonic functions. For example, Lemma 4.1 holds for plurisubharmonic functions, $PS(D)$ is closed under addition, and $u \in PS(D)$ if and only if u is plurisubharmonic in some neighborhood of every point $a \in D$. If $f \in \mathcal{O}(D)$, then $|f|^\alpha$, $\alpha > 0$, and $\log|f|$ are plurisubharmonic on D. (This follows from Proposition 4.6—notice that the restriction of f to a complex line is holomorphic where defined.)

On the other hand, Corollary 4.2. does not extend to higher dimensions. For example, if $D = \mathbb{C}^2 - \{0\}$, let $u = -\log \delta_D(z)$. For $a = (1, 0)$ and $w = (0, 1)$, $u(a + \lambda w) = -\log \delta_D(1, \lambda) = -\log\sqrt{1 + |\lambda|^2}$, and this function has a strict maximum at $\lambda = 0$, so it cannot be subharmonic (Lemma 4.5). So u is not plurisubharmonic. We shall see in §5 that the regions $D \subset \mathbb{C}^n$ for which $-\log \delta_D$ is plurisubharmonic are precisely the pseudoconvex ones.

For plurisubharmonic functions of class C^2 there is a differential characterization analogous to the one given in Proposition 4.8 for subharmonic functions.

Proposition 4.9. *Let $D \subset \mathbb{C}^n$ and suppose $u \in C^2(D)$ is real valued. Then $u \in PS(D)$ if and only if the complex Hessian of u,*

$$L_z(u; w) = \sum_{j,k=1}^{n} \frac{\partial^2 u}{\partial z_j \partial \bar{z}_k}(z) w_j \bar{w}_k,$$

is positive semidefinite on $\mathbb{C}^n$ at every point $z \in D$.

Proof. A straightforward computation gives

$$\frac{\partial^2}{\partial \lambda \partial \bar{\lambda}} u(a + \lambda w) = L_{a+\lambda w}(u; w) \tag{4.8}$$

for $w \in \mathbb{C}^n$ and $a + \lambda w \in D$. By Proposition 4.8, $u(a + \lambda w)$ is subharmonic in λ if and only if the left side in (4.8) is nonnegative. ■

It is now obvious that strictly plurisubharmonic functions, which were defined earlier in §2.7, are just a special type of plurisubharmonic function.

Corollary 4.10. *Suppose $\Omega \subset \mathbb{C}^n$ and $D \subset \mathbb{C}^m$ are open, and $F: D \to \Omega$ is holomorphic. Then $u \circ F \in PS(D)$ if $u \in PS(\Omega) \cap C^2(\Omega)$.*

Proof. A computation gives $L_a(u \circ F; w) = L_{F(a)}(u; F'(a)w)$. Now use the Proposition. ■

4.6. Smoothing and Approximation

In order to extend Corollary 4.10 to arbitrary $u \in PS(D)$, one needs to locally approximate u by smooth plurisubharmonic functions. This sort of result is very useful in many other applications as well. The technique of proof is well known in analysis; it involves a regularization of functions by certain convolution integrals. In order to get started we need to know that plurisubharmonic functions are in L^1—at least locally—with respect to $2n$-dimensional Lebesgue measure.

Lemma 4.11. *Let $D \subset \mathbb{C}^n$ be connected. If $u \in PS(D)$ and $u \not\equiv -\infty$ on D, then $u \in L^1_{\text{loc}}(D)$. In particular, $\{z \in D: u(z) = -\infty\}$ has Lebesgue measure 0.*

Proof. We first show that if $u(a) > -\infty$ at some point $a \in D$, then $u \in L^1(P(a, r))$ for every polydisc $P(a, r) \subset\subset D$. Since u is bounded from above on such a polydisc, it is enough to show $\int_{P(a,r)} u\, dV > -\infty$. By applying the submean value property in each coordinate separately, one obtains

$$u(a) \leq (2\pi)^{-n} \int_0^{2\pi} \cdots \int_0^{2\pi} u(a + \rho e^{i\theta})\, d\theta_1 \ldots d\theta_n$$

for all $\rho = (\rho_1, \ldots, \rho_n)$ with $0 \leq \rho \leq r$, where $\rho e^{i\theta} = (\rho_1 e^{i\theta_n}, \ldots, \rho_n e^{i\theta_n})$. After multiplying by $\rho_1 \ldots \rho_n\, d\rho_1 \ldots d\rho_n$ and integrating in ρ_j from 0 to r_j, $1 \leq j \leq n$, it follows that

$$-\infty < u(a) \leq [\text{vol}\, P(a, r)]^{-1} \int_{P(a,r)} u\, dV.$$

The application of the Fubini–Tonelli theorem is legitimate as u is bounded from above.

Now consider the set $E = \{a \in D: u \text{ is integrable in a neighborhood of } a\}$. E is clearly open, and we just saw that $E \neq \varnothing$. The statement proved above also implies that if $a \in D - E$, then $u(z) = -\infty$ for all z in some neighborhood of a, so $D - E$ is open as well. Since D is connected, $E = D$. ∎

Theorem 4.12. *Let $D \subset \mathbb{C}^n$ and set $D_j = \{z \in D: |z| < j \text{ and } \delta_D(z) > 1/j\}$. Suppose $u \in PS(D)$ is not identically $-\infty$ on any component of D. Then there is a sequence $\{u_j\} \subset C^\infty(D)$ with the following properties:*

(4.9) *u_j is strictly plurisubharmonic on D_j.*

(4.10) *$u_j(z) \geq u_{j+1}(z)$ for $z \in D_j$, and $\lim_{j \to \infty} u_j(z) = u(z)$ for $z \in D$.*

(4.11) *If u is also continuous, the convergence in* (4.10) *is compact on D.*

Proof. Let $\varphi \in C_0^\infty(B(0, 1))$ such that $\varphi \geq 0$, φ is radial (i.e., $\varphi(z) = \varphi(z')$ if $|z| = |z'|$), and $\int \varphi\, dV = 1$. Since $D_j \subset\subset D$, by Lemma 4.11 one has $u \in L^1(D_j)$

for each $j = 1, 2, \ldots$. The integral $v_j(z) = \int_{D_j} u(\zeta)\varphi(j(z-\zeta))j^{2n}\,dV(\zeta)$ is thus defined for $z \in \mathbb{C}^n$, and standard results give $v_j \in C^\infty$. We set $u_j(z) = v_j(z) + (1/j)|z|^2$. For $z \in D_j$, a linear change of variables gives

$$v_j(z) = \int_{|\zeta|<1} u(z - \zeta/j)\varphi(\zeta)\,dV(\zeta). \tag{4.12}$$

In order to prove (4.9) it is enough to show that $v_j \in PS(D_j)$, i.e., $v_j(a + \lambda w)$ satisfies the submean value property at $\lambda = 0$ for $a \in D_j$ and $w \in \mathbb{C}^n$, since then

$$L_a(u_j; w) = L_a(v_j, w) + (1/j)|w|^2 \geq (1/j)|w|^2.$$

But this follows easily from the corresponding property of u, as follows: for sufficiently small r one has

$$\begin{aligned}\frac{1}{2\pi}\int_0^{2\pi} v_j(a + re^{i\theta}w)\,d\theta &= \int_{|\zeta|<1}\left[\frac{1}{2\pi}\int_0^{2\pi} u(a + re^{i\theta}w - \zeta/j)\,d\theta\right]\varphi(\zeta)\,dV(\zeta)\\ &\geq \int_{|\zeta|<1} u(a - \zeta/j)\varphi(\zeta)\,dV(\zeta)\\ &= v_j(a).\end{aligned}$$

Next, observe that the integral (4.12) is invariant under substitution of ζ by $e^{it}\zeta$, $t \in \mathbb{R}$. Thus

$$v_j(z) = \int_{|\zeta|<1}\left[\frac{1}{2\pi}\int_0^{2\pi} u(z - e^{it}\zeta/j)\,dt\right]\varphi(\zeta)\,dV(\zeta). \tag{4.13}$$

By Lemma 4.7 applied to the subharmonic function $\lambda \mapsto u(z + \lambda(-\zeta))$, the inner integral in (4.13) is nondecreasing in $r = 1/j$; thus $v_j(z) \geq v_{j+1}(z)$. Also, (4.13) and the submean value property show $v_j(z) \geq u(z)\int \varphi\,dV = u(z)$. If $\varepsilon > 0$ is given, by the upper semicontinuity of u there is a ball $B(z, \delta) \subset \{\zeta \in D : u(\zeta) < u(z) + \varepsilon\}$; thus, for $j > 1/\delta$, one obtains from (4.12) and the above that $u(z) \leq v_j(z) < u(z) + \varepsilon$. This completes the proof of (4.10) for $\{v_j\}$; (4.11) follows by a similar argument. The corresponding statements for $\{u_j\}$ are then obvious. ■

We can now easily show that plurisubharmonic functions are invariant under holomorphic maps.

Theorem 4.13. *If $\Omega \subset \mathbb{C}^n$, $D \subset \mathbb{C}^m$ and $F: D \to \Omega$ is holomorphic, then $u \circ F \in PS(D)$ for every $u \in PS(\Omega)$.*

PROOF. Without loss of generality we may assume that Ω is connected and that $u \in PS(\Omega)$ is $\not\equiv -\infty$. Choose a decreasing sequence $\{u_j\}$ with $\lim u_j = u$ as in Theorem 4.12. If $D' \subset\subset D$, then $u_j \circ F$ is plurisubharmonic on D' for j sufficiently large, by the plurisubharmonicity of u_j and Corollary 4.10. Since $\{u_j \circ F\}$ decreases to $u \circ F$ as $j \to \infty$, the conclusion follows from Lemma 4.1. ■

Theorem 4.12 is usually sufficient when one deals with local properties of plurisubharmonic functions. However, when global properties are important as well, one needs to refine Theorem 4.12. We now present a result of that type which will be needed in §5.

Proposition 4.14. *Suppose u is a continuous plurisubharmonic exhaustion function for $D \subset \mathbb{C}^n$. Given a compact set K in D and $\varepsilon > 0$, there is a C^∞ strictly plurisubharmonic exhaustion function φ for D such that*

(4.14) $$u \leq \varphi \text{ on } D \text{ and } |\varphi(z) - u(z)| < \varepsilon \text{ for } z \in K.$$

Notice that the result implies that D is pseudoconvex according to the definition given in §2.10.

Proof. For $j = 0, 1, 2, \ldots$ we set $\Omega_j = \{z \in D: u(z) < j\}$. Then $\Omega_j \subset\subset D$, and by adding a suitable constant to u we may assume that $K \subset \Omega_0$. By Theorem 4.12 there are functions $u_j \in C^\infty(D)$, $j = 0, 1, 2, \ldots$, such that u_j is strictly plurisubharmonic on Ω_{j+2}, $u(z) < u_0(z) < u(z) + \varepsilon$ for $z \in \bar{\Omega}_1$, and $u(z) < u_j(z) < u(z) + 1$ for $z \in \Omega_j$ and $j \geq 1$. It follows that

(4.15) $$u_j - j + 1 < 0 \text{ on } \Omega_{j-2} \text{ and } u_j - j + 1 > 0 \text{ on } \bar{\Omega}_j - \Omega_{j-1} \qquad \text{for } j \geq 2.$$

Now choose $\chi \in C^\infty(\mathbb{R})$ with $\chi(t) = 0$ for $t \leq 0$ and $\chi(t)$, $\chi'(t)$, $\chi''(t)$ positive for $t > 0$. Then $\chi \circ (u_j - j + 1) \equiv 0$ on Ω_{j-2} and ≥ 0 otherwise. By computing the complex Hessian one sees that $\chi \circ (u_j - j + 1)$ is plurisubharmonic on Ω_{j+2}, and strictly plurisubharmonic and positive on $\bar{\Omega}_j - \Omega_{j-1}$. Finally, one inductively chooses integers $m_j \in \mathbb{N}$ so that, for $l \geq 2$,

$$\varphi_l = u_0 + \sum_{j=2}^{l} m_j \chi \circ (u_j - j + 1)$$

is strictly plurisubharmonic on Ω_l. It follows that $\varphi_l = u_0$ on Ω_0, $\varphi_l \geq u$, and $\varphi_l \equiv \varphi_{l-1}$ on Ω_{l-2}. Thus $\varphi = \lim_{l \to \infty} \varphi_l$ has all the required properties. ∎

Exercises

E.4.1. Show that the statements (4.5) and (4.6) are equivalent.

E.4.2. (i) Show that a function u is harmonic if and only if u and $-u$ are both subharmonic.
(ii) Show that a continuous function u is harmonic if and only if u satisfies the Mean Value Property (4.2).

E.4.3. Let $D \subset \mathbb{C}$. Show that if $f \in \mathcal{O}(D)$, then $|f|$, $\log|f|$, and $|f|^\alpha$, $\alpha > 0$, are subharmonic on D.

E.4.4. Let u be subharmonic on $D \subset \mathbb{C}$.

(i) Show that u^p is subharmonic for $p \geq 1$. (Hint: Use Hölder's inequality.)
(ii) More generally, show that if $\varphi: \mathbb{R} \to \mathbb{R}$ is convex and increasing, and $\varphi(-\infty) := \lim_{x \to -\infty} \varphi(x)$, then $\varphi \circ u$ is subharmonic.

E.4.5. If $z_j = x_j + iy_j$, $j = 1, 2$, show that $\varphi(z) = x_1^2 - x_2^2$ is harmonic on $\mathbb{C}^2$ but φ is not the real part of any holomorphic function (not even locally).

E.4.6. A real valued C^2 function u on the region D in $\mathbb{C}^n$ is called **pluriharmonic** on D if its complex Hessian $(\partial^2 u/\partial z_j \partial \bar{z}_k)_{j,k}$ is identically 0 on D. Show that u is pluriharmonic if and only if u is, locally, the real part of some holomorphic function. (Note: One direction is elementary; for the other one, it is convenient to use basic local results on differential forms—see Chapters III and IV.)

E.4.7. Let u be continuous on $D \subset \mathbb{C}$ and subharmonic on $D - \{a\}$. Show that u is subharmonic on D. (Hint: Consider the functions $u_\varepsilon = u + \varepsilon \log |z - a|$ for $\varepsilon > 0$.)

E.4.8. (i) Let $\Omega \subset \mathbb{R}^n$ be open and consider the tube $T(\Omega) = \{x + iy \in \mathbb{C}^n : x \in \Omega, y \in \mathbb{R}^n\}$ over Ω. Show that if $\varphi \in C^2(T(\Omega))$ depends only on x, then φ is plurisubharmonic if and only if the function $x \mapsto \varphi(x)$ is convex on Ω.
(ii) Prove (i) without assuming $\varphi \in C^2$ by using an appropriate approximation by smooth functions.

E.4.9. Let $K \subset \mathbb{C}^n$ be a pseudoconvex compact set. Show that K has a neighborhood basis of strictly pseudoconvex domains with C^∞ boundary.

E.4.10. Prove the following useful result, known as **Hopf Lemma**: Let $D \subset\subset \mathbb{R}^2$ have C^2 boundary, let U be a neighborhood of the point $p \in bD$, and suppose $u \in C(U \cap \bar{D})$ is subharmonic on $U \cap D$, $u(p) = 0$, and $u < 0$ on $U \cap D$. Let n be the unit outer normal to bD at p. Show that there is $c > 0$ such that

$$u(p - \lambda n) \leq -c\lambda$$

for all $\lambda > 0$ sufficiently close to 0. In particular, if u is differentiable at P, the outer normal derivative of u at P is positive.

(Hint: Let Δ be a disc with center $a = p - \tau n$ and radius τ, where τ is so small that $\bar{\Delta} - \{p\} \subset U \cap D$. There is $M > 0$, such that $u(z) < -M$ for $z \in \bar{\Delta}$ with $\delta_D(z) \geq \tau$. If $P_\Delta(z, \zeta)$ is the Poisson kernel for Δ, conclude that

$$u(z) \leq -M \int_{b\Delta \cap \{\zeta : \delta_D(\zeta) \geq \tau\}} P_\Delta(z, \zeta)\, d\sigma(\zeta)$$

for $z \in \Delta$, and use the explicit form of P_Δ.)

Remark. The result and the proof outlined above carry over to $D \subset\subset \mathbb{R}^n$.

§5. Characterizations of Pseudoconvexity

We shall now systematically use plurisubharmonic functions in order to present a more complete discussion of pseudoconvexity. Along the way we will introduce several new characterizations, and we will show the equivalence of the various notions of pseudoconvexity encountered in §2.

5.1. Plurisubharmonic Convexity

The following definition is analogous to other convexity conditions with respect to a family of functions, as discussed in §3.1.

Definition. A region $D \subset \mathbb{C}^n$ is called **plurisubharmonic convex** ($=$ PS-**convex**), if for every compact set $K \subset D$, its plurisubharmonic convex hull

$$\hat{K}_{PS(D)} = \{z \in D \colon u(z) \leq \sup_K u \text{ for all } u \in PS(D)\}$$

is relatively compact in D.

Remark. Since plurisubharmonic functions are not necessarily continuous, it is not clear at this point whether $\hat{K}_{PS(D)}$ is closed in D. However, this turns out to be the case if D is PS-convex (see Corollary 5.12 below).

The reader should notice that the hypothesis of the following Lemma are satisfied for pseudoconvex domains, as defined in §2.10.

Lemma 5.1. *Suppose there is a plurisubharmonic exhaustion function for the region D in $\mathbb{C}^n$. Then D is PS-convex.*

Proof. Let $u \in PS(D)$ be the given exhaustion function. If $K \subset D$ is compact, let $c = \max_K u < \infty$. Then $u \leq c$ on $\hat{K}_{PS(D)}$, so $\hat{K}_{PS(D)} \subset \{z \in D \colon u(z) \leq c\} \subset\subset D$. ∎

5.2. The Continuity Principle

Next we introduce a version of the classical "continuity principle" which describes a geometrically very intuitive analogue of linear convexity. If $\Delta \subset\subset \mathbb{C}$ is an open disc and $\varphi \colon \bar{\Delta} \to D$ is a continuous map which is holomorphic on Δ, we shall say that $\varphi(\bar{\Delta})$ is an **analytic disc S in D** and call the set $\varphi(b\Delta)$ the "**boundary**" ∂S of S.

Definition. A region D in $\mathbb{C}^n$ is said to satisfy the **continuity principle** if for every family $\{S_\alpha \colon \alpha \in I\}$ of analytic discs in D with

$$\bigcup_{\alpha \in I} \partial S_\alpha \subset\subset D, \tag{5.1}$$

it follows that

$$\bigcup_{\alpha \in I} S_\alpha \subset\subset D. \tag{5.2}$$

Lemma 5.2. *Every PS-convex domain satisfies the continuity principle.*

PROOF. If $\varphi: \bar{\Delta} \to D$ defines an analytic disc S in D and $u \in PS(D)$, then $u \circ \varphi \in PS(\Delta)$, by Theorem 4.13. By the maximum principle, $u(z) \leq \max_{\partial S} u$ for $z \in S$. It follows that $S \subset (\widehat{\partial S})_{PS(D)}$, and therefore, if D is PS-convex and (5.1) holds, one gets

$$\bigcup_\alpha S_\alpha \subset \bigcup_\alpha (\widehat{\partial S_\alpha})_{PS(D)} \subset (\widehat{\bigcup_\alpha \partial S_\alpha})_{PS(D)} \subset\subset D. \quad \blacksquare$$

By restricting oneself to special families of analytic discs one obtains different versions of the continuity principle. The interested reader may consult Behnke and Thullen [BeTh] for some of the classical versions. In particular, the proof of the following result shows that Hartogs pseudoconvexity, as defined in §2.2, is just one of these variants.

Proposition 5.3. *If D satisfies the continuity principle, then D is Hartogs pseudoconvex.*

PROOF. Suppose $(\Gamma^*, \hat{\Gamma}^*)$ is a Hartogs figure defined by the biholomorphic map $F: \hat{\Gamma} \to \hat{\Gamma}^*$, such that $\Gamma^* \subset D$. Let Δ be the open unit disc in $\mathbb{C}$; for $\tau \in \bar{\Delta}$ let S_τ be the analytic disc defined by the holomorphic map $\lambda \mapsto F(0, \ldots, 0, \tau, \lambda)$, $\lambda \in \bar{\Delta}$. The definition of Γ and the hypothesis $\Gamma^* = F(\Gamma) \subset D$ imply that $S_0 \subset D$ and $\partial S_\tau \subset D$ for all $\tau \in \bar{\Delta}$; we must show that $S_\tau \subset D$ for all $\tau \in \bar{\Delta}$. So, let $A = \{\tau \in \bar{\Delta}: S_\tau \subset D\}$. Then $A \neq \varnothing$, and clearly A is open in $\bar{\Delta}$. If the sequence $\{\tau_\nu, \nu = 1, 2, \ldots\} \subset A$ converges to $\tau \in \bar{\Delta}$, then $\bigcup_{\nu=1}^\infty \partial S_{\tau_\nu} \subset\subset D$; therefore, if D satisfies the continuity principle, it follows that $S_\tau \subset \overline{\bigcup_{\nu=1}^\infty S_{\tau_\nu}} \subset\subset D$, so that $\tau \in A$. This shows that A is closed, and hence $A = \bar{\Delta}$. $\blacksquare$

5.3. The Plurisubharmonicity of $-\log \delta_D$

We shall now prove the existence of plurisubharmonic exhaustion functions on Hartogs pseudoconvex domains. This result, which is the crucial "missing link" in our discussion of pseudoconvexity, is considerably deeper than the results proved so far in this section. It is particularly remarkable that the geometrically defined function $\varphi(z) = -\log \delta_D(z)$ turns out to be plurisubharmonic. That certain distance functions on regions of convergence of power series—the so-called "regularity radii"—lead to subharmonic functions, was already recognized by Hartogs in 1906 [Har 2] (see Exercise E.5.1.). However, only the introduction of plurisubharmonic functions by Oka and Lelong in 1941 provided the framework for realizing the full scope of Hartogs' fundamental discovery.

We first introduce a class of distance functions which generalize the "regularity radii" considered by Hartogs. If $D \subset \mathbb{C}^n$ is open and $u \in \mathbb{C}^n$ is a unit vector, we define

(5.3) $$\delta_{D,u}(z) = \sup\{\tau > 0: z + \eta u \in D \text{ for } \eta \in \mathbb{C} \text{ with } |\eta| \leq \tau\}.$$

$\delta_{D,u}(z)$ measures how large a disc in the u-direction with center at z is contained in D. Notice that $0 < \delta_{D,u}(z) \le \infty$, and that $\delta_{D,u}(z) < \infty$ if D is bounded. However, even in that case, $\delta_{D,u}$ need not be continuous. For example, if $D = P(0, (1, 2)) \cup P(0, (2, 1)) \subset \mathbb{C}^2$ and $u = (0, 1)$, then $\delta_{D,u}$ is not continuous at all points $(z_1, z_2) \in D$ with $|z_1| = 1$. On the other hand, the following properties are easily established.

Lemma 5.4. *$\delta_{D,u}$: $D \to \mathbb{R} \cup \{\infty\}$ is lower semicontinuous on D (i.e., $-\delta_{D,u}$ is upper semicontinuous), and*

$$\delta_D(z) = \inf\{\delta_{D,u}(z): u \in \mathbb{C}^n \text{ with } |u| = 1\} \tag{5.4}$$

for all $z \in D$.

The proof is left to the reader.

Proposition 5.5. *Let $D \subset \mathbb{C}^n$ be Hartogs pseudoconvex. Then $-\log \delta_{D,u}$ is plurisubharmonic on D for each unit vector $u \in \mathbb{C}^n$.*

Proof. Fix $u \in \mathbb{C}^n$ with $|u| = 1$. From Lemma 5.4 it follows that $-\log \delta_{D,u}$ is upper semicontinuous on D. We must show that for $a \in D$ and $w \in \mathbb{C}^n$ the function $\lambda \mapsto -\log \delta_{D,u}(a + \lambda w)$ is subharmonic on $D_{a,w} = \{\lambda \in \mathbb{C}: a + \lambda w \in D\}$.

In case u and w are linearly dependent, $\delta_{D,u}(a + \lambda w)$ simply measures the usual Euclidean distance from λ to the boundary of $D_{a,w} \subset \mathbb{C}$; the desired result is then already known (Corollary 4.2).

We now assume that u and w are linearly independent. Because of Theorem 4.3 it is enough to prove the following: if $r > 0$ satisfies $\{a + \lambda w: |\lambda| \le r\} \subset D$ and if g is a holomorphic polynomial with

$$-\log \delta_{D,u}(a + \lambda w) \le \operatorname{Re} g(\lambda) \tag{5.5}$$

for $|\lambda| = r$, then (5.5) holds also for $|\lambda| \le r$. Notice that (5.5) is equivalent to

$$\delta_{D,u}(a + \lambda w) \ge |e^{-g(\lambda)}|. \tag{5.6}$$

We shall use the geometric content of (5.6) for $|\lambda| = r$ to construct a suitable Hartogs figure $(\Gamma^*, \hat{\Gamma}^*)$.

By the definition of $\delta_{D,u}$, (5.6) is equivalent to

$$\{a + \lambda w + \eta u e^{-g(\lambda)}: |\eta| \le \tau\} \subset D \qquad \text{for all } 0 < \tau < 1. \tag{5.7}$$

Fix $0 < \tau < 1$ and choose $u_1, \ldots, u_{n-2} \in \mathbb{C}^n$ so that $\{u_1, \ldots, u_{n-2}, u, w\}$ is linearly independent. It follows that the map F: $\mathbb{C}^n \to \mathbb{C}^n$ defined by

$$F(z) = a + rz_n w + \tau z_{n-1} u e^{-g(rz_n)} + z_1 u_1 + \cdots + z_{n-2} u_{n-2}$$

is biholomorphic. Then

$$\Gamma_1^* = \{F(0', z_n): |z_n| \le 1\} = \{a + \lambda w: |\lambda| \le r\} \subset D,$$

and because (5.5), and hence (5.7), is assumed to hold for $|\lambda| = r$, we see that

$$\Gamma_2^* = \{F(0, \ldots, 0, z_{n-1}, z_n): |z_{n-1}| \leq 1, |z_n| = 1\} \subset D$$

as well. *If* $\Gamma^* = \Gamma_1^* \cup \Gamma_2^*$ and $\hat{\Gamma}^* = F(\hat{\Gamma})$, where $\hat{\Gamma} = \{(0, \ldots, 0, z_{n-1}, z_n): |z_{n-1}| \leq 1, |z_n| \leq 1\}$, then $(\Gamma^*, \hat{\Gamma}^*)$ is a Hartogs figure with $\Gamma^* \subset D$. Since D is assumed Hartogs pseudoconvex, it follows that $\hat{\Gamma}^* \subset D$. So (5.7) holds for $|\lambda| \leq r$ and $\tau < 1$, and therefore (5.5) holds for $|\lambda| \leq r$. ∎

The main result of this section now follows immediately.

Theorem 5.6. *If $D \subset \mathbb{C}^n$ is Hartogs pseudoconvex, then $-\log \delta_D$ is plurisubharmonic on D.*

Proof. From Lemma 5.4 it follows that

$$-\log \delta_D = \sup\{-\log \delta_{D,u}: |u| = 1\}.$$

The conclusion then follows from Proposition 5.5 and Lemma 4.1. ∎

In case D is bounded, $\varphi = -\log \delta_D$ is already an exhaustion function for D; in general, we can use the following simple fact to find a plurisubharmonic exhaustion function on an H-pseudoconvex set.

Lemma 5.7. *If $-\log \delta_D \in PS(D)$, then $\varphi = \max\{|z|^2, -\log \delta_D\}$ is a continuous plurisubharmonic exhaustion function for D.*

5.4. The Fundamental Equivalence

We now summarize the results obtained in the preceding sections.

Theorem 5.8. *The following properties are equivalent for an open set D in $\mathbb{C}^n$:*

(i) *There is a C^2 strictly plurisubharmonic exhaustion function for D (i.e., D is pseudoconvex according to the definition in* §2.10).
(ii) *There is a plurisubharmonic exhaustion function for D.*
(iii) *D is plurisubharmonic convex.*
(iv) *For every analytic disc S in D one has* $\operatorname{dist}(S, bD) = \operatorname{dist}(\partial S, bD)$.
(v) *D satisfies the continuity principle.*
(vi) *D is Hartogs pseudoconvex.*
(vii) *$-\log \delta_D$ is plurisubharmonic on D.*

Proof. We have, essentially, proved everything, except the equivalence of (iv) with the other properties. In fact, (i) ⇒ (ii) is trivial, and the sequence of implications (ii) ⇒ (iii) ⇒ (v) ⇒ (vi) ⇒ (vii) was proved in §5.1–§5.3. Given (vii), by Lemma 5.7 there is a continuous plurisubharmonic exhaustion function for D. By Proposition 4.14, we can then find a C^∞ strictly plurisubharmonic exhaustion function for D, i.e., we have (more than) (i). Regarding (iv), it is obvious that (iv) implies (v). Conversely, if $-\log \delta_D \in PS(D)$, and S is an analytic disc in D defined by the holomorphic map φ, the maximum principle applied to the subharmonic function $-\log \delta_D \circ \varphi$ implies that $-\log \delta_D(z) \leq \sup_{\partial S}(-\log \delta_D)$ for $z \in S$, and hence $\operatorname{dist}(\partial S, bD) \leq \operatorname{dist}(S, bD)$. As the reverse inequality is trivial, we see that (iv) holds. Thus (vii) $\Rightarrow$ (iv), and we are done. ■

Property (iv) exhibits once more the strong analogy between pseudoconvexity and (linear) convexity. Just observe that $D \subset \mathbb{R}^n$ is convex if and only if for every line segment $L \subset D$ one has $\operatorname{dist}(L, bD) = \operatorname{dist}(\partial L, bD)$. Classically, pseudoconvexity was identified mainly with the continuity principle and variations thereof. More recently, the emphasis has been on plurisubharmonic functions. Property (vii) gives a very deep and elegant characterization, but since δ_D is, in general, not differentiable, property (i) is often much more useful. Moreover, (i) clearly states an intrinsic complex analytic property of D, while (vii) involves the (extrinsic) Euclidean geometry. For the purposes of this book, differentiable strictly plurisubharmonic exhaustion functions provide a convenient tool for extending function theoretic results from strictly pseudoconvex domains to arbitrary pseudoconvex domains. They also play an important role in more advanced studies on abstract complex manifolds.

5.5. Some General Properties

We can now prove a strengthened version of results already noted for H-pseudoconvex domains in §2.2.

Theorem 5.9. (a) *If $\{D_\alpha, \alpha \in I\}$ is a collection of pseudoconvex domains, then the interior Ω of $\bigcap_{\alpha \in I} D_\alpha$ is pseudoconvex.*

(b) *If $D_1 \subset D_2 \subset \ldots$ is an increasing sequence of pseudoconvex domains, then $\bigcup_{\nu=1}^{\infty} D_\nu$ is pseudoconvex.*

Proof. Since (a) for *finite* collections and (b) hold for H-pseudoconvexity, Theorem 5.8 gives the desired result in this case. In order to prove (a) in general we use the characterization (iv) in Theorem 5.8. Let S be an analytic disc in Ω. Then $\operatorname{dist}(\partial S, b\Omega) \leq \operatorname{dist}(\partial S, bD_\alpha) = \operatorname{dist}(S, bD_\alpha)$ for all $\alpha \in I$; this implies $\operatorname{dist}(\partial S, b\Omega) = \operatorname{dist}(S, b\Omega)$. ■

The next result shows that pseudoconvexity is a local property of the boundary of a domain. The corresponding statement for domains of holomorphy (or holomorphically convex domains) is considerably harder—in fact, it is equivalent to the solution of the Levi problem.

Theorem 5.10. *A region $D \subset \mathbb{C}^n$ is pseudoconvex if and only if every point $\zeta \in \bar{D}$ has a neighborhood U_ζ such that $U_\zeta \cap D$ is pseudoconvex.*

Clearly the condition is a restriction only for $\zeta \in bD$.

Proof. One implication is obvious. For the other implication, we assume first that D is bounded. If $\zeta \in bD$ and $U_\zeta \cap D$ is pseudoconvex, then $-\log \delta_{U_\zeta \cap D}$ is plurisubharmonic. But $\delta_{U_\zeta \cap D}(z) = \delta_D(z)$ for $z \in D$ sufficiently close to ζ; therefore we can find a neighborhood U of bD, such that $-\log \delta_D \in PS(U \cap D)$. Since $D - U \subset\subset D$, $m = \sup\{-\log \delta_D(z): z \in D - U\}$ is finite. It follows that $\varphi = \max\{-\log \delta_D, |z|^2 + m + 1\} \in PS(D)$; clearly φ is an exhaustion function for D, and hence D is pseudoconvex, by Theorem 5.8. If D is not bounded, we apply the preceding argument to $D_\nu = D \cap B(0, \nu)$. If D is locally pseudoconvex, so is D_ν, and hence D_ν is pseudoconvex for $\nu = 1, 2, \ldots$. Since D is the increasing union of $\{D_\nu\}$, D itself is pseudoconvex, by Theorem 5.9. ■

Theorem 5.11. *Suppose $D \subset \mathbb{C}^n$ is pseudoconvex, $K \subset D$ is compact and $U \subset D$ is an open neighborhood of $\hat{K}_{PS(D)}$. Then there is a C^∞ strictly plurisubharmonic exhaustion function φ on D with*

$$\varphi(z) > 0 \text{ for } z \in D - U \quad \text{and } \varphi(z) < 0 \text{ for } z \in K. \tag{5.8}$$

The following result is an immediate consequence of this theorem.

Corollary 5.12. *If $D \subset \mathbb{C}^n$ is pseudoconvex and $K \subset D$ is compact, then*

$$\hat{K}_{PS(D)} = \hat{K}_{PS(D) \cap C^\infty(D)}. \tag{5.9}$$

In particular, $\hat{K}_{PS(D)}$ is closed in D (and hence compact).

Proof. In order to prove (5.9) it is enough to prove $\hat{K}_{PS(D) \cap C^\infty(D)} \subset \hat{K}_{PS(D)}$, the opposite inclusion being trivial. For this, note that if $a \in D - \hat{K}_{PS(D)}$, then by applying the theorem to K and $U = D - \{a\}$, one finds $\varphi \in PS(D) \cap C^\infty(D)$ with $\varphi(a) > 0$ and $\varphi < 0$ on K; thus $a \notin \hat{K}_{PS(D) \cap C^\infty(D)}$. The remaining statements are now obvious. ■

We shall see in Chapter VI, §1.8, that one even has

$$\hat{K}_{PS(D)} = \hat{K}_{\mathcal{O}(D)}, \tag{5.10}$$

a statement which obviously implies the holomorphic convexity of every pseudoconvex domain, i.e., the solution of the Levi problem. Of course, (5.10) is much harder to prove than the results we have obtained so far; the proof will require the solution of the Levi problem as well as some deep approximation theorems for holomorphic functions.

Proof of Theorem 5.11. The main point of the proof is to find a *continuous* plurisubharmonic exhaustion function φ_0 which satisfies (5.8). Given such a φ_0, by Proposition 4.14 we can then approximate φ_0 uniformly on K by a C^∞ strictly plurisubharmonic exhaustion function $\varphi \geq \varphi_0$, which will then satisfy (5.8) as well.

We fix a continuous exhaustion function $u \in PS(D)$ for D. By adding a constant we may assume that $u < 0$ on K. Set $D_c = \{z \in D: u(z) < c\}$ for $c \in \mathbb{R}$ and $L = \{z \in D - U: u(z) \leq 0\}$; then $D_c \subset\subset D$, L is compact, and $L \cap \hat{K}_{PS(D)} = \emptyset$. Therefore, if $\zeta \in L$, there is $u_\zeta \in PS(D)$ such that

$$(5.11) \qquad u_\zeta(\zeta) > 0 \text{ and } u_\zeta < 0 \text{ on } K.$$

We would like to conclude that $u_\zeta > 0$ in a neighborhood of ζ, but this is not necessarily true if u_ζ is only upper semicontinuous. Therefore, we first replace u_ζ by a *continuous* function with—essentially—the same properties. For this, we use Theorem 4.12 to approximate u_ζ from above by a decreasing sequence $\{u_j: j = 1, 2, \ldots\}$ of C^∞ plurisubharmonic functions on $D_3 \subset\subset D$. Then $u_j(\zeta) \geq u_\zeta(\zeta) > 0$ for each j, and, for $z \in K$, there is $j_z \in \mathbb{N}$, such that

$$(5.12) \qquad 0 > u_{j_z}(z) \geq u_j(z) \qquad \text{for } j \geq j_z.$$

By continuity, (5.12) will hold in a neighborhood of z, and by compactness of K we can find $j_\zeta \in \mathbb{N}$, such that (5.11) holds with u_{j_ζ} in place of u_ζ.

We can now conclude that $u_\zeta > 0$ in a neighborhood V_ζ of ζ. By compactness of L, finitely many such neighborhoods $V_{\zeta_1}, \ldots, V_{\zeta_l}$ will cover L. It follows that $v = \max\{u_{\zeta_\nu}, 1 \leq \nu \leq l\} \in PS(D_3)$ is continuous, and v satisfies $v > 0$ on L and $v < 0$ on K. Finally, we patch v and u together to obtain a global function: with $M = \sup_{\bar{D}_2} v$, we define

$$\varphi_0 = \begin{cases} \max\{v, Mu\} & \text{on } D_2 \\ Mu & \text{on } D - \bar{D}_1. \end{cases}$$

The two definitions agree on their common domain, and it follows easily that φ_0 has all the required properties. ∎

5.6. Differentiable Boundaries

In case $D \subset\subset \mathbb{C}^n$ has C^2 boundary we saw in §2.6 that the Hartogs pseudoconvexity of D implies Levi pseudoconvexity. When combined with Theorem 5.8 one thus obtains one half of the following characterization.

Theorem 5.13. *A bounded domain in $\mathbb{C}^n$ with C^2 boundary is pseudoconvex if and only if it is Levi pseudoconvex.*

Proof. We only need to show that a Levi pseudoconvex domain D is pseudoconvex. This has already been done in Lemma 2.19 in case D has boundary of class C^3. Even though the extension to C^2 boundaries looks like just a minor technical improvement, the proof is somewhat deeper, as it makes use of Theorem 5.8 and the localization of pseudoconvexity given by Theorem 5.10, as follows. If $p \in bD$, Lemma 2.13 combined with Remark 2.20 implies that there is a local C^2 defining function r on a neighborhood U of p, such that $\varphi_1 = -\log(-r) + A|z|^2$ is (strictly) plurisubharmonic on $U \cap D$ if A is sufficiently large. Obviously $\varphi_1(z) \to \infty$ as $z \to bD \cap U$. If U is chosen pseudoconvex, there is a plurisubharmonic exhaustion function φ_2 on U, and it follows that $\varphi = \max(\varphi_1, \varphi_2)$ is a plurisubharmonic exhaustion function for $U \cap D$, i.e., $U \cap D$ is pseudoconvex, by Theorem 5.8. Theorem 5.10 then implies that D is pseudoconvex. ■

An alternate proof of Theorem 5.13 can be obtained by using the plurisubharmonicity of $-\log \delta_D$ and the fact that, in case of C^2 boundary, $-\delta_D$ extends to a C^2 defining function for D in a neighborhood of bD (see Exercises E.5.9, E.5.10, and E.5.11).

Exercises

E.5.1. Let $D \subset \mathbb{C}$ and suppose $f_\nu \in \mathcal{O}(D)$ for $\nu = 0, 1, 2, \ldots$. Define the *Hartogs regularity radius* $R(z)$ of the series

$$(*) \qquad \sum_{\nu=0}^{\infty} f_\nu(z) w^\nu, \qquad w \in \mathbb{C},$$

by setting $R(z)$ equal to the radius of convergence of $(*)$ in w. Prove that $-\log R(z)$ is subharmonic on D. (This is a fundamental result of Hartogs, which is at the core of Oka's Theorem 5.6.)

E.5.2. Prove Lemma 5.4.

E.5.3. Prove Theorem 5.9 by using the fact that D is pseudoconvex if and only if $-\log \delta_D \in \mathrm{PS}(D)$.

E.5.4. Show that if $D_i \in \mathbb{C}^{n_i}$, $i = 1, 2$, are pseudoconvex, then $D_1 \times D_2$ is pseudoconvex.

E.5.5. An open set $D \subset \mathbb{C}^n$, $n > 1$, is called a **complete Hartogs domain** if D contains all points $(z', \lambda z_n)$ for $\lambda \in \mathbb{C}$ with $|\lambda| \leq 1$ whenever $z = (z', z_n) \in D$.

(i) Show that if D is a complete Hartogs domain, then there are a region $\Omega \subset \mathbb{C}^{n-1}$ and a lower semicontinuous positive function R on Ω, such that

$$(*) \qquad D = \{(z', z_n): z' \in \Omega \text{ and } |z_n| < R(z')\}.$$

(ii) Show that a complete Hartogs domain D is pseudoconvex if and only if in the representation $(*)$ Ω is pseudoconvex and $-\log R$ is plurisubharmonic on Ω.

E.5.6. (i) Let $U \subset \mathbb{C}^n$ be open and $\varphi \in \mathrm{PS}(U)$, and suppose $D = \{z \in U: \varphi(z) < 0\}$ is relatively compact in U and nonempty. Show that D is pseudoconvex.

(ii) Let $D \subset \mathbb{C}^n$ be pseudoconvex and $\varphi \in \mathrm{PS}(D)$. Show that every set $D_c = \{z \in D: \varphi(z) < c\}$, $c \in \mathbb{R}$, is either empty or an open pseudoconvex set.

E.5.7. Suppose D and D' are pseudoconvex in $\mathbb{C}^n$ and $\mathbb{C}^m$, respectively, and let $F: D \to \mathbb{C}^m$ be a holomorphic map. Show that $D_F = \{z \in D: F(z) \in D'\}$ is pseudoconvex.

E.5.8. Let $D \subset \mathbb{C}^n$ and $D' \subset \mathbb{C}^m$ be open, and let $F: D \to D'$ be a proper holomorphic map. Show:

(i) if D' is pseudoconvex, so is D.
(ii) if F is biholomorphic, then D is pseudoconvex if and only if D' is.

E.5.9. Let $D \subset \mathbb{R}^n$ have C^2 boundary and define the signed distance function $\rho = \rho_{bD}$ by $\rho(z) = -\delta_D(z)$ for $z \in D$ and $\rho(z) = \delta_{\mathbb{R}^n - \bar{D}}(z)$ for $z \notin D$. Show that there is a neighborhood U of bD, such that ρ is C^2 on U with $d\rho \neq 0$, and hence ρ is a C^2 defining function for D. (Note that it is quite easy to see that ρ is of class C^1 near bD—use the implicit function theorem; that ρ is actually C^2 requires a sharper look.)

E.5.10. Let $D \subset\subset \mathbb{C}^n$ be pseudoconvex with C^2 boundary.

(i) Use E.5.9 and Proposition 4.9 to show that there is a neighborhood U of bD such that for $z \in U \cap D$ one has

$$L_z(\rho_{bD}; t) \geq 0 \text{ for } t \in \mathbb{C}^n \text{ with } \sum_{j=1}^{n} \frac{\partial \rho_{bD}}{\partial z_j}(z) t_j = 0.$$

(ii) Deduce from (i) that D is Levi pseudoconvex.

E.5.11. Suppose $D \subset\subset \mathbb{C}^n$ has C^2 boundary and is Levi pseudoconvex. Use E.5.9 to show that $\varphi = -\log \delta_D$ is plurisubharmonic on $U \cap D$ for some neighborhood U of bD by completing the following argument: if the Leviform $L_z(\varphi; t)$ where negative for some $t \neq 0$, construct an analytic disc S in $\bar{D}$ which meets bD only at one point, so that $\delta_{D|S}$ has a *strict* minimum at P; this contradicts the Levi condition for ρ_{bD} at P.

E.5.12. Use E.5.10.(i) to prove the following result of K. Diederich and J. E. Fornaess [DiFo 1]: *If D is pseudoconvex with C^2 boundary, then $\varphi = -\delta_D^\eta e^{-K|z|}$ is strictly plurisubharmonic on $U \cap D$ for some neighborhood U of bD, provided $\eta > 0$ is sufficiently small and $K > 0$ is sufficiently large.* (The result remains true if δ_D is replaced by $-r$ in case D has C^3 boundary and r is an arbitrary C^3 defining function for D (see [Ran 5]).

E.5.13. Deduce from E.5.12 that for every bounded pseudoconvex domain with C^2 boundary there is a "*bounded* strictly plurisubharmonic exhaustion function," i.e., a strictly plurisubharmonic function φ on D, with $\varphi(z) < 0$ for $z \in D$ and $\lim \varphi(z) = 0$ as $z \to bD$. (Hint: See E.2.8.(ii).)

Remark. By a completely different proof one can show that the conclusion remains true even in case D has only a C^1 boundary (cf. N. Kerzman and J. P. Rosay [KeRo]).

Notes for Chapter II

As mentioned in the text, the early developments related to the characterization of natural boundaries for holomorphic functions owe much to the work of F. Hartogs and E. E. Levi. In 1910 Levi [Lev 1] gave a different proof of Hartogs' Theorem 1.1 based on a Laurent series expansion in the distinguished variable, and he generalized the result to meromorphic functions. Laurent series expansions in several variables were well known at the time of W. F. Osgood's book [Osg]. Theorem 1.6, at least in case the Reinhardt domain contains its center, is due to Hartogs [Har 2]. Our presentation follows R. Narasimhan [Nar 3].

The notions of Hartogs pseudoconvexity and Theorem 2.3 are due to Hartogs [Har 1]; later presentations emphasized the "continuity principle" (cf. [BeTh], §IV.1) and variants thereof. K. Oka discusses various formulations and proves their equivalence in his 9th memoir; in essence, his (C)-pseudoconvexity ([Oka], p. 149–150) coincides with Hartogs pseudoconvexity as defined here. Lemma 2.1 seems to have been noticed first by Levi [Lev 2]. The Levi condition for domains of holomorphy was discovered by Levi in 1910 in the case of two variables [Lev 1]; the extension to n variables is due to J. Krzoska ("Über die natürlichen Grenzen der analytischen Funktionen mehrerer Veränderlichen," Dissertation Greifswald, 1933). The important partial local converse (Theorem 2.15) was proved in [Lev 2]. In that paper E. E. Levi already pointed out the principal obstacle to obtaining the corresponding global result, i.e., what is now called the Levi problem (cf. §2.9). The useful Proposition 2.14 is due to J. J. Kohn [Koh 1], but strictly plurisubharmonic functions occur already in [Oka, VI]. Lemma 2.19 is a consequence of the general theory of pseudoconvexity developed by Oka and Lelong; the direct proof (for C^3 boundaries) given here is based on a construction of plurisubharmonic exhaustion functions taken from [Koh 2].

The main results in §3 are due to H. Cartan and P. Thullen [CaTh]. The terminology of Stein domains adopted here and the emphasis on compact sets have become quite standard in recent years. The proof of Theorem 3.24 is due to F. Docquier and H. Grauert [DoGr], who actually considered the more general case of Stein manifolds. The characterization of Stein manifolds in terms of strictly plurisubharmonic exhaustion functions was initiated in an influential paper by H. Grauert [Gra] in 1958, and investigated further in [DoGr] and [Nar 1]. The characterization of regions of convergence of power series by logarithmic convexity is due to F. Hartogs [Har 2]; closely related results were obtained around the same time by G. Faber [Fab]; the characterization in terms of convexity with respect to monomials is due to H. Cartan ([Car], 303–326). Subharmonic functions were first investigated by F. Hartogs [Har 2]. Plurisubharmonic functions were introduced—under the name of pseudoconvex functions—by Oka in the case of two variables in 1941 ([Oka],VI) and in the case of n variables in 1953 ([Oka],IX), and independently, by P. Lelong [Lel 1], whose terminology is the one now in use. Most of the basic properties of plurisubharmonic functions discussed in §4 are due to these authors. For a thorough discussion of the analogy between real convex functions and plurisubharmonic functions the reader should consult H. Bremermann [Bre 2].

Convexity with respect to plurisubharmonic functions was introduced by P. Lelong [Lel 3]. The plurisubharmonicity of $-\log \delta_D$ for a pseudoconvex domain D (Theorem 5.6) was first proved by K. Oka in $\mathbb{C}^2$ ([Oka],VI) but the essential idea of the proof goes back to F. Hartogs [Har 2]. Proofs in arbitrary dimension are due to Oka ([Oka], p. 155–157), P. Lelong [Lel 2], and H.

Bremermann [Bre 2]. The proof given here is based on Oka's presentation, which is the only one which starts off with Hartogs pseudoconvexity rather than with the continuity principle in terms of analytic discs. In the proofs of Proposition 4.14 and Theorem 5.11 we have followed the exposition of L. Hörmander [Hör 2].

CHAPTER III

Differential Forms and Hermitian Geometry

In this chapter we collect the technical tools from the calculus of differential forms and from complex differential geometry which will be needed in the following chapters. Section 1 deals with differentiable manifolds; the principal goal here is a thorough understanding of Stokes' Theorem in the language of differential forms. In §2 we discuss the additional structures which arise when the manifold under consideration is complex. The main topics here are the natural intrinsic complex structure on the (real) tangent space of a complex manifold M, the direct sum decomposition of the algebra of complex valued differential forms into forms of type (p, q), $0 \leq p, q \leq \dim_{\mathbb{C}} M$, and the Cauchy–Riemann complex with its associated $\bar{\partial}$-cohomology groups. In §3 we discuss the elementary aspects of Riemannian geometry in $\mathbb{C}^n$ in complex form. Of major importance for our purposes are the inner product of differential forms defined by integration over regions in $\mathbb{C}^n$, the Hodge $*$- operator, which allows us to freely go back and forth between the geometric inner product and the algebraic wedge product of forms, the various formulas for integration by parts, and the natural differential operators associated to the Cauchy–Riemann operator, i.e., the (formal) adjoint ϑ of $\bar{\partial}$ and the complex Laplacian $\square = \vartheta\bar{\partial} + \bar{\partial}\vartheta$. In this paragraph, which is more computational than the preceding ones, we consider only the case of $\mathbb{C}^n$ rather than general Hermitian manifolds; not only does this simplify matters quite a bit, but it allows us to state certain basic formulas in exact form without having to introduce the numerous error terms which occur in the general setting.

The reader familiar with some or all of the topics in §1 and §2 may safely skip the sections devoted to them, referring to them only as needed for notation and specific results. On the other hand, unless the reader is a professional complex differential geometer, he is urged to study §3 carefully before proceeding with Chapter IV.

§1. Calculus on Real Differentiable Manifolds

The principal results in Chapters IV and V are based on integration of functions and differential forms over regions D in $\mathbb{C}^n$ and over their boundaries bD. Even though all the necessary concepts and computations could be formulated explicitly in terms of the global coordinates in $\mathbb{C}^n$, we shall take the intrinsic—or coordinate-free—point of view, as it provides a deeper insight into the various formulas and operations which will be considered. Explicit computations will then appear as straightforward applications of the general concepts. Moreover, as it would be too restrictive for later applications to consider only domains with C^∞ boundaries, we shall work on differentiable manifolds of class C^k, for $1 \leq k \leq \infty$. This requires just a little bit more attention than the C^∞ case considered in many of the introductory books on differential topology. For completeness' sake we collect all the definitions and results which will be important for our purposes, but we shall skip many of the often uninspiring proofs. The reader may find more details in the books by R. Narasimhan [Nar 4] and F. Warner [War], or in the classic "Variétés Différentiables" by G. de Rham [Rha]. (The latter two concentrate on the C^∞ case.)

1.1. Differentiable Manifolds

A **differentiable manifold** M **of dimension** $n \in \mathbb{N}$ **and class** C^k, $1 \leq k \leq \infty$, is a topological Hausdorff space M together with a C^k **atlas** $\mathscr{A} = \{(U_i, \varphi_i), i \in I\}$ consisting of pairs (U_i, φ_i), where $U_i \subset M$ is open and $\varphi_i \colon U_i \to \varphi_i(U_i)$ is a homeomorphism onto an open subset of $\mathbb{R}^n$, subject to the following conditions:

(1.1) $$M = \bigcup_{i \in I} U_i;$$

(1.2) $$\varphi_j \circ \varphi_i^{-1} \colon \varphi_i(U_i \cap U_j) \to \varphi_j(U_i \cap U_j)$$

is a C^k map between open subsets of $\mathbb{R}^n$ for all $i, j \in I$ with $U_i \cap U_j \neq \varnothing$.

It is obvious from (1.2) that the inverse $\varphi_i \circ \varphi_j^{-1}$ of $\varphi_j \circ \varphi_i^{-1}$ is also of class C^k, and hence the Jacobian matrix of $\varphi_j \circ \varphi_i^{-1}$ is an invertible $n \times n$ matrix at every point $x \in \varphi_i(U_i \cap U_j)$.

One also says that the C^k atlas $\mathscr{A}$ defines a C^k **structure** on the Hausdorff space M. It is obvious that a C^k atlas is also a C^l atlas for any $1 \leq l \leq k$; hence a manifold of class C^k is also a manifold of class C^l, $1 \leq l \leq k$, in a natural way. An element (U_i, φ_i) of $\mathscr{A}$ is called a **coordinate system**. Any open set D of a C^k manifold M (with atlas $\mathscr{A}$) inherits the structure of a C^k manifold defined by the atlas $\mathscr{A}_D = \{(U_i \cap D, \varphi_{i|U_i \cap D}) \colon i \in I\}$.

Examples. (i) Every nonempty open set D of $\mathbb{R}^n$ (including $\mathbb{R}^n$ itself) has a natural structure of a C^∞ manifold of dimension n defined by the coordinate system (D, φ_D), where $\varphi_D: D \to D$ is the identity map.

(ii) Suppose $D \subset\subset \mathbb{R}^n$ has C^k boundary bD for some $1 \le k \le \infty$. Then bD carries a C^k manifold structure of dimension $n-1$ defined as follows. If $P \in bD$ and r_P is a C^k defining function for D near P, it follows from the inverse function theorem that there is an invertible C^k map $\psi_P: U_P \to V_P$ from an open neighborhood U_P of P onto an open neighborhood V_P of 0 in $\mathbb{R}^n$, such that $\psi_P = (r_P, \psi'_P)$, where $\psi'_P: U_P \to \mathbb{R}^{n-1}$ is defined by the last $n-1$ components of ψ_P. Thus $\varphi_P = \psi'_P|_{bD \cap U_P}$ is a homeomorphism of $bD \cap U_P$ onto the open set $V_P \cap \{x: x_1 = 0\}$ of $\mathbb{R}^{n-1}$, and clearly the collection $\{(bD \cap U_P, \varphi_P), P \in bD\}$ satisfies (1.1) and (1.2).

(iii) Generalizing the previous example, one can show that if $U \subset \mathbb{R}^n$ is open, $a \in U$, and $F: U \to \mathbb{R}^m$, $(0 < m \le n)$ is a C^k map such that the Jacobian matrix of F has maximal rank m at every point of $L_a(F) = \{x \in U: F(x) = F(a)\}$, then $L_a(F)$ inherits the structure of a C^k manifold of dimension $n - m$. The reader should compare this with Chapter I, §2.6, where the corresponding result was proved for holomorphic maps.

In the applications later on we shall mainly consider manifolds of the types described above. Such manifolds are examples of **submanifolds of** $\mathbb{R}^n$ (see Exercise E.1.2, or Chapter I, §2.6 for the complex version).

If $\mathscr{A}$ is a C^k atlas for M and $P \in M$, a function $f: M \to \mathbb{R}$ is **differentiable of class** C^l **at** P, $1 \le l \le k$, if $f \circ \varphi^{-1}: \varphi(U) \to \mathbb{R}$ is of class C^l in some neighborhood of $\varphi(P)$ for some coordinate system $(U, \varphi) \in \mathscr{A}$ with $P \in U$. By (1.2), this definition is independent of the particular coordinate system chosen. More generally, a map $F: M_1 \to M_2$ between two C^k manifolds is of class C^l, $1 \le l \le k$, if for any two coordinate systems (U, φ) of M_1, respectively (V, ψ) of M_2, the map $\psi \circ F \circ \varphi^{-1}$ is C^l wherever defined. F is a (C^l) **diffeomorphism** if F is a homeomorphism, and F and F^{-1} are of class C^l; if F is a diffeomorphism, then M_1 and M_2 must necessarily have equal dimension.

In practice, it is convenient to enlarge the given atlas $\mathscr{A}$ of a C^k manifold M to an atlas $\bar{\mathscr{A}}$ which is maximal with respect to (1.2); such an atlas is said to be **complete**. $\bar{\mathscr{A}}$ is determined uniquely by $\mathscr{A}$ and is obtained by adding to $\mathscr{A}$ *all* pairs (U, φ) consisting of an open subset U of M and a C^k diffeomorphism φ from U onto an open subset of $\mathbb{R}^n$. Two C^k atlases $\mathscr{A}_1$ and $\mathscr{A}_2$ define the same C^k structure on M if and only if $\bar{\mathscr{A}}_1 = \bar{\mathscr{A}}_2$.

In order to extend the construction of (differentiable) partitions of unity from $\mathbb{R}^n$ to abstract manifolds M one needs that M is paracompact. Recall that a topological space X is said to be **paracompact** if for every open covering $\mathscr{U} = \{U_i: i \in I\}$ of X there is a locally finite refinement $\mathscr{V} = \{V_j: j \in J\}$ of $\mathscr{U}$. This means that $\mathscr{V}$ is an open covering of X such that for each compact set $K \subset X$ one has $V_j \cap K \ne \varnothing$ for only finitely many indices j, and that there is a **refinement map** $\tau: J \to I$ such that $V_j \subset U_{\tau(j)}$ for all $j \in J$. It is known that a differentiable manifold is paracompact if and only if every connected component of M is countable at infinity (i.e., is the increasing union of a countable collection of compact subsets), a condition which is satisfied by all submanifolds of $\mathbb{R}^n$. From now on, all manifolds shall be assumed paracompact without explicit mention.

1.2. Tangent Vectors and Differentials

We fix a differentiable manifold M of class C^k and dimension n. We shall introduce "tangent vectors" to M as a generalization of the directional derivative of functions on $\mathbb{R}^n$. A curve in M through the point $P \in M$ is a C^1 map $\alpha: I \to M$, where I is an open interval in $\mathbb{R}$ with $0 \in I$, such that $\alpha(0) = P$. We obtain an equivalence relation in the set of all such curves by defining α **is equivalent to** β **(at** P**)** if and only if *for all* C^1 *functions* f *defined near* P *one has*

$$\frac{d}{dt}(f \circ \alpha)(0) = \frac{d}{dt}(f \circ \beta)(0). \tag{1.3}$$

The set of all equivalence classes of curves through P is called the **tangent space** T_PM of M at P. Every **tangent vector** $v \in T_PM$ defines a "directional derivative" as follows: given a C^1 function f near P, one sets

$$v(f) = \frac{d}{dt}(f \circ \alpha)(0), \tag{1.4}$$

where α is a curve through P which represents v.

It is obvious that if $v \in T_PM$ and f, g are C^1 functions near P, then

$$v(af + bg) = av(f) + bv(g) \qquad \text{for } a, b \in \mathbb{R}, \tag{1.5}$$

and

$$v(f \cdot g) = v(f)g(P) + f(P)v(g). \tag{1.6}$$

In practice, it is the action of tangent vectors on functions, given by (1.4), and the properties (1.5) and (1.6) which are used, rather than the abstract definition.

A C^1 map $F: M \to N$ between differentiable manifolds induces a map

$$dF_P: T_PM \to T_{F(P)}N \tag{1.7}$$

for every $P \in M$ by defining $dF_P(v)$ as the equivalence class of the curve $F \circ \alpha$ through $F(p)$, where α represents $v \in T_P(M)$. (The reader should check that this depends only on v!) The map dF_P is called the **differential of** F **at** p.

In case $M = \mathbb{R}^n$, there is a natural identification of $T_P\mathbb{R}^n$ with $\mathbb{R}^n$. In fact, if $\alpha: I \to \mathbb{R}^n$ is a curve in $\mathbb{R}^n$ through P, then $\alpha'(0) = d\alpha/dt(0)$ is an element of $\mathbb{R}^n$, and clearly two curves α and β are equivalent at P according to (1.3) if and only if $\alpha'(0) = \beta'(0)$. Via this identification, $T_P\mathbb{R}^n$ carries the natural structure of an n-dimensional vector space.

Theorem 1.1. *The tangent space T_PM carries a unique structure of a real n-dimensional vector space such that for every coordinate system (U, φ) around P the differential $d\varphi_P$: $T_PM \to T_{\varphi(P)}\mathbb{R}^n$ is an isomorphism.*

PROOF. If (U, φ) is a fixed coordinate system, then $d\varphi_P$: $T_PM \to T_{\varphi(P)}\mathbb{R}^n$ is clearly one-to-one and onto (with inverse $d(\varphi^{-1})_{\varphi(P)}$), so there is a unique vector space structure on T_PM which makes $d\varphi_P$ an isomorphism. In order to see that the vector space operations thus defined are independent of (U, φ), one checks that for every C^1 function f near P and every two vectors v, $w \in T_PM$ one has

$$(v + w)(f) := [(d\varphi_P)^{-1}(d\varphi_P(v) + d\varphi_P(w))](f) = v(f) + w(f),$$

and, if $a \in \mathbb{R}$, then

$$(av)(f) := [(d\varphi_P)^{-1}(a d\varphi_P(v))](f) = a(v(f)). \quad \blacksquare$$

The proof shows in particular that the vector space operations in T_PM are consistent with the interpretation of tangent vectors as differentiation operators given by (1.4).

Corollary 1.2. *The differential dF_P of a C^1 map F: $M \to N$ is $\mathbb{R}$-linear.*

Remark. Suppose $D \subset \mathbb{R}^n$ has C^k boundary bD at P. The inclusion map ι: $bD \to \mathbb{R}^n$ induces an injective map $d\iota$: $T_PbD \to T_P\mathbb{R}^n = \mathbb{R}^n$; via this map, the (abstract) tangent space T_PbD, as defined here, agrees with the (concrete) tangent space of bD defined in II,§2.3 as a subspace of $\mathbb{R}^n$.

In view of the natural identification of $T_P\mathbb{R}^n$ with $\mathbb{R}^n$, the differential dF_P of a C^1 map F: $\mathbb{R}^n \to \mathbb{R}^m$ defines a linear map $\mathbb{R}^n \to \mathbb{R}^m$; one checks that the matrix representation of dF_P if given by the Jacobian matrix of F at P. Similarly, if M is a C^1 manifold and $P \in M$, the differential df_P of a *real valued* C^1 function defined near P is naturally identified with a linear functional $T_PM \to \mathbb{R}$, i.e., one considers df_P as an element of the dual space T_P^*M of T_PM, and one has the equality $df_P(v) = v(f)$. Elements of the dual space T_P^*M are called **1-forms**, or **cotangent vectors at** P.

Finally, we consider the representation of tangent vectors and 1-forms at a point $P \in M$ with respect to a coordinate system (U, φ) with $P \in U$. We set $\varphi = (x_1, \ldots, x_n)$, where $x_1, \ldots, x_n$ are real valued C^k functions on U. If $\{e_1, \ldots, e_n\}$ is the standard basis of $\mathbb{R}^n$, the tangent vectors $(d\varphi_P)^{-1}(e_j) \in T_PM$, $1 \le j \le n$, are denoted by $(\partial/\partial x_j)_P$, or simply $\partial/\partial x_j$. This notation is motivated by the fact that for a C^1 function f near P, $(\partial/\partial x_j)_P(f)$ is the partial derivative at $\varphi(P)$ of the function $f \circ \varphi^{-1}$ with respect to the coordinate x_j. Clearly $\partial/\partial x_1$, $\ldots$, $\partial/\partial x_n$ form a basis for T_PM, and any $v \in T_PM$ has the representation

$$(1.8) \qquad v = \sum_{j=1}^{n} v(x_j)\frac{\partial}{\partial x_j} = \sum_{j=1}^{n} (dx_j)_P(v)\frac{\partial}{\partial x_j}.$$

The differentials $(dx_1)_P, \ldots, (dx_n)_P$ of the coordinate functions at P define a basis of T_P^*M, which is dual to the basis $\{\partial/\partial x_j\colon 1 \le j \le n\}$ of T_PM. If f is a C^1 function near P, one has

$$(1.9)\qquad (df)_P = \sum_{j=1}^{n} df_P\left(\frac{\partial}{\partial x_j}\right)(dx_j)_P = \sum_{j=1}^{n}\left(\frac{\partial}{\partial x_j}\right)_P(f)(dx_j)_P.$$

To simplify notation we will write dx_j instead of $(dx_j)_P$.

1.3. The Algebra of Differential Forms

We now consider the r-th **exterior power** $\Lambda^r T_P^* M$ **over** $\mathbb{R}$ of the cotangent space $T_P^* M$. By definition $\Lambda^0 T_P^* M = \mathbb{R}$, while for $r \ge 1$ it is most convenient to identify $\Lambda^r T_P^* M$ with the $\mathbb{R}$-vector space of **alternating** r**-multilinear forms on** $T_P M$, i.e., $\mathbb{R}$-multilinear maps

$$\omega\colon \underbrace{T_P M \times \cdots \times T_P M}_{r \text{ factors}} \to \mathbb{R}$$

which satisfy

$$\omega(v_{\sigma(1)}, \ldots, v_{\sigma(r)}) = \operatorname{sign} \sigma\ \omega(v_1, \ldots, v_r)$$

for all $v_1, \ldots, v_r \in T_P M$ and every permutation σ of $\{1, \ldots, r\}$[1]. In particular,

$$\omega(v_1, \ldots, v_r) = 0 \qquad \text{if } v_i = v_j \text{ for two indices } i \ne j.$$

For our purposes, the above identification may be taken as the definition of $\Lambda^r T_P^* M$. It follows that $\Lambda^1 T_P^* M = T_P^* M$ and $\Lambda^r T_P^* M = \{0\}$ for $r > \dim T_P M = \dim M$. Elements of $\Lambda^r T_P^* M$ are called r**-covectors**, or r**-forms** at P.

The direct sum

$$\mathcal{G}_P(M) = \bigoplus_{r \ge 0} \Lambda^r T_P^* M$$

is called the **exterior algebra** (or **Grassmann algebra**) of $T_P^* M$. The **wedge product** of an r-form ω and an s-form η is an $(r + s)$-form denoted by $\omega \wedge \eta$, and defined by

$$(1.10)\qquad \begin{aligned} &\omega \wedge \eta(v_1, \ldots, v_{r+s}) \\ &= \frac{1}{r!s!}\sum_{\sigma} \operatorname{sign} \sigma\ \omega(v_{\sigma(1)}, \ldots, v_{\sigma(r)})\eta(v_{\sigma(r+1)}, \ldots, v_{\sigma(r+s)}), \end{aligned}$$

where the summation is taken over all permutations σ of $\{1, \ldots, r + s\}$. Equation (1.10) and the distributive property determine the product between any two elements in $\mathcal{G}_P M$. The wedge product is associative, but not commutative; instead one has

[1] See Warner [War] for a detailed presentation of this identification.

(1.11) $$\omega \wedge \eta = (-1)^{r \cdot s} \eta \wedge \omega \qquad \text{for } \omega \in \Lambda^r \text{ and } \eta \in \Lambda^s.$$

If (U, φ) is a coordinate system near $P \in M$ with $\varphi = (x_1, \ldots, x_n)$, and if $1 \leq r \leq n$, then

$$\{dx_{j_1} \wedge \ldots \wedge dx_{j_r} : 1 \leq j_1 < \ldots < j_r \leq n\}$$

is a basis for $\Lambda^r T_P^* M$; in particular, $\dim \Lambda^r T_P^* M = \binom{n}{r}$, and every r-form ω_P at P has a unique representation

(1.12) $$\omega_P = \sum_J a_J (dx^J)_P, \qquad a_J \in \mathbb{R},$$

where the summation is over *all strictly increasing* r-tuples $J = (j_1, \ldots, j_r) \subset \{1, \ldots, n\}$, and

(1.13) $$dx^J = dx_{j_1} \wedge \ldots \wedge dx_{j_r} \qquad \text{for } J = (j_1, \ldots, j_r).$$

Notice that

$$a_J = \omega_P \left(\frac{\partial}{\partial x_{j_1}}, \ldots, \frac{\partial}{\partial x_{j_r}} \right).$$

So far we have only considered tangent vectors and r-forms at fixed points $P \in M$. We shall now examine the relevant concepts when the point P is variable.

A **vector field** on M is a map $V: M \to \bigcup_{P \in M} T_P M$,[1] which to each $P \in M$ assigns a tangent vector $V_P \in T_P M$. A coordinate system (U, φ) defines n vector fields

$$\frac{\partial}{\partial x_1}, \ldots, \frac{\partial}{\partial x_n}$$

on U, and a vector field V on U has a representation $V = \sum_{j=1}^n a_j (\partial / \partial x_j)$, where $a_1, \ldots, a_n$ are uniquely determined functions on U; V is said to be of class C^l on U if the coefficient functions $a_1, \ldots, a_n$ are C^l on U. This notion is independent of the chosen coordinate system if $0 \leq l \leq k - 1$, but not necessarily for $l = k$. Therefore, on a C^k manifold M, one only considers vector fields of class C^l for $l \leq k - 1$. One easily checks that a vector field V on an open subset D of M is of class C^l if and only if $Vf \in C^l(D)$ for every $f \in C^{l+1}(D)$, where $(Vf)(P)$ is defined by $V_P(f)$ according to (1.4).

Similarly, an r-form ω on M is given by a map $P \mapsto \omega_P \in \Lambda^r T_P^* M$ for $P \in M$. ω is of class C^l, $0 \leq l \leq k - 1$, if all the coefficients $a_J(P)$ in the local representation $\omega = \sum a_J \, dx^J$ (see (1.12)) with respect to some coordinate system are C^l functions; equivalently, ω is of class C^l on M if and only if $\omega(V_1, \ldots, V_r) \in C^l(M)$ for all C^l vector fields $V_1, \ldots, V_r$ on M.

We shall denote the space of r-forms of class C^l on M by $C_r^l(M)$. The elements of $C_r^l(M)$ are also called **differential forms of degree** r **and class** C^l **on** M. In particular, $C_0^l(M) = C^l(M)$, and $C_r^l(M) = \{0\}$ if $r > \dim M$.

[1] We do not introduce the vector bundle structure on the collection of tangent spaces, as it is not needed for our purposes.

1.4. The Exterior Derivative

The differential df of a function $f \in C^1(M)$ defines a continuous 1-form on M; we therefore have a natural linear map $d\colon C^1(M) \to C_1^0(M)$ on every C^k manifold M. Clearly $d(C^l(M)) \subset C_1^{l-1}(M)$ for every $1 \le l \le k$, and d satisfies the Leibnitz rule $d(fg) = g\,df + f\,dg$. The map d extends to the full Grassmann algebra $\mathscr{G}^1(M)$ of differential forms on M of class C^1, as follows.

Theorem 1.3. *Let M be a C^k manifold, $k \ge 2$, and let $\mathscr{G}^l(M) = \bigoplus_{r \ge 0} C_r^l(M)$, $0 \le l < k$. There is a unique linear map $d = d_M\colon \mathscr{G}^1(M) \to \mathscr{G}^0(M)$ which satisfies the following conditions.*

(1.13) *df is the differential of $f \in C^1(M)$.*

(1.14) *If $1 \le l < k$ and $0 \le r$, then $d\omega \in C_{r+1}^{l-1}(M)$ for $\omega \in C_r^l(M)$.*

(1.15) *If $f \in C^l$, $2 \le l \le k$, then $d(df) = 0$.*

(1.16) *If $\omega_1 \in C_r^1(M)$ and $\omega_2 \in C_s^1(M)$, then*

$$d(\omega_1 \wedge \omega_2) = d\omega_1 \wedge \omega_2 + (-1)^r \omega_1 \wedge d\omega_2.$$

The map d given by the theorem is called the **exterior derivative** on M.

Corollary 1.4. *The exterior derivative is a local operator, i.e., if $\omega_1 = \omega_2$ on an open set U, then $d\omega_1 = d\omega_2$ on U.*

Proof. It is enough to show that if $\omega \in C_r^1(M)$ is 0 on U, then $d\omega = 0$ on U. If $P \in U$, choose $f \in C^1(M)$ such that $f(P) = 0$ and $f = 1$ in a neighborhood of $M - U$. Then $\omega = f\omega$; by (1.16),

$$d\omega = d(f\omega) = df \wedge \omega + f \wedge d\omega,$$

which implies $(d\omega)_P = 0$, since $\omega_P = 0$ and $f(P) = 0$. ∎

Proof of Theorem 1.3. Because of Corollary 1.4, in order to prove the uniqueness statement, it is enough to consider the case where M is replaced by the open set U of a coordinate system (U, φ), $\varphi = (x_1, \ldots, x_n)$. If $\omega \in C_r^1(U)$, then $\omega = \sum_J a_J\, dx^J$ (see (1.12)), with $a_J \in C^1(U)$ for all strictly increasing r-tuples J. By (1.15), $d(dx_j) = 0$ for $1 \le j \le n$; therefore the linearity of d and repeated application of (1.16) implies that

$$d\omega = \sum_J da_J \wedge dx^J. \tag{1.17}$$

This shows that d is completely determined by its values on functions, and uniqueness follows.

Because of Corollary 1.4 and uniqueness, in order to prove the existence of a map d with the required properties, it is enough to do it on a coordinate system (U, φ). Since $\omega \in C_r^1(U)$ has a unique representation $\omega = \sum a_J\, dx^J$, one can define $d\omega \in C_{r+1}^0(U)$ by (1.17), and extend d to $\mathscr{G}^1(U)$ by linearity. One then checks that the map $d = d_U\colon \mathcal{G}^1(U) \to \mathcal{G}^0(U)$ so obtained satisfies (1.13)–(1.16). We shall verify (1.15), which is the most interesting property. If $f \in C^2(U)$, it follows from (1.9) that $df = \sum_{j=1}^n (\partial f/\partial x_j)\, dx_j$ and hence, by (1.17),

$$\begin{aligned} d(df) &= \sum_{j=1}^{n} d\left(\frac{\partial f}{\partial x_j}\right) \wedge dx_j = \sum_j \left(\sum_i \frac{\partial}{\partial x_i}\left(\frac{\partial f}{\partial x_j}\right) dx_i\right) \wedge dx_j \\ &= \sum_{i<j} \left[\frac{\partial}{\partial x_i}\left(\frac{\partial f}{\partial x_j}\right) - \frac{\partial}{\partial x_j}\left(\frac{\partial f}{\partial x_i}\right)\right] dx_i \wedge dx_j. \end{aligned} \tag{1.18}$$

But in the sum on the right in (1.18) all coefficients are 0, as $\partial/\partial x_i(\partial f/\partial x_j)$ is computed by taking the 2nd order partial derivatives of the C^2 function $f \circ \varphi^{-1}$ with respect to the coordinates x_i and x_j on $\mathbb{R}^n$. Thus $d(df) = 0$. ∎

Under suitable differentiability assumptions the property $d^2 = d \circ d = 0$ extends from functions to arbitrary forms.

Corollary 1.5. *If $2 \le l < k$, then*

$$d(d\omega) = 0 \qquad \textit{for } \omega \in C_r^l(M).$$

Proof. In a local coordinate system (U, φ) one represents $d\omega$ as in (1.17). The conclusion then follows by (1.16) and (1.15). ∎

1.5. The Pull Back

Suppose M and N are C^k manifolds and $F\colon M \to N$ is a C^1 map. The **pull back** $f \to F^*(f) = f \circ F$ defines an algebra homomorphism $F^*\colon C^1(N) \to C^1(M)$. We shall now examine how F^* extends to differential forms.

For $P \in M$, the differential $dF_P\colon T_P M \to T_{F(P)} N$ induces the transposed (or dual) map $F_P^*\colon T_{F(P)}^* N \to T_P^* M$, which extends to an algebra homomorphism $F_P^*\colon \mathcal{G}_{F(P)}(N) \to \mathcal{G}_P(M)$ of the Grassmann algebras defined by

$$(F_P^* \omega_{F(P)})(v_1, \ldots, v_r) = \omega_{F(P)}(dF_P(v_1), \ldots, dF_P(v_r)) \tag{1.19}$$

for $r > 0$, $\omega_{F(P)} \in \Lambda^r(T_{F(P)}^* N)$, and $v_1, \ldots, v_r \in T_P M$. If ω is a differential form on the open set $U \subset N$, one obtains a differential form $F^*\omega$ on $F^{-1}(U)$, called the **pull back of** ω **by** F, by setting $(F^*\omega)_P = F_P^*(\omega_{F(P)})$ for $P \in F^{-1}(U)$. The map F^* thus defined has the following natural properties.

Theorem 1.6. *Suppose M, N, and $F\colon M \to N$ are of class C^k.*

(i) *The pull back F^* is an algebra homomorphism*

$$F^*\colon \mathcal{G}^0(N) \to \mathcal{G}^0(M)$$

which satisfies $F^*(C_r^l(N)) \subset C_r^l(M)$ for $0 \le l < k$ and $r \ge 0$.

(ii) *F^* commutes with the exterior derivatives d_M and d_N on M and N, that is, if ω is of class C^l, $1 \le l < k$ (and $l = k$ if ω is of degree 0), then*

$$d_M(F^*\omega) = F^*(d_N\omega). \tag{1.20}$$

Proof. We first verify (1.20) for $\omega = f \in C^l(N)$, $1 \le l \le k$. Since $F^*f = f \circ F \in C^l(M)$, and since d_M and d_N agree on functions with the usual differentials, one obtains

$$d_M(F^*f) = d(f \circ F) = (df) \circ (dF) = F^*(df),$$

where the last equation follows from the definition (1.19) in case $\omega_{F(P)} = df_{F(P)}$.

From the pointwise definition of F^* it is clear that F^* will be an algebra homomorphism on differential forms. In order to prove the rest of (i) and (ii) it is enough to prove the corresponding *local* statements for $\omega \in C_r^l(N)$. Thus we may assume that ω is given on $U \subset N$ by $\omega = \sum a_J \, dx^J$, where $a_J \in C^l(U)$. Then

$$\begin{aligned} F^*\omega &= \sum_{J=(j_1,\dots,j_r)} F^*(a_j)F^*(dx_{j_1})\dots F^*(dx_{j_r}) \\ &= \sum (a_j \circ F) d(x_{j_1} \circ F) \wedge \dots \wedge d(x_{j_r} \circ F), \end{aligned} \tag{1.21}$$

where we have used (1.20) for $r = 0$; this shows that $F^*\omega \in C_r^l(F^{-1}(U))$. Moreover (1.21) implies

$$\begin{aligned} d_M(F^*\omega) &= \sum d(a_J \circ F) \wedge d(x_{j_1} \circ F) \wedge \dots \wedge d(x_{j_r} \circ F) \\ &= \sum F^*(da_J) \wedge F^*(dx_{j_1}) \wedge \dots \wedge F^*(dx_{j_r}) \\ &= F^*(\textstyle\sum da_J \wedge dx^J) \\ &= F^*(d_N\omega). \quad \blacksquare \end{aligned}$$

One also verifies that if $G: W \to M$ and $F: M \to N$ are maps of class C^k, then $F \circ G: W \to N$ satisfies

$$(F \circ G)^* = G^* \circ F^*. \tag{1.22}$$

1.6. Orientation and Integration

Integration of differential forms is often introduced by defining the integral over differentiable simplices and chains. This approach is useful for applications to algebraic topology, but for our purposes it suffices to define the somewhat more special concept of the integral of an n-form over an oriented n-dimensional manifold.

Definition. A C^k manifold M of dimension n is called **orientable** if there is a nowhere zero continuous n-form Ω on M (i.e., $\Omega_P \neq 0$ for all $P \in M$). Two such forms Ω_1, Ω_2 are said to define the same orientation of M if $\Omega_1 = f\Omega_2$ on M, where f is a positive continuous function on M. A manifold with a choice of orientation (given by specifying a particular nowhere zero n-form Ω) is said to be **oriented**.

Since $\Lambda^n(T_P^* M)$ is one-dimensional, it is clear that any two nowhere zero continuous n-forms Ω_1 and Ω_2 differ by a nonzero continuous function f on M. Thus, if M is connected, there are precisely two orientations on M: if one of them is defined by Ω, the other one is defined by $-\Omega$.

If M is oriented by Ω, a coordinate system (U, φ) with $\varphi = (x_1, \ldots, x_n)$ and the corresponding basis $\{dx_1, \ldots, dx_n\}$ of $T_P^* M$ are called **positively oriented** (or simply positive) if $dx_1 \wedge \ldots \wedge dx_n$ defines the same orientation on U as Ω, that is, if $\Omega_P = a(P)(dx_1 \wedge \ldots \wedge dx_n)_P$, with $a(P) > 0$ for all $P \in U$. A C^k diffeomorphism $F: M_1 \to M_2$ between oriented C^k manifolds (M_i, Ω_i), $i = 1, 2$, is said to be **orientation preserving** if $F^*\Omega_2 = a\Omega_1$, where $a > 0$ on M_1.

Examples. (i) If $(t_1, \ldots, t_n)$ are the standard coordinates of $\mathbb{R}^n$, the n-form $dt_1 \wedge \ldots \wedge dt_n$ defines an orientation on $\mathbb{R}^n$, which is called the **positive** (or natural or standard) orientation of $\mathbb{R}^n$.

(ii) Any nonempty open set $D \subset \mathbb{R}^n$ inherits the positive orientation of $\mathbb{R}^n$, and thus is also an oriented manifold. Every open set $D \subset \mathbb{R}^n$, including $\mathbb{R}^n$, will always be assumed to be positively oriented.

(iii) If $D \subset \mathbb{R}^n$ is an open set with C^k boundary, $k \geq 1$, the (positive) orientation of D induces an orientation on the C^k manifold bD as follows. Let $P \in bD$ and suppose r is a C^k defining function for D on a neighborhood U of P (thus $U \cap D = \{q \in U: r(q) < 0\}$). After shrinking U, we may assume that there is a positively oriented coordinate system φ on U with $\varphi = (r, x_2, \ldots, x_n)$. We then orient $bD \cap U$ by the $(n-1)$-form $dx_2 \wedge \ldots \wedge dx_n$. (More precisely, we consider the pull back $\imath^*(dx_2 \wedge \ldots \wedge dx_n)$, where $\imath: bD \cap U \to U$ is the inclusion map.) This orientation is independent of the defining function r and of the particular positive coordinate system φ chosen, and it can easily be extended globally to bD via partitions of unity (Exercise E.1.11). The orientation thus defined on bD is called the **positive orientation induced from** D. bD shall always be assumed to be oriented in this way. We shall write $-bD$ for the manifold bD with the opposite orientation. Notice that the manifold bD can also be viewed as the boundary of the open set $\mathbb{R}^n - \bar{D}$, and that $b(\mathbb{R}^n - \bar{D}) = -bD$ as oriented manifolds.

In order to define integration on oriented manifolds, we first consider the case of $\mathbb{R}^n$. Suppose $U \subset \mathbb{R}^n$ and $\eta \in C_n^0(U)$ has compact support in U. There is a unique continuous function f on U such that $\eta = f\, dx_1 \wedge \ldots \wedge dx_n$. We define

$$\int_U \eta = \int_U f\, dx_1 \ldots dx_n. \tag{1.23}$$

where the integral on the right is the ordinary multiple Riemann integral in $\mathbb{R}^n$. If $W \subset \mathbb{R}^n$ is open and $F: W \to F(W) = U$ is a C^1 diffeomorphism, the substitution formula for multiple integrals (see [Nar 4], §2.9, for example) gives

$$\int_{F(W)} f(x)\, dx_1 \ldots dx_n = \int_W f(F(t))|\det(dF)_t|\, dt_1 \ldots dt_n. \tag{1.24}$$

If F is orientation preserving, one has

$$\det(dF) = \det \frac{\partial(F_1, \ldots, F_n)}{\partial(t_1, \ldots, t_n)} > 0,$$

and hence the integral on the right in (1.24) is equal to

$$\int_W (f \circ F) \det(dF)_t \, dt_1 \wedge \ldots \wedge dt_n = \int_W (f \circ F) \, dF_1 \wedge \ldots \wedge dF_n$$
$$= \int_W F^*(f \, dx_1 \wedge \ldots \wedge dx_n).$$

Thus

$$\int_{F(W)} \eta = \int_W F^*\eta \tag{1.25}$$

whenever $F\colon W \to F(W)$ preserves orientation.

We now use the transformation formula (1.25) in order to define integrals of n-forms on an oriented n-manifold M. Suppose (U, φ), $\varphi = (x_1, \ldots, x_n)$, is a positive coordinate system. If $\omega \in C_n^0(U)$ has compact support in U, we define

$$\int_U \omega = \int_{\varphi(U)} (\varphi^{-1})^*\omega, \tag{1.26}$$

where the integral on the right (in $\mathbb{R}^n$!) is defined as in (1.23). If (U, ψ), $\psi = (y_1, \ldots, y_n)$, is another positive coordinate system on U, the map $F = \varphi \circ \psi^{-1}\colon \psi(U) \to \varphi(U)$ preserves orientation, and hence, by (1.25) and (1.22),

$$\int_{\varphi(U)} (\varphi^{-1})^*\omega = \int_{\psi(U)} F^*(\varphi^{-1})^*\omega = \int_{\psi(U)} (\varphi^{-1} \circ F)^*\omega = \int_{\psi(U)} (\psi^{-1})^*\omega.$$

Thus the definition (1.26) is independent of the (positive) coordinate system φ.

It is now easy to integrate globally over the oriented manifold M. Choose a countable atlas $\{(U_i, \varphi_i)\}$ for M such that each φ_i is positively oriented. (Note that for arbitrary $\varphi = (x_1, \ldots, x_n)$ either φ or $\tilde{\varphi} = (-x_1, x_2, \ldots, x_n)$ is positive!) Let $\{\chi_i\}$ be a continuous partition of unity such that each χ_i has compact support in U_i. For $\omega \in C_n^0(M)$ with compact support one defines

$$\int_M \omega = \sum_i \int_{U_i} \chi_i \omega_i. \tag{1.27}$$

One can show by standard arguments that (1.27) is independent of the positively oriented atlas and the partition of unity used in the definition (see Exercise E.1.12). Once the integral $\int_M \omega$ has been defined for continuous n-forms with compact support, one can proceed as in $\mathbb{R}^n$ and extend the integral to more general n-forms and consider integration over subsets other than M itself. We shall leave these details to the interested reader.

1.7. Stokes' Theorem

We can now discuss the version of Stokes' Theorem which we shall use later in the applications to integral representations in complex analysis.

Theorem 1.7. *Let M be an oriented C^k manifold, $k \geq 2$, of dimension n, and let $D \subset\subset M$ be an open set with C^1 boundary bD. If $\omega \in C^1_{n-1}(\bar{D})$, then*

$$\int_{bD} \omega = \int_D d\omega. \tag{1.28}$$

Remarks. The hypothesis on ω means that ω is defined on $\bar{D}$ and that the coefficients of ω with respect to any coordinate system (U, φ) are in $C^1(U \cap \bar{D})$, i.e., they have partial derivatives on $U \cap D$ which extend continuously to $U \cap \bar{D}$ (see also E.1.15). The orientation of bD is of course the one induced from D (see Example (iii) above). The integral on the left in (1.28) is, more precisely, $\int_{bD} \iota^* \omega$, where $\iota: bD \to M$ is the inclusion map.

In case $M = \mathbb{R}^n$, Theorem 1.7 is just a reformulation of the classical Gauss–Green Theorem in the language of differential forms. In particular, Theorem 1.7 contains the Fundamental Theorem of Calculus as a special case, as follows. Let $D = (\alpha, \beta)$ be an open, bounded interval in $\mathbb{R}$. Then $bD = \{\alpha, \beta\}$ as set, and $bD = \{\beta\} - \{\alpha\}$ as manifold of dimension 0 with the induced orientation. The $(n-1)$-form ω is now a function $f \in C^1([\alpha, \beta])$, and $df = f'(t)\,dt$. Thus (1.28) says

$$f(\beta) - f(\alpha) = \int_{b(\alpha,\beta)} f = \int_{(\alpha,\beta)} df = \int_\alpha^\beta f'(t)\,dt.$$

As the proof given below shows, this special case is really at the heart of the general version as well.

PROOF. By compactness of $\bar{D}$ there are finitely many positively oriented coordinate systems (U_i, φ_i), $1 \leq i \leq l$, such that $\bar{D} \subset \bigcup_{i=1}^l U_i$, and, if $U_i \cap bD \neq \emptyset$ for some i, then $\varphi_i = (r, \varphi_i')$, where $\varphi_i': U_i \to \mathbb{R}^{n-1}$, and $D \cap U_i \subset \{P \in U_i: -1 < r(P) < 0\}$. Let χ_i be of class C^k, with compact support in U_i, so that $\sum_{i=1}^l \chi_i = 1$ on $\bar{D}$. By linearity, (1.28) will follow once we prove

$$\int_{bD \cap U_i} \chi_i \omega = \int_{D \cap U_i} d(\chi_i \omega) \qquad \text{for } i = 1, \ldots, l. \tag{1.29}$$

Case 1: $U_i \cap bD \neq \emptyset$. We drop the subscript i. If $\varphi = (r, \varphi')$, then $\tilde{\varphi} = \varphi'|_{bD \cap U}$ is a positively oriented coordinate system for bD on $bD \cap U$. We set $(\varphi^{-1})^*(\chi\omega) = \sum_{j=1}^n g_j\,dt_1 \wedge \ldots \wedge [dt_j] \wedge \ldots \wedge dt_n$,[1] with coefficients $g_j \in C^1$ with compact support in $\varphi(U \cap \bar{D})$; then

[1] $[dt_j]$ means that the differential dt_j is omitted.

$$(\tilde{\varphi}^{-1})^* \imath^*(\chi\omega) = g_1(0, t_2, \ldots, t_n)\, dt_2 \wedge \ldots \wedge dt_n.$$

By (1.20),

$$(\varphi^{-1})^* d(\chi\omega) = d[(\varphi^{-1})^* \chi\omega] = \sum_{j=1}^{n} (-1)^{j-1} \frac{\partial g_j}{\partial t_j}\, dt_1 \wedge \ldots \wedge dt_n.$$

Since $\varphi(D \cap U) \subset \{-1 < t_1 < 0\}$, it follows from (1.26), (1.23), and the support condition on $g_1, \ldots, g_n$ that

$$\int_{D \cap U} d(\chi\omega) = \sum_{j=1}^{n} (-1)^{j-1} \int_{\{t \in \mathbb{R}^n : -1 < t_1 < 0\}} \frac{\partial g_j}{\partial t_j}\, dt_1 \ldots dt_n. \tag{1.30}$$

By the Fundamental Theorem of Calculus,

$$\int_{-1}^{0} \frac{\partial g_1}{\partial t_1}\, dt_1 = g_1(0, t_2, \ldots, t_n) - g_1(-1, t_2, \ldots, t_n) = g_1(0, t_2, \ldots, t_n), \tag{1.31}$$

and

$$\int_{\mathbb{R}} \frac{\partial g_j}{\partial t_j}\, dt_j = 0 \qquad \text{for } j \geq 2, \tag{1.32}$$

since g_j has compact support in t_j for $j \geq 2$. Therefore,

$$\int_{D \cap U} d(\chi\omega) = \int_{\mathbb{R}^{n-1}} g_1(0, t_2, \ldots, t_n)\, dt_2 \ldots dt_n = \int_{\tilde{\varphi}(bD \cap U)} (\tilde{\varphi}^{-1})^* \imath^*(\chi\omega)$$
$$= \int_{bD \cap U} \imath^*(\chi\omega).$$

Case 2: $U_i \cap bD = \varnothing$. We may assume that $U_i \subset D$. Since $\int_{bD \cap U_i} \chi_i\omega = 0$, (1.29) holds if $\int_{U_i} d(\chi_i\omega) = 0$. But this follows from the computations in Case 1 by replacing the region of integration on the right side of (1.30) with $\mathbb{R}^n$, and using the fact that now (1.32) holds for $j = 1$ as well. ■

In case M is compact, we may take $D = M$ in Theorem 1.7, in which case $bD = \varnothing$. One thus obtains:

Corollary 1.8. *Let M be a compact, oriented C^k manifold of dimension n. If $\omega \in C^1_{n-1}(M)$, then*

$$\int_M d\omega = 0.$$

Stokes' Theorem can easily be extended to regions $D \subset\subset M$ with **piecewise C^1 boundary**, which are defined as follows: *there are a finite covering $\{U_1, \ldots, U_l\}$ of bD and functions $r_i \in C^1(U_i)$ such that*

$$D \cap \left(\bigcup_{i=1}^{l} U_i\right) = \left\{x \in \bigcup_{i=1}^{l} U_i : r_i(x) < 0 \text{ for all } i \text{ with } x \in U_i\right\}, \tag{1.33}$$

and for every subset $\{i_1, \ldots, i_\nu\}$ *of* $\{1, \ldots, l\}$ *one has*

(1.34) $$dr_{i_1} \wedge \ldots \wedge dr_{i_\nu} \neq 0 \qquad \text{on } U_{i_1} \cap \ldots \cap U_{i_\nu}.$$

The collection $\{r_i \in C^1(U_i): i = 1, \ldots, l\}$ is called a **frame for** D. Notice that (1.34) implies that $U_{i_1} \cap \ldots \cap U_{i_\nu} = \emptyset$ for $\nu > \dim M$. Moreover, each set $\sum_i = \{x \in U_i: r_i(x) = 0\}$ is a C^1 submanifold of U_i. Condition (1.34) is often referred to by saying that the manifolds $\sum_i$, $1 \leq i \leq l$, **intersect transversally**, or that they are in **general position.** If $S_i = bD \cap \sum_i$, then $bD = \bigcup_{i=1}^{l} S_i$. The interior $\mathring{S}_i$ of S_i in $\sum_i$ is a C^1 manifold which carries the orientation induced from D. (The reader may verify that $\mathring{S}_i$ is itself an open subset of $\sum_i$ with piecewise C^1 boundary.) For $\omega \in C^1_{n-1}(\bar{D})$ one defines

$$\int_{bD} \omega = \sum_{i=1}^{l} \int_{\mathring{S}_i} \omega.$$

By suitably modifying the proof of Theorem 1.7 one shows that Stokes' Theorem remains valid, that is

$$\int_{bD} \omega = \int_D d\omega.$$

1.8. Product Manifolds

Let M_1 and M_2 be C^k manifolds of dimensions n_1 and n_2, respectively. The product space $M = M_1 \times M_2$ inherits a natural C^k manifold structure of dimension $n_1 + n_2$, as follows: if $\mathscr{A}_\nu = \{(U_i^{(\nu)}, \varphi_i^{(\nu)}): i \in I_\nu\}$ is a C^k atlas for M_ν, $\nu = 1, 2$, an atlas for M is given by

$$\{(U_{i_1}^{(1)} \times U_{i_2}^{(2)}, (\varphi_{i_1}^{(1)}, \varphi_{i_2}^{(2)})): \qquad (i_1, i_2) \in I_1 \times I_2\},$$

where $(\varphi_{i_1}^{(1)}, \varphi_{i_2}^{(2)})(P_1, P_2)$ is defined by $(\varphi_{i_1}^{(1)}(P_1), \varphi_{i_2}^{(2)}(P_2)) \in \mathbb{R}^{n_1} \times \mathbb{R}^{n_2} = \mathbb{R}^{n_1+n_2}$. The projections $\pi_\nu: M_1 \times M_2 \to M_\nu$, $\nu = 1, 2$, are C^k maps.

Assume now that M_ν is orientable, with orientation defined by $\Omega_\nu \in C^0_{n_\nu}(M_\nu)$, $\nu = 1, 2$; then $M_1 \times M_2$ is orientable as well, with the natural **product orientation** defined by $(\pi_1^*\Omega_1) \wedge (\pi_2^*\Omega_2)$.

Let $D_\nu \subset\subset M_\nu$ be open sets with C^1 boundary. (We allow $D_\nu = M_\nu$ if M_ν is compact.) Then $D_1 \times D_2 \subset\subset M_1 \times M_2$ is open with piecewise C^1 boundary (see Exercise E.1.14). As usual, bD_ν carries the orientation induced from D_ν. The manifolds $bD_1 \times D_2$ and $D_1 \times bD_2$ make up the essential part of the boundary $b(D_1 \times D_2)$ of $D_1 \times D_2$, and the orientations induced on them from $D_1 \times D_2$ correspond to $bD_1 \times D_2$ and $(-1)^{n_1} D_1 \times bD_2$, respectively. Stokes' Theorem thus takes the form

(1.35) $$\int_{D_1 \times D_2} d\omega = \int_{bD_1 \times D_2} \omega + (-1)^{n_1} \int_{D_1 \times bD_2} \omega$$

for every $\omega \in C^1_{n_1+n_2-1}(\overline{D_1 \times D_2})$. In particular, if M_1 is compact, one obtains

$$\int_{M_1 \times D_2} d\omega = (-1)^{\dim M_1} \int_{M_1 \times bD_2} \omega. \tag{1.36}$$

1.9. Double Differential Forms

Recall from §1.3 that an r-form $\omega_P \in \Lambda^r(T_P^* M)$ is an alternating r-multilinear map $T_P M \times \cdots \times T_P M \to \mathbb{R}$. By analogy, if A is an $\mathbb{R}$-algebra, one may consider alternating r-multilinear maps $T_P M \times \cdots \times T_P M \to A$; such maps are called **A-valued r-forms at P**, or **r-forms with coefficients in A**. As in the case of $\mathbb{R}$-valued forms, one can then define the Grassman algebra of A-valued forms on the manifold M.

We shall need this concept in the special case where $A = \mathscr{G}^0(N)$ is the Grassman algebra of continuous differential forms on another manifold N. Differential forms on M with values in $\mathscr{G}^0(N)$ are also called **double differential forms** on $M \times N$, for reasons which will become clear shortly. Let $(U, (x_1, \ldots, x_m))$ and $(V, (y_1, \ldots, y_n))$ be local coordinate systems for M and N, respectively. A double form ω on $U \times N$ of **bidegree** (r, s) can be written as $\omega_x = \sum_J a_J(x)\, dx^J$ for $x \in U$, where the summation is over all strictly increasing r-tuples $J = (j_1, \ldots, j_r) \subset \{1, \ldots, m\}$, and where

$$a_J(x) = \omega_x\left(\frac{\partial}{\partial x_{j_1}}, \ldots, \frac{\partial}{\partial x_{j_r}}\right)$$

is an s-form on N for each $x \in U$.

On the neighborhood V we have $[a_J(x)]_y = \sum_K a_{JK}(x, y)\, dy^K$, the summation being over all strictly increasing s-tuples $K = (k_1, \ldots, k_s) \subset \{1, \ldots, n\}$, and $a_{JK}(x, y) \in \mathbb{R}$. Therefore ω is represented on $U \times V$ by

$$\omega_{(x,y)} = \sum_{J,K} (a_{JK}(x, y)\, dy^K)\, dx^J. \tag{1.37}$$

One says that ω is of class C^l on $U \times V$, if $a_{JK} \in C^l(U \times V)$ for all J, K. This definition is invariant under coordinate changes if M and N are of class $\geq l + 1$.

It is convenient to introduce the wedge product $dy_k \wedge dx_j$ between differentials on N and on M, with the convention that

$$dy_k \wedge dx_j = dx_j \wedge dy_k \qquad \text{for } j, k. \tag{1.38}$$

We shall write $dy^K\, dx^J$ in (1.37) as $dy^K \wedge dx^J = dy_{j_1} \wedge \ldots \wedge dy_{j_s} \wedge dx_{i_1} \wedge \ldots \wedge dx_{i_r}$. Thus $dy^K \wedge dx^J = dx^J \wedge dy^K$, and we may write (1.37) as

$$\omega = \sum_{J,K} a_{JK}\, dx^J \wedge dy^K.$$

One must be careful to distinguish double forms of bidegree (r, s) from $(r + s)$-forms on the manifold $M \times N$.

It is now obvious that a double form on $M \times N$ can also be viewed as a differential form on N with values in the differential forms on M.

The various operations on differential forms defined on M and on N can be applied to double forms in a straightforward manner. For example, if $F: M_1 \to M_2$ is a C^1 map, the pullback $F^*\omega$ of a double form ω on $M_2 \times N$ of bidegree (r, s) is a double form on $M_1 \times N$ of bidegree (r, s); the exterior derivative d_x $(x \in M)$ on M, respectively d_y $(y \in N)$ on N, sends double forms of bidegree (r, s) to forms of bidegree $(r + 1, s)$, respectively $(r, s + 1)$, and F^* commutes with d_x and d_y. Also, the wedge product between two arbitrary double forms is well defined, keeping in mind the commutativity relations (1.38). If the bidegrees of ω and ω' are (r, s), respectively (r', s'), then $\omega \wedge \omega'$ has bidegree $(r + r', s + s')$, and one has

$$\omega \wedge \omega' = (-1)^{rr' + ss'} \omega' \wedge \omega. \tag{1.39}$$

If M is oriented, a double form on $M \times N$ can be integrated over suitable subsets of M, the result being a differential form on N. In particular, Stokes' Theorem remains valid in the following form: *if $D \subset\subset M$ has piecewise C^1 boundary and ω is a double form on $\bar{D} \times N$ of class C^1 and bidegree $(m - 1, s)$, $m = \dim M$, then*

$$\int_{bD} \omega = \int_D d_x \omega, \tag{1.40}$$

where both sides of (1.40) *are now s-forms on N.*

Remark. Since ω is of class C^1, it follows by standard calculus techniques that the integral $\int_{bD} \omega$ is of class C^1 on N. Equation (1.40) then implies that the integral on the right is of class C^1 as well, something which could not easily be proved directly, as the coefficients of $d_x\omega$ are not necessarily of class C^1.

Exercises

E.1.1. Let M_1 be the manifold $\mathbb{R}$ with atlas $\{(\mathbb{R}, id)\}$ and let M_2 be $\mathbb{R}$ with the atlas $\{(\mathbb{R}, \varphi)\}$, where $\varphi(x) = x^{1/3}$. Show that M_1 and M_2 are C^∞ manifolds which are diffeomorphic, but that the two C^∞ structures defined on $\mathbb{R}$ are not identical, i.e., the identity map $id: M_1 \to M_2$ is not a diffeomorphism.

E.1.2. Let M be an n-dimensional manifold of class C^k. A subset $N \subset M$ is called a **submanifold of** M if for every $P \in N$ there is a coordinate system (U, φ) for M with $P \in U$, such that $\varphi(P) = 0$ and if $\varphi = (x_1, \ldots, x_n)$, then $U \cap N = \{q \in U: x_1(q) = \cdots = x_m(q) = 0\}$ for some $m \in \mathbb{N}$ with $0 \leq m \leq n$.

(i) Show that if N is a submanifold of M, then N carries a natural structure of a C^k manifold such that the inclusion $\iota: N \to M$ is a *regular* C^k map (i.e., the differential $(d\iota)_P$ is injective for every $P \in N$.

(ii) Let $D \subset \mathbb{R}^n$ be open and suppose $F: D \to \mathbb{R}^m$ is a map of class C^k, such that its differential dF_P has maximal rank at every point $P \in D$. Show that for fixed $P \in D$, the level set

$$L_P(F) = \{x \in D: F(x) = F(P)\}$$

is a submanifold of class C^k of D of dimension $\max(0, n - m)$.

E.1.3. Suppose $F: M \to N$ is a C^1 map between two C^1 manifolds. Verify that for $v \in T_P M$ the definition $dF_P(v) = [F \circ \alpha] \in T_{F(P)}N$ is independent of the curve α chosen to represent v.

E.1.4. Let $F: M \to N$ and $G: N \to W$ be C^1 maps between C^1 manifolds M, N, and W.

(i) Prove the "chain rule"

$$d(G \circ F)_P = (dG)_{F(P)} \circ (dF)_P$$

for all $P \in M$.

(ii) By specializing (i) to $M = \mathbb{R}^{n_1}$, $N = \mathbb{R}^{n_2}$, and $W = \mathbb{R}^{n_3}$ and using the standard coordinates, obtain the usual chain rule in matrix form.

(iii) Have you obtained a "new" trivial proof of the chain rule from calculus? Explain!

E.1.5. Let t be the standard coordinate for $\mathbb{R}$, and suppose $f: M \to \mathbb{R}$ is a C^1 function on the C^1 manifold M. Show that for all $v \in T_P M$, $(df)_P(v) = v(f)\partial/\partial t$.

E.1.6. (i) Prove that the wedge product defined by (1.10) is associative.
(ii) Prove (1.11).

E.1.7. Let M be a C^k manifold. Prove that a vector field V on M is of class $C^l (l < k)$, as defined in §1.3, if and only if Vf is of class C^l for every $f \in C^{l+1}(M)$.

E.1.8. Suppose $\dim M = n$ and $\omega_1, \ldots, \omega_n$ is a basis for $\Lambda^1 T_P^* M$. Show that if $\eta_j = \sum_{i=1}^n a_{ij}\omega_i$, $1 \le j \le n$, with $a_{ij} \in \mathbb{R}$, then

$$\eta_1 \wedge \ldots \wedge \eta_n = \det([a_{ij}])\omega_1 \wedge \ldots \wedge \omega_n.$$

E.1.9. Fill in the missing parts in the proof of Theorem 1.3.

E.1.10. (i) Prove that $F_P^*: \mathscr{G}_{F(P)}(N) \to \mathscr{G}_P(M)$ defined in (1.19) defines an $\mathbb{R}$-algebra homomorphism.
(ii) If $G: W \to M$ and $F: M \to N$ are C^1 maps, show that

$$(F \circ G)^* = (G)^* \circ (F^*)$$

on $\mathscr{G}^0(N)$.

E.1.11. (i) Let $D \subset \mathbb{R}^n$ be a domain with C^1 boundary. Prove in detail that bD is (globally) orientable.
(ii) Show that a C^1 manifold M is orientable if and only if there is an atlas $\{(U_i, \varphi_i)\}$ for M, such that all C^1 maps $\varphi_j \circ \varphi_i^{-1}$ preserve the natural orientation of $\mathbb{R}^n$.

E.1.12. Prove in detail that the definition of $\int_M \omega$ in (1.27) is independent of the covering $\{U_i\}$ and the choice of partition of unity $\{\chi_i\}$.

E.1.13. Fill in the details in the proof of Stokes' Theorem for domains with piecewise differentiable boundary (cf. the end of §1.7).

E.1.14. Suppose $D_1 \subset \mathbb{R}^n$ and $D_2 \subset \mathbb{R}^m$ are regions with C^1 boundary.

(i) Show that $D_1 \times D_2$ is a region with piecewise C^1 boundary.
(ii) Generalize (i) to the case where D_1 and D_2 have piecewise C^1 boundary.

E.1.15. Let M be an n-dimensional manifold of class C^k and suppose $D \subset\subset M$ is an open set with C^k boundary. Show that every $f \in C^k(\bar{D})$ has an extension $\tilde{f} \in C^k(M)$ (i.e., $\tilde{f}(x) = f(x)$ for $x \in \bar{D}$) by proving the following steps.

(i) Use partitions of unity and special local coordinate systems to reduce the general case to the following special case: Let $D = \{x \in \mathbb{R}^n : |x| < 1 \text{ and } x_n < 0\}$; show that every $f \in C^k(\bar{D})$ whose support is a compact subset of $B = \{x \in \mathbb{R}^n : |x| < 1\}$ has an extension $\tilde{f} \in C_0^k(B)$.
(ii) Show that the $(k + 1) \times (k + 1)$ matrix (a_{ij}), with $a_{ij} = (-j)^i$ for $1 \le j \le k + 1$ and $0 \le i \le k$ is nonsingular. (Hint: this is the Vandermonde matrix of $-1, -2, \ldots, -(k + 1)$.)
(iii) By (ii) there are numbers $c_j \in \mathbb{R}$, $1 \le j \le k + 1$, such that $\sum_{j=1}^{k+1} c_j(-j)^i = 1$ for $0 \le i \le k$. Define

$$(E_k f)(x', x_n) = \begin{cases} f(x', x_n) & \text{if } x_n \le 0 \\ \sum_{j=1}^{k+1} c_j f(x', -jx_n) & \text{if } x_n > 0 \end{cases}$$

for $x = (x', x_n) \in B$. Verify that $\tilde{f} = E_k f$ is the desired extension in the special case stated in (i).

Remark. Notice that the operator E_k in (iii) is linear, and that $|E_k f|_{k,B} \lesssim |f|_{k,D}$. By doing step (i) carefully, one therefore can show that also in the general case the extension can be defined by a linear operator which is bounded in C^k norm.

§2. Complex Structures

In this section we consider some of the additional features of the calculus of differential forms on a manifold M which arise in the presence of a complex structure. Even though later on we shall primarily deal with the case $M = \mathbb{C}^n$, it is instructive to consider the related abstract concept of a **complex manifold**. Besides $\mathbb{C}^n$, other examples of such manifolds already familiar to us are the complex submanifolds of $\mathbb{C}^n$, discussed in I.§2.5, and Stein manifolds, which were briefly mentioned in II.§3.7. (See also E.2.3.)

2.1. Complex Manifolds

The following definition is analogous to the one of differentiable manifold in §1.1. A **complex manifold** M **of dimension** n is a topological Hausdorff space M together with a **complex atlas** $\mathscr{A} = \{(U_i, \varphi_i) : i \in I\}$ of pairs (U_i, φ_i) consisting of open subsets U_i of M and homeomorphisms φ_i of U_i onto open subsets of $\mathbb{C}^n$ such that $M = \bigcup_{i \in I} U_i$, and for all $i, j \in I$ with $U_i \cap U_j \neq \emptyset$, $\varphi_j \circ \varphi_i^{-1}$:

$\varphi_i(U_i \cap U_j) \to \varphi_j(U_i \cap U_j)$ is a biholomorphic map between open subsets of $\mathbb{C}^n$. $\mathscr{A}$ is also called an **atlas for the complex structure of** M.

Holomorphic functions and holomorphic maps between complex manifolds are defined as in case of C^k manifolds in terms of the coordinate systems in an atlas. We denote by $\mathcal{O}(M)$ the space of holomorphic functions on M. A biholomorphic map $\varphi: U \to \varphi(U) \subset \mathbb{C}^n$ from an open subset U of a complex manifold onto an open set in $\mathbb{C}^n$ is called a **holomorphic** (or **complex**) **coordinate system**.

It is clear that a complex manifold M of dimension n carries a natural C^∞ manifold structure of dimension $2n$; any holomorphic coordinate system $\varphi = (z_1, \ldots, z_n)$ induces a C^∞-coordinate system $(x_1, y_1, \ldots, x_n, y_n)$, where $z_j = x_j + iy_j$ for $1 \le j \le n$. Therefore, the topics discussed in §1 apply to complex manifolds. Moreover, it is natural to now consider *complex valued* functions and differential forms of M. (This really has nothing to do with the *complex* structure of M, and it could be done on any differentiable manifold.) A tangent vector $v \in T_P M$ acts on complex valued functions by setting $v(f) = v(\operatorname{Re} f) + iv(\operatorname{Im} f)$, where $f = \operatorname{Re} f + i \operatorname{Im} f$ is the decomposition of f into real and imaginary part. A complex valued r-form ω at P is an alternating r-multilinear map: $T_P M \times \cdots \times T_P M \to \mathbb{C}$. Any such ω has a unique decomposition $\omega = \operatorname{Re} \omega + i \operatorname{Im} \omega$, with $\operatorname{Re} \omega$, $\operatorname{Im} \omega$ real valued r-forms at P. From now on, $C_r^k(U)$ will denote the space of *complex valued* r-forms of class C^k on U, and $\mathscr{G}^k(U) = \bigoplus_{r \ge 0} C_r^k(U)$.

For greater clarity we shall denote the space of complex valued 1-forms at $P \in M$, i.e., $\mathbb{R}$-linear maps $T_P M \to \mathbb{C}$, by $\mathbb{C}T_P^* M$, in order to distinguish it from the space of real valued 1-forms $T_P^* M$.[1] $\mathbb{C}T_P^* M$ is a vector space of complex dimension $2n$ over $\mathbb{C}$. Correspondingly, we consider the complexification $\mathbb{C}T_P M$ of $T_P M$; in terms of the action of tangent vectors on functions, this simply means that we define λv for $\lambda \in \mathbb{C}$ and $v \in T_P M$ by $(\lambda v)(f) = \lambda[v(f)]$. Every $v \in \mathbb{C}T_P M$ has a unique representation $v = v_1 + iv_2$, with $v_1, v_2 \in T_P M$. The conjugation operator in $\mathbb{C}T_P M$ is defined by $\overline{v_1 + iv_2} = v_1 - iv_2$. $\mathbb{C}T_P M$ is a complex vector space of dimension $2n$. A complex valued 1-form ω: $T_P M \to \mathbb{C}$ extends in the obvious way to a $\mathbb{C}$-linear map $\omega^{\mathbb{C}}: \mathbb{C}T_P M \to \mathbb{C}$ by setting $\omega^{\mathbb{C}}(v_1 + iv_2) = \omega(v_1) + i\omega(v_2)$. In the following, we shall extend all 1-forms in this way and write again ω instead of $\omega^{\mathbb{C}}$. One easily checks that $\mathbb{C}T_P^* M$ and $\mathbb{C}T_P M$ are now naturally dual to each other as $\mathbb{C}$-vector spaces.

In terms of a holomorphic coordinate system $(z_1, \ldots, z_n)$ near P, with $z_j = x_j + iy_j$, one has the basis $\{dx_1, dy_1, \ldots, dx_n, dy_n\}$ for $\mathbb{C}T_P^* M$ over $\mathbb{C}$ and the corresponding dual basis

$$\left\{\frac{\partial}{\partial x_1}, \frac{\partial}{\partial y_1}, \ldots, \frac{\partial}{\partial x_n}, \frac{\partial}{\partial y_n}\right\}$$

for $\mathbb{C}T_P M$. More useful for *complex* analysis are the basis $\{dz_1, d\bar{z}_1, \ldots,$

[1] $\mathbb{C}T_P^* M$ can be thought of as the complexification of the $\mathbb{R}$-vector space $T_P^* M$, defined for an arbitrary $\mathbb{R}$-vector space V by the tensor product $\mathbb{C} \otimes_{\mathbb{R}} V$.

$dz_n, d\bar{z}_n\}$ for $\mathbb{C}T_P^*M$ and the corresponding dual basis for $\mathbb{C}T_PM$, denoted by

$$\left\{\frac{\partial}{\partial z_1}, \frac{\partial}{\partial \bar{z}_1}, \ldots, \frac{\partial}{\partial z_n}, \frac{\partial}{\partial \bar{z}_n}\right\};$$

one has

$$dz_j(\partial/\partial z_k) = \delta_{jk} \qquad \text{and } d\bar{z}_j(\partial/\partial z_k) = 0$$

for $1 \leq j, k \leq n$, and analogous formulas hold for $\partial/\partial \bar{z}_k$. Depending on the choice of basis, the differential df_P of a C^1 function f at P can be represented either by

$$df_P = \sum_{j=1}^n \frac{\partial f}{\partial x_j}(P)\, dx_j + \sum_{j=1}^n \frac{\partial f}{\partial y_j}(P)\, dy_j \tag{2.1}$$

or by

$$df_P = \sum_{j=1}^n \frac{\partial f}{\partial z_j}(P)\, dz_j + \sum_{j=1}^n \frac{\partial f}{\partial \bar{z}_j}(P)\, d\bar{z}_j. \tag{2.2}$$

Notice that (2.2) is just a formal consequence of the definition of dual basis. A simple computation now gives

$$\frac{\partial}{\partial z_j} = \frac{1}{2}\left(\frac{\partial}{\partial x_j} + \frac{1}{i}\frac{\partial}{\partial y_j}\right); \qquad \frac{\partial}{\partial \bar{z}_j} = \frac{1}{2}\left(\frac{\partial}{\partial x_j} - \frac{1}{i}\frac{\partial}{\partial y_j}\right) \tag{2.3}$$

for $1 \leq j \leq n$; that is, we recover the expressions which, in I.§1.2, were taken as the definition of the differential operators $\partial/\partial z_j$ and $\partial/\partial \bar{z}_j$.

2.2. The Complex Structure on T_PM

In the preceding section we associated to T_PM the complex vector space $\mathbb{C}T_PM$. This can be done for any differentiable manifold, and it does not mean that we have made T_PM into a complex vector space. We shall now show that for a *complex* manifold M, T_PM itself carries a natural structure of a complex vector space.

Let us first consider the case $M = \mathbb{C}^n$. As a C^∞ manifold, $\mathbb{C}^n = \mathbb{R}^{2n}$ in a natural way. Thus $T_P\mathbb{C}^n = T_P\mathbb{R}^{2n} = \mathbb{R}^{2n}$, and the latter can again be identified with $\mathbb{C}^n$. The resulting identification of $T_P\mathbb{C}^n$ with $\mathbb{C}^n$ defines a complex vector space structure on $T_P\mathbb{C}^n$. For greater clarity we shall denote the corresponding operator of multiplication by i by the symbol J. $J\colon T_P\mathbb{C}^n \to T_P\mathbb{C}^n$ is an $\mathbb{R}$-linear map which satisfies $J^2 = -$identity; in terms of the standard basis

$$\left\{\frac{\partial}{\partial x_1}, \frac{\partial}{\partial y_1}, \ldots, \frac{\partial}{\partial x_n}, \frac{\partial}{\partial y_n}\right\}$$

of $T_P\mathbb{C}^n = T_P\mathbb{R}^{2n}$ one has

$$J(\partial/\partial x_j) = \partial/\partial y_j, \qquad J(\partial/\partial y_j) = -\partial/\partial x_j,\ 1 \leq j \leq n. \tag{2.4}$$

The scalar product of $a + ib \in \mathbb{C}$ and $v \in T_P\mathbb{C}^n$ is then given by $(a + ib)v = av + bJ(v)$. With this notation, Theorem I.1.2 states that a C^1 function f satisfies the Cauchy–Riemann equations at P if and only if $df_P(Jv) = i\,df_P(v)$ for $v \in T_P\mathbb{C}^n$. (At this point the reader may wish to look over Section I.1.2 again.) This relationship between the complex structure of $T_P\mathbb{C}^n$ and holomorphic functions implies that the operator J is independent of the Euclidean coordinates of $\mathbb{C}^n$, and thus J can be introduced on complex manifolds. The following theorem makes this precise.

Theorem 2.1. *Let M be a complex manifold. For every $P \in M$ there is a unique $\mathbb{R}$-linear map $J = J_P\colon T_PM \to T_PM$ such that for all functions f holomorphic at P one has*

$$df_P(Jv) = i\,df_P(v) \qquad \text{for all } v \in T_PM. \tag{2.5}$$

Moreover, $J^2 = -\text{identity}$, and the definition $(a + ib)v = av + b(Jv)$ for $a + ib \in \mathbb{C}$ and $v \in T_PM$ turns T_PM into a vector space over $\mathbb{C}$.

Proof. We first prove uniqueness. Let $(z_1, \ldots, z_n)$ be a holomorphic coordinate system near P with underlying real coordinates $(x_1, y_1, \ldots, x_n, y_n)$; then $dx_j = \operatorname{Re} dz_j$ and $dy_j = \operatorname{Im} dz_j$ for $1 \le j \le n$. We shall determine $J(\partial/\partial x_k)$ by computing the coefficients in the representation

$$J(\partial/\partial x_k) = \sum_{j=1}^n \left[dx_j\left(J\frac{\partial}{\partial x_k}\right)\right]\frac{\partial}{\partial x_j} + \sum_{j=1}^n \left[dy_j\left(J\frac{\partial}{\partial x_k}\right)\right]\frac{\partial}{\partial y_j}. \tag{2.6}$$

Since z_j is holomorphic at P and $dz_j(\partial/\partial x_k) = \delta_{jk}$, (2.5) implies

$$dx_j\left(J\frac{\partial}{\partial x_k}\right) = \operatorname{Re} dz_j\left(J\frac{\partial}{\partial x_k}\right) = \operatorname{Re} i\,dz_j\left(\frac{\partial}{\partial x_k}\right) = \operatorname{Re} i\delta_{jk} = 0,$$

and

$$dy_j\left(J\frac{\partial}{\partial x_k}\right) = \operatorname{Im} dz_j\left(J\frac{\partial}{\partial x_k}\right) = \operatorname{Im} i\,dz_j\left(\frac{\partial}{\partial x_k}\right) = \delta_{jk}.$$

Therefore $J(\partial/\partial x_k) = \partial/\partial y_k$. A similar computation gives $J(\partial/\partial y_k) = -\partial/\partial x_k$. Thus J is determined uniquely by (2.5).

To prove the existence of J, one defines J by (2.4) with respect to some holomorphic coordinates. By what we had seen in $\mathbb{C}^n$, J has all the required properties. The uniqueness statement implies that this definition is independent of the choice of coordinates. ∎

Remark. Equation (2.5) may be written as $(Jv)(f) = iv(f)$ for all $v \in T_PM$. Separating real and imaginary parts, this is equivalent to

$$v(\operatorname{Re} f) = (Jv)(\operatorname{Im} f) \qquad \text{for all } v \in T_PM. \tag{2.7}$$

The reader will recognize (2.7) as a coordinate-free formulation of the classical Cauchy–Riemann equations.

We now establish the connection between the complex structures of $T_P M$ and $\mathbb{C}T_P M$. Since $J^2 = -\text{identity}$, the eigenvalues of J are i and $-i$. Therefore, in order to diagonalize J, one is forced to consider the natural extension of J to a $\mathbb{C}$-linear map $J: \mathbb{C}T_P M \to \mathbb{C}T_P M$. Denote by $T_P^{1,0}M$ and $T_P^{0,1}M$ the eigenspaces of J corresponding to the eigenvalues $+i$ and $-i$, respectively. Then $T_P^{0,1}M = \overline{T_P^{1,0}M}$, and

$$\mathbb{C}T_P M = T_P^{1,0}M \oplus T_P^{0,1}M. \tag{2.8}$$

One easily checks that for a holomorphic coordinate system $(z_1, \ldots, z_n)$ near P the vectors $\partial/\partial z_1, \ldots, \partial/\partial z_n$ at P form a basis for $T_P^{1,0}M$, Since $Jw = iw$ if and only if $w \in T_P^{1,0}M$, it is exactly on the space $T_P^{1,0}M$ that the intrinsic complex structure operator J of $T_P M$ can be realized by simple scalar multiplication. $T_P^{1,0}M$ is called the space of **(tangent) vectors of type** $(1, 0)$. Correspondingly, $T_P^{0,1}M$ is called the space of **vectors of type** $(0, 1)$.

Remark. The complex vector spaces $T_P M$ (with its natural complex structure defined by J) and $T_P^{1,0}M$ are naturally isomorphic via the map given by $\partial/\partial x_j \mapsto \partial/\partial z_j$ for $1 \le j \le n$. For a *holomorphic* function f one has

$$\left(\frac{\partial}{\partial z_j}\right) f = \frac{1}{2}\left(\frac{\partial}{\partial x_j} + \frac{1}{i}\frac{\partial}{\partial y_j}\right) f = \frac{\partial f}{\partial x_j},$$

which equals the complex derivative of f with respect to z_j; therefore, the $\mathbb{C}$-vector spaces $T_P M$, $T_P^{1,0}M$, and the space of "complex derivations on the ring of holomorphic functions at P" are often identified without specific mentioning, something which can be a source of confusion for the non-expert. The reader interested in more details should consult Chapter 0 in [Wu].

2.3. Forms on Type (p, q)

Let M be a complex manifold of dimension n. The intrinsic complex structure of $T_P M$ induces a natural decomposition of $\mathbb{C}T_P^* M$, as follows. Define the space of 1-forms of type $(1, 0)$ at P by

$$\Lambda^{1,0}(T_P^* M) = \{\omega \in \mathbb{C}T_P^* M : \omega(Jv) = i\omega(v) \qquad \text{for all } v \in \mathbb{C}T_P M\}.$$

Notice that every 1-form ω at P is $\mathbb{C}$-linear as a map $\mathbb{C}T_P M \to \mathbb{C}$, but ω is $\mathbb{C}$-linear on $T_P M$ (with complex structure defined by J) precisely when ω is of type $(1, 0)$. The differentials of *holomorphic* functions are of type $(1, 0)$, and if $(z_1, \ldots, z_n)$ are holomorphic coordinates near P, $\{dz_1, \ldots, dz_n\}$ defines a basis for $\Lambda^{1,0}(T_P^* M)$. The conjugate space $\overline{\Lambda^{1,0}(T_P^* M)}$, with basis $\{d\bar{z}_1, \ldots, d\bar{z}_n\}$, is denoted by $\Lambda^{0,1}(T_P^* M)$ and is called the space of forms of type $(0, 1)$ at P. One has the direct sum decomposition

$$\mathbb{C}T_P^* M = \Lambda^{1,0}(T_P^* M) \oplus \Lambda^{0,1}(T_P^* M). \tag{2.9}$$

It is straightforward to extend (2.9) to forms of higher degree. Any complex valued r-form ω at P is a linear combination of r-forms $\omega_1 \wedge \ldots \wedge \omega_r$, with $\omega_j \in \mathbb{C}T_P^* M$. According to (2.9) one can write $\omega_j = \omega_j' + \omega_j''$, with $\omega_j' \in \Lambda^{1,0}$ and $\omega_j'' \in \Lambda^{0,1}$, and it follows that ω can be written as a linear combination of terms $\eta_1 \wedge \ldots \wedge \eta_r$, where η_j is either of type (1, 0) or of type (0, 1) for $1 \le j \le r$. The r-form ω is said to be of **type** (p, q), $p + q = r$, if ω can be written as a linear combination of r-forms $\omega_{i_1} \wedge \ldots \wedge \omega_{i_p} \wedge \bar{\omega}_{j_1} \wedge \ldots \wedge \bar{\omega}_{j_q}$, where all ω_ν are of type (1, 0) (and thus the last q factors are of type (0, 1)). We denote the space of forms at P of type (p, q) by $\Lambda_P^{p,q}$, and we shall denote by $C_{p,q}^k(M)$ the subspace of $C_{p+q}^k(M)$ consisting of those forms which are of type (p, q) at every point. It follows that

$$(2.10) \qquad C_r^k(M) = \bigoplus_{p+q=r} C_{p,q}^k(M)$$

and

$$(2.11) \qquad \mathscr{G}^k(M) = \bigoplus_{p,q \ge 0} C_{p,q}^k(M).$$

Notice that $C_{p,q}^k(M) = \{0\}$ if p or $q > n = \dim_{\mathbb{C}} M$.

If $(z_1, \ldots, z_n)$ are holomorphic coordinates on U, then $dz_j \in C_{1,0}^\infty(U)$ for $1 \le j \le n$; and a (p, q)-form $\omega \in C_{(p,q)}^k(U)$ has a unique representation

$$(2.12) \qquad \omega = \sum_{\substack{1 \le i_1 < \ldots < i_p \le n \\ 1 \le j_1 < \ldots < j_q \le n}} a_{i_1 \cdots i_p j_1 \cdots j_q} \, dz_{i_1} \wedge \ldots \wedge dz_{i_p} \wedge d\bar{z}_{j_1} \wedge \ldots \wedge d\bar{z}_{j_q}$$

with coefficients $a_{i_1 \ldots i_p j_i \ldots j_q} \in C^k(U)$. We shall use the more compact notation

$$(2.13) \qquad \omega = \sum_{I,J} a_{IJ} \, dz^I \wedge d\bar{z}^J$$

for (2.12), where it is understood that the summation is over *all strictly increasing* p-tuples I and q-tuples J in $\{1, \ldots, n\}$.

2.4. The $\bar{\partial}$-Operator

According to (2.2) or (2.9) the differential df of a function $f \in C^1(U)$ has a unique decomposition

$$(2.14) \qquad df = \partial f + \bar{\partial} f$$

where ∂f is of type (1, 0) and $\bar{\partial} f$ is of type (0, 1). This decomposition $d = \partial + \bar{\partial}$ of the exterior derivative on functions generalizes to arbitrary forms. First we observe that

$$(2.15) \qquad dC_{p,q}^1(M) \subset C_{p+1,q}^0(M) \oplus C_{p,q+1}^0(M).$$

Equation (2.15) is a local statement: if $\omega \in C_{p,q}^1(M)$ is given locally by (2.13), it follows that

$$d\omega = \sum_{I,J} da_{IJ} \wedge dz^I \wedge d\bar{z}^J = \sum (\partial a_{IJ} + \bar{\partial} a_{IJ}) \wedge dz^I \wedge d\bar{z}^J,$$

which clearly proves (2.15). One now defines $\partial\omega$ and $\bar\partial\omega$ as the components of $d\omega$ of type $(p+1, q)$ and $(p, q+1)$, respectively. In local coordinates one has

$$(2.16)\quad \partial\omega = \sum_{I,J} \partial a_{IJ} \wedge dz^I \wedge d\bar z^J \qquad \text{and } \bar\partial\omega = \sum_{I,J} \bar\partial a_{IJ} \wedge dz^I \wedge d\bar z^J.$$

The operators

$$\partial\colon C^1_{p,q}(M) \to C^0_{p+1,q}(M) \qquad \text{and } \bar\partial\colon C^1_{p,q}(M) \to C^0_{p,q+1}(M)$$

thus defined on $C^1_{p,q}(M)$ are extended to the full Grassman algebra $\mathscr{G}^1(M)$ by linearity.

Proposition 2.2. *The operators ∂ and $\bar\partial$ on a complex manifold M have the following properties*

(a) $d = \partial + \bar\partial$ *on* $\mathscr{G}^1(M)$.
(b) $\partial\circ\partial = 0$, $\bar\partial\circ\bar\partial = 0$, $\partial\circ\bar\partial + \bar\partial\circ\partial = 0$ on $\mathscr{G}^2(M)$.
(c) ∂ *and* $\bar\partial$ *commute with the pullback under holomorphic maps.*

Proof. (a) is obvious since $d = \partial + \bar\partial$ on $C^1_{p,q}$ for all $p, q \geq 0$. For (b), we recall that $d\circ d = 0$ on $\mathscr{G}^2(M)$; combined with (a), one obtains

$$(2.17)\quad 0 = (\partial + \bar\partial)\circ(\partial + \bar\partial)\omega = (\partial\circ\partial)\omega + (\partial\circ\bar\partial + \bar\partial\circ\partial)\omega + (\bar\partial\circ\bar\partial)\omega.$$

If ω is of type (p, q), the three components on the right in (2.17) are of type $(p+2, q)$, $(p+1, q+1)$, and $(p, q+2)$, respectively; since their sum is 0, they must be 0 individually, proving (b) on $C^2_{p,q}$, and hence in general. For (c), let N be another complex manifold and let $F\colon N \to M$ be holomorphic. If $(z_1, \ldots, z_n)$ are local holomorphic coordinates on M, the $F^*z_j = z_j\circ F$ is holomorphic; hence $F^*(dz_j) = d(F^*z_j)$ is of type (1, 0) and $F^*(d\bar z_j)$ is of type (0, 1). By using the local representation (2.13) it now follows that

$$(2.18)\qquad F^*(C^k_{p,q}(M)) \subset C^k_{p,q}(N) \qquad \text{for all } p, q \geq 0 \text{ and } k \geq 0.$$

From Theorem 1.6 one obtains

$$(2.19)\quad \partial(F^*\omega) + \bar\partial(F^*\omega) = d(F^*\omega) = F^*(d\omega) = F^*(\partial\omega) + F^*(\bar\partial\omega).$$

Now take $\omega \in C^1_{p,q}$ in (2.19) and compare types on both sides by using (2.18); it follows that

$$\partial(F^*\omega) = F^*(\partial\omega) \qquad \text{and } \bar\partial(F^*\omega) = F^*(\bar\partial\omega). \quad \blacksquare$$

2.5. $\bar\partial$-Cohomology Groups

The inhomogeneous system of Cauchy–Riemann equations introduced in I.§1.2 can now be reformulated as follows: *given $D \subset \mathbb{C}^n$ and $u = \sum_{j=1}^n u_j\, d\bar z_j \in C^1_{0,1}(D)$, find $f \in C^1(D)$ such that $\bar\partial f = u$.* The necessary integrability condition $\partial u_j/\partial\bar z_k = \partial u_k/\partial\bar z_j$ for $1 \leq j, k \leq n$ is equivalent to $\bar\partial u = 0$.

More generally, one can formulate the corresponding $\bar{\partial}$-equation (or generalized Cauchy–Riemann equations) for (p, q)-forms on any complex manifold M: *given* $u \in C^\infty_{(p,q)}(M)$, $q \geq 1$, *one wants to find a solution* $\omega \in C^\infty_{(p,q-1)}(M)$ *of*

$$\bar{\partial}\omega = u \qquad \text{on } M. \tag{2.20}$$

Since $\bar{\partial}^2 = 0$, in order to solve (2.20) it is necessary that $\bar{\partial}u = 0$. It turns out that this condition is also sufficient for the existence of a *local* solution of (2.20). If fact, it is elementary to solve (2.20) with $\bar{\partial}u = 0$ on polydiscs in $\mathbb{C}^n$. (This is the Lemma of Bochner, Dolbeault, and Grothendieck. See Exercise E.2.4 and the Notes to Chapter IV for some additional comments.) However, the question of *existence of global solutions* of (2.20), even for $M = D \subset \mathbb{C}^n$, is much deeper and depends on global complex analytic properties of M.

The situation is analogous to the one for the equation $d\omega = u$, where u is a given r-form on a differentiable manifold M. The condition $du = 0$ is necessary for solvability, and it is sufficient for the existence of *local* solutions (the Poincaré Lemma). However, the existence of *global* solutions depends on global topological properties of M.

Formally, this kind of question is conveniently described in the language of complexes and their cohomology groups. Given a C^∞ manifold M, one introduces the space of **d-closed r-forms** $Z_r(M) = \{\omega \in C^\infty_r(M)\colon d\omega = 0\}$, $r \geq 0$, and the space of **d-exact forms** $B_r(M) = dC^\infty_{r-1}(M)$ for $r \geq 1$ and $\{0\}$ otherwise. Since $d^2 = 0$, $B_r(M)$ is a subspace of $Z_r(M)$. The quotient spaces

$$H^r_d(M) = Z_r(M)/B_r(M), \qquad r = 1, 2, \ldots,$$

measure the obstructions to the solvability of $d\omega = u$ on M. $H^r_d(M)$ is called the **d-cohomology group** (or **de Rham cohomology** group) of M of degree r.

The definition of the groups $H^r_d(M)$ clearly involves the differentiable structure of M; it is therefore quite remarkable that these groups in fact depend only on the topology of M (de Rham's Theorem [Rha]). More precisely, de Rham established a natural isomorphism between $H^r_d(M)$ and the r-th cohomology group $H^r(M, \mathbb{C})$ of M with coefficients in $\mathbb{C}$, as defined in algebraic topology.

Similarly, one may consider the cohomology groups of the $\bar{\partial}$-complex on a complex manifold M. Let $Z_{p,q}(M) = \{\omega \in C^\infty_{p,q}(M)\colon \bar{\partial}\omega = 0\}$ be the space of **$\bar{\partial}$-closed (p, q)-forms** and $B_{p,q}(M) = \bar{\partial}C^\infty_{p,q-1}(M)$ for $q \geq 1$ and $\{0\}$ for $q = 0$ the space of **$\bar{\partial}$-exact (p, q)-forms**. The quotient space

$$H^{p,q}_{\bar{\partial}}(M) = Z_{p,q}(M)/B_{p,q}(M) \tag{2.21}$$

is called the **$\bar{\partial}$-cohomology group of M of type (p, q)**. Notice that $H^{0,0}_{\bar{\partial}}(M) = \mathcal{O}(M)$. If $p = 0$, we shall write $H^q_{\bar{\partial}}(M)$ instead of $H^{0,q}_{\bar{\partial}}(M)$; this is the case of greatest interest for $M = D \subset \mathbb{C}^n$. The solvability of the $\bar{\partial}$-equation (2.20) can now be rephrased as the vanishing of the corresponding $\bar{\partial}$-cohomology group.

We shall see later that information about the groups $H^q_{\bar{\partial}}(M)$ for $q \geq 1$ provides a powerful tool for solving many problems in complex analysis. In Chapter V we will solve the $\bar{\partial}$-equation on strictly pseudoconvex domains in

$\mathbb{C}^n$ by means of integral formulas. A crucial step will involve showing first that $\dim_{\mathbb{C}} H^q_{\bar\partial} < \infty$ for $q \geq 1$ in this case. Moreover, we shall see in Chapter VI, §2, that an open set $D \subset \mathbb{C}^n$ is a Stein domain (i.e., a domain of holomorphy) if and only if $H^q_{\bar\partial}(D) = 0$ for $q \geq 1$.

It is convenient to also introduce the **$\bar\partial$-cohomology groups $H^{p,q}_{\bar\partial}(K)$ of compact sets $K \subset M$.** Denote by $C^\infty_{(p,q)}(K)$ the space of (p, q)-forms which are defined and of class C^∞ on some open neighborhood of K, with the understanding that two elements $\omega_1, \omega_2 \in C^\infty_{(p,q)}(K)$ are equal if they agree on some neighborhood of K,[1] and set $Z_{p,q}(K) = \{\omega \in C^\infty_{p,q}(K): \bar\partial\omega = 0$ on some neighborhood of $K\}$, and $B_{p,q}(K) = \bar\partial C^\infty_{p,q-1}(K)$ for $q \geq 1$ and $\{0\}$ otherwise. One then defines

$$H^{p,q}_{\bar\partial}(K) = Z_{p,q}(K)/B_{p,q}(K). \tag{2.22}$$

Finally, we also mention the **$\bar\partial$-cohomology groups $H^{p,q}_{\bar\partial,c}(M)$ with compact support**, defined as follows. Denote by $\mathscr{D}_{p,q}(M)$ the space of $C^\infty_{p,q}$ forms on M with compact support. Then, for $q \geq 1$, one has

$$H^{p,q}_{\bar\partial,c}(M) = (Z_{p,q} \cap \mathscr{D}_{p,q})/\bar\partial\mathscr{D}_{p,q-1}, \tag{2.23}$$

and

$$H^{p,0}_{\bar\partial,c}(M) = Z_{p,0} \cap \mathscr{D}_{p,0}.$$

We leave it to the reader to check that $H^{p,0}_{\bar\partial,c}(M) = \{0\}$ if M has no compact component. We write $H^q_{\bar\partial,c}$ for $H^{0,q}_{\bar\partial,c}$. We shall see in Chapter IV, §2, that $H^1_{\bar\partial,c}(\mathbb{C}^n) = 0$ for $n \geq 2$ (but *not* for $n = 1$!), a fact which is very closely related to the remarkable extension properties of holomorphic functions in several variables.

Exercises

E.2.1. Prove that a complex submanifold M of the region D in $\mathbb{C}^n$, as defined in I,§2.6, carries a natural structure of a complex manifold.

E.2.2. Show that every complex manifold is orientable.

E.2.3. A complex manifold M is said to be a **Stein manifold** if M has a countable basis for open sets and if M satisfies the properties (i), (ii), and (iii) in E.II.3.11.

(i) Show that every closed submanifold N of a Stein manifold M is also Stein.
(ii) Show that an open subset D of a Stein manifold M is itself a Stein manifold if and only if D is holomorphically convex (i.e., $\mathscr{O}(D)$-convex).

Remark. It is known that every Stein manifold M is biholomorphically equivalent to a closed complex submanifold of some $\mathbb{C}^n$. See [Hör 2], §5.3, for a proof of this result due to R. Remmert (Habilitationsschrift, Münster 1958), R. Narasimhan (Amer. J. Math. **82**(1960), 917–934) and E. Bishop (Amer. J. Math. **83**(1961), 209–242).

[1] In the language of sheaves, $C^\infty_{p,q}(K)$ is the space of sections over K of the sheaf of germs of C^∞ (p, q)-forms on M (see VI, §4 and §6).

E.2.4. Let K be a compact polydisc in $\mathbb{C}^n$. Prove that $H^1_{\bar{\partial}}(K) = 0$ by completing the following outline:

(i) Suppose $n = 1$. Show that if $u = f d\bar{z} \in C^\infty_{0,1}(U)$ has compact support in a neighborhood U of K, then

$$Tu(z) = -\frac{1}{2\pi i}\int \frac{f(\zeta)\, d\bar{\zeta} \wedge d\zeta}{\zeta - z}$$

satisfies $\bar{\partial}(Tu) = \partial(Tu)/\partial\bar{z}\, d\bar{z} = u$, and if u depends holomorphically on some parameters, so does Tu. (This is a classical one variable result; it is also a special case of Corollary IV.1.11.)

(ii) For the general case, use induction on k, where $u = \sum_{j=1}^k u_j\, d\bar{z}_j$. Use $\bar{\partial}u = 0$ to show that $u - \bar{\partial}(T(\chi u_k))$, where χ is a suitable cutoff function, involves only $d\bar{z}_1, \ldots, d\bar{z}_{k-1}$ in some neighborhood of K.

E.2.5. Let K be a compact polydisc in $\mathbb{C}^n$. Modify the ideas used in E.2.4 to show that $H^q_{\bar{\partial}}(K) = 0$ for $q \geq 1$.

E.2.6. Let M be a complex manifold and let $P \in M$. Show that

$$\Lambda^{p,q}_P(M) \cap \Lambda^{p',q'}_P(M) = \{0\}$$

if $p \neq p'$ or $q \neq q'$. (Hint: Use the fact that ω is of type (p, q) if and only if ω is a $(p + q)$-multilinear map on $\mathbb{C}T_PM$ which is $\mathbb{C}$-linear in p entries, and conjugate $\mathbb{C}$-linear in the remaining q entries.)

E.2.7. By Theorem I.2.12 there are no nontrivial *compact* complex submanifolds of $\mathbb{C}^n$. The Riemann sphere is the simplest compact complex manifold of dimension 1. Construct higher dimensional analogs as follows. Define

$$\mathbb{P}_n(\mathbb{C}) = \{\text{complex lines in } \mathbb{C}^{n+1} \text{ through } 0\},$$

with the natural topology as a quotient space of $\mathbb{C}^{n+1} - \{0\}$. For $0 \leq j \leq n$, let

$$U_j = \big\{L \in \mathbb{P}_n(\mathbb{C})\colon L = \{\lambda(a_1, \ldots, a_j, 1, a_{j+1}, \ldots, a_n)\colon \lambda \in \mathbb{C}\} \text{ for some } a = (a_1, \ldots, a_n) \in \mathbb{C}^n\big\}.$$

Show that there are natural homeomorphisms $\varphi_j\colon U_j \to \mathbb{C}^n$, and that $\{(U_j, \varphi_j)\colon 0 \leq j \leq n\}$ is a complex atlas for $\mathbb{P}_n(\mathbb{C})$ which makes $\mathbb{P}_n(\mathbb{C})$ into a compact complex manifold of dimension n. $\mathbb{P}_n(\mathbb{C})$ is called the **n-dimensional complex projective space**.

§3. Hermitian Geometry in $\mathbb{C}^n$

3.1. Riemannian Structures

A **Riemannian structure** (or **Riemannian metric**) on a C^k manifold M is given by an inner product $\langle\cdot,\cdot\rangle_P$ on T_PM for each $P \in M$, which depends smoothly on P in the following sense: if V and W are C^{k-1} vector fields on $U \subset M$, then $P \mapsto \langle V_P, W_P\rangle_P$ is a C^{k-1} function on U.

The most important example for us is $\mathbb{R}^n$. Via the natural identification $T_P\mathbb{R}^n = \mathbb{R}^n$, each tangent space $T_P\mathbb{R}^n$ inherits the standard Euclidean inner product from $\mathbb{R}^n$, with smooth dependence on P. With respect to this Riemannian metric, the global vector fields $\partial/\partial x_1, \ldots, \partial/\partial x_n$ corresponding to the standard coordinates $(x_1, \ldots, x_n)$ of $\mathbb{R}^n$ define an orthonormal basis for $T_P\mathbb{R}^n$ at each point P.

Riemannian structures exist on every C^k manifold M. Locally one can define the inner product as in $\mathbb{R}^n$ via a fixed coordinate system, and the local structures can be patched together via partitions of unity (see Exercise E.3.4).

Given a Riemannian metric on M, the inner product on T_PM induces in a natural way a Hermitian inner product on the full Grassman algebra $\mathscr{G}_P(M)$ of complex valued forms at P, which is determined uniquely by the following property: if $\{v_1, \ldots, v_n\}$ is an orthonormal basis for T_PM, and if $\{\omega_1, \ldots, \omega_n\}$ is the corresponding dual basis of T_P^*M, then

$$\{\omega_{j_1} \wedge \ldots \wedge \omega_{j_r}: 1 \leq j_1 < \ldots < j_r \leq n \text{ and } 1 \leq r \leq n\}.$$

is orthonormal in $\mathscr{G}_P(M)$. In particular, two forms of different degree are orthogonal.

For $M = \mathbb{R}^n$, with its standard Riemannian structure, it follows that the product of two r-forms $\varphi = \sum_J a_J\, dx^J$ and $\psi = \sum_J b_J\, dx^J$ at P is given by

$$\langle \varphi, \psi \rangle_P = \sum_J a_J \bar{b}_J,$$

where, as usual, the summation is over all strictly increasing r-tuples $J \subset \{1, \ldots, n\}$.

3.2. The Hermitian Structure on $\mathbb{C}^n$

The identification on $\mathbb{C}^n$ with $\mathbb{R}^{2n}$ defines a natural Riemannian structure on $\mathbb{C}^n$. The (real) inner product $\langle \cdot, \cdot \rangle_P$ on $T_P\mathbb{C}^n$ extends in the obvious way to a (complex valued) Hermitian inner product on the complexification $\mathbb{C}T_P\mathbb{C}^n$, which we still denote by $\langle \cdot, \cdot \rangle_P$. Thus $\langle av, bw \rangle_P = a\bar{b}\langle v, w \rangle_P$ for $a, b \in \mathbb{C}$ and $v, w \in T_P\mathbb{C}^n$. A simple computation shows that the basis $\{\partial/\partial z_1, \partial/\partial \bar{z}_1, \ldots, \partial/\partial z_n, \partial/\partial \bar{z}_n\}$ of $\mathbb{C}T_P\mathbb{C}^n$ is orthogonal and satisfies

$$(3.1) \qquad \langle \partial/\partial z_j, \partial/\partial z_j \rangle_P = \langle \partial/\partial \bar{z}_j, \partial/\partial \bar{z}_j \rangle_P = 1/2 \qquad \text{for } 1 \leq j \leq n;$$

thus $\|\partial/\partial z_j\| = 1/\sqrt{2}$.

In particular, $\mathbb{C}T_P\mathbb{C}^n = T_P^{1,0}\mathbb{C}^n \oplus T_P^{0,1}\mathbb{C}^n$ is an *orthogonal* decomposition.

Remark. The procedure just described gives the Hermitian structure on $\mathbb{C}T_P\mathbb{C}^n$ in terms of the Riemannian structure of $\mathbb{R}^{2n}$. Clearly one could do the same thing for any complex manifold M with a given Riemannian structure on the underlying C^∞ manifold, but, in general, the decomposition $\mathbb{C}T_P = T_P^{1,0} + T_P^{0,1}$ would not be orthogonal, and the Hermitian structure on $\mathbb{C}T_PM$

so obtained would not reflect the complex structure of M. For this reason, in complex differential geometry one mainly considers **Hermitian structures** (or **metrics**) which are defined by directly specifying a Hermitian inner product either on $T_P^{1,0}$ or on $T_P M$ (with its natural complex structure defined by J). Since in this book we only need the Hermitian structure of $\mathbb{C}^n$, we do not elaborate this more intrinsic complex point of view. (See Exercise E.3.1 for more details in case of $\mathbb{C}^n$.)

As in the real case considered in §3.1, the Hermitian product on $\mathbb{C}T_P\mathbb{C}^n$ induces a Hermitian inner product on the Grassman algebra of complex valued froms at P. In terms of the natural coordinates $(z_1, \ldots, z_n)$ on $\mathbb{C}^n$, the product of two (p, q)-forms $\varphi = \sum_{I,J} a_{IJ}\, dz^I \wedge d\bar{z}^J$ and $\psi = \sum_{I,J} b_{IJ}\, dz^I \wedge d\bar{z}^J$ at P is given by

$$\langle \varphi, \psi \rangle_P = 2^{p+q} \sum_{I,J} a_{IJ} \overline{b_{IJ}}, \tag{3.2}$$

and any two forms of different type are orthogonal. The norm of a form φ at P is defined by $\|\varphi\| = \sqrt{\langle \varphi, \varphi \rangle_P}$. The factor 2^{p+q} in (3.2) is a consequence of

$$\langle dz_j, dz_j \rangle = \langle d\bar{z}_j, d\bar{z}_j \rangle = 2 \qquad \text{for } 1 \le j \le n, \tag{3.3}$$

which is the statement dual to (3.1).

3.3. Volume Forms and Global Inner Products

Let M be an n-dimensional C^k manifold with a Riemannian structure. A **volume form** dV on M is a real, continuous n-form on M with $\|dV_P\| = 1$ for all $P \in M$. A volume form clearly defines an orientation of M, and, conversely, if M is oriented, there is a unique volume form on M which defines the given orientation of M. The volume form of $\mathbb{C}^n = \mathbb{R}^{2n}$ is given by $dV = dx_1 \wedge dy_1 \wedge \ldots \wedge dx_n \wedge dy_n$, where, as usual, $z_j = x_j + iy_j$. The following representations of dV for $\mathbb{C}^n$ are used as well:

$$dV = \left(\frac{1}{2}\right)^n dz_1 \wedge d\bar{z}_1 \wedge \ldots \wedge dz_n \wedge d\bar{z}_n; \tag{3.4}$$

$$dV = \frac{1}{n!}\tau^n, \qquad \text{where } \tau = \frac{i}{2}\sum_{j=1}^n dz_j \wedge d\bar{z}_j; \tag{3.5}$$

$$dV = \frac{(-1)^{n(n-1)/2}}{(2i)^n} d\bar{z}_1 \wedge \ldots \wedge d\bar{z}_n \wedge dz_1 \wedge \ldots \wedge dz_n. \tag{3.6}$$

A volume form dV on M allows to integrate functions f over M by setting $\int_M f = \int_M f\, dV$. In particular, $\operatorname{vol}(M) = \int_M 1 = \int_M dV$ is the **volume of** M (which may by ∞!). Moreover, by integrating the pointwise inner product of two forms one obtains the following important *global inner product* between differential forms.

Definition. The inner product $(\varphi, \psi)_M$ of two differential forms on M is given by

$$(\varphi, \psi)_M = \int_M \langle \varphi, \psi \rangle_P \, dV \tag{3.7}$$

whenever the integral makes sense.

It is clear that (3.7) defines a Hermitian inner product on the space $\mathscr{G}_c^0(M)$ of continuous differential forms with compact support in M. The associated norm $\sqrt{(\varphi, \varphi)_M}$ is denoted by $\|\varphi\|_{L^2(M)}$ or $\|\varphi\|_M$; when M is clear from the context one sometimes omits the symbol M.

By taking the completion with respect to $\|\cdot\|_M$ of the subspace of r-forms of $\mathscr{G}_c^0(M)$ one obtains the Hilbert space $L_r^2(M)$ of **square integrable r-forms on M**. Similarly, on a *complex* manifold with Hermitian structure there are the spaces $L_{p,q}^2(M)$ of **square integrable (p, q)-forms**. One has the orthogonal decomposition

$$L_r^2(M) = \bigoplus_{p+q=r} L_{p,q}^2(M). \tag{3.8}$$

In case $M = D$ is an open subset of $\mathbb{C}^n$, the global coordinates of $\mathbb{C}^n$ can be used to obtain a more concrete description of $L_{p,q}^2(D)$, as follows. A (p, q)-form φ on D has a unique representation

$$\varphi = \sum_{I,J} a_{IJ} \, dz^I \wedge d\bar{z}^J, \tag{3.9}$$

and

$$(\varphi, \varphi)_D = \int_D \langle \varphi_\zeta, \varphi_\zeta \rangle_\zeta \, dV = 2^{p+q} \sum_{I,J} \int_D |a_{IJ}(\zeta)|^2 \, dV; \tag{3.10}$$

thus φ is in $L_{p,q}^2(D)$ if and only if the coefficients a_{IJ} are functions in $L^2(D)$ (with respect to Lebesgue measure on $\mathbb{R}^{2n}$).

The global representation (3.9) allows to define spaces $\mathscr{F}_{p,q}(D)$ of **(p, q)-forms with coefficients in $\mathscr{F}$** for any other function space $\mathscr{F}(D)$ by

$$\mathscr{F}_{p,q}(D) = \left\{ \varphi = \sum_{I,J} a_{IJ} \, dz^I \wedge d\bar{z}^J : a_{IJ} \in \mathscr{F}(D) \quad \text{for all } I, J \right\}.$$

In particular, this procedure defines the space $L_{p,q}^s(D)$ of (p, q)-forms with coefficients in $L^s(D)$ for any $1 \le s \le \infty$. Notice that the inner product $(\varphi, \psi)_D$ is defined whenever $\varphi \in L_{p,q}^s(D)$, $\psi \in L_{p,q}^{s'}(D)$, and $1/s + 1/s' = 1$.

If $D \subset \mathbb{R}^n$ has C^k boundary, $k \ge 1$, the boundary bD inherits a Riemannian structure from $\mathbb{R}^n$ by restricting the inner product in $T_P\mathbb{R}^n$ to $T_P bD \subset T_P\mathbb{R}^n$. There is thus a unique volume element on bD, denoted by dS, which defines the induced orientation of bD. dS is also called the **surface element** of bD. By the Riesz representation theorem there is a unique positive Borel measure $\sigma = \sigma_{bD}$ on bD such that

$$\int_{bD} f \, dS = \int_{bD} f \, d\sigma \qquad \text{for all } f \in C_c(bD); \tag{3.11}$$

$d\sigma$ is called the **surface measure** or $(n-1)$**-dimensional Lebesgue measure induced on** bD. The following Lemma is an immediate consequence of the definitions.

Lemma 3.1. *Suppose r is a C^k defining function for $D \subset \mathbb{R}^n$ near $P \in bD$, such that $\|dr_P\| = 1$. If $\omega_2, \ldots, \omega_n$ are real 1-forms at P such that $dr_P, \omega_2, \ldots, \omega_n$ is a positively oriented orthonormal basis of $T_P^*(\mathbb{R}^n)$, then $dS_P = \iota^*(\omega_2 \wedge \ldots \wedge \omega_n)$, where $\iota\colon bD \to \mathbb{R}^n$ is the inclusion map.*

In Lemma 3.4 we will give a formula for the surface element dS which is more practical for computations.

3.4. The $*$-Operator

We now discuss a very useful operator on forms on an oriented Riemannian manifold M which transforms inner products of forms into wedge products.

For convenience of notation we first introduce the generalization of the Kronecker δ-symbol to ordered subsets of $N = \{1, \ldots, n\}$. For $A \subset N$ we denote by $|A|$ the cardinality of A and by A' the complement of A in N, with the order of A' induced by the natural order of N. If A, B are two ordered subsets of N one sets

(3.8) $\varepsilon_B^A = \operatorname{sign} \pi$ *if* $A = B$ *as sets and* π *is a permutation which takes* A *into* B, and $\varepsilon_B^A = 0$ *in all other cases.*

It follows that $\varepsilon_B^A = \varepsilon_A^B$, $\varepsilon_B^A \varepsilon_C^B = \varepsilon_C^A$, and $\varepsilon_{BA}^{AB} = (-1)^{rs}$ if $|A| = r$ and $|B| = s$. Also note that $\varepsilon_B^A = 0$ if $|A| \neq |B|$.

Theorem 3.2. *Suppose* $\dim M = n$ *and let dV_P be a volume form at $P \in M$. There is a unique $\mathbb{C}$-linear map $*\colon \mathcal{G}_P(M) \to \mathcal{G}_P(M)$ with the following properties.*

(3.12) $$*\varphi \in \Lambda^{n-r}\mathbb{C}T_P^*M \qquad \text{for } \varphi \in \Lambda^r\mathbb{C}T_P^*M.$$

(3.13) $$* \text{ is real, that is } *\bar{\varphi} = \overline{*\varphi}.$$

(3.14) $$**\varphi = (-1)^{(n-r)r}\varphi \qquad \text{for } \varphi \in \Lambda^r\mathbb{C}T_P^*M.$$

(3.15) $$*1 = dV_P \qquad \text{and } *dV_P = 1.$$

(3.16) $$\psi \wedge *\bar{\varphi} = \langle \psi, \varphi \rangle_P \, dV_P \qquad \text{for all } \psi, \varphi \in \mathcal{G}_P(M).$$

Proof. We first prove uniqueness. Choose an orthonormal basis $\omega_1, \ldots, \omega_n$ of $\mathbb{C}T_P^*M$ such that $\omega_1 \wedge \ldots \wedge \omega_n = dV_P$. By linearity of $*$ it is enough to verify that the properties stated in the theorem determine $*\omega^J$ for each r-tuple $J \subset N$. By (3.15) it is enough to consider $1 \le r \le n$. By (3.12), $*\omega^J$ is an $(n-r)$-form, thus $*\bar{\omega}^J = \sum_{|K|=n-r} a_K \omega^K$, where $a_K \in \mathbb{C}$ and the summation is over all strictly increasing $(n-r)$-tuples $K \subset N$. Fix such a K. Then $\omega^{K'} \wedge *\bar{\omega}^J = a_K \omega^{K'} \wedge \omega^K$

$= a_K \varepsilon_N^{K'K}\, dV_P$, and by (3.16), $\omega^{K'} \wedge *\bar{\omega}^J = \langle \omega^{K'}, \omega^J \rangle\, dV_P = \varepsilon_{K'}^J\, dV_P$. It follows that $a_K = \varepsilon_{K'}^J \varepsilon_N^{K'K} = \varepsilon_N^{JK}$, which implies

$$*\omega^J = \varepsilon_N^{JJ'} \bar{\omega}^{J'}. \tag{3.17}$$

To prove the existence of $*$, choose any orthonormal basis $\omega_1, \dots, \omega_n$ with $\omega_1 \wedge \dots \wedge \omega_n = dV_P$, define $*$ by (3.17) and extend it to the whole Grassman algebra by $\mathbb{C}$-linearity. It is then straightforward to verify that $*$ has all the required properties. The uniqueness part implies that the definition of $*$ is independent of the choice of orthonormal basis. ∎

We now specialize to $M = \mathbb{C}^n$, with the standard global coordinates $(z_1, \dots, z_n)$.

Lemma 3.3. *The $*$-operator on $\mathbb{C}^n$ has the following additional properties.*

(3.18) $\quad *(\Lambda_P^{p,q}(\mathbb{C}^n)) \subset \Lambda_P^{n-q,n-p}(\mathbb{C}^n) \qquad$ *for all $0 \le p, q \le n$ and $P \in \mathbb{C}^n$.*

(3.19) $\quad **\varphi = (-1)^{p+q}\varphi \qquad$ *for $\varphi \in \Lambda_P^{p,q}(\mathbb{C}^n)$.*

If $J \subset N = \{1, \dots, n\}$ and $|J| = q$, then

$$*dz^J = \frac{(-1)^{q(q-1)/2}}{2^{n-q} i^n} dz^J \wedge \left(\bigwedge_{\nu \in J'} d\bar{z}_\nu \wedge dz_\nu \right). \tag{3.20}$$

Proof. If φ is a (p, q) form, then $\langle \psi, \varphi \rangle \neq 0$ only for ψ of type (p, q). From (3.16) it follows that $*\bar{\varphi}$ is of type $(n-p, n-q)$, i.e., $*\varphi$ is of type $(n-q, n-p)$. This proves (3.18). Equation (3.19) is an immediate consequence of (3.14). For (3.20), notice that $dz_1 \wedge d\bar{z}_1 \wedge \dots \wedge dz_n \wedge d\bar{z}_n = (-1)^{q(q-1)/2} dz^J \wedge d\bar{z}^J \wedge (dz \wedge d\bar{z})^{J'}$, and that, by (3.16), $dz^J \wedge \overline{*dz^J} = 2^q dV$. Using the expression (3.4) for dV it follows that

$$dz^J \wedge \overline{*dz^J} = 2^q \left(\frac{i}{2}\right)^n (-1)^{q(q-1)/2} dz^J \wedge d\bar{z}^J \wedge (dz \wedge d\bar{z})^{J'}.$$

This implies

$$\overline{*dz^J} = \frac{i^n (-1)^{q(q-1)/2}}{2^{n-q}} d\bar{z}^J \wedge (dz \wedge d\bar{z})^{J'},$$

and (3.20) follows by conjugation. ∎

Next we use the $*$-operator in order to represent the surface element dS for the boundary bD of an open set D in $\mathbb{R}^n$.

Lemma 3.4. *Let r be a C^k defining function for D, $k \ge 1$, and let $\iota\colon bD \to \mathbb{R}^n$ be the inclusion map. Then*

$$dS_P = \frac{\iota^*(*dr_P)}{\|dr_P\|} \qquad \text{for } P \in bD. \tag{3.21}$$

PROOF. Fix $P \in bD$ and choose $\omega_2, \ldots, \omega_n \in T_P^*\mathbb{R}^n$, such that $dr_P/\|dr_P\|, \omega_2, \ldots, \omega_n$ is a positively oriented orthonormal basis. By (3.17), $*(dr_P/\|dr_P\|) = \omega_2 \wedge \ldots \wedge \omega_n$. The result now follows from Lemma 3.1. ∎

Notice that by applying (3.17) to the orthonormal basis $dx_1, \ldots, dx_n$, it follows that

$$*dr = \sum_{j=1}^{n} (-1)^{j-1} \frac{\partial r}{\partial x_j} dx_1 \wedge \ldots \wedge [dx_j] \wedge \ldots \wedge dx_n.$$

Thus the expression (3.21) is easily computable.

Corollary 3.5. *If D is open in $\mathbb{C}^n$ with C^k defining function r, then*

$$dS = \frac{2}{\|dr\|} \iota^*(*\partial r) \qquad \text{on } bD. \tag{3.22}$$

PROOF. We use (3.21) and compute $*dr = *(\partial r + \bar{\partial} r) = *\partial r + \overline{*\partial r}$. By (3.20),

$$\begin{aligned} *\partial r &= \sum_{j=1}^{n} \frac{1}{2^{n-1} i^n} \frac{\partial r}{\partial z_j} dz_j \wedge \left(\bigwedge_{\nu \neq j} (d\bar{z}_\nu \wedge dz_\nu) \right) \\ &= \frac{1}{i} \partial r \wedge \frac{\tau^{n-1}}{(n-1)!}, \end{aligned} \tag{3.23}$$

where $\tau = 1/2i \sum_{j=1}^n d\bar{z}_j \wedge dz_j$. Hence

$$*dr = \left(\frac{1}{i} \partial r - \frac{1}{i} \bar{\partial} r \right) \frac{\tau^{n-1}}{(n-1)!}.$$

Notice that $\iota^*(dr) = 0$ implies $\iota^*(\partial r) = -\iota^*(\bar{\partial} r)$; therefore this formula for $*dr$ implies

$$\iota^*(*dr) = 2\iota^* \left(\frac{1}{i} \partial r \wedge \frac{\tau^{n-1}}{(n-1)!} \right).$$

By (3.23) we thus have $\iota^*(*dr) = 2\iota^*(*\partial r)$, and the result follows from (3.21). ∎

Remark. The pullback ι^* is often not explicitly indicated in formulas like (3.21) or (3.22), since for $\varphi \in \Lambda_P(\mathbb{R}^n)$, $\iota^*\varphi$ can just be viewed as the restriction of φ to $T_P bD$.

3.5. Integration by Parts and the Adjoint of $\bar{\partial}$

We shall now use Stokes' Theorem and the $*$-operator in order to derive an "integration by parts" formula for $\bar{\partial}$.

Let $\varphi \in C^1_{p,q}(\mathbb{C}^n)$ and $\psi \in C^1_{p,q+1}(\mathbb{C}^n)$ be two forms and suppose at least one of them has compact support. The inner product $(\bar{\partial}\varphi, \psi)_{\mathbb{C}^n}$ is then defined, and since $\alpha = \varphi \wedge *\bar{\psi}$ is of type $(n, n-1)$, one has $\bar{\partial}\alpha = d\alpha$. Therefore

$$
\begin{aligned}
(\bar{\partial}\varphi, \psi) &= \int \bar{\partial}\varphi \wedge *\bar{\psi} \\
&= \int \bar{\partial}(\varphi \wedge *\bar{\psi}) - \int (-1)^{p+q}\varphi \wedge \bar{\partial}*\bar{\psi} \\
&= \int d(\varphi \wedge *\bar{\psi}) - \int \varphi \wedge **\bar{\partial}*\bar{\psi} \qquad \text{(use Lemma 3.3!)} \\
&= \int \varphi \wedge \overline{*(-*\partial*\psi)} \\
&= (\varphi, -*\partial*\psi),
\end{aligned}
\tag{3.24}
$$

where we have used Stokes' Theorem applied to a region D with C^1 boundary containing the support of $\varphi \wedge *\bar{\psi}$ in order to obtain $\int_D d(\varphi \wedge *\bar{\psi}) = 0$.

We have thus proved that the operator $\vartheta = -*\partial*$ satisfies

$$(\bar{\partial}\varphi, \psi) = (\varphi, \vartheta\psi) \tag{3.25}$$

if φ or ψ has compact support. ϑ is called the **formal adjoint of** $\bar{\partial}$. It is clear that ϑ is a first order differential operator which maps forms of type $(p, q+1)$ to forms of type (p, q) and is 0 on forms of type $(p, 0)$.

Lemma 3.6. *If*

$$\psi = \sum_{I,J} \psi_{IJ}\, dz^I \wedge d\bar{z}^J \in C^1_{p,q+1}(\mathbb{C}^n),$$

then

$$\vartheta\psi = -2(-1)^p \sum_{I,K}\left(\sum_{k,J} \varepsilon^J_{kK} \frac{\partial \psi_{IJ}}{\partial z_k}\right) dz^I \wedge d\bar{z}^K, \tag{3.26}$$

where the summation is over all k between 1 *and n, and over all strictly increasing p-tuples I, q-tuples K, and* $(q+1)$*-tuples J in* $\{1, \ldots, n\}$.

Proof. Rather than computing ϑ from $-*\partial*$, which is messy, we carry out the integration by parts in more explicit form. Let

$$\varphi = \sum_{I,K} \varphi_{IK}\, dz^I \wedge d\bar{z}^K \in C^1_{p,q}(\mathbb{C}^n)$$

have compact support. Since

$$\bar{\partial}\varphi = \sum_{I,K} \bar{\partial}\varphi_{IK} \wedge dz^I \wedge d\bar{z}^K = (-1)^p \sum_{I,J}\left(\sum_{k,K} \varepsilon^J_{kK} \frac{\partial \varphi_{IK}}{\partial \bar{z}_k}\right) dz^I \wedge d\bar{z}^J,$$

one obtains

$$(\bar{\partial}\varphi, \psi) = 2^{p+q+1}(-1)^p \sum_{I,J}\sum_{k,K} \varepsilon^J_{kK} \int \frac{\partial \varphi_{IK}}{\partial \bar{z}_k} \overline{\psi_{IJ}}\, dV. \tag{3.27}$$

Standard integration by parts in $\mathbb{R}^{2n}$ gives

$$(3.28)\qquad \int \frac{\partial g}{\partial \bar{z}_k} \bar{f}\, dV = -\int g \overline{\left(\frac{\partial f}{\partial z_k}\right)} dV, \qquad 1 \le k \le n,$$

for $g, f \in C^1(\mathbb{C}^n)$ and g with compact support. By applying (3.28) in (3.27) and interchanging the order of summation, one obtains

$$(\bar{\partial}\varphi, \psi) = -2^{p+q+1}(-1)^p \sum_{I,K} \int \varphi_{IK} \overline{\left(\sum_{k,J} \varepsilon^J_{kK} \frac{\partial \psi_{IJ}}{\partial z_k}\right)} dV = (\varphi, \omega),$$

where ω denotes the form on the right side of (3.26). We thus see that $(\varphi, \vartheta\psi) = (\bar{\partial}\varphi, \psi) = (\varphi, \omega)$ holds for all φ with compact support. This implies that $\vartheta\psi = \omega$. ∎

Next we examine how (3.25) has to be modified if φ and ψ do not have compact support.

Lemma 3.7. *Suppose $D \subset\subset \mathbb{C}^n$ has piecewise C^1 boundary and $\varphi \in C^1_{p,q}(\bar{D})$, $\psi \in C^1_{p,q+1}(\bar{D})$. Then*

$$(3.29)\qquad (\bar{\partial}\varphi, \psi)_D = (\varphi, \vartheta\psi)_D + \int_{bD} \varphi \wedge *\bar{\psi}$$

and

$$(3.30)\qquad (\vartheta\psi, \varphi)_D = (\psi, \bar{\partial}\varphi)_D - \int_{bD} \bar{\varphi} \wedge *\psi.$$

Proof. One proceeds as in (3.24), with all integrals taken over D instead of $\mathbb{C}^n$. Stokes' Theorem now gives

$$\int_D d(\varphi \wedge *\bar{\psi}) = \int_{bD} \varphi \wedge *\bar{\psi},$$

and (3.29) follows. Equation (3.30) follows from (3.29) by conjugation. ∎

3.6. The Complex Laplacian

In Riemannian geometry one considers the formal adjoint δ of the exterior derivative d and the **Laplace–Beltrami operator** $\Delta = \delta d + d\delta$. For $\mathbb{R}^n$ one obtains (see Exercise E.3.8)

$$(3.31)\qquad \Delta f = \delta d f = -\sum_{j=1}^{n} \frac{\partial^2 f}{\partial x_j^2} \qquad \text{if } f \in C^2(\mathbb{R}^n).$$

The appearance of the minus sign in (3.31) is one of the reasons why the Laplace operator Δ in $\mathbb{R}^n$ is often defined as the *negative* of $\sum_{j=1}^n \partial^2/\partial x_j^2$. From now on we shall take Δ as in (3.31).

In analogy to the Laplace–Beltrami operator, in complex differential geometry one considers the **complex Laplacian** $\square$ defined by

$$\square = \vartheta\bar{\partial} + \bar{\partial}\vartheta. \tag{3.32}$$

We shall compute $\square$ explicitly in $\mathbb{C}^n$.

Lemma 3.8. *If $f \in C^2(\mathbb{C}^n)$, then*

$$\square f = -2\sum_{j=1}^{n} \frac{\partial^2 f}{\partial z_j \partial \bar{z}_j} = \frac{1}{2}\Delta f.$$

PROOF. Since $\vartheta = 0$ on functions,

$$\square f = \vartheta\bar{\partial} f = \vartheta\left(\sum_{j=1}^{n} (\partial f/\partial \bar{z}_j)\, d\bar{z}_j\right).$$

The first equation now follows from Lemma 3.6, and the second equation then follows from

$$\frac{\partial^2}{\partial z_j \partial \bar{z}_j} = 1/4\left(\frac{\partial^2}{\partial x_j^2} + \frac{\partial^2}{\partial y_j^2}\right) \qquad \text{if } z_j = x_j + iy_j. \quad \blacksquare$$

Warning. The formula $\square = 1/2\Delta$ does not generalize to arbitrary Hermitian manifolds, though it remains true in the important case of Kähler manifolds (see, for example, [MoKo]).

Next we compute $\square$ on forms of positive degree. Surprisingly, in $\mathbb{C}^n$, $\square$ is a diagonal operator with respect to the standard global basis for (p, q)-forms. Precisely, one has the following result.

Lemma 3.9. *If*

$$\varphi = \sum_{I,J} \varphi_{IJ}\, dz^I \wedge d\bar{z}^J \in C^2_{p,q}(\mathbb{C}^n),$$

then

$$\square\varphi = \sum_{I,J} (\square\varphi_{IJ})\, dz^I \wedge d\bar{z}^J.$$

PROOF. In order to simplify notation we will assume that $p = 0$. This is really no restriction, since both $\bar{\partial}$ and ϑ do not act on differentials dz^I. Thus, let $\varphi = \sum_J \varphi_J\, d\bar{z}^J$; by using (3.26), one computes

$$\bar{\partial}\varphi = \sum_L \left(\sum_{j,J} \varepsilon^L_{jJ} \frac{\partial \varphi_J}{\partial \bar{z}_j}\right) d\bar{z}^L,$$

$$\vartheta\bar{\partial}\varphi = -2\sum_M \left(\sum_{\substack{j,J\\k,L}} \varepsilon^L_{kM}\, \varepsilon^L_{jJ} \frac{\partial^2 \varphi_J}{\partial z_k \partial \bar{z}_j}\right) d\bar{z}^M,$$

and similarly,

$$\bar{\partial}\vartheta\varphi = -2\sum_M \left(\sum_{\substack{k,J \\ j,K}} \varepsilon^M_{jK} \varepsilon^J_{kK} \frac{\partial^2 \varphi_J}{\partial \bar{z}_j \partial z_k} \right) d\bar{z}^M.$$

It follows that

$$\square\varphi = \vartheta\bar{\partial}\varphi + \bar{\partial}\vartheta\varphi = -2\sum_M \left(\sum_{j,k=1}^n C^M_{jk} \frac{\partial^2 \varphi_J}{\partial z_k \partial \bar{z}_j} \right) d\bar{z}^M, \tag{3.33}$$

where

$$C^M_{jk} = \sum_{J,K,L} (\varepsilon^L_{kM} \varepsilon^L_{jJ} + \varepsilon^M_{jK} \varepsilon^J_{kK}), \tag{3.34}$$

the summation being, as usual, over all strictly increasing tuples of the appropriate size. We now examine C^M_{jk}. If $j = k$, the only nonzero terms in (3.34) arise for $J = M$. By considering the two possibilities $j \in M$ and $j \notin M$ one sees that $C^M_{jj} = 1$. If $j \neq k$, nonzero terms in (3.34) arise only if $j \in M$. If $M = \{j\} \cup Q$, where $|Q| = q - 1$, then $\varepsilon^L_{kM} \varepsilon^L_{jJ} \neq 0$ only if $J = \{k\} \cup Q$ and $L = \{j, k\} \cup Q$ as sets; moreover $\varepsilon^M_{jK} \varepsilon^J_{kK} \neq 0$ only if $M = \{j\} \cup K$ and $J = \{k\} \cup K$ as sets. Thus only the term with $K = Q$ matters in (3.34), and therefore

$$C^M_{jk} = \varepsilon^{jJ}_{kM} + \varepsilon^M_{jQ} \varepsilon^J_{kQ} \qquad \text{if } j \neq k. \tag{3.35}$$

Since $\varepsilon^M_{jQ} \varepsilon^J_{kQ} = \varepsilon^{kM}_{kjQ} \varepsilon^{jJ}_{jkQ} = -\varepsilon^{jJ}_{kM}$, (3.35) implies that $C^M_{jk} = 0$. From (3.33) we now obtain

$$\square\varphi = -2\sum_M \left(\sum_{j=1}^n \frac{\partial^2 \varphi_M}{\partial z_j \partial \bar{z}_j} \right) d\bar{z}^M,$$

which gives the desired result in view of Lemma 3.8. ■

The formula $\square = 1/2\Delta$ in $\mathbb{C}^n$, which relates the complex Laplacian, and hence the Cauchy–Riemann operator $\bar{\partial}$, to the usual Laplace operator, can be viewed as a several-variable substitute for the intimate connection between holomorphic and harmonic functions in *one* complex variable. In the next chapter we shall exploit this relationship in order to derive from Green's formula certain integral representation formulas in $\mathbb{C}^n$ which are, in some sense, natural generalizations of the Cauchy Integral Formula in $\mathbb{C}^1$.

Exercises

E.3.1. Via the natural identification of $T_P\mathbb{C}^n$ with $\mathbb{C}^n$, the standard Hermitian inner product on $\mathbb{C}^n$ induces a Hermitian inner product $\langle\langle \cdot, \cdot \rangle\rangle$ on $T_P\mathbb{C}^n$.

(i) Show that

$$\left\langle\left\langle \frac{\partial}{\partial x_j}, \frac{\partial}{\partial x_k} \right\rangle\right\rangle = \delta_{jk}$$

and

$$\left\langle\!\left\langle \frac{\partial}{\partial x_j}, \frac{\partial}{\partial y_k} \right\rangle\!\right\rangle = -i\delta_{jk}.$$

(ii) Show that for $v, w \in T_P\mathbb{C}^n$ one has $\langle\!\langle v, Jw \rangle\!\rangle = -i\langle\!\langle v, w \rangle\!\rangle$ and $\langle\!\langle Jv, w \rangle\!\rangle = i\langle\!\langle v, w \rangle\!\rangle$.

(iii) Show that $\operatorname{Re}\langle\!\langle \cdot, \cdot \rangle\!\rangle$ defines the standard Riemannian structure on $T_P\mathbb{C}^n = T_P\mathbb{R}^{2n}$.

(iv) More generally, if $\langle \cdot, \cdot \rangle$ is a Hermitian inner product on a complex vector space V, show that $\operatorname{Re}\langle \cdot, \cdot \rangle$ defines a (real) inner product on the underlying $\mathbb{R}$-vector space.

E.3.2. Let dV be the volume form of $\mathbb{C}^n$.

(i) Show that if $\{\omega_1, \ldots, \omega_n\}$ is an orthonormal basis of $\Lambda_P^{1,0}(\mathbb{C})$, then

$$dV_P = i^n \omega_1 \wedge \bar{\omega}_1 \wedge \ldots \wedge \omega_n \wedge \bar{\omega}_n.$$

(ii) If $F = (f_1, \ldots, f_n)$ is a holomorphic map at P, show that

$$\left(\frac{i}{2}\right)^n df_1 \wedge d\bar{f}_1 \wedge \ldots \wedge df_n \wedge d\bar{f}_n(P) = |\det F'(P)|^2\, dV_P.$$

E.3.3. Prove Lemma 3.1.

E.3.4. Let M be a C^k manifold of dimension n. Complete the details of the following construction of a Riemannian metric on M.

(i) By paracompactness, choose a locally finite atlas $\{(U_i, \varphi_i), i \in I\}$ for M. Use $\varphi_i\colon U_i \to \varphi_i(U_i) \subset \mathbb{R}^n$ to define a Riemannian metric $g_i(\cdot, \cdot)$ on U_i.

(ii) Let $\{\chi_i\colon i \in I\}$ be a C^k partition of unity subordinate the covering $\{U_i\colon i \in I\}$ of M. Show that $\langle \cdot, \cdot \rangle = \sum \chi_i g_i(\cdot, \cdot)$ is a Riemannian metric on M.

E.3.5. In analogy to formula (3.20) in $\mathbb{C}^n$, compute

$$* \, dz^I \wedge d\bar{z}^J \wedge (dz \wedge d\bar{z})^K$$

for $I, J, K \subset N = \{1, \ldots, n\}$ with $I \cap K = \varnothing$ and $J \cap K = \varnothing$.

E.3.6. Compute $\vartheta = -*\partial*$ in $\mathbb{C}^n$ directly by using the formula for $*$ obtained in E.3.5.

E.3.7. Let $D \subset\subset \mathbb{C}^n$ have C^1 boundary and let r be a C^1 defining function. Define the $(1, 0)$-form ω_n near bD by $\omega_n = \partial r/\|\partial r\|$.

(i) Show that if $P \in bD$, there are a neighborhood U of P and $\omega_1, \ldots, \omega_{n-1} \in C^0_{1,0}(U)$, such that $\omega_1, \ldots, \omega_n$ defines an orthonormal basis for each space $\Lambda_a^{1,0}(\mathbb{C}^n)$ for $a \in U$.

(ii) Let $\psi \in C^1_{p,q}(\bar{D})$ have compact support in U. Show that one has

$$(\bar{\partial}\varphi, \psi)_D = (\varphi, \vartheta\psi) \qquad \text{for every } \varphi \in C^1_{p,q-1}(\bar{D})$$

if and only if in the representation $\psi = \sum_{I,J} \psi_{IJ}\omega^I \wedge \bar{\omega}^J$ valid on $U \cap \bar{D}$ ($\omega_1, \ldots, \omega_n$ as in (i)) one has $\psi_{IJ} = 0$ on $U \cap bD$ whenever $n \in J$. (Note: if $J = (j_1, \ldots, j_q)$, $\bar{\omega}^J := \bar{\omega}_{j_1} \wedge \ldots \wedge \bar{\omega}_{j_q}$, etc.)

E.3.8. Let d be the exterior derivative on $\mathbb{R}^n$.

(i) Show that $\delta = -*d*$ satisfies

$$(d\varphi, \psi) = (\varphi, \delta\psi)$$

for all forms $\varphi \in C^1_{r-1}(\mathbb{R}^n)$ and $\psi \in C^1_r(\mathbb{R}^n)$ with compact support.

(ii) Compute an explicit formula for δ on 1-forms.

(iii) Show that

$$\Delta f := \delta\, df = -\sum_{j=1}^{n} \frac{\partial^2 f}{\partial x_j^2}$$

for every C^2 function f.

Notes for Chapter III

Since the topics of this chapter are not properly part of complex function theory, we limit our comments to a few suggestions for further reading. An excellent reference for §1 and §2 is R. Narasimhan's book [Nar 4] which, of course, contains much more than is needed here. G. deRham's book [Rha] is still the classic for those aspects of Riemannian Geometry which do not involve curvature. There are numerous good texts on Differential Geometry, but Hermitian Geometry, i.e., *complex* differential geometry, is a fairly new and difficult field which is developing very rapidly; unfortunately, a clear and comprehensive introduction to this subject, suitable for the nonexpert, still needs to be written. In the meantime, the reader may consult the short and very readable notes by S.S. Chern [Che], the excellent survey by H. Wu [Wu], and the books by J. Morrow and K. Kodaira [MoKo] and by R.O. Wells, Jr. [Wel]. A very good self-contained discussion of the existence and regularity theory for the non-elliptic boundary value problem associated to the complex Laplacian is given by G. Folland and J.J. Kohn [FoKo].

CHAPTER IV

Integral Representations in $\mathbb{C}^n$

In this chapter we develop the basic machinery of integral representations of functions and differential forms in $\mathbb{C}^n$ as it relates to the Cauchy–Riemann operator. These representations have their roots in potential theory, the link being the relationship between the complex Laplacian $\square$ and the ordinary Laplacian Δ established in Chapter III, §3.6.

In §1 we use a simple integration by parts and the Hodge $*$-operator in order to derive from the Newtonian potential the kernels relevant for complex analysis—the so-called Bochner–Martinelli–Koppelman kernels—and to establish the corresponding basic representation formula for differential forms on domains with piecewise C^1 boundary. In case of one complex variable this result coincides with the classical generalized Cauchy (or Cauchy–Green) Integral Formula for C^1 functions due to D. Pompeiu. The far-reaching applications of this formula (in $\mathbb{C}^1$) are mainly due to the fact that the Cauchy kernel is holomorphic in the parameter z, and thus it can be readily used to construct global holomorphic functions with many useful special properties. This fact is no longer true for the kernels of Bochner, Martinelli, and Koppelman in two or more variables, which suggests, once again, that potential theoretic methods are of more limited use in several complex variables than in case of one variable (see also Chapter II, §4.5). Integral representations really became a useful tool for global problems in complex analysis only after learning how to construct kernels holomorphic in the parameter near the boundary of sufficiently general classes of domains.

Nevertheless, the potential theoretic kernels have interesting applications: we use them in §2.1 to obtain a simple proof of the Hartogs Extension Theorem, and in §2.4 we discuss results on $\bar{\partial}$-cohomology with compact supports. We also use them to prove a version of Hartogs' Theorem, due to S. Bochner, in which the functions to be extended are only defined and smooth on the boundary bD and satisfy the "tangential Cauchy–Riemann equations".

This result is obtained in §2.3 in its sharpest form—i.e., if $f \in C^k(bD)$, then the extension is in $\mathcal{O}(D) \cap C^k(\bar{D})$—after a careful analysis of the "jump behavior" of the Bochner–Martinelli transform in §2.2. The results in §2.2 and §2.3 are somewhat more delicate and technical than the rest of this chapter, and the reader may skip the details without loss of continuity. In fact, none of the material in §2 is used later in the book.

From the point of view of integral representations the major difference between the case of *one* complex variable and the general case is the fact that in one variable there is essentially only one kernel—the Cauchy kernel—while in several variables one has great freedom to modify, by a basically algebraic procedure, the original potential theoretic kernels. This seems to have been noticed first by J. Leray [Ler 1] in 1956, and it was eventually generalized by W. Koppelman [Kop 1, 2] in 1967; shortly thereafter G.M. Henkin [Hen 1, 2], E. Ramirez [Ram], and H. Grauert and I. Lieb [GrLi] combined these methods with deep global results from the classical theory of several complex variables in order to construct appropriate generalizations of the Cauchy kernel and of the Cauchy–Green formula on strictly pseudoconvex domains. In §3 of this chapter we discuss a general version of the results of Leray and Koppelman which is of fundamental importance for the following chapters, and we give a first indication of the power of these methods by considering the geometrically simple case of convex domains. This sets the stage for the solution of the fundamental global problems on strictly pseudoconvex domains in Chapter V.

In §4 we discuss a completely different type of integral representation for holomorphic functions due to S. Bergman, the roots of which are based in abstract Hilbert space theory. These results will be used again only in Chapter VII, §7, where we will obtain detailed analytic information about the Bergman kernel which is needed for the study of boundary regularity of biholomorphic maps in VII.§8.

§1. The Bochner–Martinelli–Koppelman Formula

In this section we discuss a complex version of Green's Formula and the related integral representation formulas of Bochner, Martinelli, and Koppelman. We shall first deal with the case of functions in order to familiarize the reader with an important special case. The general case of differential forms of positive degree, which will be fundamental for our discussion later on, will then appear as a natural and easy generalization.

1.1. A Complex Green's Formula

The reader may be familiar with the following version of Green's formula in potential theory (see Exercise E.1.1): *if* $n \geq 2$ and $f \in C^2(\mathbb{R}^n)$ *has compact support, then*

$$f(y) = \int_{\mathbb{R}^n} \Delta f(x) G^{(n)}(x, y)\, dV(x), \tag{1.1}$$

where $\Delta = -\sum \partial^2/\partial x_j^2$ (see III.§3.6) and $G^{(n)}$ is the Newtonian potential

$$G^{(n)}(x, y) = \begin{cases} -\dfrac{1}{\sigma_1} \log |x - y| & \text{for } n = 2 \\ \dfrac{|x - y|^{2-n}}{(n-2)\sigma_{n-1}} & \text{for } n > 2, \end{cases}$$

with σ_{n-1} denoting the surface area of the unit sphere S^{n-1} in $\mathbb{R}^n$.

In $\mathbb{C}^n$ the complex Laplacian $\Box$ satisfies $\Box = 1/2\Delta$ (Lemma III.3.8), so, if one sets $\Gamma^{(n)} = 2G^{(2n)}$, (1.1) trivially implies

$$f(z) = \int_{\mathbb{C}^n} \Box f(\zeta) \Gamma^{(n)}(\zeta, z)\, dV(\zeta) = (\Box f, \Gamma^{(n)}(\cdot, z)) \tag{1.2}$$

for $f \in C_0^2(\mathbb{C}^n)$.

Since the area σ_{2n-1} of the unit sphere in $\mathbb{C}^n = \mathbb{R}^{2n}$ is given by

$$\sigma_{2n-1} = \frac{2\pi^n}{(n-1)!},$$

one has the following more explicit formula for $\Gamma^{(n)}$ on $\mathbb{C}^n \times \mathbb{C}^n$:

$$\Gamma^{(n)}(\zeta, z) = \begin{cases} -\dfrac{1}{2\pi} \log |\zeta - z|^2 & \text{for } n = 1. \\ \dfrac{(n-2)!}{2\pi^n} |\zeta - z|^{2-2n} & \text{for } n \geq 2. \end{cases} \tag{1.3}$$

Remark on notation. Here and in the following we shall always denote the variable of integration by ζ and think of z as a parameter in the integral. All operators on functions and forms will act in ζ unless indicated otherwise by subscripts. For example, $\bar\partial \Gamma = \sum (\partial \Gamma/\partial \bar\zeta_j)\, d\bar\zeta_j$, and $\bar\partial_z \Gamma = \sum (\partial \Gamma/\partial \bar z_j)\, d\bar z_j$. Also, we shall write Γ instead of $\Gamma^{(n)}$, and set $|\zeta - z|^2 = \beta$.

More significant for our purposes than (1.2) is the following equivalent representation formula.

Theorem 1.1. *The kernel $\Gamma = \Gamma^{(n)}$ defined on $\mathbb{C}^n \times \mathbb{C}^n$ by (1.3) satisfies*

$$f(z) = (\bar\partial f, \bar\partial \Gamma(\cdot, z))_{\mathbb{C}^n} = \int \bar\partial f \wedge *\partial \Gamma \tag{1.4}$$

for all $f \in C_0^1(\mathbb{C}^n)$.

The equivalence of (1.2) and (1.4) for $f \in C_0^2$ involves just an integration by parts (see III.3.5), since

$$(\square f, \Gamma) = (\vartheta\bar{\partial} f, \Gamma) = (\bar{\partial} f, \bar{\partial}\Gamma). \tag{1.5}$$

The reader should check that (1.5) holds in spite of the singularity of Γ at $\zeta = z$ (see Exercise E.1.2). Thus Theorem 1.1 is a consequence of Green's formula (1.1). Because of the fundamental importance of (1.4) we shall now give a complete proof of Theorem 1.1, independent of Green's formula.

We need the following properties of Γ.

Lemma 1.2. (i) $\square\Gamma(\zeta, z) = 0$ *for* $\zeta \neq z$; (ii) $-\int_{bB(z,\varepsilon)} *\partial\bar{\Gamma}(\cdot, z) = 1$ *for all* $z \in \mathbb{C}^n$ *and* $\varepsilon > 0$.

Proof. (i) is a straightforward computation which is left to the reader (use Lemma III.3.8). For (ii), it follows from (1.3) that

$$\partial\bar{\Gamma} = -\sigma_{2n-1}^{-1}\beta^{-n}\partial\beta,$$

where $\beta = |\zeta - z|^2$. Since $\beta = \varepsilon^2$ on $bB(z, \varepsilon)$, we obtain

$$-\int_{bB} *\partial\bar{\Gamma} = \sigma_{2n-1}^{-1}\varepsilon^{-2n}\int_{bB} *\partial\beta. \tag{1.6}$$

Now $r(\zeta) = \beta(\zeta, z) - \varepsilon^2$ is a defining function for $B(z, \varepsilon)$ with $\partial\beta = \partial r$ and $\|dr\| = 2\varepsilon$ on $bB(z, \varepsilon)$. Thus it follows from Corollary III.3.5 that

$$\int_{bB(z,\varepsilon)} *\partial\beta = \int_{bB(z,\varepsilon)} *\partial r = \varepsilon\int_{bB(z,\varepsilon)} dS = \varepsilon^{2n}\sigma_{2n-1}. \tag{1.7}$$

Equations (1.6) and (1.7) imply the desired result. ∎

Proof of Theorem 1.1. In order to integrate by parts in $(\bar{\partial} f, \bar{\partial}\Gamma(\cdot, z))$ we must first remove a ball $B(z, \varepsilon)$ from the region of integration. Because f has compact support in $\mathbb{C}^n$, Lemma III.3.7 gives

$$\begin{aligned}(\bar{\partial} f, \bar{\partial}\Gamma)_{\mathbb{C}^n - B(z,\varepsilon)} &= (f, \vartheta\bar{\partial}\Gamma)_{\mathbb{C}^n - B(z,\varepsilon)} + \int_{-bB(z,\varepsilon)} f \wedge *\partial\bar{\Gamma} \\ &= -\int_{bB(z,\varepsilon)} f \wedge *\partial\bar{\Gamma},\end{aligned} \tag{1.8}$$

where we have used $\vartheta\bar{\partial}\Gamma = \square\Gamma = 0$ outside $B(z, \varepsilon)$.

We now let ε go to zero in (1.8). Since $\bar{\partial}\Gamma$ is locally integrable in $\mathbb{C}^n$, we have $\lim_{\varepsilon\to 0}(\bar{\partial} f, \bar{\partial}\Gamma)_{\mathbb{C}^n - B(z,\varepsilon)} = (\bar{\partial} f, \bar{\partial}\Gamma)_{\mathbb{C}^n}$. For the other side of (1.8) notice that Lemma 1.2 (ii) implies

$$-\int_{bB(z,\varepsilon)} f(\zeta) \wedge *\partial\bar{\Gamma}(\zeta, z) = f(z) - \int_{bB(z,\varepsilon)} [f(\zeta) - f(z)]*\partial\bar{\Gamma}(\zeta, z). \tag{1.9}$$

Since $\sup_{bB(z,\varepsilon)}|f(\zeta) - f(z)| \to 0$ as $\varepsilon \to 0$ (continuity of f at z is all that is needed here), and $|\partial\bar{\Gamma}| \leq C\varepsilon^{-(2n-1)}$ on $bB(z, \varepsilon)$, a simple estimation of the integral on the right in (1.9) implies

$$\lim_{\varepsilon \to 0} -\int_{bB(z,\varepsilon)} f \wedge *\partial\bar{\Gamma} = f(z). \quad \blacksquare \tag{1.10}$$

If one drops the support condition in Theorem 1.1, one obtains the following integral representation formula for functions.

Corollary 1.3. *Suppose $D \subset\subset \mathbb{C}^n$ has piecewise C^1 boundary. Then*

$$f(z) = -\int_{bD} f \wedge *\partial\bar{\Gamma}(\cdot, z) + (\bar{\partial} f, \bar{\partial}\Gamma(\cdot, z))_D$$

for $f \in C^1(\bar{D})$ and $z \in D$.

Proof. In the proof of Theorem 1.1 one replaces the region of integration $\mathbb{C}^n - B(z, \varepsilon)$ by $D - B(z, \varepsilon)$, where ε is so small that $B(z, \varepsilon) \subset\subset D$. According to Lemma III.3.7, there is now an additional boundary integral $\int_{bD} f \wedge *\partial\bar{\Gamma}$ on the right side of (1.8), and the desired result follows as before. $\blacksquare$

1.2. The Bochner–Martinelli Kernel

The double form

$$K_0(\zeta, z) = -*\partial\bar{\Gamma}(\zeta, z),$$

of type $(n, n-1)$ in ζ and type $(0, 0)$ in z, is called the **Bochner–Martinelli kernel** (for functions). We note that $K_0(\zeta, z)$ is real analytic on $\mathbb{C}^n \times \mathbb{C}^n - \{\zeta = z\}$. Furthermore, K_0 is $\bar{\partial}$-closed.

Lemma 1.4. *The Bochner-Martinelli kernel satisfies $\bar{\partial}_\zeta K_0 = 0$ on $\mathbb{C}^n \times \mathbb{C}^n - \{\zeta = z\}$.*

Proof. For $\zeta \neq z$, $\bar{\partial}_\zeta K_0 = -\bar{\partial}_\zeta * \partial_\zeta \bar{\Gamma} = *\Box\bar{\Gamma} = 0$. $\blacksquare$

Let us calculate K_0 more explicitly. First, if $n = 1$,

$$K_0(\zeta, z) = \frac{1}{2\pi} \frac{*d\zeta}{\zeta - z} = \frac{1}{2\pi i} \frac{d\zeta}{\zeta - z}.$$

So K_0 is simply the Cauchy kernel! Corollary 1.3 is nothing else but the classical (generalized) Cauchy integral formula:

$$f(z) = \frac{1}{2\pi i} \int_{bD} \frac{f(\zeta)\, d\zeta}{\zeta - z} - \frac{1}{2\pi i} \int_D \frac{\partial f}{\partial \bar{\zeta}}(\zeta) \frac{d\bar{\zeta} \wedge d\zeta}{\zeta - z}. \tag{1.11}$$

If $n > 1$, one can write $K_0(\zeta, z)$ in several ways.

Lemma 1.5. *With $\beta = |\zeta - z|^2$, one has the following representations for $K_0 = -*\partial\bar{\Gamma}$:*

(a) $$K_0 = \frac{(n-1)!}{(2\pi i)^n} \beta^{-n} \sum_{j=1}^{n} (\bar{\zeta}_j - \bar{z}_j)\, d\zeta_j \wedge \left(\bigwedge_{\nu \neq j} d\bar{\zeta}_\nu \wedge d\zeta_\nu \right);$$

(b) $$K_0 = \frac{(n-1)!}{(2\pi i)^n} (-1)^{n(n-1)/2} \beta^{-n} \sum_{j=1}^{n} (-1)^{j-1} (\bar{\zeta}_j - \bar{z}_j)\, d\bar{\zeta}_1 \wedge \ldots [d\bar{\zeta}_j] \wedge \ldots d\bar{\zeta}_n \wedge d\zeta^N;$$

(c) $$K_0 = \frac{1}{(2\pi i)^n} \beta^{-n} \cdot \partial\beta \wedge (\bar{\partial}\partial\beta)^{n-1};$$

(d) $$K_0 = \frac{1}{(2\pi i)^n} \frac{\partial\beta}{\beta} \wedge \left(\bar{\partial} \frac{\partial\beta}{\beta} \right)^{n-1}.$$

Proof. By calculating $\partial\Gamma$, one obtains

$$K_0 = \frac{(n-1)!}{2\pi^n} \beta^{-n} \sum_{j=1}^{n} (\bar{\zeta}_j - \bar{z}_j) * d\zeta_j,$$

and by III, (3.20),

$$* d\zeta_j = 2 \cdot (2i)^{-n}\, d\zeta_j \wedge \left(\bigwedge_{\nu \neq j} d\bar{\zeta}_\nu \wedge d\zeta_\nu \right),$$

this gives (a). (b) follows from (a) by a permutation of the differentials. For (c), observe that $\bar{\partial}\partial\beta = \sum_{j=1}^{n} d\bar{\zeta}_j \wedge d\zeta_j$, so

$$(\bar{\partial}\partial\beta)^{n-1} = (n-1)! \sum_{j=1}^{n} \bigwedge_{\nu \neq j} d\bar{\zeta}_\nu \wedge d\zeta_\nu.$$

Therefore,

$$\partial\beta \wedge (\bar{\partial}\partial\beta)^{n-1} = (n-1)! \sum_{j=1}^{n} (\bar{\zeta}_j - \bar{z}_j)\, d\zeta_j \wedge \left(\bigwedge_{\nu \neq j} d\bar{\zeta}_\nu \wedge d\zeta_\nu \right),$$

which shows that (c) is equivalent to (a). Finally, by the quotient rule, $\bar{\partial}(\partial\beta/\beta) = (\bar{\partial}\partial\beta)/\beta - (\bar{\partial}\beta/\beta) \wedge (\partial\beta/\beta)$; so $\partial\beta/\beta \wedge \bar{\partial}(\partial\beta/\beta) = (\partial\beta/\beta) \wedge \bar{\partial}\partial\beta/\beta$, and (d) follows. ∎

Corollary 1.3 includes as a special case the following important representation formula for holomorphic functions (S. Bochner [Boc], E. Martinelli [Mar 1]).

Proposition 1.6. (The Bochner–Martinelli Integral Formula). *Let $D \subset\subset \mathbb{C}^n$ be a domain with piecewise C^1 boundary. Then, for $f \in C(\bar{D}) \cap \mathcal{O}(D)$,*

(1.12) $$\int_{bD} f(\zeta) K_0(\zeta, z) = \begin{cases} f(z) & \text{for } z \in D \\ 0 & \text{for } z \notin \bar{D} \end{cases}.$$

Proof. Assume first that $f \in C^1(\bar{D})$ and $\bar{\partial} f = 0$ on D. If $z \in D$, the result follows from Corollary 1.3. For $z \notin \bar{D}$, note that

$$d(fK_0) = \bar{\partial}(fK_0) = f \wedge \bar{\partial}K_0 = 0$$

on D, by Lemma 1.4, so the result follows by applying Stokes' Theorem. For the general case, apply the Proposition to a suitable exhaustion of D by domains $D_k \subset\subset D$, $k = 1, 2, \ldots$, and pass to the limit $k \to \infty$. ∎

Note that in case $n = 1$, formula (1.12) is just the usual Cauchy integral formula. However, this generalization to several variables has the serious drawback that the kernel $K_0(z, \zeta)$ is not holomorphic for $n > 1$, so if f is a continuous function on bD, $\int_{bD} fK_0(\cdot, z)$ does not define, in general, a holomorphic function on $\mathbb{C}^n \setminus bD$ if $n > 1$. The question arises if, and under what circumstances, one can find a reproducing kernel which is holomorphic in z. It turns out that this question is intimately related with function theoretic properties of the domain D. The case of convex domains will be discussed in §3.2, and strictly pseudoconvex domains will be discussed in Chapter VII, §1 and §3.

It is of interest to understand better how the kernel $K_0(\zeta, z)$ nevertheless gives rise to a holomorphic expression $\int_{bD} fK_0(\cdot, z)$ when f is holomorphic; that is, how can one see explicitly that $\int_{bD} fK_0(\cdot, z)$ is independent of $\bar{z}$? Fix $p \in D$; for $\zeta \in bD$ and z in a suitable neighborhood U of p, one obviously can write $K_0(\zeta, z) = \tilde{K}_0(\zeta, z, \bar{z})$, where $\tilde{K}_0(\zeta, z, w)$ is holomorphic in $(z, w) \in U \times \bar{U}$. Therefore, the function $H(z, w) = \int_{bD} f\tilde{K}_0(\cdot, z, w)$ is holomorphic in $U \times \bar{U}$, and, by (1.12), $H(z, w) = f(z)$ on the hyperplane $w = \bar{z}$. By Exercise E.I.1.13, this implies $H(z, w) \equiv f(z)$ in $U \times \bar{U}$. In particular, for $z \in U$,

$$f(z) = H(z, \bar{p}) = \frac{(n-1)!}{(2\pi i)^n} \int_{bD} f(\zeta) \frac{\sum_{j=1}^{n} (\bar{\zeta}_j - \bar{p}_j)\, d\zeta_j \wedge \left(\bigwedge_{\nu \neq j} d\bar{\zeta}_\nu \wedge d\zeta_\nu \right)}{\sum_{j=1}^{n} (\zeta_j - z_j)(\bar{\zeta}_j - \bar{p}_j)},$$

where now the kernel is indeed independent of $\bar{z}$.

1.3. A Representation in Terms of $\bar{\partial}$ and ϑ

We now consider the analog of the representation given by Corollary 1.3 for forms of type $(0, q)$, $q > 0$. We begin by finding analogs of $\Gamma = \Gamma^{(n)}$ and of the representation $f = (\Box f, \Gamma)$ (cf. (1.2)) for such forms. This is very easy, since by Lemma III.3.9 the complex Laplacian $\Box$ in $\mathbb{C}^n$ is a diagonal operator.

Let $f = \sum f_J \, d\bar{z}^J$ be a $(0, q)$-form with coefficients $f_J \in C_0^2(\mathbb{C}^n)$. Then

$$\begin{aligned} f(z) &= \sum f_J(z)\, d\bar{z}^J = \sum (\Box f_J, \Gamma(\cdot, z))\, d\bar{z}^J \\ &= 2^{-q} \left(\Box f, \sum_J \Gamma(\cdot, z)\, d\bar{\zeta}^J \wedge dz^J \right), \end{aligned}$$

where we have used Lemma III.3.9.

Let us define the double form Γ_q on $\mathbb{C}^n \times \mathbb{C}^n - \{\zeta = z\}$ by

$$\Gamma_q(\zeta, z) = 2^{-q} \Gamma(\zeta, z) \sum_{|J|=q} d\bar{\zeta}^J \wedge dz^J$$

for $0 \le q \le n$. Also set $\Gamma_{-1} \equiv 0$. The above formula then states

(1.13) $\quad f(z) = (\Box f, \Gamma_q(\cdot, z)) \qquad$ *for all* $f \in C^2_{0,q}(\mathbb{C}^n)$ *with compact support.*

Notice that $\Gamma_0 = \Gamma$ and that $\Box\Gamma_q(\zeta, z) = 0$ for $\zeta \ne z$ (use Lemma III.3.9 and Lemma 1.2(i)). Furthermore, the singularity of Γ shows that every first order derivative of Γ_q is locally in L^1. Let us introduce the notation

$$Q_D(f, g) = (\bar\partial f, \bar\partial g)_D + (\vartheta f, \vartheta g)_D$$

for $(0, q)$-forms f, g, where it is understood that the integrals should exist.

Theorem 1.7. *Let $D \subset\subset \mathbb{C}^n$ be a domain with piecewise C^1 boundary. Fix $0 \le q \le n$, and let $f \in C^1_{0,q}(\bar D)$. Then*

$$f(z) = Q_D(f, \Gamma_q(\cdot, z)) - \int_{bD} f \wedge *\partial\bar\Gamma_q + \int_{bD} \overline{\vartheta\Gamma_q} \wedge *f, \qquad z \in D.$$

Remark. For $q = 0$, the above representation reduces precisely to the one given by Corollary 1.3, as $\vartheta \equiv 0$ on functions.

Proof. First we consider $f \in C^2_{0,q}(\bar D)$. Fix $z \in D$ and write $f = f_1 + f_2$, where $f_1 \in C^2_{0,q}$ has compact support in D, and $f_2 \equiv 0$ on $\overline{B(z, \varepsilon)} \subset\subset D$, for some $\varepsilon > 0$. Integrating by parts once in (1.13) with $f = f_1$, one obtains

$$\begin{aligned} f(z) &= f_1(z) = (\Box f_1, \Gamma_q)_D = ((\vartheta\bar\partial + \bar\partial\vartheta)f_1, \Gamma_q)_D \\ &= Q_D(f_1, \Gamma_q) = Q_D(f, \Gamma_q) - Q_{D-B}(f_2, \Gamma_q). \end{aligned} \tag{1.14}$$

Integrating by parts again (use Lemma III.3.7, and note that $f_2 = 0$ on bB!), gives

$$\begin{aligned} Q_{D-B}(f_2, \Gamma_q) &= \int_{bD} f_2 \wedge *\partial\bar\Gamma_q + (f_2, \vartheta\bar\partial\Gamma_q)_{D-B} - \int_{bD} \overline{\vartheta\Gamma_q} \wedge *f_2 \\ &\quad + (f_2, \bar\partial\vartheta\Gamma_q)_{D-B} \\ &= \int_{bD} f \wedge *\partial\bar\Gamma_q - \int_{bD} \overline{\vartheta\Gamma_q} \wedge *f. \end{aligned}$$

The last equality holds since $f_2 = f$ on bD and $\Box\Gamma_q = 0$ for $\zeta \ne z$. Combining this result with (1.14) gives the Theorem for f of class C^2. If f is of class C^1, one approximates f in C^1 norm on $\bar D$ by forms of class C^2, and uses a limit argument. ■

1.4. The Basic Representation in Terms of $\bar\partial$

Suppose f is a $C^1_{0,q}$ form with compact support. By Theorem 1.7,

$$f(z) = Q_{\mathbb{C}^n}(f, \Gamma_q(\cdot, z)) = (\bar\partial f, \bar\partial\Gamma_q) + (\vartheta f, \vartheta\Gamma_q). \tag{1.15}$$

If $q = 0$, the part involving ϑ vanishes. Superficially, this appears just as a matter of definition, since $\vartheta \equiv 0$ on functions. But there is a deeper explanation. Suppose, in addition, that $\bar{\partial} f = 0$; then f is a holomorphic function with compact support, hence $f \equiv 0$. Thus, for functions with compact support, the 0-function is characterized by $\bar{\partial} f = 0$. So it is indeed reasonable to expect that nontrivial functions f with compact support can be described completely by $\bar{\partial} f$.

However, if $q \geq 1$, there are nontrivial $\bar{\partial}$-closed forms f with compact support; so obviously $\bar{\partial} f$ is not sufficient to describe $(0, q)$-forms f when $q > 0$. But Equation (1.15) shows that $\bar{\partial}$ together with its formal adjoint ϑ completely describes forms with compact support. This is a remarkable property for the differential operator $\bar{\partial}$! In order to appreciate this, let us first consider the following abstract setting. Let T and S be bounded linear operators between Hilbert spaces H_i, $i = 1, 2, 3$,

$$H_1 \xrightarrow{T} H_2 \xrightarrow{S} H_3,$$

and assume $S \circ T = 0$, i.e., $R_T = \text{range } T \subset \ker S$. It is a standard result that one has the orthogonal decomposition

$$\ker T^* \oplus \bar{R}_T = H_2.$$

Since $\bar{R}_T \subset \ker S$, one obtains

$$\ker S = (\ker S \cap \ker T^*) \oplus \bar{R}_T.$$

So, if

$$\ker S \cap \ker T^* = \{0\}, \tag{1.16}$$

it follows that

$$\ker S = \bar{R}_T. \tag{1.17}$$

These abstract results remain valid for densely defined, closed linear operators (which may be unbounded). So, arguing formally, without specifying the Hilbert spaces and the domains of the operators, the above can be applied to $T = \bar{\partial}$ on $(0, q-1)$-forms and $S = \bar{\partial}$ on $(0, q)$-forms, $q > 0$. Formally, $T^* = \vartheta$, so that Equation (1.15) implies (1.16) and hence (1.17), at least when restricting oneself to compactly supported $(0, q)$-forms. So, the basic formula (1.15) leads one to suspect that a $\bar{\partial}$-closed $(0, q)$-form f with compact support lies in the "closure of the range of $\bar{\partial}$", or, even better, is actually $\bar{\partial}$-exact. We will see shortly that this is indeed correct. On the other hand, the above discussion also shows that the main difficulty in completely understanding the $\bar{\partial}$-operator without support conditions will involve an analysis of the boundary integrals in Theorem 1.7.

The following, surprisingly simple lemma is the key to making precise what has been suggested above.

Lemma 1.8. *Let* $0 \leq q \leq n$. *Then*

(a)
$$\vartheta_\zeta \Gamma_q = \partial_z \Gamma_{q-1},$$

(b) $$\partial_\zeta \bar{\Gamma}_{q-1} = \vartheta_z \bar{\Gamma}_q$$

on $\mathbb{C}^n \times \mathbb{C}^n - \{\zeta = z\}$.

PROOF. This is a straightforward computation. Since $\Gamma_{-1} \equiv 0$, it is enough to consider $q \geq 1$. Using the definitions of Γ_q, ϑ, and ∂, one obtains

$$\vartheta_\zeta \Gamma_q = -2\frac{-(n-1)!}{2^{q+1}\pi^n}\beta^{-n}\sum_{j,K,J}\varepsilon^J_{jK}\frac{\partial\beta}{\partial\zeta_j}d\bar{\zeta}^K \wedge dz^J$$

and

$$\partial_z \Gamma_{q-1} = \frac{-(n-1)!}{2^q\pi^n}\beta^{-n}\sum_{j,K,J}\varepsilon^J_{jK}\frac{\partial\beta}{\partial z_j}d\bar{\zeta}^K \wedge dz^J.$$

Since $\partial\beta/\partial\zeta_j = (\bar{\zeta}_j - \bar{z}_j) = -\partial\beta/\partial z_j$, part (a) follows. (b) follows from (a) by interchanging z and ζ. ■

We now apply this result to obtain a new version of Theorem 1.7.

Proposition 1.9. *Let $D \subset\subset \mathbb{C}^n$ be a domain with piecewise C^1 boundary. Let $0 \leq q \leq n$, and $f \in C^1_{0,q}(\bar{D})$. Then*

(a) $$(f, \bar{\partial}\Gamma_{q-1}(\cdot, z))_D \text{ is of class } C^1 \text{ for } z \in D;$$

(b) $$f(z) = -\int_{bD} f \wedge *\partial\bar{\Gamma}_q(\cdot, z) + (\bar{\partial}f, \bar{\partial}\Gamma_q(\cdot, z))_D$$
$$+ \bar{\partial}_z(f, \bar{\partial}\Gamma_{q-1}(\cdot, z))_D, \qquad z \in D.$$

PROOF. By Lemma 1.8.,

$$(1.18)\qquad \begin{aligned}(\vartheta f, \vartheta\Gamma_q(\cdot, z))_D &= (\vartheta f, \partial_z\Gamma_{q-1}(\cdot, z))_D\\ &= \bar{\partial}_z(\vartheta f, \Gamma_{q-1}(\cdot, z))_D.\end{aligned}$$

The interchange of differentiation and integration is legitimate, as the first order derivatives of Γ_{q-1} are in L^1_{loc}. Furthermore, the expression $(\vartheta f, \Gamma_{q-1}(\cdot, z))_D$ is of class C^1 on D. Integrating by parts,

$$(1.19)\qquad (\vartheta f, \Gamma_{q-1}(\cdot, z))_D = -\int_{bD} \bar{\Gamma}_{q-1}(\cdot, z) \wedge *f + (f, \bar{\partial}\Gamma_{q-1}(\cdot, z))_D.$$

Clearly the boundary integral is C^1 (even real analytic) on $\mathbb{C}^n - bD$, so the inner product on the right side is in $C^1(D)$. This proves (a). (A different proof of this will be given in §1.6.)

Now apply $\bar{\partial}_z$ to (1.19), and use Lemma 1.8. in the boundary integral to obtain, together with (1.18),

$$(\vartheta f, \vartheta\Gamma_q(\cdot, z))_D = -\int_{bD} \overline{\vartheta\Gamma_q}(\cdot, z) \wedge *f + \bar{\partial}_z(f, \bar{\partial}\Gamma_{q-1}(\cdot, z))_D.$$

After substituting the last equation into the formula in Theorem 1.7, two of the boundary integrals cancel, and the proof is complete. ∎

It is now quite natural to introduce the following generalization of the Bochner–Martinelli kernel $K_0(\zeta, z)$:

Definition. For $0 \le q \le n$, the Bochner–Martinelli–Koppelman (BMK) kernel K_q for $(0, q)$-forms is defined by

$$K_q(\zeta, z) = -*_\zeta \partial_\zeta \overline{\Gamma}_q(\zeta, z).$$

One also sets $K_{-1} = 0$.

Note that $K_n \equiv 0$. K_q is a real analytic double differential form on $\mathbb{C}^n \times \mathbb{C}^n - \{\zeta = z\}$, of type $(n, n-q-1)$ in ζ and type $(0, q)$ in z.

We now reformulate Proposition 1.9, using the new terminology.

Theorem 1.10. (The Bochner–Martinelli–Koppelman formula [Kop 2]). *Let $D \subset\subset \mathbb{C}^n$ be a domain with piecewise C^1 boundary. Let $0 \le q \le n$. Then every $f \in C^1_{0,q}(\overline{D})$ is represented on D by*

$$f(z) = \int_{bD} f \wedge K_q(\cdot, z) - \int_D \bar{\partial} f \wedge K_q(\cdot, z) - \bar{\partial}_z \int_D f \wedge K_{q-1}(\cdot, z),$$

where $\int_D f \wedge K_{q-1}(\cdot, z) \in C^1_{0,q-1}(D)$.

We state some immediate consequences of Theorem 1.10.

Corollary 1.11.

(a) *If $1 \le q \le n-1$, $f \in C^1_{0,q}(\mathbb{C}^n)$ with compact support, and $\bar{\partial} f = 0$, then there is a $u \in C^1_{0,q-1}(\mathbb{C}^n)$, such that $\bar{\partial} u = f$ on $\mathbb{C}^n$.*

(b) *If $D \subset\subset \mathbb{C}^n$, with piecewise C^1 boundary, and $f \in C^1_{0,n}(\overline{D})$, there is $u \in C^1_{0,n-1}(D)$, such that $\bar{\partial} u = f$ on D.*

Proof. Theorem 1.10 shows in both cases that

$$u(z) = -\int_D f \wedge K_{q-1}(\cdot, z) = (f, \bar{\partial}\Gamma_{q-1}(\cdot, z))_D$$

will do ($D = \mathbb{C}^n$ in a)).

Remark. In case $n = 1$, Theorem 1.10 for $q = 1$ gives the familiar solution formula for the Cauchy–Riemann operator $\partial/\partial \bar{z}$ in $\mathbb{C}$:

$$\frac{\partial}{\partial \bar{z}}\left(-\frac{1}{2\pi i}\int_D f(\zeta)\frac{d\bar{\zeta} \wedge d\zeta}{\zeta - z}\right) = f(z), \qquad z \in D.$$

Here, as well as in Corollary 1.11(b), the hypotheses on f and D can be relaxed considerably (see Exercise E.1.5).

We will see in §2.4 that in Corollary 1.11(a) one can even find a solution of $\bar{\partial} u = f$ with *compact support*.

1.5. Explicit Formulas for K_q

As we have seen, the BMK kernel $K_q = -*\partial\bar{\Gamma}_q$ arises in a natural manner from potential theory. The formula above exhibits this origin well, and it is useful for establishing properties of the BMK kernels. For example, Lemma 1.4 generalizes as follows.

Lemma 1.12. *For* $0 \le q \le n$ *one has*

$$\bar{\partial}_\zeta K_q = (-1)^q \bar{\partial}_z K_{q-1}$$

on $\mathbb{C}^n \times \mathbb{C}^n - \{\zeta = z\}$.

PROOF. By Lemma 1.8(a),

$$\begin{aligned}\bar{\partial}_z K_{q-1} &= -*_\zeta \bar{\partial}_z \partial_\zeta \bar{\Gamma}_{q-1} = -*_\zeta \overline{\partial_\zeta \partial_z \Gamma_{q-1}} \\ &= -*_\zeta \overline{\partial_\zeta \vartheta_\zeta \Gamma_q} = -*_\zeta \overline{\bar{\partial}_\zeta \vartheta_\zeta \Gamma_q} \\ &= *_\zeta \overline{\vartheta_\zeta \bar{\partial}_\zeta \Gamma_q},\end{aligned}$$

where we have used $\square \Gamma_q = 0$ in the last equality. Since $**f = (-1)^q f$ for $(n, n-q)$-forms f,

$$\begin{aligned}*_\zeta \bar{\vartheta}_\zeta \partial_\zeta \bar{\Gamma}_q &= -*_\zeta (*_\zeta \bar{\partial}_\zeta *_\zeta) \partial_\zeta \bar{\Gamma}_q \\ &= (-1)^q \bar{\partial}_\zeta (-*_\zeta \partial_\zeta \bar{\Gamma}_q) = (-1)^q \bar{\partial}_\zeta K_q. \quad \blacksquare\end{aligned}$$

In order to establish regularity properties of the kernels K_q and in order to show the connection with other types of kernels to be discussed in §3, additional representations of K_q are useful (cf. Lemma 1.5).

Lemma 1.13. *With* $\beta = |\zeta - z|^2$, *the following representations are valid for the* BMK *kernel* K_q *for* $0 \le q \le n$:

(a)
$$K_q = \frac{(n-1)!}{2^{q+1}\pi^n} \beta^{-n} \sum_{\substack{j,J \\ |L|=q+1}} \varepsilon^L_{jJ} (\bar{\zeta}_j - \bar{z}_j)(*d\zeta^L) \wedge d\bar{z}^J.$$

(b)
$$K_q = \frac{(-1)^{q(q-1)/2}}{(2\pi i)^n} \binom{n-1}{q} \beta^{-n} \partial_\zeta \beta \wedge (\bar{\partial}_\zeta \partial_\zeta \beta)^{n-q-1} \wedge (\bar{\partial}_z \partial_\zeta \beta)^q.$$

(c) *With* $B = \partial\beta/\beta$,

$$K_q = \frac{(-1)^{q(q-1)/2}}{(2\pi i)^n} \binom{n-1}{q} B \wedge (\bar{\partial}_\zeta B)^{n-q-1} \wedge (\bar{\partial}_z B)^q.$$

PROOF. Using the definitions and calculating $\partial_\zeta \bar{\Gamma}_q$ immediately gives (a). Next, note that by Lemma III.3.3, if $|L| = q + 1$, then

$$*d\zeta^L = \frac{(-1)^{(q+1)q/2}}{2^{n-q-1} i^n} d\zeta^L \wedge \bigwedge_{\nu \in L} (d\bar{\zeta}_\nu \wedge d\zeta_\nu),$$

and $\varepsilon^L_{jJ}\, d\zeta^L = d\zeta_j \wedge d\zeta^J$. Furthermore,

$$(\bar{\partial}_\zeta \partial_\zeta \beta)^{n-q-1} = (n - q - 1)! \sum_{|L'|=n-q-1} \bigwedge_{\nu \in L'} (d\bar{\zeta}_\zeta \wedge d\zeta_\nu),$$

and

$$(\bar{\partial}_z \partial_\zeta \beta)^q = (-1)^q q! \sum_{|J|=q} \bigwedge_{\nu \in J} (d\zeta_\nu \wedge d\bar{z}_\nu).$$

Putting all this together, one obtains (b) as a consequence of (a). Finally, (c) follows from (b) by the quotient rule, as in the proof of Lemma 1.5. ∎

1.6. Regularity Properties

Let D be an open set in $\mathbb{R}^n$. For $0 < \alpha < 1$, the Lipschitz norm $|f|_{\alpha, D}$ of order α of a function $f: D \to \mathbb{C}$ is defined by

$$|f|_{\alpha, D} = \sup_{x \in D} |f(x)| + \sup_{\substack{x, x' \in D \\ x \neq x'}} \frac{|f(x) - f(x')|}{|x - x'|^\alpha}.$$

One defines the Lipschitz space $\Lambda_\alpha(D)$ of order α by

$$\Lambda_\alpha(D) = \{f | f: D \to \mathbb{C}, |f|_{\alpha, D} < \infty\}.$$

One verifies easily that $\Lambda_\alpha(D)$ with the norm $|\cdot|_{\alpha, D}$ is a Banach space. A function $f \in \Lambda_\alpha(D)$ is bounded and uniformly continuous on D:

$$|f(x) - f(x')| \leq |f|_{\alpha, D} |x - x'|^\alpha, \qquad x, x' \in D.$$

We also use the notation $|f|_{0, D} = \sup_{x \in D} |f(x)|$. If the set D is clear from the context, we will often simply write $|f|_\alpha$ instead of $|f|_{\alpha, D}$. If $D \subset \mathbb{C}^n$, and $f = \sum_J f_J\, d\bar{z}^J$ is a $(0, q)$-form on D, we set $|f|_\alpha = \sum |f_J|_\alpha$ for any $0 \leq \alpha < 1$.

For a $(0, q + 1)$-form f on D, we denote by $\mathbf{K}^D_q f$ the $(0, q)$-form defined by

$$\mathbf{K}^D_q f(z) = \int_D f(\zeta) \wedge K_q(\zeta, z),$$

provided the integral makes sense. $\mathbf{K}^D_q f$ is certainly well defined for all $z \in \mathbb{C}^n$ if D is bounded and $f \in L^\infty_{0, q+1}(D)$; furthermore $\mathbf{K}^D_q f$ is real analytic on $\mathbb{C}^n - \bar{D}$.

Theorem 1.14. *Let D be a bounded domain in $\mathbb{C}^n$. Then*

(a) *for $0 < \alpha < 1$, $\mathbf{K}^D_q$ defines a bounded linear transformation $L^\infty_{0, q+1}(D) \to \Lambda_{\alpha, (0, q)}(D)$, i.e., there is a constant $c_\alpha < \infty$, such that*

$$|\mathbf{K}_q^D f|_\alpha \leq c_\alpha |f|_0 \qquad \textit{for all } f \in L^\infty_{0,q+1}(D);$$

(b) *for* $k = 0, 1, 2, \ldots, \infty$, *if* $f \in C^k_{0,q+1}(D) \cap L^\infty_{0,q+1}(D)$, *then* $\mathbf{K}_q^D f \in C^k_{0,q}(D)$.

Proof. Using Lemma 1.13(a), one sees that up to a constant factor, the coefficient $(\mathbf{K}_q^D f)_J$ is given by

$$\sum \varepsilon^L_{jJ} \int_D f_L(\zeta) \frac{\bar{\zeta}_j - \bar{z}_j}{|\zeta - z|^{2n}} \, dV(\zeta).$$

So the theorem is an immediate consequence of the following real variable lemma. ∎

Lemma 1.15. *Let D be a bounded open set in $\mathbb{R}^n$, $n \geq 2$. For $1 \leq j \leq n$ and $f \in L^\infty(D)$, define*

$$\mathbf{T}_j f(x) = \int_D f(y) \frac{y_j - x_j}{|y - x|^n} \, dV(y).$$

Then

(a) *for $0 < \alpha < 1$ there is a constant c_α, such that*

$$|\mathbf{T}_j f|_{\alpha, D} \leq c_\alpha |f|_{0,D} \qquad \textit{for all } f \in L^\infty(D);$$

(b) $\mathbf{T}_j f \in C^k(D)$ *if* $f \in C^k(D) \cap L^\infty(D)$, $\qquad k = 0, 1, 2, \ldots \infty$.

Proof. It is enough to consider $\mathbf{T} = \mathbf{T}_1$. Direct estimation shows $|\mathbf{T}f|_0 \lesssim |f|_0$.[1] Choose $R < \infty$, so that $D \subset B(0, R)$. Fix $x, x' \in B(0, R)$, $x \neq x'$; set $|x - x'| = d$, and let $p = (x + x')/2$. With $T(s) = s_1/|s|^n$, one has

$$|\mathbf{T}f(x) - \mathbf{T}f(x')| \leq |f|_{0,D} \int_{B(0,R)} |T(y - x) - T(y - x')| \, dV(y).$$

Decompose

$$\int_{B(0,R)} |T(y - x) - T(y - x')| = I_1 + I_2,$$

where

$$I_1 = \int_{B(0,R) \cap B(p,2d)} |T(y - x) - T(y - x')|,$$

and

$$I_2 = \int_{B(0,R) - B(p,2d)} |T(y - x) - T(y - x')|.$$

[1] The notation $A \lesssim B$ for two expressions A and B which may depend on various quantities, means that there is a constant c, independent of the quantities under consideration, such that $A \leq cB$. This notation will be used often in estimations.

Then

$$I_1 \le \int_{B(x,3d)} |T(y-x)| + \int_{B(x',3d)} |T(y-x')| \lesssim d.$$

For I_2, note that by the Mean Value Theorem

$$|T(y-x) - T(y-x')| \lesssim |x-x'| \sup_{t\in[x,x']} |y-t|^{-n},$$

where $[x, x']$ denotes the line segment connecting x with x'. Since for $t \in [x, x']$, $B(0, R) - B(p, 2d) \subset B(t, 2R) - B(t, d)$, it follows that

$$I_2 \lesssim d \sup_{t\in[x,x']} \int_{d\le|y-t|\le 2R} |y-t|^{-n}\, dV(y) \lesssim d + d|\log d|.$$

We have thus proved that

$$|\mathbf{T}f(x) - \mathbf{T}f(x')| \lesssim |x-x'|(1 + |\log|x-x'||)$$

for $x, x' \in B(0, R)$. This implies (a).

For part (b), fix $p \in D$ and $\varepsilon > 0$, with $B(p, \varepsilon) \subset\subset D$. Decompose $f = f_1 + f_2$, where $f_1 \in C_0^k(D)$, and $f_2 = 0$ on $B(p, \varepsilon)$. Then $\mathbf{T}f = \mathbf{T}f_1 + \mathbf{T}f_2$, and $\mathbf{T}f_2 \in C^\infty(B(p, \varepsilon))$. After the change of variables $s = y - x$,

$$\mathbf{T}f_1(x) = \int_{\mathbb{R}^n} f_1(y)T(y-x)\, dV(y) = \int_{\mathbb{R}^n} f_1(s+x)T(s)\, dV(s).$$

One can now differentiate with respect to x under the integral sign up to order k and, after changing back the variables, one obtains $D^\alpha(\mathbf{T}f_1) = \mathbf{T}(D^\alpha f_1)$ for any multi-index α, $|\alpha| \le k$. This shows that $\mathbf{T}f_1 \in C^k(\mathbb{R}^n)$, and therefore $\mathbf{T}f \in C^k(B(p, \varepsilon))$. Since p is arbitrary, the proof is complete. ∎

Exercises

E.1.1. Give a direct real variable proof of Green's Formula (1.1) in the text by completing the following steps.

(i) Show that if $f \in C^1(\mathbb{R}^n)$ has compact support, then

$$(df, dG^{(n)}(\cdot, y))_{\mathbb{R}^n} = f(y) \qquad \text{for } y \in \mathbb{R}^n.$$

(Hint: Modify the proof of Theorem 1.1.)

(ii) Integrate by parts in (i) and use $\Delta = \delta d$ on functions where δ is the formal adjoint of d.

E.1.2. Prove in detail that if $f \in C^2(\mathbb{C}^n)$ has compact support, then

$$(\Box f, \Gamma) = (\bar\partial f, \bar\partial \Gamma).$$

(Hint: Consider integration over $\mathbb{C}^n - B(z, \varepsilon)$ first, and then let $\varepsilon \to 0$.)

E.1.3. Generalize E.1.2 to $(0, q)$-forms f with compact support, that is, prove

$$(\Box f, \Gamma_q) = (\bar\partial f, \bar\partial \Gamma_q) + (\vartheta f, \vartheta \Gamma_q).$$

(Careful: There is a singularity at $\zeta = z$!)

E.1.4. Show that $\square\Gamma(\zeta,z) = 0$ for $\zeta \neq z$.

E.1.5. Let D be a bounded domain in $\mathbb{C}^n$ and suppose $f \in C^1_{0,n}(D)$ has coefficients in $L^1(D)$. Prove that $u = \mathbf{K}^D_{n-1} f \in C^1_{0,n-1}(D)$, and that $\bar\partial u = f$ on D.

E.1.6. Let $D \subset\subset \mathbb{C}^n$ have piecewise C^1 boundary and let $f \in C^1_{0,q}(\bar{D})$, $0 \leq q \leq n$. Show that

$$\int_{bD} f \wedge K_q - \int_D \bar\partial f \wedge K_q - \bar\partial_z \int_D f \wedge K_{q-1} = 0$$

for $z \in \mathbb{C}^n - \bar{D}$.

E.1.7. Find an analog of the BMK integral representation (Theorem 1.10) valid for forms of type (p, q), $0 \leq p, q \leq n$. (Hint: Find a kernel $\Gamma_{p,q}$ such that

$$f(z) = (\square f, \Gamma_{p,q}(\cdot, z))$$

for all $f \in C^2_{p,q}(\mathbb{C}^n)$ with compact support, and define $K_{p,q} = -*\partial\bar\Gamma_{p,q}$.)

E.1.8. (i) Show that the Bochner–Martinelli kernel $K_0(\zeta, z)$ is harmonic in z for $z \neq \zeta$.
(ii) Suppose $D \subset\subset \mathbb{C}^n$. Show that $\mathbf{K}^D_0 f$ is harmonic on $\mathbb{C}^n - \bar{D}$ for $f \in C_{0,1}(\bar{D})$.

E.1.9. Suppose $D \subset\subset \mathbb{C}^n$ is a connected region with C^1 boundary. Use Proposition 1.6 to show that if $f \in A(D)$ is zero on an open, nonempty subset of bD, then $f \equiv 0$ on D.

E.1.10. Let $D \subset\subset \mathbb{C}^n$ have C^1 boundary. Prove that a form $f \in C^1_{0,q}(\bar{D})$ satisfies

$$f = \int_{bD} f \wedge K_q - \int_D \bar\partial f \wedge K_q$$

if and only if $\vartheta f \equiv 0$ on D.

§2. Some Applications

2.1. The Hartogs Extension Theorem

In II.§1 we saw examples of domains D with the property that every holomorphic function on D has a holomorphic extension to a strictly larger domain. We now discuss a result of this type which is particularly striking.

Theorem 2.1 (Hartogs [Har 1]). *Let D be a bounded domain in $\mathbb{C}^n$ with connected boundary bD. Assume that $n > 1$. Then every $f \in \mathcal{O}(bD)$ can be extended to a holomorphic function on $\bar{D}$, i.e., there is $F \in \mathcal{O}(\bar{D})$, such that $F = f$ on bD.*

Proof. Let U be an open and connected neighborhood of bD, such that f is holomorphic on U. Choose domains $D_1 \subset\subset D \subset\subset D_2$ with connected piecewise C^1 boundaries, such that $\bar{D}_2 - D_1 \subset U$. By Proposition 1.6, if f has a holomorphic extension F to D_2, F is determined uniquely on D_2 by

$$F(z) = \int_{bD_2} f(\zeta) K_0(\zeta, z). \tag{2.1}$$

So, define F by (2.1) on $\mathbb{C}^n \backslash bD_2$. For $n = 1$, F is clearly holomorphic and this remains true for $n > 1$ as well, even though K_0 is no longer holomorphic in z. This follows from Lemma 1.12 and Stokes' Theorem:

$$\bar{\partial}_z F = \int_{bD_2} f \wedge \bar{\partial}_z K_0 = -\int_{bD_2} f \bar{\partial}_\zeta K_1 = -\int_{bD_2} d(fK_1) = 0. \tag{2.2}$$

What additional conditions imply that $F|D_2$ is an extension of f? Clearly this cannot be true in general: a necessary condition is that $F(z) = \int_{bD_2} fK_0(\cdot, z) = 0$ for $z \notin \bar{D}_2$ (cf. Proposition 1.6, or the Cauchy integral theorem for $n = 1$), and this is certainly false in $\mathbb{C}^1$ if f is only holomorphic on bD_2. The main point of the proof is thus to show that the above necessary condition holds in $\mathbb{C}^n$ for $n > 1$! This is seen as follows.

It follows from (2.1) and Lemma 1.5(a), that $\lim_{|z| \to \infty} F(z) = 0$. Therefore, writing $z = (z', z_n)$ with $z' \in \mathbb{C}^{n-1}$, if $|z'|$ is so large that $(\{z'\} \times \mathbb{C}) \cap \bar{D}_2 = \emptyset$, $F(z', z_n)$ is an entire function in z_n, which vanishes at ∞. By Liouville's Theorem, $F(z', z_n) \equiv 0$ for $|z'|$ large; so, by the identity theorem, $F \equiv 0$ on the unbounded component $\mathbb{C}^n - \bar{D}_2$ of $\mathbb{C}^n - bD_2$!

It is now easy to complete the proof of the theorem. By applying Proposition 1.6 to the domain $D_2 - \bar{D}_1$, one obtains

$$f(z) = \int_{bD_2} fK_0(\cdot, z) - \int_{bD_1} fK_0(\zeta, z), \qquad z \in D_2 - \bar{D}_1.$$

Here the second integral is zero for $z \in \mathbb{C}^n \backslash \bar{D}_1$, by the argument just given, with bD_1 in place of bD_2. So

$$f(z) = \int_{bD_2} fK_0(\cdot, z) = F(z) \qquad \text{for } z \in D_2 - \bar{D}_1,$$

which shows that F is indeed a holomorphic *extension* of f to D_2. ∎

2.2. The Jump Formula for the *BM* Transform

The last part of the preceding proof shows that, under suitable hypotheses, the Bochner–Martinelli transform

$$\mathbf{K}^{bD} f(z) = \int_{bD} f \wedge K_0(\cdot, z)$$

defines a continuous extension of f to $\bar{D}$. We now want to study the boundary behavior of the *BM* transform in more detail. As a motivation, we first consider the case of a C^1 function f on bD, assuming bD of class C^1 as well.

Fix a defining function r for D, and set

$$D_\varepsilon = \{z : r(z) < \varepsilon\}.$$

We can assume that f is defined and of class C^1 on $\bar{D}$. Applying Corollary 1.3 to $D - \bar{D}_\varepsilon$, where $\varepsilon < 0$, one obtains

$$f(z) = \int_{bD} fK_0(\cdot, z) - \int_{bD_\varepsilon} fK_0(\cdot, z) - \int_{D \backslash \bar{D}_\varepsilon} \bar{\partial} f \wedge K_0(\cdot, z), \qquad z \in D \backslash \bar{D}_\varepsilon. \tag{2.3}$$

Fix $p \in bD$, and let $z \to p$ in (2.3); then let $\varepsilon \to 0^-$, obtaining

$$f(p) = \lim_{\substack{z \in D \\ z \to p}} \mathbf{K}^{bD} f(z) - \lim_{\varepsilon \to 0^-} \mathbf{K}^{bD_\varepsilon} f(p).$$

It appears plausible that

$$\lim_{\varepsilon \to 0^-} \mathbf{K}^{bD_\varepsilon} f(p) = \lim_{\substack{z \to p \\ z \notin \bar{D}}} \mathbf{K}^{bD} f(z).$$

So, with a little bit of hand waving, we have established that the "jump" of $\mathbf{K}^{bD} f$ on bD is given by $f|_{bD}$. The argument just given required f of class C^1, or at least $\bar{\partial} f$ bounded on $D \backslash D_\varepsilon$ for some ε. In fact, the jump relation for $\mathbf{K}^{bD} f$ holds if f is only continuous on bD, as stated in the following result, but the proof requires a more careful analysis.

Proposition 2.2. *Let U be an open set in $\mathbb{C}^n$, and let M be a real C^1 submanifold of U of dimension $2n - 1$, such that $U \backslash M$ has two connected components U^+ and U^-. Assume that M carries the orientation induced from U^+. Let $f \in C_0(M)$, and denote by $\mathbf{K}^M f(z)$ the Bochner–Martinelli transform $\int_M fK_0(\cdot, z)$. Then $\mathbf{K}^M f|_{U^+}$ has a continuous extension F^+ to $\overline{U^+}$ if and only if $\mathbf{K}^M f|_{U^-}$ has a continuous extension F^- to $\overline{U^-}$. Furthermore, if either extension exists, then*

$$f(z) = F^+(z) - F^-(z) \qquad \text{for } z \in M.$$

See Exercise E.2.2 for related results.

Proof. Fix $p \in M$ and let $r > 0$, with $B(p, 2r) \subset U$. Decompose $f = f_0 + f_1$, where f_0 and f_1 are continuous, $\operatorname{supp} f_1 \subset B(p, 2r)$, and $f_0 \equiv 0$ on $B(p, r) \cap M$. Since $\mathbf{K}^M f_0$ is clearly continuous on $B(p, r)$, it is enough to prove the theorem for $\mathbf{K}^M f_1$ in a neighborhood of p. We can thus assume that $p = 0$ and $f \in C_0(M)$ has compact support in $B = B(0, 2r)$; $r > 0$ is chosen so small that the following hold:

(i) bB intersects M transversally, so that $D = B \cap U^+$ has piecewise C^1 boundary $(M \cap B) \cap (bB \cap U^+)$;
(ii) if $\mathbf{n}_z$ is the unit normal to M at $z \in M$ pointing to the U^+ side, there is $c < 1$, such that

$$|(\zeta - z, \mathbf{n}_z)| \leq c|\zeta - z| \qquad \text{for } \zeta, z \in B \cap M. \tag{2.4}$$

Let $M_r = \{z \in M : |z| \le r\}$. We will prove that $F = \mathbf{K}^M f$ satisfies

$$\lim_{t \to 0^+} [F(z + t\mathbf{n}_z) - F(z - t\mathbf{n}_z)] = f(z) \tag{2.5}$$

uniformly for $z \in M_r$.

The proof of the theorem is then immediate. Suppose $F|_{U^+ \cap B}$ has a continuous extension F^+ to $\bar{U}^+ \cap B$; by (2.5)

$$\lim_{t \to 0^+} F(z - t\mathbf{n}_z) = F^+(z) - f(z)$$

uniformly for $z \in M_r$. This limit defines $F^-(z)$ on M_r and it is a simple exercise to verify that F^- is continuous on $\{z \in \bar{U}^- : |z| \le r\}$.

In order to prove (2.5), extend f by 0 to all of bD, so that $F = \mathbf{K}^{bD} f$. Proposition 1.6, applied to the constant function $g \equiv f(z)$, z fixed in M_r, gives

$$\int_{bD} f(z) K_0(\cdot, z + t\mathbf{n}_z) = \begin{cases} f(z) & \text{for } t > 0 \\ 0 & \text{for } t < 0 \end{cases}$$

provided $|t| \le t_0$ for some small $t_0 > 0$. Therefore, if $0 < t < t_0$, $G(z, t) = F(z + t\mathbf{n}_z) - F(z - t\mathbf{n}_z) - f(z)$ is given by

$$G(z, t) = \int_{bD} [f(\zeta) - f(z)][K_0(\zeta, z + t\mathbf{n}_z) - K_0(\zeta, z - t\mathbf{n}_z)], \tag{2.6}$$

and (2.5) is equivalent to

$$\lim_{t \to 0^+} G(z, t) = 0 \textit{ uniformly for } z \in M_r. \tag{2.7}$$

The crucial estimate is contained in the following

Claim. *There is a constant* $C < \infty$ *such that*

$$\int_{bD} |K_0(\zeta, z + t\mathbf{n}_z) - K_0(\zeta, z - t\mathbf{n}_z)|\, d\sigma(\zeta) \le C \quad \textit{for } z \in M_r, 0 < t \le t_0. \tag{2.8}$$

Assume the claim for a moment. Given $\varepsilon > 0$, choose $\eta > 0$, such that $|f(\zeta) - f(z)| < \varepsilon/C$ for $|\zeta - z| < \eta$. By dividing the region of integration in (2.6) into the pieces $bD \cap B(z, \eta)$ and $bD - B(z, \eta)$, and using (2.8) in the obvious estimation, one obtains

$$|G(z, t)| \le \varepsilon + 2 \sup_{bD} |f| \int_{bD - B(z,\eta)} |K_0(\zeta, z + t\mathbf{n}_z) - K_0(\zeta, z - t\mathbf{n}_z)|\, d\sigma(\zeta).$$

The integral on the right now tends to 0 uniformly in $z \in M_r$ as $t \to 0^+$; therefore one can find $\delta > 0$, such that $|G(z, t)| \le \varepsilon + \varepsilon$ for $0 < t < \delta$. This proves (2.7), and hence the theorem.

We now prove the claim. A straightforward computation shows that for $t > 0$ the integrand $Q(t)$ in (2.8) is estimated by

$$Q(t) \lesssim \frac{t}{|\zeta - (z + t\mathbf{n}_z)|^{2n}} + \frac{\big||\zeta - (z - t\mathbf{n}_z)|^{2n} - |\zeta - (z + t\mathbf{n}_z)|^{2n}\big|}{|\zeta - (z - t\mathbf{n}_z)|^{2n-1}|\zeta - (z + t\mathbf{n}_z)|^{2n}}.$$

Since $\zeta, z \in M \cap B$, one can use (2.4) to obtain

$$|\zeta - z \pm t\mathbf{n}_z|^2 \geq (1 - c)[|\zeta - z|^2 + t^2].$$

It then follows that

$$Q(t) \lesssim \frac{t}{(|\zeta - z|^2 + t^2)^n}, \qquad t > 0.$$

For $z \in M_r$ and any γ with $0 < \gamma < r$,

$$\int_{bD} Q = \int_{bD - B(z,\gamma)} Q + \int_{M \cap B(z,\gamma)} Q.$$

Here the first integral on the right is bounded uniformly in $z \in M_r$ and $t > 0$; by introducing suitable local coordinates, the second integral is bounded by

$$I(t) = \int_{x \in \mathbb{R}^{2n-1}, |x| \leq R} \frac{t\, dV(x)}{(|x|^2 + t^2)^n}$$

for some fixed constant R. Finally, the substitution $x = ty$ leads to

$$I(t) = \int_{y \in \mathbb{R}^{2n-1}; |y| \leq R/t} \frac{t \cdot t^{2n-1}\, dV(y)}{(t^2|y|^2 + t^2)^n} \leq \int_{\mathbb{R}^{2n-1}} \frac{dV(y)}{(|y|^2 + 1)^n} < \infty,$$

so that (2.8) follows. ∎

Let D be a bounded domain in $\mathbb{C}^n$ with connected C^1 boundary bD. The geometric hypotheses of Proposition 2.2 are satisfied for $M = bD$, $U^+ = D$, and $U^- = \mathbb{C}^n \setminus \bar{D}$. One thus obtains the following.

Corollary 2.3. *Let $f \in C(bD)$, and suppose that*

$$\int_{bD} f K_0(\cdot, z) = 0 \qquad \text{for } z \notin \bar{D}. \tag{2.9}$$

Then $\mathbf{K}^{bD} f|_D$ extends to a continuous function F on the closure of D and

$$F(z) = f(z) \qquad \text{for } z \in bD.$$

PROOF. Simply apply Proposition 2.2, observing that (2.9) implies the existence of the continuous extension F^-, and $F^-(z) = 0$ for $z \in bD$. ∎

2.3. The Extension of CR-Functions from the Boundary.

Corollary 2.3 allows us to eliminate from Theorem 2.1 (and its proof) the hypothesis that f is holomorphic in a *neighborhood* of bD. All that is needed is that f satisfies the "tangential" Cauchy–Riemann equations on bD, a concept which we now make precise. Let D be a domain in $\mathbb{C}^n$ with C^1 boundary. Recall from Chapter II, §2.4, that the complex tangent space $T_p^{\mathbb{C}}(bD)$ of bD at $p \in bD$ is given by $\{t \in \mathbb{C}^n : \sum (\partial r/\partial z_j)(p) t_j = 0\}$ for some C^1 defining function r for D. By identifying $t \in T_p^{\mathbb{C}}(bD)$ with the $(1, 0)$ tangent vector $v_t = \sum_{j=1}^n t_j(\partial/\partial z_j)$ of $\mathbb{C}^n$ at p, one obtains the space $T_p^{1,0}(bD)$ of $(1, 0)$ **tangent vectors to** bD. Clearly $T_p^{1,0}(bD)$ is a complex subspace of $T_p^{1,0}(\mathbb{C}^n)$ of dimension $n - 1$, and also of the complexified tangent space $\mathbb{C}T_p bD$ of bD at p; its conjugate

$$T_p^{0,1}(bD) := \overline{T_p^{1,0}(bD)}$$

is called the space of **tangential Cauchy–Riemann operators at** $p \in bD$.

Definition. Let U be an open subset of bD. A function $f \in C^1(U)$ is called a **CR-function** (for Cauchy–Riemann function) on U if for all $p \in U$ one has

$$v(f) = 0 \qquad \text{for all } v \in T_p^{0,1}(bD). \tag{2.10}$$

Remarks.

(1) It is clearly enough to check (2.10) for a basis of $T_p^{0,1}(bD)$.
(2) If $V \subset \mathbb{C}^n$ is open, the restriction of any $f \in \mathcal{O}(V)$ to $V \cap bD$ is a CR-function.
(3) If $f \in \mathcal{O}(D) \cap C(\bar{D})$ and $f|_{bD} \in C^1(bD)$, then $f|_{bD}$ is a CR-function (see Exercise E.2.3).

A CR-function f on bD satisfies the full set of Cauchy–Riemann equations on bD in the following sense.

Lemma 2.4. *Suppose $1 \le k \le \infty$, and let D be a bounded domain with boundary of class C^k. Let $f \in C^k(bD)$ be a* CR*-function. There are a neighborhood U of bD and an extension $\tilde{f} \in C^{k-1}(U)$ of f, such that*

(i) $$\tilde{f}|_{bD} = f|_{bD};$$

(ii) *Any derivative of $\tilde{f}$ of order $k - 1$ is differentiable at every point $z \in bD$, and $D^\alpha \tilde{f}$ is continuous on bD for $|\alpha| = k$.*

(iii) $$\bar{\partial}\tilde{f}(z) = 0 \qquad \textit{for } z \in bD.$$

Proof. For $p \in bD$, let $L_1(p), \ldots, L_{n-1}(p)$ be a basis of $T_p^{1,0}(bD)$. Let $L_n(p) \neq 0$ be orthogonal to $T_p^{1,0}(bD)$. The extension $\tilde{f}$ must be chosen, so that $\bar{L}_n(z)\tilde{f} = 0$ for $z \in bD$. This condition completely determines the Taylor expansion of $\tilde{f}$ up to order 1 (this is also called the $1 -$ jet of $\tilde{f}$) at every point in bD! In fact, if r is a C^k defining function of D on a neighborhood U, we can choose

$$L_n(z) = \sum_{j=1}^n \frac{\partial r}{\partial \bar{z}_j}(z) \frac{\partial}{\partial z_j}\bigg|_z, \qquad z \in U.$$

Then $(L_n - \bar{L}_n)r = 0$, so that $L_n(z) - \bar{L}_n(z)$ is a tangent vector in $\mathbb{C}T_z bD$.

Therefore, if $\bar{L}_n(z)\tilde{f} = 0$, one obtains $L_n(z)\tilde{f} = [L_n(z) - \bar{L}_n(z)]\tilde{f} = [L_n(z) - \bar{L}_n(z)]f$; so $v\tilde{f}$ is indeed determined for all $v \in \mathbb{C}T_z\mathbb{C}^n$, $z \in bD$.

Without loss of generality, we can assume that $(\bar{L}_n r)(z) = 1$ on bD and that f is given as a C^k function on U. Then we define

$$\tilde{f}(z) = f(z) - r(z)(\bar{L}_n f)(z), \qquad z \in U.$$

Clearly $\tilde{f}$ is of class C^{k-1} and satisfies (i); (ii) follows by a simple computation (see also II(2.8)), and (iii) follows as well, since $\bar{L}_n\tilde{f} = \bar{L}_n f - (\bar{L}_n r)\bar{L}_n f = 0$ on bD, and $\bar{L}_j f = \bar{L}_j\tilde{f} = 0$ for $j < n$. ∎

We can now state a generalization of the Hartogs Extension Theorem.

Theorem 2.5. *Let D be a bounded domain in $\mathbb{C}^n$ with connected C^1 boundary. Assume that $n > 1$. Then every* CR*-function on bD of class C^1 can be extended continuously to a holomorphic function F on D.*

Moreover, if bD is of class C^m and $f \in C^k(bD)$, where $1 \le k \le m$, then $F \in \mathcal{O}(D) \cap C^k(\bar{D})$.

For any $k = 0, 1, 2, \ldots, \infty$, the space $\mathcal{O}(D) \cap C^k(\bar{D})$ will be denoted by $A^k(D)$; one also writes $A(D)$ for $A^0(D)$. $A^k(D)$ with norm $|\cdot|_{k,D}$ is clearly a Banach space if $k < \infty$.

Proof. Let $f \in C^1(bD)$ be a CR-function, and let $\tilde{f}$ be the extension given by Lemma 2.4. As in the proof of Theorem 2.1, it follows that $F = \mathbf{K}^{bD}f = \mathbf{K}^{bD}\tilde{f}$ is holomorphic on $\mathbb{C}^n - bD$ and ($n > 1$!) $F \equiv 0$ on $\mathbb{C}^n - \bar{D}$. The fact that $F|_D$ has continuous boundary values given by f now follows from Corollary 2.3.

Next we show that F is actually in $C^1(\bar{D})$. Since $\partial F/\partial \bar{z}_j = 0$ on D, it is enough to show that $\partial F/\partial z_j$ has a continuous extension to $\bar{D}$ for $j = 1, \ldots, n$. Since K_0 is of type $(n, n-1)$ in ζ and $\bar{\partial}$-closed, we can write, for a fixed j, $K_0 = d\zeta_j \wedge K^{(j)}$, where $K^{(j)}$ is of type $(n-1, n-1)$, $\bar{\partial}_\zeta K^{(j)} = 0$, and the coefficients of $K^{(j)}$ are functions of $\zeta - z$. Therefore,

$$\frac{\partial K_0}{\partial z_j} = -d\zeta_j \wedge \frac{\partial K^{(j)}}{\partial \zeta_j} = -\partial_\zeta K^{(j)} = -d_\zeta K^{(j)};$$

from this it follows by Stokes' Theorem that on $\mathbb{C}^n - bD$ one has

$$\begin{aligned} \frac{\partial F}{\partial z_j} &= -\int_{bD} \tilde{f}\, d_\zeta K^{(j)} = \int_{bD} (d_\zeta \tilde{f}) \wedge K^{(j)} \\ &= \int_{bD} \frac{\partial \tilde{f}}{\partial \zeta_j}\, d\zeta_j \wedge K^{(j)} = \int_{bD} \frac{\partial \tilde{f}}{\partial \zeta_j} K_0, \end{aligned} \tag{2.11}$$

i.e., $\partial/\partial z_j$ commutes with the Bochner–Martinelli transform $\mathbf{K}^{bD}$! Since $\partial F/\partial z_j = 0$ outside $\bar{D}$, Corollary 2.3 implies that $\partial F/\partial z_j|_D$ has continuous boundary values $\partial\tilde{f}/\partial\zeta_j$ on bD.

The general case follows by an easy induction. Let $1 \le k < m$, and suppose we already proved that $\mathbf{K}^{bD}g|_D$ extends to a function in $C^k(\bar{D})$ for any CR-function $g \in C^k(bD)$. Let f be a CR-function on bD of class C^{k+1}, set $F = \mathbf{K}^{bD}f$, and let $\tilde{f}$ be the extension of f given by Lemma 2.4. Then $g_j = (\partial\tilde{f}/\partial\zeta_j)|_{bD} \in C^k(bD)$, and, by (2.11), $\mathbf{K}^{bD}g_j = \partial F/\partial z_j$. Since $\partial F/\partial z_j \in A(D)$, it follows that g_j is a CR-function on bD (see Exercise E.2.3). By inductive assumption, $\partial F/\partial z_j \in C^k(\bar{D})$ for $1 \le j \le n$, and hence $F \in C^{k+1}(\bar{D})$. ■

2.4. $\bar{\partial}$-Cohomology with Compact Support

The Hartogs Extension Theorem is closely related to the solution of $\bar{\partial}$ with compact support. If f is a compactly supported $\bar{\partial}$-closed (0, 1)-form on $\mathbb{C}^n$, there is a function u on $\mathbb{C}^n$, such that $\bar{\partial}u = f$ (cf. Corollary 1.11). The function u is holomorphic outside of a ball $B \supset \operatorname{supp} f$, hence, if $n > 1$, Hartogs' Theorem gives $\tilde{u} \in \mathcal{O}(\mathbb{C}^n)$, with $u = \tilde{u}$ on $\mathbb{C}^n \backslash B$. Therefore $v = u - \tilde{u}$ solves $\bar{\partial}v = f$ and v has *compact* support. In terms of $\bar{\partial}$-cohomology, this can be stated as follows (see III.§2.5, for the notation).

Lemma 2.6. *If* $n > 1$, *then*

$$H^1_{\bar{\partial},c}(\mathbb{C}^n) = 0.$$

Lemma 2.6 can also be proved directly, without using the Hartogs Extension Theorem. For example, the solution $u = -\int_{\mathbb{C}^n} f \wedge K_0$ of $\bar{\partial}u = f$ given by Corollary 1.11 vanishes at ∞, and is holomorphic outside $\operatorname{supp} f$, so, if $n > 1$, $u \equiv 0$ on the unbounded component of $\mathbb{C}^n - \operatorname{supp} f$ (cf. the *proof* of Theorem 2.1). Another solution operator is described in Exercise E.2.4. Moreover, the Hartogs Extension Theorem is a simple consequence of $H^1_{\bar{\partial},c}(\mathbb{C}^n) = 0$ (see Exercise E.2.5).

We now consider the cohomology groups $H^q_{\bar{\partial},c}(\mathbb{C}^n)$ for $q > 1$. First we observe that $H^n_{\bar{\partial},c}(\mathbb{C}^n) \neq \{0\}$. In fact, let $f = f_N\, d\bar{z}^N \in \mathcal{D}_{0,n}(\mathbb{C}^n)$ and suppose

$$f = \bar{\partial}u \qquad \text{for some } u \in \mathcal{D}_{0,n-1}(\mathbb{C}^n). \tag{2.12}$$

An application of Stokes' Theorem to a large ball $B \supset \operatorname{supp} u$ gives

$$0 = \int_{bB} u \wedge d\zeta^N = \int_B \bar{\partial}u \wedge d\zeta^N = \int_{\mathbb{C}^n} f_N\, d\bar{\zeta}^N \wedge d\zeta^N.$$

So, if $f_N \in C_0^\infty(\mathbb{C}^n)$ satisfies $\int f_N\, dV \neq 0$, (2.12) cannot hold. In particular, $H^1_{\bar{\partial},c}(\mathbb{C}^1) \neq 0$; this gives another "explanation" for the failure of the Hartogs Extension Theorem in $\mathbb{C}^1$.

Lemma 2.6 generalizes as follows.

Theorem 2.7. *Let* n *be* ≥ 1; *then*

$$H^q_{\bar{\partial},c}(\mathbb{C}^n) = 0$$

for $q = 0, 1, \ldots, n - 1$.

Proof. The case $q = 0$ is obvious, and the case $q = 1 \le n - 1$ was discussed above. The case $1 < q \le n - 1$ is done similarly, except the extension of holomorphic functions is replaced by extension of $\bar{\partial}$-closed $(0, q - 1)$-forms.

Let $f \in \mathscr{D}_{0,q}(\mathbb{C}^n)$ and $\bar{\partial} f = 0$. By Corollary 1.11, $u = -\int f \wedge K_{q-1}$ solves $\bar{\partial} u = f$. So $\bar{\partial} u = 0$ outside of a ball $B = B(0, R) \supset \operatorname{supp} f$. Suppose there is an extension $\tilde{u} \in C^\infty_{0,q-1}(\mathbb{C}^n)$ of u, $\tilde{u} = u$ outside of a large ball, with $\bar{\partial}\tilde{u} = 0$ on $\mathbb{C}^n$. Then $g = u - \tilde{u} \in \mathscr{D}_{0,q-1}$ and $\bar{\partial} g = f$.

In order to find the extension $\tilde{u}$, we use the following result.

(2.13)
If $0 < R < R_2$ *and* $K = \{z : R \le |z| \le R_2\}$,
then $H^q_{\bar{\partial}}(K) = 0$ *for* $1 \le q \le n - 2$.

Equation (2.13) will be proved in §3.5, after we have developed further integral representation formulas. Assuming (2.13), let us return to the $(0, q - 1)$-form u. Since $1 \le q - 1 \le n - 2$ in this case and $\bar{\partial} u = 0$ on a neighborhood of K, by (2.13) there is $v \in C^\infty_{0,q-2}(K)$ with $\bar{\partial} v = u$ in a neighborhood of K. Choose $\chi \in C^\infty(\mathbb{C}^n)$ with $\chi \equiv 0$ on $B(0, R)$ and $\chi \equiv 1$ on $\mathbb{C}^n - B(0, R_2)$. Then $\tilde{u}$ defined by

$$\tilde{u} = \begin{cases} \bar{\partial}(\chi v) & \text{on } \overline{B(0, R_2)} \\ u & \text{on } \mathbb{C}^n - B(0, R_2) \end{cases}$$

defines the desired extension of u. ∎

Exercises

E.2.1. Let $D \subset \mathbb{C}^n$ have C^1 boundary and suppose r is a C^1 defining function for D. Show that $f \in C^1(bD)$ is a CR-function if and only if $\bar{\partial}\tilde{f} \wedge \bar{\partial} r = 0$ on bD for any C^1 extension $\tilde{f}$ of f to a neighborhood of bD.

E.2.2. In the setting of Proposition 2.2, show that if $f \in \Lambda_\alpha(M)$ for some $0 < \alpha < 1$, then $\mathbf{K}^M f|_{U^\pm}$ has continuous extensions $F^\pm$ to $\overline{U^\pm}$, which satisfy

$$F^\pm(z) = \pm\frac{1}{2} f(z) + \lim_{\varepsilon \to 0^+} \int_{\zeta \in M; |\zeta - z| \ge \varepsilon} f(\zeta) K_0(\zeta, z)$$

for $z \in M$. (See also [HaLa].)

E.2.3. Let $D \subset\subset \mathbb{C}^n$ have C^1 boundary.

(i) Show that if $f \in A^1(D) = \mathcal{O}(D) \cap C^1(\bar{D})$, then $f|_{bD}$ is a CR-function. (Hint: Use Exercise E.III.1.15.)

(ii) Show that if $f \in A(D)$ and $f|_{bD} \in C^1(bD)$, then $f|_{bD}$ is a CR-function, and hence $f \in A^1(D)$, by Theorem 2.5. (Hint: By translation, approximate f locally near $p \in bD$ by functions holomorphic in a neighborhood of p in bD.)

E.2.4. Suppose $f \in C^1_{0,1}(\mathbb{C}^n)$ has compact support and $\bar{\partial} f = 0$ on $\mathbb{C}^n$. Define

$$Tf(z', z_n) = -\frac{1}{2\pi i} \int_{\mathbb{C}} \frac{f(z', \zeta)\, d\bar{\zeta} \wedge d\zeta}{\zeta - z_n}.$$

(i) Show that $Tf \in C^1(\mathbb{C}^n)$ and that $\bar{\partial}(Tf) = f$. (Hint: Use Corollary 1.11 for $n = 1$.)

(ii) Show that if $n > 1$, then Tf has compact support.

E.2.5. Show that $H^1_{\bar\partial,c}(\mathbb{C}^n) = 0$ for $n > 1$ implies the Hartogs Extension Theorem, as follows: if $K \subset D$ is compact and $f \in \mathcal{O}(D - K)$, choose $\chi \in C^\infty(D)$ such that $\chi \equiv 0$ in a neighborhood of K and $\chi \equiv 1$ near bD, so that χf gives a C^∞ extension of f to D. Now use $H^1_{\bar\partial,c}(\mathbb{C}^n) = 0$ to modify χf suitably.

§3. The General Homotopy Formula

3.1. Cauchy–Fantappiè Forms of Order 0

We now discuss a method which allows to construct integral representations of a more general type than those presented in §1. In order to present the main idea more clearly, we first deal with the case of functions. The reproducing properties of the Bochner–Martinelli kernel K_0 depend on two basic facts: (i) $\bar\partial_\zeta K_0 = 0$ for $\zeta \neq z$, and (ii) K_0 has the "right" singularity at $\zeta = z$. In §1.2 we proved (i) by using the relationship between K_0 and the fundamental solution for the complex Laplacian $\square$. However, (i) holds for a much wider class of forms whose algebraic properties are similar to those of K_0.

Lemma 3.1. *Let $W(\zeta) = \sum_{j=1}^n w_j(\zeta)\, d\zeta_j$ be a $C^2_{1,0}$ form on a set $U \subset \mathbb{C}^n$. Suppose there is $z \notin U$, such that*

$$\langle W(\zeta), \zeta - z\rangle = \sum_{j=1}^n w_j(\zeta)(\zeta_j - z_j) = 1 \qquad \text{for } \zeta \in U. \tag{3.1}$$

(This is consistent with the usual notation if $\zeta - z$ is identified with the $(1, 0)$ vector $\sum_{j=1}^n (\zeta_j - z_j)(\partial/\partial\zeta_j)$ at ζ.) Then the $(n, n-1)$-form

$$\Omega_0(W) = (2\pi i)^{-n} W \wedge (\bar\partial W)^{n-1}$$

satisfies

$$d\Omega_0(W) = \bar\partial_\zeta \Omega_0(W) = 0 \text{ on } U.$$

Remark. By Lemma 1.3(d), $K_0 = \Omega_0(\partial\beta/\beta)$, where $\beta = |\zeta - z|^2$, and the $(1, 0)$-form $B = \partial\beta/\beta$ satisfies (3.1).

Proof of Lemma 3.1. It is immediate that

$$\begin{aligned}(2\pi i)^n\, d\Omega_0(W) &= (2\pi i)^n \bar\partial\Omega_0(W) = (\bar\partial W)^n \\ &= \left(\sum_{j=1}^n \bar\partial w_j \wedge d\zeta_j\right)^n \\ &= n!(\bar\partial_\zeta w_1 \wedge d\zeta_1) \wedge \cdots \wedge (\bar\partial_\zeta w_n \wedge d\zeta_n).\end{aligned} \tag{3.2}$$

Applying $\bar\partial$ to (3.1) one obtains

$$\sum_{j=1}^n \bar\partial w_j(\zeta_j - z_j) = 0.$$

So, for $\zeta \neq z$, the set $\{\bar{\partial}_\zeta w_1(\zeta), \ldots, \bar{\partial}_\zeta w_n(\zeta)\}$ is linearly dependent, which implies that (3.2) is 0. ■

Addendum to Lemma 3.1. Lemma 3.1 remains valid if W is only of class C^1, provided $d\Omega_0(W) = 0$ is interpreted in the sense of distributions (or better "currents": see de Rham [Rha]). More concretely, if M is a C^∞ manifold of dimension n, and $\varphi \in C_r(M)$ and $\psi \in C_{r+1}(M)$ are continuous forms on M, the statement "$d\varphi = \psi$ on M in the sense of distributions" implies that Stokes' Theorem remains valid in the following form: if $D \subset\subset M$ has piecewise C^1 boundary and $f \in C_s^1(\bar{D})$, where $s = n - (r + 1)$, then

$$\int_{bD} f \wedge \varphi = \int_D df \wedge \varphi + (-1)^s \int_D f \wedge \psi.$$

The reader who wants to see a detailed proof of $d\Omega_0(W) = 0$ in the above sense when W is only of class C^1 should work out Exercise E.3.1.

From now on we will assume that W is of class C^1. This will allow us to keep the differentiability assumptions in later applications at a minimum. If this makes the reader uncomfortable, he may simply increase the order of differentiability by one in whatever follows.

Definition. A generating form W on $U \subset \mathbb{C}^n$ (for the point z) is a $C^1_{1,0}$ form on U which satisfies (3.1). The $(n, n-1)$-form

$$\Omega_0(W) = (2\pi i)^{-n} W \wedge (\bar{\partial}_\zeta W)^{n-1}$$

is called the **Cauchy–Fantappiè** ($= CF$) **form** (of order 0) generated by W.

We denote by $B = \partial\beta/\beta$ the generating form for the Bochner–Martinelli kernel.

The following simple relationship holds for CF forms:

$$\Omega_0(gW) = g^n \Omega_0(W) \tag{3.3}$$

for any C^1 function g. The proof is immediate by observing that

$$gW \wedge \bar{\partial}_\zeta(gW) = gW \wedge (\bar{\partial}g \wedge W + g\bar{\partial}W) = gW \wedge g\bar{\partial}W.$$

Note that in case $n = 1$ the condition (3.1) determines W uniquely, namely $W = (\zeta - z)^{-1}\, d\zeta$, while for $n \geq 2$ there obviously are infinitely many solutions $(w_1, \ldots, w_n)$ of (3.1). This explains the special role of the Cauchy kernel in dimension 1, while in higher dimensions there is a multitude of possible choices. And each Cauchy–Fantappiè form $\Omega_0(W)$ with the "right" singularity leads to an integral representation formula, as follows.

Proposition 3.2. *Let $D \subset\subset \mathbb{C}^n$ be a domain with piecewise C^1 boundary. Let $z \in D$ and suppose that $W \in C^1_{1,0}(\bar{D} - \{z\})$ is a generating form for z such that $W = \partial\beta/\beta$ on a ball $B(z, \varepsilon) - \{z\}$. Then every $f \in C^1(\bar{D})$ satisfies*

$$(3.4)\qquad f(z) = \int_{bD} f\Omega_0(W) - \int_D \bar\partial f \wedge \Omega_0(W).$$

The proof is left as an exercise for the reader (see E.3.3).

Example 3.3. Let $B = B(0, 1) \subset \mathbb{C}^n$. For $1 \le j \le n$ and $|\zeta| > |z|$ define

$$s_j(\zeta, z) = \frac{\bar\zeta_j}{\sum_{j=1}^n \bar\zeta_j(\zeta_j - z_j)}, \qquad S(\zeta, z) = \sum_{j=1}^n s_j\, d\zeta_j;$$

then S satisfies (3.1). Let $0 < a < 1$ and choose $\chi \in C^\infty(\mathbb{C}^n)$, such that $\chi \equiv 1$ on bB and $\chi \equiv 0$ on $B(0, a)$. For $z \in B(0, a)$ the (1, 0)-form

$$W(\zeta, z) = \chi(\zeta)S(\zeta, z) + (1 - \chi(\zeta))B(\zeta, z)$$

satisfies the hypothesis of Proposition 3.2. Therefore (3.4) implies

$$(3.5)\qquad f(z) = \int_{bB} f\Omega_0(S) \qquad \text{if } \bar\partial f = 0 \text{ on } B.$$

This formula no longer involves the function χ, so (3.5) holds for all $z \in B$. Moreover, the kernel $\Omega_0(S)$ is holomorphic in z! The Bochner–Martinelli kernel does not have this property, unless $n = 1$. $\Omega_0(S)$ is called the Szegö kernel for the unit ball (see E.3.5). Of course $\Omega_0(S) = K_0$ in dimension 1.

The obvious question now is, for which domains D can one find a *holomorphic* reproducing kernel like $\Omega_0(S)$. In general, this will not be possible, at least not by the above method. In fact, if for $\zeta \in bD$ one could find a solution $(w_1(\zeta, z), \ldots, w_n(\zeta, z))$ of (3.1) which is holomorphic in $z \in D$, then one of the w_j's must become singular at ζ. Hence D would have to be a domain of holomorphy. We will see in the next section that (3.5) generalizes to arbitrary *convex* domains. Later, in VII.§1 and §3, we will consider holomorphic reproducing kernels on strictly pseudoconvex domains.

3.2. The Cauchy Integral for Convex Domains

A simple method for constructing a CF form $\Omega_0(W)$ which is holomorphic in $z \in D$ for $\zeta \in bD$ is to find a double form α on $bD \times D$, of type (1, 0) in ζ, which is holomorphic in $z \in D$ and such that $g = \langle \alpha, \zeta - z\rangle \neq 0$ for $z \in D$. Then $W(\cdot, z) = \alpha/g(\cdot, z)$ will generate such a CF form. This can easily be done for convex domains, as follows.

Let $D \subset\subset \mathbb{C}^n$ be convex with C^2 boundary, and let $r \in C^2$ be a defining function for D on a neighborhood U of bD. For $\zeta \in bD$, the condition $\langle \partial r(\zeta), \zeta - z\rangle = 0$ characterizes the complex tangent space $T_\zeta^{\mathbb{C}}(bD)$. From the convexity of D one obtains

$$\langle \partial r(\zeta), \zeta - z\rangle \neq 0 \qquad \text{for } \zeta \in bD, z \in D.$$

Therefore,

$$C^{(r)}(\zeta, z) = \partial r(\zeta) \cdot \langle \partial r(\zeta), \zeta - z \rangle^{-1} \tag{3.6}$$

is a generating form on bD for $z \in D$, and the CF form $\Omega_0(C^{(r)})$ is holomorphic in z!

Theorem 3.4. *Let $D \subset\subset \mathbb{C}^n$ be a convex domain with C^2 defining function r. Then*

$$f(z) = \frac{1}{(2\pi i)^n} \int_{bD} f(\zeta) \frac{\partial r(\zeta) \wedge (\bar{\partial}\partial r(\zeta))^{n-1}}{\langle \partial r(\zeta), \zeta - z \rangle^n} \tag{3.7}$$

for $f \in A(D)$ and $z \in D$.

Proof. As in Example 3.3, one shows that

$$f(z) = \int_{bD} f \Omega_0(C^{(r)}(\cdot, z)) \tag{3.8}$$

for $f \in A^1(D)$. Another proof of (3.8), independent of Proposition 3.2, will follow from a more general result in the next section. Since D is convex, one can approximate $F \in A(D)$ uniformly by functions in $\mathcal{O}(\bar{D})$, so (3.8) holds for $f \in A(D)$. Finally, using (3.3), one obtains

$$\begin{aligned} \Omega_0(C^{(r)}) &= \langle \partial r, \zeta - z \rangle^{-n} \Omega_0(\partial r) \\ &= \langle \partial r, \zeta - z \rangle^{-n} \cdot (2\pi i)^{-n} \partial r \wedge (\bar{\partial}\partial r)^{n-1}. \quad \blacksquare \end{aligned}$$

Formula (3.7) is called the Cauchy Integral Formula for the convex domain D. It is another natural generalization of the classical one-variable formula, which is quite different from the Bochner–Martinelli Formula. In (3.7) the kernel is holomorphic in z, but we had to pay a price: the new kernel depends on the domain D, while the BM kernel is "universal".

Remark. Even though the $(n, n-1)$-form $\Omega_0(C^{(r)})$ depends on the choice of the defining function r, only the "tangential" component of $\Omega_0(C^{(r)})$ (i.e., its pullback to bD) enters into (3.7), and the latter is independent of r! This is seen as follows. Let $\iota: bD \to \mathbb{C}^n$ be the inclusion. Suppose $\tilde{r}$ is another C^2 defining function for D. Then $\tilde{r} = h \cdot r$, where $h > 0$ on bD. For $\zeta \in bD$,

$$\partial \tilde{r} = h \partial r \qquad \text{and} \qquad \bar{\partial}\partial \tilde{r} = h \bar{\partial}\partial r + \bar{\partial} h \wedge \partial r - \partial h \wedge \bar{\partial} r.$$

Since $\iota^*(\bar{\partial} r) = -\iota^*(\partial r)$, one obtains

$$\iota^*(\bar{\partial}\partial \tilde{r}) = h \iota^*(\bar{\partial}\partial r) + \iota^*(dh) \wedge \iota^*(\partial r),$$

and therefore

$$\iota^* \Omega_0(\partial \tilde{r}) = h^n \iota^* \Omega_0(\partial r).$$

Also, for $\zeta \in bD$, $\langle \partial \tilde{r}(\zeta), \zeta - z \rangle^n = h^n \langle \partial r(\zeta), \zeta - z \rangle^n$. It follows that

$$\iota^*\Omega_0(C^{(\tilde{r})}) = \iota^*\Omega_0(C^{(r)}).$$

So, for any convex domain $D \subset \mathbb{C}^n$ with C^2 boundary, the Cauchy kernel

$$C_D(\zeta, z) = \iota^*\Omega_0(C^{(r)})$$

is an intrinsically defined differential form on bD with coefficients in $\mathcal{O}(D)$.

3.3. The General Homotopy Formula

In Example 3.3 the patching function χ depended on the choice of $B(0, a)$; by choosing a more complicated patching function depending on ζ *and* z one can eliminate this dependence. So for a convex domain D, one can pass from the representation (3.7) for $f \in A(D)$ to the general Cauchy integral formula (3.4) for $f \in C^1(\bar{D})$. However, for the estimation of the kernel it is technically simpler to start directly from the Bochner–Martinelli formula (Corollary 1.3) and to introduce a linear homotopy over bD. We now carry out this construction for forms of type $(0, q)$, q arbitrary.

We first need to make some technical definitions. Given a C^k manifold M and a C^∞ manifold N we denote by $C^{0,\infty}(M \times N)$ the space of those functions $f \in C^0(M \times N)$, all of whose partial derivatives with respect to local coordinates of N are continuous on $M \times N$, and we define inductively, for $l = 1, 2, \ldots, k$,

$$C^{l,\infty}(M \times N) = \{f \in C^l(M \times N) \colon f \text{ and } d_x f \in C^{l-1,\infty}(M \times N)\}.$$

Clearly, if $g \in C^l(M)$ and $h \in C^\infty(N)$, then f defined on $M \times N$ by $f(x, y) = g(x)h(y)$ is in $C^{l,\infty}(M \times N)$. If U and D are open in $\mathbb{C}^n$, we denote by $C^{l,\infty}_{p,q;s,t}(U \times D)$ the space of double forms on $U \times D$ of type (p, q) in $\zeta \in U$ and type (s, t) in $z \in D$ whose coefficients are in $C^{l,\infty}(U \times D)$. We write $C^{l,\infty}_{p,q}$ instead of $C^{l,\infty}_{p,q;0,0}$. We will need to consider such forms defined for $(\zeta, z) \in bD \times D$, where D has boundary of class C^k. The statement $W \in C^{k,\infty}_{p,q}(bD \times D)$ refers to a double differential form W on a subset of $\mathbb{C}^n \times \mathbb{C}^n$, i.e., we have $W(\zeta, z) \in \Lambda^{p,q}_\zeta(\mathbb{C}^n)$ for $\zeta \in bD$ and $z \in D$.[1] The regularity condition means that the coefficients of W with respect to the standard global frame on $\mathbb{C}^n$ are in $C^{k,\infty}(bD \times D)$. Without loss of generality such a form $W \in C^{k,\infty}_{p,q}(bD \times D)$ can be extended as a form in $C^{k,\infty}_{p,q}(\Omega)$, where Ω is a neighborhood of $bD \times D$ (Ω is not necessarily a product!), so that operators like d and $\bar{\partial}$ can be applied to W. Whenever a form $W \in C^{k,\infty}_{p,q}(bD \times D)$ appears in an integral over the manifold bD it is understood that the pullback $\iota^* W$ under the inclusion $\iota \colon bD \to \mathbb{C}^n$ must be used, although we shall not indicate this explicitly in the formulas.

Let us now assume that $D \subset\subset \mathbb{C}^n$ has C^1 boundary and that $W \in C^{1,\infty}_{1,0}(bD \times D)$

[1] Notice that for differential forms on the manifold $bD \subset \mathbb{C}^n$ it is not clear how to define the type: for example, if r is a defining function for D, then $\iota^*(\partial r) = -\iota^*(\bar{\partial} r) \neq 0$ on bD.

is a generating form (in ζ) for each $z \in D$. Let $I = [0, 1]$ be the unit interval. The **homotopy form** $\hat{W}$ on $(bD \times I) \times D$ associated to W is defined by

$$\hat{W}(\zeta, \lambda, z) = \lambda W(\zeta, z) + (1 - \lambda) B(\zeta, z),$$

where $B = \partial\beta/\beta$ is the generating form for the BMK kernel.

Obviously $\hat{W}$ is still a generating form, i.e.,

$$\sum_{j=1}^{n} \hat{w}_j(\zeta, \lambda, z)(\zeta_j - z_j) = 1 \qquad \text{on } (bD \times I) \times D, \tag{3.9}$$

where $\hat{W} = \sum \hat{w}_j \, d\zeta_j$.

The exterior derivative $d_{\zeta,\lambda}$ on $\mathbb{C}^n \times I$ and on $bD \times I$ decomposes into $d_\zeta + d_\lambda$; similarly, we write $\bar{\partial}_{\zeta,\lambda} = \bar{\partial}_\zeta + d_\lambda$.

For example, if $f = \sum f_J \, d\bar{\zeta}^J + f_\lambda \, d\lambda$, then

$$\bar{\partial}_{\zeta,\lambda} f = \bar{\partial}_\zeta\left(\sum f_J \, d\bar{\zeta}^J\right) + \bar{\partial}_\zeta f_\lambda \wedge d\lambda + \sum \frac{\partial f_J}{d\lambda} \, d\lambda \wedge d\bar{\zeta}^J.$$

Definition. Let W be a generating form with coefficients in $C^{1,\infty}(bD \times D)$ and let $\hat{W}$ be the associated homotopy form. For $-1 \le q \le n$, the **Cauchy–Fantappiè kernel $\Omega_q(\hat{W})$ of order q generated by** $\hat{W}$ is defined by

$$\Omega_q(\hat{W}) = \frac{(-1)^{q(q-1)/2}}{(2\pi i)^n} \binom{n-1}{q} \hat{W} \wedge (\bar{\partial}_{\zeta,\lambda} \hat{W})^{n-q-1} \wedge (\bar{\partial}_z \hat{W})^q \tag{3.10}$$

for $0 \le q \le n-1$, and 0 otherwise. $\Omega_q(W)$ is defined in the same way, with W instead of $\hat{W}$.

Remark. $\Omega_q(\hat{W})$ is a double form on $(bD \times I) \times D$, of degree $2n - q - 1$ in (ζ, λ) and type $(0, q)$ in z. Equation (3.3) holds for CF forms of order $q \ge 0$. Also, if $\mu_\lambda \colon \mathbb{C}^n \to \mathbb{C}^n \times I$ is defined by $\mu_\lambda(\zeta) = (\zeta, \lambda)$, then

$$\mu_0^* \Omega_q(\hat{W}) = \Omega_q(B) = K_q, \qquad \text{and} \qquad \mu_1^* \Omega_q(\hat{W}) = \Omega_q(W) \text{ on } bD \times D. \tag{3.11}$$

Lemma 3.1 generalizes as follows.

Lemma 3.5. *Let $W \in C^{1;\infty}_{1,0}(bD \times D)$ be a generating form. Then, for any $0 \le q \le n$,*

$$d_{\zeta,\lambda} \Omega_q(\hat{W}) = \bar{\partial}_{\zeta,\lambda} \Omega_q(\hat{W}) = (-1)^q \bar{\partial}_z \Omega_{q-1}(\hat{W}), \tag{3.12}$$

and an analogous relation holds with W in place of $\hat{W}$.

Proof. The right side is a continuous form; so, unless W is of class $C^{2,\infty}$, (3.12) has to be interpreted in the distribution sense (cf. the addendum to Lemma 3.1).

From (3.9) it follows that, at a fixed point $(\zeta, \lambda, z) \in (bD \times I) \times D$, the $(1, 0)$-forms $\partial/\partial\bar{\zeta}_k \hat{W}$, $\partial/\partial\bar{z}_k \hat{W}$, and $\partial/\partial\lambda \hat{W}$, $1 \le k \le n$, which arise as coefficients of $d\bar{\zeta}_k$, $d\bar{z}_k$, resp. $d\lambda$ in $\bar{\partial}_{\zeta,\lambda} \hat{W}$ and $\bar{\partial}_z \hat{W}$, lie in the subspace

$$H_{\zeta,z} = \left\{ \sum_{j=1}^{n} a_j \, d\zeta_j : \sum_{j=1}^{n} a_j(\zeta_j - z_j) = 0 \right\}$$

of $\Lambda_\zeta^{1,0}$. Since $\dim_{\mathbb{C}} H_{\zeta,z} = n - 1$, one has $\Lambda^n H_{\zeta,z} = \{0\}$. This implies the vanishing of any n-fold product of the above coefficients, and hence one obtains

$$(\bar{\partial}_{\zeta,\lambda} \hat{W})^{n-s} \wedge (\bar{\partial}_z \hat{W})^s = 0 \qquad \text{for any } s, 0 \leq s \leq n.$$

Therefore, denoting the numerical factor in (3.10) by $c_{n,q}$,

$$\begin{aligned} d_{\zeta,\lambda}\Omega_q(\hat{W}) &= \bar{\partial}_{\zeta,\lambda}\Omega_q(\hat{W}) \\ &= c_{n,q}(-q)\hat{W} \wedge (\bar{\partial}_{\zeta,\lambda}\hat{W})^{n-q-1} \wedge \bar{\partial}_{\zeta,\lambda}\bar{\partial}_z \hat{W} \wedge (\bar{\partial}_z \hat{W})^{q-1}, \end{aligned}$$

and

$$\begin{aligned} (-1)^q \bar{\partial}_z \Omega_{q-1}(\hat{W}) = (-1)^q c_{n,q-1}(n-q)\hat{W} \wedge \bar{\partial}_z \bar{\partial}_{\zeta,\lambda} \hat{W} \\ \wedge (\bar{\partial}_{\zeta,\lambda}\hat{W})^{n-q-1} \wedge (\bar{\partial}_z \hat{W})^{q-1}. \end{aligned}$$

Since $\bar{\partial}_z \bar{\partial}_{\zeta,\lambda} \hat{W} = \bar{\partial}_{\zeta,\lambda}\bar{\partial}_z \hat{W}$ (we are dealing with *double* forms) and $\bar{\partial}_{\zeta,\lambda}\hat{W}$ is of even degree in (ζ, λ) and in z, (3.12) follows from the above by checking that $-qc_{n,q} = (-1)^q(n-q)c_{n,q-1}$.

The remaining statement follows by pulling back (3.12) under the map $\mu_1 : \zeta \mapsto (\zeta, 1)$. ■

We can now prove a general representation formula in terms of Cauchy–Fantappiè forms.

Theorem 3.6. *Let $W \in C^{1;\infty}_{1,0}(bD \times D)$ be a generating form for points z in the domain $D \subset\subset \mathbb{C}^n$ with C^1 boundary and set $\hat{W} = \lambda W + (1 - \lambda) B$ on $(bD \times I) \times D$. For $1 \leq q \leq n$, define the linear operator*

$$\mathbf{T}_q^W : C_{0,q}(\bar{D}) \to C_{0,q-1}(D)$$

by

$$\mathbf{T}_q^W f = \int_{bD \times I} f \wedge \Omega_{q-1}(\hat{W}) - \int_D f \wedge K_{q-1}, \tag{3.13}$$

and set $\mathbf{T}_0^W = \mathbf{T}_{n+1}^W \equiv 0$. Then the following hold:

(a) *For $k = 0, 1, 2, \ldots, \infty$, if $f \in C^k_{0,q}(D) \cap C_{0,q}(\bar{D})$, then $\mathbf{T}_q^W f \in C^k_{0,q-1}(D)$.*
(b) *For $0 \leq q \leq n$, if $f \in C^1_{0,q}(\bar{D})$, then*

$$f = \int_{bD} f \wedge \Omega_q(W) + \bar{\partial}\mathbf{T}_q^W f + \mathbf{T}_{q+1}^W \bar{\partial} f \qquad \text{on } D. \tag{3.14}$$

Remark. In case $q = n$, $\Omega_n(W) \equiv 0$, and $f \wedge \Omega_{n-1}(\hat{W})$ is an (n, n)-form in ζ; therefore

$$\int_{bD \times I} f \wedge \Omega_{n-1}(\hat{W}) \equiv 0,$$

and the representation (3.14) simplifies to $f = -\bar\partial \int_D f \wedge K_{n-1}$, which is known from Theorem 1.10.

PROOF OF 3.6. Since the first integral in (3.13) is C^∞ in z, part (a) follows from Theorem 1.14. By the Remark above, it is enough to prove (b) for $q < n$. We use the representation given by Theorem 1.10 and modify the boundary integral as follows. By (3.11),

$$\begin{aligned}
&\int_{bD} f \wedge K_q - \int_{bD} f \wedge \Omega_q(W) \\
&= -\left\{\int_{bD} f \wedge \mu_1^*\Omega_q(\hat W) - \int_{bD} f \wedge \mu_0^*\Omega_q(\hat W)\right\} \\
&= (-1)^{\dim bD} \int_{bD\times bI} f \wedge \Omega_q(\hat W) \\
&= \int_{b(bD\times I)} f \wedge \Omega_q(\hat W)
\end{aligned}$$

(see III.§1.8). On the other hand, by Stokes' Theorem and Lemma 3.5,

$$\begin{aligned}
\int_{b(bD\times I)} f \wedge \Omega_q(\hat W) &= \int_{bD\times I} \bar\partial f \wedge \Omega_q(\hat W) + (-1)^q \int_{bD\times I} f \wedge \bar\partial_{\zeta,\lambda}\Omega_q(\hat W) \\
&= \int_{bD\times I} \bar\partial f \wedge \Omega_q(\hat W) + \bar\partial_z \int_{bD\times I} f \wedge \Omega_{q-1}(\hat W).
\end{aligned}$$

Therefore,

$$\begin{aligned}
\int_{bD} f \wedge K_q = \int_{bD} f \wedge \Omega_q(W) + \bar\partial_z \int_{bD\times I} f \wedge \Omega_{q-1}(\hat W) \\
+ \int_{bD\times I} \bar\partial f \wedge \Omega_q(\hat W).
\end{aligned} \tag{3.15}$$

Substituting (3.15) into the representation in Theorem 1.10 gives (3.14). ∎

Theorem 3.6 shows that in order to obtain a solution operator for the equation $\bar\partial u = f$ when $\bar\partial f = 0$, it is enough to find a generating form W such that $\Omega_q(W) \equiv 0$ on $bD \times D$. In the next two sections we discuss some simple situations where this can be done without much further work.

3.4. The Solution of $\bar\partial$ on Convex Domains

We first consider a *convex* domain D in $\mathbb{C}^n$ with C^2 defining function r. Let $C^{(r)}$ be the generating form defined by (3.6), and let $C_D = \Omega_0(C^{(r)})$ be the Cauchy kernel for D. Applying Theorem 3.6 in case $q = 0$ gives the following higher dimensional version of the general Cauchy integral formula for a function $f \in C^1(\bar{D})$:

$$(3.16)\quad f(z) = \int_{bD} f C_D(\cdot, z) + \int_{bD \times I} \bar{\partial} f \wedge \Omega_0(\hat{C}^{(r)}) - \int_D \bar{\partial} f \wedge K_0 \qquad \text{for } z \in D.$$

This formula includes Theorem 3.4 as a special case.

Furthermore, since $C^{(r)}$ is holomorphic in $z \in D$, one obtains

$$\Omega_q(C^{(r)}) \equiv 0 \qquad \text{if } q \geq 1.$$

Theorem 3.6 now gives the following result.

Proposition 3.7. *The operator* $\mathbf{T}_q^{(r)}: C_{0,q}(\bar{D}) \to C_{0,q-1}(D)$ *defined by*

$$\mathbf{T}_q^{(r)} f = \int_{bD \times I} f \wedge \Omega_{q-1}(\hat{C}^{(r)}) - \int_D f \wedge K_{q-1}$$

satisfies

$$\bar{\partial}(\mathbf{T}_q^{(r)} f) = f$$

on D *if* $f \in C^1_{0,q}(\bar{D})$ *and* $\bar{\partial} f = 0$.

Corollary 3.8. *Let* K *be a compact convex set in* $\mathbb{C}^n$. *Then*

$$H^q_{\bar{\partial}}(K) = 0 \qquad \textit{for } q \geq 1.$$

Proof. Let $q \geq 1$ and suppose $f \in C^\infty_{0,q}(K)$ is defined and $\bar{\partial}$-closed on the open neighborhood U of K. Since K is convex, there is a smoothly bounded open convex neighborhood D, $K \subset D \subset\subset U$, with defining function r. By Theorem 3.6(a) and Proposition 3.7, $u = \mathbf{T}_q^{(r)} f \in C^\infty_{0,q-1}(D)$ and $\bar{\partial} u = f$ on D. ∎

3.5. The Solution of $\bar{\partial}$ on Spherical Shells

We now consider another method for achieving $\Omega_q(W) = 0$. If W is holomorphic in ζ, then $\Omega_q(W) = 0$ for $q \leq n - 2$! This works well for *concave* boundaries by simply switching the variables ζ and z in the construction for convex boundaries. We discuss in detail the simple case of spherical shells $G = G(R_1, R_2) = \{z \in \mathbb{C}^n: R_1 < |z| < R_2\}$ with $0 < R_1 < R_2$; more general situations are considered in Exercise E.3.7.

Note that $bG = S_2 - S_1$, where $S_i = \{z: |z| = R_i\}$, $i = 1, 2$. On $S_2 \times G$ we can take the generating form $C^{(r)}$, with $r(\zeta) = |\zeta|^2 - R_2$. Observe that $C^{(r)}$ is nonsingular for all (ζ, z) with $|\zeta| > |z|$, hence on $G \times S_1$. If $C^{(r)}(\zeta, z) = \sum_{j=1}^n c_j(\zeta, z)\, d\zeta_j$, define $C^{(r)}_\#(\zeta, z) = -\sum c_j(z, \zeta)\, d\zeta_j$. Then $C^{(r)}_\# \in C^{\infty;\infty}_{1,0}(S_1 \times G)$, and $C^{(r)}_\#$ is a generating form which is holomorphic in ζ. Therefore, if

$$W_G = \begin{cases} C^{(r)} & \text{on } S_2 \times G \\ C_{\#}^{(r)} & \text{on } S_1 \times G \end{cases},$$

it follows that

$$\Omega_q(W_G) = 0 \text{ on } bG \times G \qquad \text{for } 1 \le q \le n-2.$$

An application of Theorem 3.6 gives

Proposition 3.9. *With G and W_G as defined above, the operator $\mathbf{T}_q$ given by*

$$\mathbf{T}_q f = \int_{bG \times I} f \wedge \Omega_{q-1}(\hat{W}_G) - \int_G f \wedge K_{q-1} \tag{3.17}$$

for $q \ge 1$, satisfies

$$\bar{\partial}(\mathbf{T}_q f) = f \qquad \textit{for } 1 \le q \ne n-1 \tag{3.18}$$

if $f \in C^1_{0,q}(\bar{G})$ and $\bar{\partial} f = 0$.

Corollary 3.10. *If $K = \{R_1 \le |z| \le R_2\} \subset \mathbb{C}^n$ with $0 < R_1 \le R_2 < \infty$, then*

$$H^q_{\bar{\partial}}(K) = 0 \qquad \textit{for } q \ne 0, n-1.$$

The proof is immediate (see the proof of Corollary 3.8).

Remark 3.11. The above result was needed in the proof of Theorem 2.7. In fact, this shows that Corollary 3.10, and hence also (3.18), is false for $q = n - 1$. Otherwise the proof of Theorem 2.7 would hold for $q = n$, implying $H^n_{\bar{\partial},c}(\mathbb{C}^n) = 0$, which we know is false!

For $q = 0 \le n - 2$, $\Omega_q(C_{\#}^{(r)}) = 0$ on $S_1 \times G$. Therefore, Theorem 3.6 shows that if $f \in A(G)$, then

$$f = \int_{S_2} f \wedge \Omega_0(C^{(r)}) \qquad \text{on } G.$$

Here the integral is holomorphic on $B(0, R_2)$! We have thus obtained another proof of the Hartogs extension theorem for the case of a ball.

Exercises

E.3.1. Let U be open in $\mathbb{C}^n$ and suppose $W \in C^1_{1,0}(U)$ is a generating form for the point $z \notin U$ (i.e., (3.1) holds). Show that $d\Omega_0(W) = 0$ in the sense of distributions by proving the following steps.

(i) Let $V \subset\subset U$ be open. Then there is a sequence $\{W_\nu: \nu = 1, 2, \ldots\} \subset C^2_{1,0}(V)$, such that W_ν satisfies (3.1) on V and such that $\lim_{\nu\to\infty} W_\nu = W$ in C^1 norm over V.

(ii) Suppose $D \subset\subset U$ has piecewise C^1 boundary and let $f \in C^1(\bar{D})$. If $\{W_\nu\}$ is chosen as in (i), where $D \subset\subset V \subset\subset U$, show that

$$\int_{bD} f\Omega_0(W_\nu) = \int_D df \wedge \Omega_0(W_\nu)$$

for all $\nu = 1, 2, \ldots$.

(iii) Let $\nu \to \infty$ in (ii) to conclude that

$$\int_{bD} f\Omega_0(W) = \int_D df \wedge \Omega_0(W).$$

E.3.2. Let $W \in C^{1;\infty}_{1;0}(U \times D)$ and $f \in C^{1,\infty}(U \times D)$, where U and D are in $\mathbb{C}^n$. Show that

$$\Omega_q(fW) = f^n\Omega_q(W).$$

E.3.3. Prove Proposition 3.2. (Hint: Use Stokes' Theorem on $D - B(z, \varepsilon)$ and Lemma 3.1, and let $\varepsilon \to 0$.)

E.3.4. Let $D \subset\subset \mathbb{C}^n$ be a convex open set. Show that every $f \in A(D)$ is the uniform limit on $\bar{D}$ of a sequence of functions in $\mathcal{O}(\bar{D})$.

E.3.5. Let $\Omega_0(S)$ be the kernel on $bB \times B$ defined in Example 3.3, where B is the unit ball in $\mathbb{C}^n$.

(i) If $r = |\zeta|^2 - 1$ is the standard defining function for the unit ball, show that

$$\Omega_0(S) = [1 - (z, \zeta)]^{-n}\Omega_0(\partial r) \text{ on } bB \times B.$$

(ii) Show that

$$\iota^*\Omega_0(\partial r) = \frac{(n-1)!}{2\pi^n}\, dS,$$

where $\iota\colon bB \hookrightarrow \mathbb{C}^n$ is the inclusion and dS is the surface element on bB. (Hint: Use Corollary III.3.5; this result will be generalized in Lemma VII.3.8.)

(iii) (i) and (ii) imply that the operator $\mathbf{S}\colon L^1(bB) \to \mathcal{O}(B)$ defined by $\mathbf{S}f = \int_{bB} f\Omega_0(S)$ is given by

$$(*) \qquad \mathbf{S}f(z) = \frac{(n-1)!}{2\pi^n}\int_{bB} f(\zeta)\frac{dS_\zeta}{[1 - (z, \zeta)]^n}.$$

Let $H^2(bB)$ denote the closure in $L^2(bB)$ of the restriction to bB of functions in $A(B)$. Show that if $f \in H^2(bB)$, then $\tilde{f} = \mathbf{S}f \in \mathcal{O}(B)$ is an "extension" of f, with the property that if $f_r \in \mathcal{O}(\bar{B})$ is defined by $f_r(z) = \tilde{f}(rz)$ for $z \in \bar{B}$, then $\lim_{r\to 1} \|f_r - f\|_{L^2(bB)} = 0$.

Remark. One can show that $(*)$ defines the orthogonal projection from $L^2(bB)$ onto the closed subspace $H^2(bD)$. This projection is called the Szegö projection. (See [Rud 3] or [SteE 2] for more details.)

E.3.6. Let $D \subset\subset \mathbb{C}^n$ be convex with C^2 boundary. Define $r(z) = -\delta_D(z)$ for $z \in \bar{D}$ and $r(z) = \delta_{\mathbb{C}^n - \bar{D}}(z)$ for $z \notin \bar{D}$. Show that

$$D_\varepsilon = \{z\colon r(z) < \varepsilon\}$$

is a convex domain with C^2 boundary provided $|\varepsilon|$ is sufficiently small. (Hint: Use E.II.5.9.)

E.3.7. Let $D_1 \subset\subset D_2$ be two bounded convex domains in $\mathbb{C}^n$ and let $K = \bar{D}_2 - D_1$. Show that $H^q_{\bar\partial}(K) = 0$ for $q \geq 1$ with $q \neq n - 1$.

§4. The Bergman Kernel

In contrast to the concrete integral representations which we have discussed in the preceding paragraphs we shall now use abstract Hilbert space techniques in order to obtain a new type of integral representation for holomorphic functions. Such methods were introduced in complex analysis already in the 1920s by Stefan Bergman. Bergman developed his theory in an attempt to deal with the classification problem for domains in $\mathbb{C}^n$ up to biholomorphic equivalence, a problem which is particularly significant if $n > 1$, because there is no analogue of the Riemann Mapping Theorem in that case (recall Theorem I.2.7). The relevant abstract kernel—the so-called "Bergman kernel"—can be defined quite easily for arbitrary domains, but it is very difficult to obtain concrete representations for it and to study its behavior at the boundary, except in special cases like a ball or polydisc; thus the Bergman kernel was of limited use for a long time. However, major progress has been made during the last decade, especially in case of strictly pseudoconvex domains, and the Bergman kernel is now used extensively in the study of boundary regularity of biholomorphic maps. In the present paragraph we present some of the more elementary results about this kernel. Some of its deeper properties on strictly pseudoconvex domains and applications to biholomorphic maps will be discussed in Chapter VII.

4.1. The Bergman Kernel

Given $D \subset \mathbb{C}^n$, we consider the Hilbert space $L^2(D)$, with inner product

$$(f, g) = (f, g)_D = \int_D f\bar{g}\, dV \quad \left(= \int_D f \wedge *\bar{g}\right).$$

By Corollary I.1.10, $\mathscr{H}^2 = \mathscr{H}^2(D) := \mathcal{O}L^2(D)$ is a closed subspace of $L^2(D)$, and hence is itself a Hilbert space. In order to ensure that $\mathscr{H}^2(D)$ is nontrivial, we shall henceforth assume that D is bounded. By Corollary I.1.7, for each $a \in D$, the evaluation map

$$\tau_a \colon \mathscr{H}^2 \to \mathbb{C} \qquad \text{defined by } \tau_a(f) = f(a),$$

is a *bounded* linear functional on $\mathscr{H}^2(D)$. Therefore, by the Riesz Representation Theorem, there is a unique element in $\mathscr{H}^2(D)$, which we denote by $K_D(\cdot, a)$, such that

$$f(a) = \tau_a(f) = (f, K_D(\cdot, a)) = \int_D f(\zeta)\overline{K_D(\zeta, a)}\, dV(\zeta) \tag{4.1}$$

for all $f \in \mathscr{H}^2$. The function

$$K_D \colon D \times D \to \mathbb{C} \qquad \text{with } K_D(\cdot, a) \in \mathscr{H}^2(D),$$

thus defined is called the **Bergman kernel for** D.

Notice that for all $a \in D$ one has

$$|\tau_a(f)| = |f(a)| \leq C\delta_D^{-n}(a) \| f \|_{L^2(D)} \qquad \text{for } f \in \mathscr{H}^2(D),$$

with a constant C which depends only on the dimension n. To see this, apply the Cauchy estimate (I.1.20) with $\alpha = 0$ to the polydisc $P = P(a, r)$, where $r = \delta_D(a)/\sqrt{n}$, and use the inequality $\| f \|_{L^1(P)} \leq (\text{vol } P)^{1/2} \| f \|_{L^2(P)}$. Since

$$\| K_D(\cdot, a) \|_{L^2} = \| \tau_a \| = \sup\{|f(a)|: f \in \mathscr{H}^2, \| f \| \leq 1\},$$

the Bergman kernel K_D therefore satisfies the estimate

$$\| K_D(\cdot, a) \|_{L^2} \leq C\delta_D^{-n}(a) \qquad \textit{for } a \in D. \tag{4.2}$$

Next we verify a fundamental symmetry property of K_D.

Lemma 4.1. *The Bergman kernel K_D satisfies*

$$K_D(\zeta, z) = \overline{K_D(z, \zeta)} \qquad \text{for all } \zeta, z \in D, \tag{4.3}$$

and hence $K_D(\zeta, z)$ is conjugate holomorphic in z.

Proof. By definition, for fixed $z \in D$ we have $K_D(\cdot, z) \in \mathscr{H}^2$; so we may apply (4.1) with $f = K_D(\cdot, z)$, and $a = \zeta \in D$, giving

$$\begin{aligned} K_D(\zeta, z) &= (K_D(\cdot, z), K_D(\cdot, \zeta)) \\ &= \overline{(K_D(\cdot, \zeta), K_D(\cdot, z))} \\ &= \overline{K_D(z, \zeta)}, \end{aligned}$$

where, in the last equality, we have again used (4.1), but this time with $f = K(\cdot, \zeta)$ and $a = z$. ■

Notice that by (4.3) we can rewrite (4.1) in the form

$$f(z) = \int_D f(\zeta) K_D(z, \zeta)\, dV(\zeta) \qquad \textit{for all } f \in \mathscr{H}^2 \textit{ and } z \in D. \tag{4.4}$$

4.2. Representation by Orthonormal Basis

By Lemma 4.1 the function $H(\zeta, z) = K_D(\zeta, \bar{z})$, defined on $\Omega = D \times \bar{D} = \{(\zeta, \bar{z}) \in \mathbb{C}^{2n}: \zeta, z \in D\}$, is separately holomorphic in ζ and z. In order to see that K_D is jointly continuous in (ζ, z), one could thus apply Hartogs' theorem on separate analyticity to the function H, implying that H is holomorphic on Ω. But for the special function considered here, a more elementary proof of the latter fact is available, which is based on an interesting representation of the Bergman kernel in terms of an orthonormal basis for $\mathscr{H}^2(D)$.

Lemma 4.2. *For every compact set $K \subset D$ there is a constant $C_K < \infty$ such that for every orthonormal basis $\{\varphi_j, j = 1, 2, \ldots\}$ of $\mathscr{H}^2$ one has*

$$\sup_{z \in K} \sum_{j=1}^{\infty} |\varphi_j(z)|^2 \le C_K. \tag{4.5}$$

PROOF. Let $K \subset D$ be compact. Then $\operatorname{dist}(K, bD) > 0$, and it follows from (4.2) that there is a constant C_K such that

$$\| K_D(\cdot, z) \|_{L^2(D)} \le C_K \tag{4.6}$$

for all $z \in K$. Given an orthonormal basis $\{\varphi_j : j = 1, 2, \ldots\}$ for $\mathscr{H}^2(D)$, the function $K_D(\cdot, z) \in \mathscr{H}^2$ has the representation

$$K_D(\zeta, z) = \sum_{j=1}^{\infty} (K_D(\cdot, z), \varphi_j)\varphi_j(\zeta), \tag{4.7}$$

the series converging in $\mathscr{H}^2$, and hence also compactly in $\mathscr{O}(D)$. Moreover, one has

$$\sum_{j=1}^{\infty} |(K_D(\cdot, z), \varphi_j)|^2 = \| K_D(\cdot, z) \|_{L^2(D)}^2. \tag{4.8}$$

Since by (4.1) one has

$$(K_D(\cdot, z), \varphi_j) = \overline{(\varphi_j, K_D(\cdot, z))} = \overline{\varphi_j(z)} \qquad \text{for } j = 1, 2, \ldots, \tag{4.9}$$

the desired result follows from (4.8) and (4.6). ■

Theorem 4.3. *For any orthonormal basis $\{\varphi_j, j = 1, 2, \ldots\}$ for $\mathscr{H}^2(D)$ one has the representation*

$$K_D(\zeta, z) = \sum_{j=1}^{\infty} \varphi_j(\zeta)\overline{\varphi_j(z)} \qquad \text{for all } (\zeta, z) \in D \times D, \tag{4.10}$$

with uniform convergence on compact subsets of $D \times D$.

PROOF. The representation (4.10) follows directly from (4.7) and (4.9), since convergence in $\mathscr{H}^2$ implies pointwise convergence. For the remaining statement, it is enough to prove uniform boundedness of the partial sums $\sum_{j=1}^{m} |\varphi_j(\zeta)| \, |\varphi_j(z)|$, $m = 1, 2, \ldots$, on $K \times K$ for an arbitrary compact set $K \subset D$. Compact convergence then follows by a normality argument. By Lemma 4.2, $\{|\varphi_j(z)|, j = 1, 2, \ldots\} \in l^2$ for each fixed $z \in D$; so by the Cauchy–Schwarz inequality (in l^2) and (4.5), one has

$$\sum_{j=1}^{\infty} |\varphi_j(\zeta)| \, |\varphi_j(z)| \le \left(\sum_{j=1}^{\infty} |\varphi_j(\zeta)|^2 \right)^{1/2} \left(\sum_{j=1}^{\infty} |\varphi_j(z)|^2 \right)^{1/2} \le C_K \tag{4.11}$$

for all $\zeta, z \in K$. ■

Corollary 4.4. *$K_D(\zeta, \bar{z})$ is holomorphic in $(\zeta, z) \in D \times \bar{D}$, and hence $K_D \in C^\infty(D \times D)$.*

4.3. The Bergman Projection

Since $\mathscr{H}^2(D)$ is a *closed* subspace of $L^2(D)$, there is an orthogonal projection operator

$$\mathbf{P} = \mathbf{P}_D: L^2(D) \to \mathscr{H}^2(D).$$

$\mathbf{P}$ is a bounded Hermitian operator of norm 1 which satisfies $\mathbf{P}f = f$ for $f \in \mathscr{H}^2$. We shall now see that the abstract operator $\mathbf{P}_D$—called the **Bergman projection**—is given by integration against the Bergman kernel.

Proposition 4.5. *The Bergman projection* $\mathbf{P}_D: L^2(D) \to \mathscr{H}^2(D)$ *satisfies*

$$(4.12) \qquad (\mathbf{P}_D f)(z) = (f, K_D(\cdot, z)) = \int_D f(\zeta) K_D(z, \zeta)\, dV(\zeta)$$

for all $f \in L^2(D)$ *and* $z \in D$.

PROOF. Given $f \in L^2(D)$, we apply the reproducing property (4.1) to $\mathbf{P}f \in \mathscr{H}^2$, giving

$$(4.13) \qquad \mathbf{P}f(z) = (\mathbf{P}f, K_D(\cdot, z)),$$

Since $\mathbf{P}$ is Hermitian and $K_D(\cdot, z) \in \mathscr{H}^2$, we obtain

$$(4.14) \qquad (\mathbf{P}f, K_D(\cdot, z)) = (f, \mathbf{P}K_D(\cdot, z)) = (f, K_D(\cdot, z)),$$

and (4.12) follows. ■

4.4. The Bergman Kernel for the Ball

We shall now obtain an explicit formula for the Bergman kernel K_B of the unit ball $B = B(0, 1)$ as a simple consequence of the Cauchy Integral formula for convex domains discussed in §3.2. The function $r = |z|^2 - 1$ is a C^∞ defining function for the (convex) unit ball B. Since $\partial r(\zeta) = \sum_{j=1}^m \bar{\zeta}_j\, d\zeta_j$, one has

$$(4.15) \qquad \langle \partial r(\zeta), \zeta - z \rangle = \sum_{j=1}^n \bar{\zeta}_j(\zeta_j - z_j) = |\zeta|^2 - (z, \zeta),$$

which equals $1 - (z, \zeta)$ for $\zeta \in bB$. By Theorem 3.4,

$$(4.16) \qquad f(z) = (2\pi i)^{-n} \int_{bB} f(\zeta) \frac{\partial r(\zeta) \wedge [\bar{\partial}\partial r(\zeta)]^{n-1}}{[1 - (z, \zeta)]^n}$$

for $f \in A(B)$ and $z \in B$. For fixed $z \in B$, the kernel $C(\zeta, z)$ in (4.16) is clearly smooth in $\zeta \in \bar{B}$; so if $f \in A^1(B)$, we may apply Stokes' Theorem in (4.16), obtaining

$$(4.17) \qquad f(z) = (2\pi i)^{-n} \int_B f(\zeta) \bar{\partial}_\zeta C(\zeta, z).$$

We now compute

$$\bar{\partial}_\zeta C(\zeta, z) = \frac{(\bar{\partial}\partial r(\zeta))^n}{[1 - (z, \zeta)]^n} + n\frac{\sum_{k=1}^n z_k\, d\bar{\zeta}_k \wedge \partial r(\zeta) \wedge (\bar{\partial}\partial r(\zeta))^{n-1}}{[1 - (z, \zeta)]^{n+1}}$$
$$= \frac{n! \bigwedge_{j=1}^n (d\bar{\zeta}_j \wedge d\zeta_j)}{[1 - (z, \zeta)]^n} + n\frac{(\sum_{k=1}^n z_k\bar{\zeta}_k)(n-1)! \bigwedge_{j=1}^n (d\bar{\zeta}_j \wedge d\zeta_j)}{[1 - (z, \zeta)]^{n+1}}$$
$$= \frac{n! \bigwedge_{j=1}^n (d\bar{\zeta}_j \wedge d\zeta_j)}{[1 - (z, \zeta)]^{n+1}}.$$

Since

$$\frac{1}{(2i)^n}\bigwedge_{j=1}^n (d\bar{\zeta}_j \wedge d\zeta_j) = \left(\frac{i}{2}\right)^n \bigwedge_{j=1}^n (d\zeta_j \wedge d\bar{\zeta}_j) = dV(\zeta),$$

by III.(3.4), we obtain

$$\frac{1}{(2\pi i)^n}\bar{\partial}_\zeta C = \frac{n!}{\pi^n}\frac{1}{[1 - (z, \zeta)]^{n+1}}\, dV. \tag{4.18}$$

Let us introduce the function

$$G(\zeta, z) = \frac{n!}{\pi^n}\frac{1}{[1 - (z, \zeta)]^{n+1}}. \tag{4.19}$$

It follows from (4.17) and (4.18) that

$$f(z) = \int_B f(\zeta)G(\zeta, z)\, dV = (f, \overline{G(\cdot, z)})_B \qquad \text{for } z \in B. \tag{4.20}$$

The proof of (4.20) had assumed that $f \in A^1(B)$, but (4.20) holds for arbitrary $f \in \mathscr{H}^2(B)$ by a simple limit argument: if $f \in \mathscr{H}^2(B)$ and $0 < t < 1$, then $f_t(z) = f(tz)$ defines a function $f_t \in \mathcal{O}(\bar{B})$ for which (4.20) holds, and since $f_t \to f$ in $\mathscr{H}^2$ as $t \to 1$, one obtains (4.20) for f as well. By (4.19), for fixed $z \in B$, the function $\overline{G(\zeta, z)} = G(z, \zeta)$ is holomorphic on $\bar{B}$ in ζ, and hence is in $\mathscr{H}^2(B)$. Since by (4.20) it satisfies the necessary reproducing property (4.1), it follows that $\overline{G(\zeta, z)}$ must coincide with the Bergman kernel $K_B(\zeta, z)$ of the ball.

Let us summarize the result of our computations.

Theorem 4.6. *The Bergman kernel K_B for the unit ball $B = \{z \in \mathbb{C}^n : |z| < 1\}$ is given by*

$$K_B(\zeta, z) = \frac{n!}{\pi^n}[1 - (\zeta, z)]^{-(n+1)}.$$

4.5. Product Domains

Theorem 4.6 for $n = 1$ gives the Bergman kernel for the unit disc $\Delta = \{z \in \mathbb{C} : |z| < 1\}$. The following general result immediately implies a formula for the Bergman kernel on the unit polydisc $\Delta^n = P(0, 1)$.

Theorem 4.7. *Suppose $D_i \subset\subset \mathbb{C}^{n_i}$, $i = 1, 2$, are bounded domains with Bergman kernels K_{D_1} and K_{D_2}. Then the Bergman kernel K_D for the product domain $D = D_1 \times D_2$ is given by*

$$K_D((\zeta_1, \zeta_2), (z_1, z_2)) = K_{D_1}(\zeta_1, z_1)K_{D_2}(\zeta_2, z_2) \tag{4.21}$$

for all (ζ_1, ζ_2) and $(z_1, z_2) \in D_1 \times D_2$.

PROOF. Let G denote the function on the right in (4.21). It is clear that $G(\cdot, (a_1, a_2)) \in \mathscr{H}^2(D)$ for each fixed $(a_1, a_2) \in D$; moreover, the reproducing property

$$f(a_1, a_2) = (f, G(\cdot, (a_1, a_2)))_D = \int_{D_1 \times D_2} f\bar{G}(\cdot, (a_1, a_2))\, dV$$

is an easy consequence of Fubini's Theorem and the corresponding reproducing properties for K_{D_1} and K_{D_2}. It follows that $G \equiv K_D$. ■

Corollary 4.8. *The Bergman kernel K_{Δ^n} for the unit polydisc Δ^n in $\mathbb{C}^n$ is given by*

$$K_{\Delta^n}(\zeta, z) = \frac{1}{\pi^n} \prod_{j=1}^{n} \frac{1}{(1 - \zeta_j \bar{z}_j)^2}.$$

4.6. The Transformation Formula

We shall now study the behavior of the Bergman kernel under biholomorphic maps. The simple transformation formula given in Theorem 4.9 below is of great importance in the study of biholomorphic maps, as we shall see in Chapter VII. A differential geometric interpretation is discussed in Exercise E.4.3.

Theorem 4.9. *Suppose $F: D_1 \to D_2$ is a biholomorphic map between bounded domains in $\mathbb{C}^n$. Then*

$$K_{D_1}(\zeta, z) = \det F'(\zeta) K_{D_2}(F(\zeta), F(z))\, \overline{\det F'(z)} \tag{4.22}$$

for all $\zeta, z \in D_1$.

PROOF. The substitution formula for integrals and Lemma I.2.1 imply that

$$\int_{D_2} |f(w)|^2\, dV(w) = \int_{D_1} |f \circ F(\zeta)|^2 |\det F'(\zeta)|^2\, dV(\zeta). \tag{4.23}$$

Hence the map $T_F: f \mapsto (f \circ F) \det F'$ establishes an isometric isomorphism from $L^2(D_2)$ to $L^2(D_1)$, with inverse $T_{F^{-1}}$, which restricts to an isomorphism $\mathscr{H}^2(D_2) \to \mathscr{H}^2(D_1)$. Suppose $f \in \mathscr{H}^2(D_1)$, fix $z \in D_1$, and let $w = F(z)$. By applying the reproducing property of $K_{D_2}(\cdot, w)$ to the function $T_{F^{-1}} f = (f \circ F^{-1}) \det(F^{-1})'$ one obtains

$$(4.24) \qquad (T_{F^{-1}}f, K_{D_2}(\cdot, w))_{D_2} = (T_{F^{-1}}f)(w) = f(z)[\det F'(z)]^{-1}.$$

Since T_F is an isometry,

$$(4.25) \qquad (T_{F^{-1}}f, K_{D_2}(\cdot, w))_{D_2} = (f, T_F K_{D_2}(\cdot, w))_{D_1}.$$

From (4.24) and (4.25) one obtains

$$f(z) = (f, \det F'(\cdot)K_{D_2}(F(\cdot), F(z)) \det F'(z))_{D_1},$$

which shows that the function on the right side of (4.22), which is clearly in $\mathscr{H}^2(D_1)$ as a function of ζ, has the required reproducing property, and hence must agree with $K_{D_1}(\zeta, z)$. ■

Theorem 4.9 implies a transformation formula for the Bergman projection $\mathbf{P}_D$ under biholomorphic maps.

Corollary 4.10. *If $F: D_1 \to D_2$ is biholomorphic, then*

$$(4.26) \qquad \mathbf{P}_{D_1}(\det F' \; f \circ F) = \det F' \; (\mathbf{P}_{D_2}(f) \circ F)$$

for all $f \in L^2(D_2)$.

Proof. Recall the isometry $T_F: L^2(D_2) \to L^2(D_1)$ from the proof of Theorem 4.9. The left side of (4.26) is $\mathbf{P}_{D_1}(T_F f)$. By Proposition 4.5, if $z \in D_1$, then

$$\mathbf{P}_{D_1}(T_F f)(z) = (T_F f, K_{D_1}(\cdot, z)).$$

Now (4.22) implies $K_{D_1}(\cdot, z) = [T_F K_{D_2}(\cdot, F(z))]\, \overline{\det F'(z)}$, so

$$\begin{aligned} \mathbf{P}_{D_1}(T_F f)(z) &= \det F'(z)(T_F f, T_F K_{D_2}(\cdot, F(z))) \\ &= \det F'(z)(f, K_{D_2}(\cdot, F(z))) \qquad \text{(since } T_F \text{ is an isometry)} \\ &= \det F'(z)(\mathbf{P}_{D_2} f)(F(z)), \end{aligned}$$

where the last equality follows again from Proposition 4.5. ■

Remark 4.11. The transformation formula (4.26) for the Bergman projection (but not the transformation formula (4.22) for the Bergman kernel) remains true for *proper* holomorphic maps $F: D_1 \to D_2$ (i.e., $F^{-1}(K)$ is compact for K compact in D_2) which are not necessarily biholomorphic. This important fact was discovered only recently by S. Bell [Bel 3]; it has become a fundamental tool in the study of boundary regularity of proper holomorphic maps (see the Notes to Chapter VII).

Exercises

E.4.1. Let $D \subset \mathbb{C}^n$ be bounded.

(i) Show that $k_D(z) = K_D(z, z) > 0$ for $z \in D$.
(ii) Prove that $\log k_D$ is strictly plurisubharmonic on D.

E.4.2. (*The Bergman metric*). Let $D \subset\subset \mathbb{C}^n$.

(i) Show that the map $H_z^D: T_z\mathbb{C}^n \times T_z\mathbb{C}^n \to \mathbb{C}$ defined by

$$(*) \qquad H_z^D(u, v) = \sum_{j,k=1}^{n} \frac{\partial^2 \log k_D}{\partial z_j \partial \bar{z}_k}(z) u_j \bar{v}_k, \qquad z \in D,$$

defines a Hermitian inner product on $T_z\mathbb{C}^n$ for each $z \in D$.

(ii) If $V^{(1)}$ and $V^{(2)}$ are C^∞ vector fields on $\Omega \subset D$, then $h(z) := H_z^D(V_z^{(1)}, V_z^{(2)})$ is C^∞ on Ω.

Remark. (i) and (ii) are the defining properties of a **Hermitian structure (or metric)** on D; this concept generalizes in the obvious way to any complex manifold. The particular Hermitian metric H^D defined by (*) is called the **Bergman metric** on D.

(iii) Let H^D be any Hermitian metric on D. Show that $G^D := \operatorname{Re} H^D$ defines a Riemannian metric on D.

E.4.3. Let $F: D_1 \to D_2$ be a biholomorphic map between bounded domains in $\mathbb{C}^n$. Prove that F is an isometry with respect to the Bergman metrics H^{D_1} and H^{D_2} on D_1 and D_2, i.e., for all $u, v \in T_z\mathbb{C}^n$, $z \in D_1$, one has $H_z^{D_1}(u, v) = H_{F(z)}^{D_2}(dF(u), dF(v))$.

E.4.4. (i) Compute the Bergman metric for the unit ball $B(0, 1)$ in $\mathbb{C}^n$.

(ii) Verify that in case $n = 1$ the Bergman metric on the open unit disc Δ agrees with the classical Poincaré metric on Δ.

E.4.5. Compute an exact formula for the Bergman kernel for a ball $B(a, R)$ with center $a \in \mathbb{C}^n$ and radius $R > 0$.

E.4.6. (i) Show that for all $z \in D$ one has

$$K_D(z, z) = \sup\{|f(z)|^2 : f \in \mathscr{H}^2(D) \text{ with } \|f\| \le 1\}.$$

(ii) Show that for each $z \in D$ there is an extremal function $h_z \in \mathscr{H}^2(D)$ with $\|h_z\| = 1$ and $K_D(z, z) = |h_z(z)|^2$.

(iii) Show that $K_D(z, z) \ge 1/\operatorname{vol}(D)$.

E.4.7. Show that $|K_D(\zeta, z)| \le C_n \delta_D^{-n}(\zeta)\delta_D^{-n}(z)$ for all $\zeta, z \in D$, where the constant C_n depends only on the dimension n.

E.4.8. Suppose $D_1 \subset D_2$ are open in $\mathbb{C}^n$. Prove that

$$K_{D_2}(z, z) \le K_{D_1}(z, z) \qquad \text{for } z \in D_1.$$

E.4.9. Suppose $D \subset\subset \mathbb{C}^n$ has boundary of class C^2. Show that the estimate in E.4.7 holds in the strengthened form

$$(*) \qquad |K_D(\zeta, z)| \le C\delta_D^{-(n+1)/2}(\zeta)\delta_D^{-(n+1)/2}(z)$$

for $z, \zeta \in D$, where C is independent of ζ and z. (Hint: Prove $K_D(z, z) \lesssim \delta_D^{-n-1}(z)$ by showing it first for a ball and then by applying this together with E.4.8 to a family of balls of fixed positive radius which are contained in D and whose boundaries meet bD.)

Notes for Chapter IV

The importance of integral representations in complex analysis was recognized at least as early as 1831, when Cauchy discovered the famous formula which carries his name. Its generalization to several variables in case of polydiscs (Theorem I.1.3) is almost as old, and it has played a fundamental role in the local theory from the very beginning, just as in the case of one variable. However, it took much longer to develop Cauchy-type integral formulas for more general domains, which would then lead to substantial *global* applications.

The first major result seems to be due to A. Weil, who in 1932 announced a Cauchy integral formula for *polynomial* polyhedra in $\mathbb{C}^2$ (the proof was published in 1935 [Wei]). Weil used this integral formula to prove the first generalization of the Runge Approximation Theorem to several variables—this result is now known as the Oka–Weil Theorem. We will say more about it in VI.§1.3, and in the Notes for Chapter VI. Essentially the same integral formula was obtained independently in 1934 by S. Bergman [Ber 1]. A special feature of the Bergman–Weil formula is that integration is only over a very small subset of the topological boundary of the polyhedron, the so-called *distinguished boundary*; we have already noticed this for the Cauchy integral on polydiscs in Chapter I. The extension of the Bergman–Weil integral to general *analytic* polyhedra requires a global decomposition for holomorphic functions on domains of holomorphy, a result now known as Hefer's Theorem (see Theorem V.2.2) and which was only published in 1950 [Hef]. However, already in 1940, K. Oka, based on his earlier work, had been able to circumvent this difficulty by introducing a technical modification in the Bergman–Weil integral, thus obtaining an integral representation formula on analytic polyhedra in $\mathbb{C}^2$ [Oka, V]. This modification of the Bergman–Weil integral was one of several major ingredients in Oka's 1942 solution of the Levi problem in $\mathbb{C}^2$ [Oka, VI]. Once Hefer's Theorem was available, Oka's modification became obsolete. Surprisingly, Hefer's Theorem is a rather elementary consequence of the solution of the Cousin I problem[1] obtained by Oka already in 1936. This seems to be the one occasion where Oka's work did not follow the direct route. In 1952 F. Sommer [Som] systematically investigated the Bergman–Weil integral on analytic polyhedra and established a connection between it and the formula of E. Martinelli and S. Bochner discussed in §1 (see also below), where integration is over the full topological boundary. Integral representations on *analytic polyhedra* are rather difficult to use in applications, and today they are of interest mainly to specialists, but we hope that the above remarks will convince the reader of their central role in the early developments of global function theory in several variables.

Of much greater importance—at least from today's perspective, and certainly

[1] See Chapter VI, §4, for a discussion of this problem.

for the present book—are generalizations of the Cauchy Integral Formula to smoothly bounded domains, especially in case of a strictly pseudoconvex boundary. The first such formula (Proposition 1.6) was discovered in 1938 by E. Martinelli [Mar 1], and, independently, by S. Bochner [Boc] in 1941. Bochner's paper was published in 1943. In a footnote added in proof, Bochner makes reference to a recently published paper of E. Martinelli [Mar 2], in which the same integral formula appears, and he points out that he had already lectured on this formula in a course at Princeton University in the Winter of 1940–41, and that the result was later incorporated in a 1941 Princeton Dissertation. Clearly, at that time Bochner had only learned of Martinelli's 1943 paper and not of the earlier 1938 paper. Thus there seems to be no question about Martinelli's priority (notice that Henkin and Leiterer [HeLe] refer to Proposition 1.6 as the Martinelli–Bochner formula). However, as our presentation follows Bochner's potential theoretic proof, it seems justified to give equal credit to both authors and retain the commonly used alphabetic order.

The more general representation formula for C^1 functions f, which involves a correction term depending on $\bar{\partial} f$ (Corollary 1.3), appeared in 1967 as a special case of Theorem 1.10, due to W. Koppelman [Kop 2], but it probably was known earlier. It is implicitly contained in [Boc], at least for harmonic f. Certainly in case $n = 1$ it is *much* older: it seems to have appeared first in 1912 in a paper of D. Pompeiu ("Sur une classe de fonctions d'une variable complexe...," Rend. Circ. Matem. Palermo **35** (1913), 277–281), but it remained largely forgotten in classical complex analysis until the early 1950s, when A. Grothendieck and P. Dolbeault used it to solve the inhomogeneous Cauchy–Riemann equation $\partial u/\partial \bar{z} = f$ in one variable (in our presentation this is now a special case of Corollary 1.11(b)). More significantly, Grothendieck and Dolbeault used this result to then prove—by an inductive procedure—the solvability of the equation $\bar{\partial} u = f$ for arbitrary $\bar{\partial}$-closed (p, q)-forms on polydiscs in *several* variables (see Exercises E.III.2.4 and E.III.2.5 for an outline). In particular, this gives local existence for solutions for the $\bar{\partial}$-equation. This result, widely known as the Dolbeault–Grothendieck Lemma, is of fundamental importance in the representation of analytic sheaf cohomology groups in terms of the $\bar{\partial}$-complex due to P. Dolbeault [Dol] (see Chapter VI, §6.3). In this book the *local* solvability of $\bar{\partial} u = f$ follows from the solution of $\bar{\partial}$ on convex sets (Corollary 3.8). Quite surprisingly, it seems to have been overlooked that the inversion of $\partial/\partial \bar{z}$ in $\mathbb{C}^1$ by the Cauchy kernel, and its application by induction to solve $\bar{\partial} u = f$ on polydiscs in $\mathbb{C}^n$ was already discovered by S. Bochner in 1943 ([Boc], Theorem 11) for the case of a real analytic $(0, 1)$-form f—this was the case of interest to Bochner. It therefore seems appropriate to refer to this result as the Bochner–Dolbeault–Grothendieck Lemma.

The relationship between the $\bar{\partial}$-complex and sheaf cohomology groups discovered in the 1950s marks the beginning of the systematic use of differential forms in complex analysis, but it took surprisingly long until concrete integral representations were introduced for them. The first result of this type (Theorem 1.10) was obtained in 1967 by W. Koppelman [Kop 2]; however, because of his untimely death, Koppelman was not able to publish any proofs. Proofs were eventually obtained around 1970, first by I. Lieb [Lie 2], then by N. Øvrelid [Øvr 1], and subsequently by many others. The simple proof given here is the natural extension to differential forms of Bochner's proof for the case of holomorphic functions, and was discovered by I. Lieb and R.M. Range [LiRa 2].

Lemma 1.15, the essential step in the proof of the regularity Theorem 1.14, is a standard result in real analysis; the proof given here follows [Ker 1].

The proof of the Hartogs Extension Theorem given in §2.1 is due to E. Martinelli [Mar 2]. A similar proof was given by S. Bochner [Boc]. A careful reading of the latter proof shows that all that is needed for the function f which is to be extended is that f is a CR-function, though Bochner did not explicitly state the result in this form. A precise statement of the extension theorem for CR-functions was given by L. Hörmander [Hör 2], who credits Bochner. A different proof of Theorem 2.1, based on the solution of $\bar{\partial}$ with compact support (see E.2.5), was found in 1960 by L. Ehrenpreis [Ehr]. Hörmander's proof of Bochner's Theorem is based on the same idea. Bochner's 1943 proof of Hartogs' Theorem involves the jump formula for the BM-transform, which he only proves for the case of harmonic functions. The more delicate case of *continuous* functions (Proposition 2.2) was proved first for $n = 1$ in 1908 by J. Plemelj ("Ein Ergänzungssatz zur Cauchyschen Integraldarstellung analytischer Funktionen, Randwerte betreffend," Monatsh. Math. und Phys. **19** (1908), 205–210), whose arguments were extended to several variables by R. Harvey and B. Lawson in 1975 [HaLa]. The proof of Theorem 2.5 given here is essentially the one of Harvey and Lawson [HaLa], stripped of all references to distributions and currents. The interested reader should consult that paper for further references related to this circle of ideas.

The Cauchy Integral Formula for arbitrary generating forms (Theorem 3.6 in case $q = 0$ and f holomorphic) is due to J. Leray [Ler 1, 2], who introduced the terminology of Cauchy–Fantappiè forms. Related results were obtained by W. Koppelman [Kop 1]. The Cauchy Integral for convex domains (Theorem 3.4) was stated first by Leray in 1959 [Ler 2]; in the special case of a ball it seems to be due to L.K. Hua [Hua], who used a computation based on complete orthonormal systems of monomials.

Cauchy–Fantappiè forms of order $q > 0$ were introduced by W. Koppelman [Kop 2], who stated Lemma 3.5 for such forms as well as the fundamental result that any two CF forms of order $q \geq 1$ differ by a sum $\bar{\partial}_\zeta A_q + \bar{\partial}_z A_{q-1}$. These results are, essentially, equivalent to Theorem 3.6, with the operators $\mathbf{T}_q^W$ written in different form. The first detailed proof of Koppelman's results was given by I. Lieb [Lie 2] in 1970. For $q = 0$, the version of Theorem 3.6 stated here was proved first in 1970 by G.M. Henkin [Hen 2]. A closely related result was proved at the same time by H. Grauert and I. Lieb [GrLi]. For $q \geq 1$, the proof of Theorem 3.6 given here is due to R.M. Range and Y.T. Siu

[RaSi]. Besides I. Lieb's original proof and the Range–Siu proof, other proofs of results like Theorem 3.6 are due to P.L. Poljakov [Pol 1] and N. Øvrelid [Øvr 1]. The idea of switching variables near the strictly *concave* boundary in order to solve $\bar{\partial}$ on spherical shells by integral kernels (Proposition 3.9) is due to M. Hortmann [Hor], who considered regions more general than the simple case considered here.

Most of the material in §4 is due to S. Bergman. The reader interested in a comprehensive account of Bergman's theory should consult [Ber 2]. The *modern* theory of the Bergman kernel begins in 1965 with L. Hörmander's paper [Hör 1]. Additional contributions were then made by K. Diederich [Die] and N. Kerzman [Ker 2], and a complete description of the Bergman kernel near the boundary of strictly pseudoconvex domains was eventually achieved in 1974 in C. Fefferman's fundamental paper [Fef]. We will discuss some of these developments in Chapter VII. The Bergman kernel for the ball can be computed in many different ways besides the method used here, for example, by explicitly summing up the expansion in terms of the orthonormal basis given by the normalized monomials (see [Kra 2]), or by using the mean value formula, the transivity of the group of automorphisms of the ball and the transformation formula (4.22) (see [Hua] or [SteE 2]).

CHAPTER V

The Levi Problem and the Solution of $\bar{\partial}$ on Strictly Pseudoconvex Domains

In the preceding chapter we developed a general integral representation formula for differential forms, and we saw, in the case of convex domains, some of its major applications whenever there is a generating form which is *globally* holomorphic in the parameter z. In this chapter we apply these techniques to a strictly pseudoconvex domain D. Here the geometric information is only *local*, and there is no simple way to find a globally holomorphic generating form.

In §1 we use the Levi polynomial of a strictly plurisubharmonic defining function r for D—in essence, this contains the information that D is locally biholomorphically equivalent to a convex domain—to construct a generating form L_D on $bD \times D$ which is holomorphic in z for z close to $\zeta \in bD$; that is, near the singularities of L_D. Not much is gained at this point for the CF form $\Omega_q(L_D)$ in case $q = 0$; but if $q \geq 1$, then $\Omega_q(L_D) \equiv 0$ near the (potential) singularities, and this has far-reaching consequences! For most of this chapter we therefore concentrate on the case $q \geq 1$. The discussion of $\Omega_0(L_D)$ and of globally holomorphic Cauchy-type kernels is postponed to Chapter VII, when we will have available the global solution of $\bar{\partial}$ on $\bar{D}$. By applying Theorem IV.3.6 in case $q \geq 1$ to the generating form L_D, one easily obtains a *compact* extension operator $\mathbf{E}_q$ for $\bar{\partial}$-cohomology classes which, by the finiteness theorem for compact operators in Banach spaces, implies that there are at most *finitely* many obstructions to the solution of $\bar{\partial}u = f$ on D. By a classical argument of H. Grauert, this fact readily implies the solution of the Levi problem, i.e., that D is Stein.

Via the extension operator $\mathbf{E}_q$, the problem of solving $\bar{\partial}u = f$ on D is reduced to the case where f is a $\bar{\partial}$-closed $(0, q)$ form on a strictly pseudoconvex neighborhood D_δ of $\bar{D}$. In §2 we solve this problem by constructing a generating form holomorphic in $z \in \bar{D}$ on a suitable neighborhood of $\bar{D}$, which by

Theorem IV.3.6 will then produce the required integral solution operator for $\bar{\partial}$. The main idea is as follows. By IV.§3.2, we know how to handle *convex* regions. Since by the solution of the Levi problem $\bar{D}$ is Stein, there is a neighborhood basis for $\bar{D}$ consisting of analytic polyhedra. We therefore can apply a fundamental idea of K. Oka to simplify the geometry by embedding a suitable analytic polyhedron into a polydisc in a higher dimensional $\mathbb{C}^N$. We are thus, essentially, reduced to the convex case! After pulling back the global data from an appropriate region in $\mathbb{C}^N$ to $\bar{D} \subset \mathbb{C}^n$, we obtain the desired generating form by applying a decomposition theorem for holomorphic functions on a Stein neighborhood of $\bar{D}$. The proof of this latter result is obtained by a simple, but crucial technical modification of the classical proof of H. Hefer, needed to take care of the possibly finitely many obstructions to the solution of $\bar{\partial}u = f$ on $\bar{D}$.

These results imply the fundamental vanishing theorem $H^q_{\bar{\partial}}(\bar{D}) = 0$ for $q \geq 1$ for a strictly pseudoconvex domain D. Moreover, the solution of $\bar{\partial}$ on D is given by a rather explicit *integral operator*. In §3 we estimate the relevant kernels and prove that the solution operator is bounded from L^∞ into the Hölder space $\Lambda_{1/2}$. This fractional gain in regularity, which is optimal except for regions in the complex plane, is a characteristic feature of the $\bar{\partial}$-equation which has led to the concept of *subellipticity* and other important developments in the theory of partial differential equations. Even though these estimates will not be needed until Chapter VII, we have included them here so as to provide a direct route to a result which is very useful in many applications. In Chapter VII, §5, we will discuss variants of the solution operator for $\bar{\partial}$ and prove estimates in L^p norms.

§1. A Parametrix for $\bar{\partial}$ on Strictly Pseudoconvex Domains

In this paragraph D will always denote a bounded strictly pseudoconvex domain and r will be a strictly plurisubharmonic C^{k+2} defining function for D in a neighborhood U of bD, where k is some integer ≥ 0. At first, we do not assume $dr \neq 0$.

1.1. A Special Cauchy–Fantappiè Form

For $\zeta \in U$, the Levi polynomial $F(\zeta, z) = F^{(r)}$ of r is defined by

$$F(\zeta, z) = \sum_{j=1}^{n} \frac{\partial r}{\partial \zeta_j}(\zeta)(\zeta_j - z_j) - \frac{1}{2} \sum_{j,k=1}^{n} \frac{\partial^2 r}{\partial \zeta_j \partial \zeta_k}(\zeta)(\zeta_j - z_j)(\zeta_k - z_k).$$

By Proposition II.2.16 there are positive numbers ε and c, such that

$$2 \operatorname{Re} F(\zeta, z) \geq r(\zeta) - r(z) + 2c|\zeta - z|^2 \tag{1.1}$$

for $\zeta \in U$ and $|z - \zeta| \leq \varepsilon$. If $k = 0$, F is only continuous in ζ, while generating forms must be at least C^1. We therefore modify F as follows (this could be avoided by assuming $k \geq 1$, but it is of interest to keep the differentiability assumptions minimal). We fix a neighborhood $U' \subset\subset U$ of bD and, for $j, k = 1, \ldots, n$, we choose $\varphi_{jk} \in C^\infty(U)$, such that

$$\left|\varphi_{jk} - \frac{\partial^2 r}{\partial \zeta_j \partial \zeta_k}\right|_{U'} \leq c/2n^2. \tag{1.2}$$

Then

$$F^{\#}(\zeta, z) = \sum_{j=1}^{n} \frac{\partial r}{\partial \zeta_j}(\zeta)(\zeta_j - z_j) - \frac{1}{2}\sum_{j,k=1}^{n} \varphi_{jk}(\zeta)(\zeta_i - z_j)(\zeta_k - z_k) \tag{1.3}$$

satisfies

$$2 \operatorname{Re} F^{\#}(\zeta, z) \geq r(\zeta) - r(z) + c|\zeta - z|^2 \tag{1.4}$$

for $\zeta \in U'$ and $|z - \zeta| \leq \varepsilon$, $F^{\#} \in C^{k+1,\infty}(U' \times \mathbb{C}^n)$, and $F^{\#}$ is holomorphic in z.

Choose $\chi \in C^\infty(\mathbb{C}^n \times \mathbb{C}^n)$, such that $0 \leq \chi \leq 1$ and

$$\chi = \begin{cases} 1 & \text{for } |\zeta - z| \leq \varepsilon/2 \\ 0 & \text{for } |\zeta - z| \geq \varepsilon. \end{cases}$$

It follows that

$$\Phi = \chi F^{\#} + (1 - \chi)|\zeta - z|^2 \tag{1.5}$$

has the following properties for $\zeta \in U'$:

$$2 \operatorname{Re} \Phi(\zeta, z) \geq r(\zeta) - r(z) + c|z - \zeta|^2 \qquad \text{for } |z - \zeta| \leq \varepsilon/2; \tag{1.6}$$

$$2 \operatorname{Re} \Phi(\zeta, z) \geq r(\zeta) + c\varepsilon^2/8 \qquad \text{for } |z - \zeta| \geq \varepsilon/2 \text{ and } r(z) \leq c\varepsilon^2/8. \tag{1.7}$$

Fix $0 < \delta \leq c\varepsilon^2/8$, such that

$$D_\delta = D \cup \{z \in U : r(z) < \delta\}$$

is a neighborhood of $\bar{D}$. One then has

$$|\Phi(\zeta, z)| \geq \delta \qquad \text{for } \zeta \in bD \text{ and all } z \in \bar{D}_\delta \text{ with } |z - \zeta| \geq \varepsilon/2. \tag{1.8}$$

Finally, Φ has a decomposition $\Phi = \sum_{j=1}^{n} P_j(\zeta_j - z_j)$, where

$$P_j(\zeta, z) = \chi\left[\frac{\partial r}{\partial \zeta_j}(\zeta) - \frac{1}{2}\sum_{k=1}^{n} \varphi_{jk}(\zeta_k - z_k)\right] + (1 - \chi)(\bar{\zeta}_j - \bar{z}_j). \tag{1.9}$$

It follows that

$$L_D \stackrel{\text{f}}{=} \sum_{j=1}^{n} (P_j/\Phi)\, d\zeta_j.$$

is a generating form with coefficients in $C^{k+1,\infty}(bD \times D)$. We now summarize the basic properties of the associated CF form $\Omega_q(L_D)$.

Proposition 1.1. *The form $\Omega_0(L_D)$ is defined on $bD \times \bar{D} - \{\zeta = z\}$, with coefficients in $C^{k,\infty}$. Furthermore, $\Omega_0(L_D)$ is holomorphic in z for $z \in \bar{D}$ with $0 \neq |z - \zeta| < \varepsilon/2$.*

For $q \geq 1$, $\Omega_q(L_D)$ is nonsingular on $bD \times \bar{D}_\delta$, with coefficients in $C^{k,\infty}(bD \times \bar{D}_\delta)$. In particular, if D has C^{k+2} boundary and if $f \in C^0_{0,q}(bD)$, then the form

$$\int_{bD} f \wedge \Omega_q(L_D)$$

is in $C^\infty_{0,q}(\bar{D}_\delta)$.

PROOF. Since $\chi \equiv 1$ for $|z - \zeta| \leq \varepsilon/2$, the construction of L_D shows that its coefficients are holomorphic in z on $|z - \zeta| < \varepsilon/2$, provided $\Phi(\zeta, z) \neq 0$. By (1.6) and (1.7), $\Phi(\zeta, z) \neq 0$ for $r(\zeta) = 0$ and $z \in \bar{D} - \{\zeta\}$. This proves the statements about $\Omega_0(L_D)$. If $q \geq 1$, one obtains $\Omega_q(L_D) \equiv 0$ for $(\zeta, z) \in bD \times D$ with $|\zeta - z| \leq \varepsilon/2$. This fact and (1.8) show that $\Omega_q(L_D)$ can be extended without singularities to $bD \times \bar{D}_\delta$. The last statement is then obvious. ■

We now assume that $dr \neq 0$ on bD. For $q \geq 1$ we define the linear operator

$$\mathbf{E}_q: C^0_{0,q}(bD) \to C^\infty_{0,q}(\bar{D}_\delta)$$

by

$$\mathbf{E}_q f = \int_{bD} f \wedge \Omega_q(L_D). \tag{1.10}$$

$\mathbf{E}_q$ is a "smoothing operator", that is, the output $\mathbf{E}_q f$ is more regular than the input f.

We now apply Theorem IV.3.6 to the generating form L_D. It follows that the operator $\mathbf{T}_q = \mathbf{T}_q^{L_D}: C^0_{0,q}(\bar{D}) \to C^0_{0,q-1}(D)$ defined by IV(3.13) in terms of L_D satisfies

$$f = \bar{\partial}\mathbf{T}_q f + \mathbf{E}_q f, \qquad q \geq 1, \tag{1.11}$$

for $f \in C^1_{0,q}(\bar{D})$ with $\bar{\partial} f = 0$. So $\mathbf{T}_q$ does not solve the $\bar{\partial}$-equation exactly, but only up to a smoothing term $\mathbf{E}_q f$. However, we will see that this is sufficient for important applications. Classically, an operator $\mathbf{T}_q$ with the above property is called a *parametrix* (for $\bar{\partial}$).

1.2. The Extension of $\bar{\partial}$-Cohomology Classes

The operator $\mathbf{E}_q$ defined by (1.10) has another important property.

Lemma 1.2. *If $q \geq 1$ and $f \in C^1_{0,q}(\bar{D})$ satisfies $\bar{\partial} f = 0$ on D, then*

$$\bar{\partial}(\mathbf{E}_q f) = 0 \text{ on } \bar{D}_\delta. \tag{1.12}$$

PROOF. By Proposition 1.1 and by Lemma IV.3.5, if $q \geq 1$, $\bar{\partial}_z \Omega_q(L_D) = (-1)^{q+1} \bar{\partial}_\zeta \Omega_{q+1}(L_D)$ on $bD \times \bar{D}_\delta$. By continuity, $\bar{\partial} f = 0$ on bD, so an application of Stokes' Theorem gives

$$\bar{\partial}_z(\mathbf{E}_q f) = (-1)^{q+1} \int_{bD} f \wedge \bar{\partial}_\zeta \Omega_{q+1}(L_D)$$

$$= \int_{bD} \bar{\partial} f \wedge \Omega_{q+1}(L_D) = 0. \quad \blacksquare$$

Hence, if $f \in C^1_{0,q}(\bar{D})$ and $\bar{\partial} f = 0$, then $\mathbf{E}_q f$ extends the $\bar{\partial}$-cohomology class of f on D to $\bar{D}_\delta$. A different version of the extension property of $\mathbf{E}_q$ is as follows (this result can be omitted without loss of continuity).

Proposition 1.3. *The homomorphism*

$$\rho_q^*: H^q_{\bar{\partial}}(\bar{D}_\delta) \to H^q_{\bar{\partial}}(\bar{D})$$

induced by the restriction map $\rho: C^\infty_{0,q}(\bar{D}_\delta) \to C^\infty_{0,q}(\bar{D})$ *is surjective for* $q \geq 1$.

Proof. There is $\eta_0 > 0$, such that

$$\mathbf{E}^\eta_q f = \int_{bD_\eta} f \wedge \Omega_q(L_D)$$

is well defined for $0 \leq \eta \leq \eta_0$ and $f \in C_{0,q}(bD_\eta)$, and $\mathbf{E}^\eta_q f \in C^\infty_{0,q}(\bar{D}_\delta)$. So, if $[f] \in H^q_{\bar{\partial}}(\bar{D})$, choose $0 < \eta \leq \eta_0$, such that $f \in C^\infty_{0,q}(\bar{D}_\eta)$ and $\bar{\partial} f = 0$ on D_η. Define $\mathbf{T}^{(\eta)}_q$ with respect to D_η as before (note that one can take $L_{D_\eta} = L_D$); then

$$f = \bar{\partial}\mathbf{T}^{(\eta)}_q f + \mathbf{E}^{(\eta)}_q f \qquad \text{on } D_\eta,$$

which shows that $\rho_q^*[\mathbf{E}^{(\eta)}_q f] = [f]$. $\blacksquare$

1.3. The Finiteness Theorem

For $q \geq 1$, define

$$Z^1_q = Z^1_q(\bar{D}) = \{f \in C^1_{0,q}(\bar{D}): \bar{\partial} f = 0 \text{ on } D\}.$$

Equipped with the norm $|\ |_{1,D}$, Z^1_q is a Banach space.

Theorem 1.4. *For* $q \geq 1$, *the operator* $\bar{\partial}\mathbf{T}_q$ *defines a bounded linear transformation*

$$\bar{\partial}\mathbf{T}_q: Z^1_q \to Z^1_q$$

whose range has finite codimension.

Proof. By (1.11),

$$\bar{\partial}\mathbf{T}_q = Id - \mathbf{E}_q \tag{1.13}$$

on Z^1_q, and by Lemma 1.2, the restriction of $\mathbf{E}_q$ to Z^1_q maps Z^1_q into $C^\infty_{0,q}(\bar{D}) \cap \ker \bar{\partial} \subset Z^1_q$.

Furthermore, differentiation under the integral sign shows that any partial derivative of $\mathbf{E}_q f$ is uniformly estimated by $|f|_{0,bD}$. An application of the

Ascoli–Arzela Theorem then shows that $\mathbf{E}_q: Z_q^1 \to Z_q^1$ is a *compact* operator. By basic Banach space theory it follows that the range of $Id - \mathbf{E}_q$ has finite codimension in Z_q^1. (See [Rud 2], Chapter 4.) ∎

1.4. The Levi Problem

The Finiteness Theorem 1.4 leads to a simple solution of the Levi problem.

Theorem 1.5. *Let $D \subset\subset \mathbb{C}^n$ be a strictly pseudoconvex domain defined by a strictly plurisubharmonic function r on a neighborhood U of bD such that $L = D \cup \{z \in U: r(z) \leq 0\}$ is compact. Then D is holomorphically convex, and hence a domain of holomorphy. In particular, the conclusion holds for every bounded strictly pseudoconvex domain with C^2 boundary.*

Proof. By Proposition II.3.4, it is enough to show that for each $\zeta \in bD$ there is $h = h_\zeta \in \mathcal{O}(D)$, such that

$$\lim_{\substack{z \to \zeta \\ z \in D}} |h_\zeta(z)| = \infty.$$

Fix $\zeta \in bD$ and let $v(z) = \Phi(\zeta, z)$, where Φ was defined by (1.5). Then v is C^∞ on a neighborhood of L, $v(\zeta) = 0$, and $v(z) \neq 0$ on $L - \{\zeta\}$ by (1.6) and (1.7). Furthermore, v is holomorphic for $|z - \zeta| < \varepsilon/2$, and therefore there is $\eta > 0$ such that for each $j = 1, 2, \ldots, f_j = \bar{\partial}(1/v^j)$ can be trivially extended across ζ to the *strictly pseudoconvex* neighborhood $D_\eta = D \cup \{z \in U: r(z) < \eta\}$ of L. So we can view $\{f_1, f_2, \ldots\}$ as a linearly independent set in $Z_1^1(\bar{G})$, where G is strictly pseudoconvex with C^2 boundary, $L \subset G \subset\subset D_\eta$ (use Corollary II.2.23). By Theorem 1.4 (applied to G instead of D) there are constants $c_1, \ldots, c_l \in \mathbb{C}$, $c_l \neq 0$, and a (0, 1)-form $g \in Z_1^1(\bar{G})$, such that

$$\sum_{j=1}^{l} c_j f_j = \bar{\partial} \mathbf{T}_1 g \qquad \text{on } G. \tag{1.14}$$

Since $D \subset\subset G$,

$$\sup_{z \in D} |\mathbf{T}_1 g(z)| < \infty.$$

Therefore, the function $h \in C^\infty(L - \{\zeta\})$ defined by

$$h = \sum_{j=1}^{l} c_j/v^j - \mathbf{T}_1 g$$

satisfies, for $z \in D$,

$$\lim_{z \to \zeta} |h| \geq \lim_{z \to \zeta} |c_l/v^l| - \sup_D |\mathbf{T}_1^{(\eta)} g| = \infty,$$

and, by (1.14), $\bar{\partial} h = 0$ on D. ∎

Corollary 1.6. *A compact set $K \subset \mathbb{C}^n$ is a Stein compactum (i.e., an intersection of holomorphically convex domains) if and only if K has a neighborhood basis of* *strictly pseudoconvex domains with C^2 boundary. In particular, the closure of a strictly pseudoconvex domain with C^2 boundary, or, more generally, any pseudoconvex compactum is a Stein compactum.*

PROOF. The "easy" part was proved in Corollary II.3.25. The "hard" part is an immediate consequence of Theorem 1.5. ∎

EXERCISES

E.1.1. For $m \in \mathbb{N}^+$ set $D^m = \{(z_1, z_2) \in \mathbb{C}^2: |z_1|^2 + |z_2|^{2m} < 1\}$ and let $r^{(m)} = 1 - |z_1|^2 - |z_2|^{2m}$ be a defining function for D^m.

(i) Show that D^m is convex, but not *strictly* convex except for $m = 1$.

(ii) If

$$\Phi^{(m)} = \sum_{j=1}^{2} \frac{\partial r^{(m)}}{\partial \zeta_j}(\zeta)(\zeta_j - z_j),$$

show that there is a constant $c > 0$ such that

$$2 \operatorname{Re} \Phi^{(m)}(\zeta, z) \geq -r(z) + c|\zeta - z|^{2m}$$

for $\zeta \in bD^m$ and $z \in \bar{D}^m$.

E.1.2. Prove that if $D \subset\subset \mathbb{C}^n$, the space $Z_q^1(\bar{D}) = \{f \in C_{0,q}^1(\bar{D}): \bar{\partial} f = 0 \text{ on } D\}$ with norm $|f|_{1,D} := \sum_J |f_J|_{1,D}$ for $f = \sum_J f_J d\bar{z}^J$ is a Banach space.

§2. A Solution Operator for $\bar{\partial}$

2.1. Extension of Holomorphic Functions from Hyperplanes

In order to construct a solution operator for $\bar{\partial}$ on strictly pseudoconvex domains we will need a decomposition theorem. The following proposition, which is of independent interest, contains the crucial analytic part of that result.

Proposition 2.1. *Let $K \subset \mathbb{C}^n$ be a Stein compactum and set $K_1 = K \cap \{z \in \mathbb{C}^n: z_1 = 0\}$. Then for every $f \in \mathcal{O}(K_1)$ there is a holomorphic function $F \in \mathcal{O}(K)$, such that*

$$F(0, z') = f(0, z') \tag{2.1}$$

for all $(0, z') = (0, z_2, \ldots, z_n)$ in a neighborhood of K_1.

Remark 2.2. An analogous result holds for $\bar{\partial}$-closed $(0, q)$-forms in a neighborhood of K_1, $q \geq 1$ (see Exercise E.2.1).

PROOF OF 2.1. Given $f \in \mathcal{O}(K_1)$, we choose a neighborhood W of K such that f is defined and holomorphic on $W_1 = W \cap \{z: z_1 = 0\}$. Let $\pi: \mathbb{C}^n \to \mathbb{C}^n$ be defined by $\pi(z_1, z') = (0, z')$. Since W_1 and $W - \pi^{-1}(W_1)$ are disjoint and closed (in W), there is $\chi \in C^\infty(W)$ such that $\chi \equiv 1$ in a neighborhood of W_1 and $\chi \equiv 0$ in a neighborhood of $W - \pi^{-1}(W_1)$. For $j = 1, 2, \ldots$

$$\alpha_j = \bar\partial[\chi(f \circ \pi)]/z_1^j \tag{2.2}$$

is a well defined $\bar\partial$-closed $C^\infty_{0,1}$ form on W.

Since K is Stein, there is a strictly pseudoconvex domain D with C^2 boundary, such that $K \subset D \subset\subset W$. Hence $\{\alpha_1, \alpha_2, \ldots\}$ is a linearly independent set in $Z_1^1(\bar D)$, and by Theorem 1.4, there are constants $c_1, \ldots, c_l \in \mathbb{C}$, $c_l \neq 0$, and a form $g \in Z_1^1(\bar D)$, such that

$$\sum_{j=1}^{l} c_j \alpha_j = \bar\partial \mathbf{T}_1 g. \tag{2.3}$$

Without loss of generality we may assume $c_l = 1$. Equations (2.2) and (2.3) then imply that

$$F = \sum_{j=1}^{l} c_j z_1^{l-j} \chi \cdot (f \circ \pi) - z_1^l(\mathbf{T}_1 g)$$

is holomorphic on D and satisfies (2.1), since $\chi \equiv 1$ in a neighborhood of K_1. ■

2.2. Hefer's Decomposition Theorem

We will now prove the following.

Theorem 2.2. *Let K be a Stein compactum in $\mathbb{C}^n$. Given $f \in \mathcal{O}(K)$, there are holomorphic functions $Q_j \in \mathcal{O}(K \times K)$, $1 \le j \le n$, such that*

$$f(\zeta) - f(z) = \sum_{j=1}^{n} Q_j(\zeta, z)(\zeta_j - z_j) \tag{2.4}$$

for $(\zeta, z) \in K \times K$.

The proof will be an easy consequence of the following result.

Lemma 2.3. *Let $K \subset \mathbb{C}^n$ be a Stein compactum. Let $1 \le k \le n$ and define*

$$K_k = \{z \in K : z_1 = \cdots = z_k = 0\}.$$

Suppose $f \in \mathcal{O}(K)$ satisfies $f(z) = 0$ for all z in a neighborhood of K with $z_1 = \cdots = z_k = 0$. Then there are holomorphic functions $g_1, \ldots, g_k \in \mathcal{O}(K)$, such that

$$f(z) = \sum_{j=1}^{k} z_j g_j. \tag{2.5}$$

Proof of 2.3. We use induction. Lemma 2.3 is certainly true for $k = 1$ and arbitrary $n \geq 1$: simply take $g_1 = f/z_1$. Suppose the Lemma has been proved for $k - 1$ and all $n \geq k$. Let $K^\# = K \cap \{z \in \mathbb{C}^n : z_1 = 0\}$. Then $K^\#$ is a Stein compactum in $\mathbb{C}^{n-1}$, $f \in \mathcal{O}(K^\#)$, and $f(0, z') = 0$ for $z' = (z_2, \ldots, z_n)$ with $z_2 = \cdots = z_k = 0$ in a neighborhood of $K^\#$. By inductive hypotheses, there are $g_j^\# \in \mathcal{O}(K^\#)$ with

$$f(0, z') = \sum_{j=2}^{k} z_j g_j^\#(z'). \tag{2.6}$$

By Proposition 2.1, there are functions $g_j \in \mathcal{O}(K)$ with $g_j(0, z') = g_j^\#(z')$, $2 \leq j \leq k$. Define

$$F(z) = f(z) - \sum_{j=2}^{k} z_j g_j(z). \tag{2.7}$$

Then $F \in \mathcal{O}(K)$ and, by (2.6), $F(0, z') = 0$ in a neighborhood of $K^\#$. By the inductive beginning $k = 1$, there is $g_1 \in \mathcal{O}(K)$ with $F = z_1 g_1$. Introducing this into (2.7) gives (2.5). ■

Proof of Theorem 2.2. The set $K \times K$ is a Stein compactum in $\mathbb{C}^{2n}$. Let $F \in \mathcal{O}(K \times K)$ be defined by $F(\zeta, z) = f(\zeta) - f(z)$. Introduce new holomorphic coordinates $u = (u_1, \ldots, u_{2n})$ in $\mathbb{C}^{2n}$ by $u_j = \zeta_j - z_j$ and $u_{n+j} = z_j$, $1 \leq j \leq n$. Then $F = 0$ on

$$K \times K \cap \{u \in \mathbb{C}^{2n} : u_1 = \cdots = u_n = 0\}.$$

Now apply Lemma 2.3 to F and go back to the original coordinates (ζ, z). ■

2.3. The Solution of $\bar{\partial}$ on a Stein Compactum

The results of IV.§3 have shown that the main difficulty for constructing a solution operator for $\bar{\partial}$ is the existence of a generating form which is globally holomorphic in the parameter z. We now construct such a generating form on suitable neighborhoods of a Stein compactum. There are two ingredients: first we use a classical idea of K. Oka [Oka, I] to simplify the geometry by going to higher dimension; and second, we must use Hefer's Theorem 2.2, whose proof is based on the Finiteness Theorem in §1.3.

Proposition 2.4. *Let $G \subset\subset \mathbb{C}^n$ be holomorphically convex, and suppose $K \subset G$ is compact. Then there are neighborhoods $V_0 \subset\subset V \subset\subset G$ of K, V with smooth boundary, and a generating $(1, 0)$-form $W \in C^{\infty,\infty}_{1,0}(bV \times V_0)$ which is holomorphic in $z \in V_0$.*

Proof. Replacing K by $\hat{K}_{\mathcal{O}(G)}$, we may assume that K is holomorphically convex in G. Let $\omega \subset\subset G$ be a neighborhood of K. Using Proposition II.3.10, one finds an analytic polyhedron

$$A = \{z \in \omega \colon |h_k(z)| < 1, \qquad 1 \le k \le N\},$$

defined by functions $h_k \in \mathcal{O}(G)$, $1 \le k \le N$, such that $K \subset A \subset\subset \omega$. Let Δ^N denote the open unit polydisc in $\mathbb{C}^N$. The restriction of $H = (h_1, \ldots, h_N)$: $G \to \mathbb{C}^N$ to A defines a proper holomorphic map

$$H_{|A} \colon A \to \Delta^N.$$

Since $H(K)$ is a compact subset of the convex set Δ^N, there is a smoothly bounded open *convex* neighborhood $U \subset\subset \Delta^N$ of $H(K)$, with a smooth defining function $\rho \colon \Delta^N \to \mathbb{R}$, i.e., $U = \{t \in \Delta^N \colon \rho(t) < 0\}$ and $d\rho \neq 0$ on bU. It follows that

$$\langle \partial\rho(\eta), \eta - t \rangle \neq 0 \qquad \text{for } (\eta, t) \in bU \times U.$$

Therefore, the function $\Phi \colon A \times A \to \mathbb{C}$ defined by

$$\Phi(\zeta, z) = \sum_{k=1}^{N} \frac{\partial\rho}{\partial\eta_k}(H(\zeta))(h_k(\zeta) - h_k(z)) \tag{2.8}$$

satisfies $\Phi(\zeta, z) \neq 0$ for $\zeta \in (H_{|A})^{-1}(bU)$ and $z \in K$. By continuity there are open neighborhoods $V_0 \subset\subset V \subset\subset A$ of K, V with smooth boundary, such that

$$\Phi(\zeta, z) \neq 0 \qquad \text{for } (\zeta, z) \in bV \times V_0. \tag{2.9}$$

Since $h_k \in \mathcal{O}(\bar{A})$ for $1 \le k \le N$ and $\bar{A}$ is a Stein compactum, Theorem 2.2 gives functions $Q_{jk} \in \mathcal{O}(\bar{A} \times \bar{A})$, $1 \le j \le n$, such that

$$h_k(\zeta) - h_k(z) = \sum_{j=1}^{n} Q_{jk}(\zeta, z)(\zeta_j - z_j).$$

Therefore one has the decomposition

$$\Phi(\zeta, z) = \sum_{j=1}^{n} P_j(\zeta, z)(\zeta_j - z_j) \tag{2.10}$$

where

$$P_j(\zeta, z) = \sum_{k=1}^{N} \frac{\partial\rho}{\partial\eta_k}(H(\zeta))Q_{jk}(\zeta, z), \qquad 1 \le j \le n. \tag{2.11}$$

Finally, we define

$$W = \sum_{j=1}^{n} (P_j/\Phi)\, d\zeta_j.$$

It follows from (2.9), (2.10), and (2.11) that W has all the required properties. ∎

We can now easily prove the following.

Theorem 2.5. *Let G be a neighborhood of the Stein compactum $K \subset \mathbb{C}^n$. Then there are neighborhoods $V_0 \subset\subset V \subset\subset G$ of K and linear operators*

$$\mathbf{T}_q^{V, V_0} \colon C_{0,q}(\bar{V}) \to C_{0,q-1}(V_0) \qquad (1 \le q \le n)$$

with the following properties:

(i) *for* $k = 1, 2, \ldots$, $\mathbf{T}_q^{V,V_0} f \in C_{0,q-1}^k(V_0)$ *if* $f \in C_{0,q}(\bar V) \cap C_{0,q}^k(V)$;
(ii) $\bar\partial(\mathbf{T}_q^{V,V_0} f) = f$ *on* V_0 *if* $\bar\partial f = 0$ *on* V.
(iii) *If* M *is a* C^l *manifold and* $f \in C_{0,q}^{k,l}(\bar V \times M)$, *then* $\mathbf{T}_q^{V,V_0} f \in C_{0,q-1}^{k,l}(V_0 \times M)$.

Proof. Without loss of generality we may assume that G is holomorphically convex. Apply Theorem IV.3.6 with $D = V$ to the generating form $W \in C_{1,0}^{1;\infty}(bV \times V_0)$ given by Proposition 2.4, restricting z to $V_0 \subset\subset V$. The operator $\mathbf{T}_q^{V,V_0}$ is defined by $\mathbf{T}_q^W$ as in IV.(3.13). Since $\Omega_q(W) \equiv 0$ on $bV \times V_0$ for $q \geq 1$, conclusions (i) and (ii) follow immediately; (iii) follows by standard theorems on the dependence of integrals on parameters. ■

Corollary 2.6. *Let* K *be a Stein compactum in* $\mathbb{C}^n$. *Then*

$$H_{\bar\partial}^q(K) = 0 \qquad \textit{for } q \geq 1.$$

2.4. A Solution Operator on Strictly Pseudoconvex Domains

The solution operator $\mathbf{T}_q^{V,V_0}$ for $\bar\partial$ given by Theorem 2.5 has the drawback that it gives a solution only on the smaller subset V_0 of V. However, in applications involving the boundary behavior of holomorphic functions, it is important to have solutions of $\bar\partial u = f$ with good estimates up to the boundary of the domain under consideration. This we can achieve for strictly pseudoconvex domains with C^2 boundary by combining Theorem 2.5 with the extension of $\bar\partial$-cohomology classes.

Theorem 2.7. *Let* $D \subset\subset \mathbb{C}^n$ *be strictly pseudoconvex with* C^2 *boundary. For* $1 \leq q \leq n$ *there are linear operators*

$$\mathbf{S}_q = \mathbf{S}_q^{(D)} \colon C_{0,q}(\bar D) \to C_{0,q-1}(D)$$

with the following properties:

(i) *for* $k = 0, 1, 2, \ldots$, *if* $f \in C_{0,q}(\bar D) \cap C_{0,q}^k(D)$, *then* $\mathbf{S}_q f \in C_{0,q-1}^k(D)$;
(ii) *there is a constant* $C > 0$, *such that* $|\mathbf{S}_q f|_{1/2,D} \leq C|f|_{\bar D}$;
(iii) *if* $f \in C_{0,q}^1(\bar D)$ *and* $\bar\partial f = 0$, *then* $\bar\partial(\mathbf{S}_q f) = f$.

Remark. Aside from the fact that $\mathbf{S}_q$ solves the $\bar\partial$-equation on D, the key result is the estimate (ii), which gives information about the regularity of the solution up to the boundary. (ii) expresses the property that $\mathbf{S}_q$ is "smoothing of order $\frac{1}{2}$". It turns out that this is the best possible gain in smoothness for solutions of $\bar\partial$ (see §2.5), unless $q = n$. Historically, the "$\frac{1}{2}$-estimate for $\bar\partial$" has been the first example of so-called "subelliptic estimates" (cf. J.J. Kohn [Koh 1]), and it has had a profound influence on the theory of partial differential equations. Classically, the $\frac{1}{2}$-estimate was formulated in terms of Sobolev norms (these measure the L^2 norm of derivatives of a function) rather than in terms of Lipschitz norms.

PROOF OF THEOREM 2.7. The construction in §1.1 gives the parametrix $\mathbf{T}_q = \mathbf{T}_q^{L_D}$ and the extension operator $\mathbf{E}_q: C_{0,q}(\bar D) \to C^\infty_{0,q}(\bar D_\delta)$, such that (1.11) holds. Since $K = \bar D$ is a Stein compactum (Corollary 1.6), one can apply Theorem 2.5 with $G = D_\delta$ to obtain neighborhoods V_0 and V of $\bar D$ with $\bar D \subset V_0 \subset\subset V \subset\subset D_\delta$, and the operator $\mathbf{T}_q^{V,V_0}$. We now define

$$\mathbf{S}_q = \mathbf{T}_q + \mathbf{T}_q^{V,V_0} \circ \mathbf{E}_q. \tag{2.12}$$

It follows from the known properties of the operators involved (Theorem IV.3.6, Theorem 2.5) that $\mathbf{S}_q: C_{0,q}(\bar D) \to C_{0,q-1}(D)$ and that (i) holds. If $\bar\partial f = 0$ on D, then $\bar\partial(\mathbf{E}_q f) = 0$ on D_δ by Lemma 1.2; so, by Theorem 2.5 and (1.11),

$$\bar\partial(\mathbf{S}_q f) = \bar\partial(\mathbf{T}_q f) + \mathbf{E}_q f = f;$$

this proves (iii). The proof of the estimate (ii) is somewhat involved, so we will prove it separately in §3. ∎

The estimate (ii) will only be used in Chapter VII, where we also prove more general estimates for a solution operator for $\bar\partial$ closely related to $\mathbf{S}_q$. So the reader may skip §3 without loss of continuity, until reference is made to it.

2.5. Sharp Bounds for Lipschitz Estimates for $\bar\partial$

For $q = n$, the operator $\mathbf{E}_n \equiv 0$, so $\mathbf{S}_n f = \mathbf{T}_n f = -\int_D f \wedge K_{n-1}$. By Theorem IV.1.14,

$$|\mathbf{S}_n f|_\alpha \le C_\alpha |f|_D$$

for any $\alpha < 1$. So it appears reasonable to ask whether one could improve the estimate (ii) for $\mathbf{S}_q$ in Theorem 2.7. We now discuss an example which shows that this is not possible if $n > q \ge 1$.

For $m = 1, 2, \ldots$ consider the domain $D^m = \{(z_1, z_2) \in \mathbb{C}^2: |z_1|^2 + |z_2|^{2m} < 1\}$ and define $v: D^m \to \mathbb{C}$ by $v(z) = \bar z_2/\log(z_1 - 1)$, where we use the principal branch for the logarithm. It follows that the (0, 1)-form

$$g = \bar\partial v = d\bar z_2/\log(z_1 - 1)$$

is $\bar\partial$-closed and bounded on D^m.

Lemma 2.8. *Suppose $u \in \Lambda_\alpha(D^m)$ satisfies $\bar\partial u = g$ on D^m. Then $\alpha \le 1/2m$.*

PROOF. For $0 < d < 1/2$, the integral

$$I(d) = \int_{|z_2| = d^{1/2m}} [u(1 - d, z_2) - u(1 - 2d, z_2)]\, dz_2 \tag{2.13}$$

is well defined, and, if $u \in \Lambda_\alpha(D^m)$, one obtains

$$|I(d)| \lesssim d^\alpha \cdot d^{1/2m} \tag{2.14}$$

by direct estimation. On the other hand, $\bar{\partial}(u - v) = 0$, so $u = v + h$, with $h \in \mathcal{O}(D^m)$. By Cauchy's Integral Theorem we can replace u by v in the integral (2.13). Therefore

$$\begin{aligned} I(d) &= \left[\frac{1}{\log(-d)} - \frac{1}{\log(-2d)}\right] \int_{|z_2| = d^{1/2m}} \bar{z}_2 \, dz_2 \\ &= \left[\frac{1}{\log(-d)} - \frac{1}{\log(-2d)}\right] 2\pi i \cdot d^{1/m}. \end{aligned} \tag{2.15}$$

If $\alpha > 1/2m$, (2.14) and (2.15) lead to a contradiction as $d \to 0$. ■

In particular, for $m = 1$, D^1 is the unit ball in $\mathbb{C}^2$, which is strictly pseudoconvex. So the $\frac{1}{2}$-estimate in Theorem 2.7 is the best possible. If $m > 1$, D^m fails to be strictly pseudoconvex at all boundary points $(e^{i\theta}, 0)$, so Theorem 2.7 does not apply. Geometrically speaking, Lemma 2.8 shows that the estimates for solutions of $\bar{\partial}$ must get worse as the boundary of the domain flattens out—more precisely, as the order of contact at $P \in bD$ between the boundary and the complex tangent space $T_P^{\mathbb{C}}(bD)$ increases.

The example above settles the case $n = 2$ and $q = 1$. The construction can be modified to take care of the general case $1 \leq q \leq n - 1$ as well (see Exercise E.2.4).

Exercises

E.2.1. Let $K \subset \mathbb{C}^n$ be a Stein compactum and set $K_1 = \{z \in K: z_1 = 0\}$. Let $\mu: \mathbb{C}^{n-1} \to \mathbb{C}^n$ be given by $\mu(z') = (0, z')$. Show that for every $f \in C^\infty_{0,q}(\mu^{-1}(K_1))$ with $\bar{\partial} f = 0$ in a neighborhood of $\mu^{-1}(K_1)$ in $\mathbb{C}^{n-1}$ there is $F \in C^\infty_{0,q}(K)$ with $\bar{\partial} F = 0$ in a neighborhood of K, such that $\mu^* F = f$ on $\mu^{-1}(K_1)$. (Hint: Generalize the proof of Proposition 2.1.)

E.2.2. Carry out the details of the proof of part (iii) of Theorem 2.5.

E.2.3. Let Δ^2 be the unit polydisc in $\mathbb{C}^2$.

(i) Show that there is $g \in C^\infty_{0,1}(\Delta^2)$ with bounded coefficients on Δ^2 and $\bar{\partial} g = 0$ on Δ^2, such that the equation $\bar{\partial} u = g$ has no solution on Δ^2 with coefficients in $\Lambda_\alpha(\Delta^2)$ for any $\alpha > 0$.

(ii) Modify the example in (i) to find g as in (i) so that $\bar{\partial} u = g$ has no solution u on Δ^2 whose coefficients extend continuously to the closure $\bar{\Delta}^2$.

(iii) Find a bounded convex domain with C^∞ boundary in $\mathbb{C}^2$ which exhibits the same property stated for Δ^2 in (i).

E.2.4. For $m \in \mathbb{N}^+$ set $G^m = \{(z_1, z') \in \mathbb{C}^n: |z_1|^2 + |z'|^{2m}\}$. Show that if $1 \leq q \leq n - 1$, then there is $g \in C^\infty_{0,q}(G^m)$ with bounded coefficients on G^m and $\bar{\partial} g = 0$, such that if $u \in C^\infty_{0,q-1}(G^m)$ has coefficients in $\Lambda_\alpha(G^m)$ and u solves $\bar{\partial} u = g$, then $\alpha \leq 1/2m$.

§3. The Lipschitz $\frac{1}{2}$-Estimate

3.1. A Criterion for Lipschitz Functions

In order to prove that a function belongs to a Lipschitz space Λ_α we will use the following elementary real variable fact.

Lemma 3.1. *Let $D \subset\subset \mathbb{R}^n$ be a bounded domain with C^1 boundary. Suppose $g \in C^1(D)$ and that for some $0 < \alpha < 1$ there is a constant c_g, such that*

$$|dg(x)| \le c_g \delta_D(x)^{\alpha-1}, \qquad x \in D.$$

Then $g \in \Lambda_\alpha(D)$. Furthermore, there are a compact set $K \subset D$ and a constant C, both depending only on D and α, such that

$$|g|_{\alpha,D} \le C[c_g + |g|_K]. \tag{3.1}$$

Proof. The main point of the proof is contained in the following special case. *Let $\delta > 0$ and*

$$U(\delta) = \{(x_1, x') \in \mathbb{R}^N : 0 < x_1 < \delta, \qquad |x'| < \delta\},$$

and suppose $g \in C^1(U(\delta))$ satisfies

$$|dg(x)| \le c_g x_1^{\alpha-1} \tag{3.2}$$

for $x \in U(\delta)$. Then there is C_1 depending only on α and δ such that

$$|g(x) - g(y)| \le C_1 c_g |x-y|^\alpha \tag{3.3}$$

for $x, y \in U(\delta/2)$ with $|x-y| \le \delta/2$.

Let $x, y \in U(\delta/2)$ and set $d = |x-y| \le \delta/2$. Then

$$\begin{aligned} |g(x_1, x') - g(x_1+d, x')| &\le \int_{x_1}^{x_1+d} \left|\frac{\partial g}{\partial x_1}(t, x')\right| dt \\ &\le c_g \int_{x_1}^{x_1+d} t^{\alpha-1}\, dt \qquad \text{(by (3.2))} \\ &\le C_2 c_g d^\alpha. \end{aligned} \tag{3.4}$$

Moreover, by the Mean Value Theorem and (3.2),

$$|g(x_1+d, x') - g(y_1+d, y')| \le c_g d \cdot d^{\alpha-1} = c_g d^\alpha. \tag{3.5}$$

Since

$$\begin{aligned} |g(x) - g(y)| \le\ & |g(x_1, x') - g(x_1+d, x')| \\ & + |g(x_1+d, x') - g(y_1+d, y')| + |g(y_1+d, y') - g(y_1, y')|, \end{aligned}$$

(3.3) follows from the estimates (3.4) and (3.5).

Returning to the general case, since bD is of class C^1, a standard compactness argument and local C^1 coordinate changes, together with the special case considered above, lead to the following statement.

There are $a > 0$ and C_3, depending only on D and α, such that

$$|g(x) - g(y)| \leq C_3 c_g |x - y|^\alpha \qquad \text{for } |x - y| \leq a. \tag{3.6}$$

In particular, with $K = \{x \in D: \delta_D(x) \geq a\}$, one obtains

$$|g|_D \leq |g|_K + C_3 c_g a^\alpha < \infty. \tag{3.7}$$

Hence, for $|x - y| > a$,

$$\Delta_g(x, y) = \frac{|g(x) - g(y)|}{|x - y|^\alpha} \leq 2|g|_D a^{-\alpha}.$$

Together with (3.6) this implies

$$\sup_{x, y \in D} \Delta_g(x, y) \leq \max(C_3 c_g, 2|g|_D a^{-\alpha}),$$

So $|g|_{\alpha, D} \lesssim c_g + |g|_D$; now use (3.7) to obtain (3.1). ∎

3.2. The Principal Part of S_q

Since the kernel of $\mathbf{E}_q$ is nonsingular, it follows that $|\mathbf{E}_q f|_{0, D_\delta} \lesssim |f|_{0, bD}$. Recall that

$$\mathbf{T}_q^{V, V_0} g = \int_{bV \times I} g \wedge \Omega_{q-1}(\hat{W}) - \int_V g \wedge K_{q-1}.$$

Let $J_1(g)$ denote the first integral and $J_2(g)$ the second. Notice that the integrand in J_1 is in $C^{0, \infty}((bV \times I) \times V_0)$, and that $D \subset\subset V_0$; so one has an estimate $|J_1(g)|_{\alpha, D} \lesssim |g|_{bV}$ for any $\alpha > 0$. By Theorem IV.1.14, $|J_2(g)|_{\alpha, D} \lesssim |g|_D$ for any $0 < \alpha < 1$. It follows that

$$|\mathbf{T}_q^{V, V_0} \mathbf{E}_q f|_{\alpha, D} \lesssim |f|_D$$

for any $\alpha < 1$. So, in order to prove the estimate (ii) in Theorem 2.7, it is enough, by (2.12), to prove it for $\mathbf{T}_q$.

The definition of $\mathbf{T}_q$ in IV.(3.13) shows that it is enough to prove

$$\left| \int_{bD \times I} f \wedge \Omega_{q-1}(\hat{L}_D) \right|_{1/2, D} \lesssim |f|_{bD}, \tag{3.8}$$

the other integral in $\mathbf{T}_q$ being of the type J_2 above.

We only need to consider the case $0 \leq q - 1 \leq n - 2$, otherwise the integral in (3.8) is 0. In the following, we replace $q - 1$ by q. We decompose

$$\Omega_q(\hat{L}_D) = \Omega_q^{(1)} \wedge d\lambda + \Omega_q^{(0)},$$

where $\Omega_q^{(0)}$ is of degree 0 in λ. Only the component with $d\lambda$ contributes to (3.8). Using $\bar{\partial}_{\zeta, \lambda} \hat{L}_D = d\lambda \wedge (\partial/\partial\lambda \hat{L}_D) + \bar{\partial}_\zeta \hat{L}_D$ one obtains

$$\Omega_q^{(1)} = (-1)^{q+1}(n-q-1)\hat{L}_D \wedge \left(\frac{\partial}{\partial \lambda}\hat{L}_D\right) \wedge R_q^\lambda(\hat{L}_D),$$

where

$$R_q^\lambda(\hat{L}_D) = \frac{(-1)^{q(q-1)/2}}{(2\pi i)^n}\binom{n-1}{q}(\bar{\partial}_\zeta \hat{L}_D)^{n-q-2} \wedge (\bar{\partial}_z \hat{L}_D)^q. \tag{3.9}$$

Recall that $\hat{L}_D = \lambda L_D + (1-\lambda)B$. So

$$\begin{aligned}\hat{L}_D \wedge \frac{\partial}{\partial \lambda}\hat{L}_D &= \hat{L}_D \wedge (L_D - B)\\ &= -L_D \wedge B.\end{aligned}$$

Therefore

$$\Omega_q^{(1)} = (-1)^q(n-q-1)L_D \wedge B \wedge R_q^\lambda(\hat{L}_D). \tag{3.10}$$

We now eliminate the parameter λ by integrating over I.

Lemma 3.2. *For $0 \le q \le n-2$ and any $f \in C_{0,q+1}(bD)$ one has*

$$\int_{bD \times I} f \wedge \Omega_q(\hat{L}_D) = \int_{bD} f \wedge A_q(L_D, B), \tag{3.11}$$

where the double form $A_q(L_D, B)$ is given by

$$A_q(L_D, B) = \sum_{j=0}^{n-q-2}\sum_{k=0}^{q} a_q^{j,k} A_q^{j,k}(L_D, B) \tag{3.12}$$

with numerical constants $a_q^{j,k}$ and

$$A_q^{j,k}(L, B) = L \wedge B \wedge (\bar{\partial}_\zeta L)^j \wedge (\bar{\partial}_\zeta B)^{n-q-2-j} \wedge (\bar{\partial}_z L)^k \wedge (\bar{\partial}_z B)^{q-k}.$$

Proof. The integral on the left in (3.11) equals $\int_{bD \times I} f \wedge \Omega_q^{(1)} \wedge d\lambda$. Expanding (3.9) by multilinearity, (3.10) implies

$$\Omega_q^{(1)} = \sum_{j=0}^{n-q-2}\sum_{k=0}^{q} b_q^{j,k}(\lambda) A_q^{j,k}(L_D, B),$$

with certain polynomials $b_q^{j,k}$ in λ. Lemma 3.2 then follows by setting $a_q^{j,k} = \int_0^1 b_q^{j,k}(\lambda)\, d\lambda$. ∎

Recall that $L_D = P/\Phi$ with $P = \sum P_j\, d\zeta_j$ (see §1.1) and $B = \partial\beta/\beta$. A straightforward computation gives

$$\begin{aligned}A_q^{j,k}(L_D, B) &= \frac{P \wedge \partial_\zeta \beta \wedge (\bar{\partial}_\zeta P)^j \wedge (\bar{\partial}_\zeta \partial_\zeta \beta)^{n-q-2-j} \wedge (\bar{\partial}_z P)^k \wedge (\bar{\partial}_z \partial_\zeta \beta)^{q-k}}{\Phi^{j+k+1} \cdot \beta^{n-(j+k+1)}}\\ &= \frac{A_q^{j,k}(P, \partial_\zeta \beta)}{\Phi^{j+k+1}\beta^{n-(j+k+1)}}.\end{aligned} \tag{3.13}$$

The heart of the Lipschitz $\frac{1}{2}$-estimate is contained in the following result.

Proposition 3.3. *For any* $0 \leq q \leq n-2$ *the double form* $A_q(L_D, B) = \sum_{|J|=q} A_{q,J}\, d\bar{z}^J$ *defined in Lemma 3.2 satisfies*

$$\int_{bD} \left| d_z A_{q,J}(\cdot, z) \right| \lesssim \delta_D(z)^{-1/2} \tag{3.14}$$

for $z \in D$ *and any* q*-tuple* J.

We will prove Proposition 3.3 in the next section. We now show how (3.14) implies the Lipschitz $\frac{1}{2}$-estimate for the solution operator $\mathbf{S}_{q+1}$. For $f \in C_{0,q+1}(bD)$ define

$$\mathbf{A}_q f(z) := \sum_{|J|=q} (\mathbf{A}_q f)_J\, d\bar{z}^J = \sum_J \int_{bD} f \wedge A_{q,J}(\cdot, z) \wedge d\bar{z}^J = \int_{bD} f \wedge A_q(L_D, B).$$

By differentiation under the integral sign and the obvious estimate, (3.14) implies

$$|d_z(\mathbf{A}_q f)_J(z)| \lesssim |f|_{0,bD} \cdot \delta_D(z)^{-1/2}.$$

We now apply Lemma 3.1. Note that if $K \subset D$ is compact, then $|(\mathbf{A}_q f)_J|_{0,K} \lesssim |f|_{0,bD}$ by trivial estimation. Hence one obtains

$$|\mathbf{A}_q f|_{1/2,D} = \sum_J |(\mathbf{A}_q f)_J|_{1/2,D} \lesssim |f|_{bD}$$

uniformly for all $f \in C_{0,q}(bD)$. Combined with (3.11), this proves the estimate (3.8), and we are done.

3.3. Integral Estimates

We now prove Proposition 3.3. Because of (3.12) it is enough to prove (3.14) for $A_{q,J}^{j,k}$ instead of $A_{q,J}$. Differentiation of (3.13) with respect to z gives

$$\begin{aligned} d_z A_{q,J}^{j,k} &= \frac{N_1}{\Phi^{j+k+1} \beta^{n-(j+k+1)}} + \frac{d_z\Phi \wedge \partial_\zeta \beta \wedge N_2}{\Phi^{j+k+2} \cdot \beta^{n-(j+k+1)}} \\ &\quad + \frac{d_z\beta \wedge \partial_\zeta \beta \wedge N_3}{\Phi^{j+k+1} \beta^{n-(j+k)}}, \end{aligned} \tag{3.15}$$

where N_j, $j = 1, 2, 3$ are forms with coefficients in $C^{0,\infty}(bD \times \bar{D})$, so that $|N_j| \lesssim 1$. Also, note that $|d_z\Phi| \lesssim 1$. Since by (1.6) and (1.7) one has $|\Phi| \gtrsim |\zeta - z|^2$ for $(\zeta, z) \in bD \times \bar{D}$, (3.15) implies

$$\begin{aligned} |d_z A_{q,J}^{j,k}(\zeta, z)| &\lesssim \frac{1}{|\Phi||\zeta - z|^{2n-2}} + \frac{|\zeta - z|}{|\Phi|^2 |\zeta - z|^{2n-2}} \\ &\quad + \frac{|\zeta - z|^2}{|\Phi||\zeta - z|^{2n}} \end{aligned} \tag{3.16}$$

for $(\zeta, z) \in bD \times \bar{D}$ and all j, k with $0 \leq j \leq n - q - 2$ and $0 \leq k \leq q$. Since $|\Phi|^{-1} = |\Phi|^{-2} \cdot |\Phi| \lesssim |\Phi|^{-2} |\zeta - z|$, it follows from (3.16) that

$$|d_z A_{q,J}^{j,k}(\zeta, z)| \lesssim \frac{1}{|\Phi|^2|\zeta - z|^{2n-3}}. \tag{3.17}$$

The crucial fact now is that in suitable local coordinates the function Φ vanishes only to first order in one of the coordinates. This "non isotropic" behavior of Φ reflects the splitting of the tangent space $T_\zeta bD$ into the complex tangent space $T_\zeta^{\mathbb{C}} bD$ and a real one-dimensional line F_ζ, where F_ζ has the property that the complex line $\mathbb{C}F_\zeta$ contains the normal vector to bD at ζ. We now make this precise.

Since $|\Phi| \gtrsim |\text{Re}\,\Phi| + |\text{Im}\,\Phi|$ and $|r(z)| \gtrsim \delta_D(z)$, one obtains from (1.6) that

$$|\Phi(\zeta, z)| \gtrsim |\text{Im}\,\Phi(\zeta, z)| + \delta_D(z) + |z - \zeta|^2 \tag{3.18}$$

for all $(\zeta, z) \in bD \times \bar{D}$ with $|\zeta - z| \leq \varepsilon/2$. We now show that $\text{Im}\,\Phi(\zeta, z)$ can be used as a local coordinate on bD.

Lemma 3.4. *There are positive constants M, a, and $\eta \leq \varepsilon/2$, and, for each z with $\delta_D(z) \leq a$, there is a C^1 coordinate system $(t_1, \ldots, t_{2n}) = t = t(\zeta, z)$ on $B(z, \eta)$, such that the following hold:*

$$t_1(\zeta, z) = r(\zeta) \quad \textit{and } t(z, z) = (r(z), 0, 0, \ldots, 0), \tag{3.19}$$

$$t_2(\zeta, z) = \text{Im}\,\Phi(\zeta, z), \tag{3.20}$$

$$|t(\zeta, z)| < 1 \quad \textit{for } \zeta \in B(z, \eta), \tag{3.21}$$

$$|J_{\mathbb{R}}(t(\cdot, z))| \leq M \quad \textit{and } |\det J_{\mathbb{R}}(t(\cdot, z))| \geq 1/M. \tag{3.22}$$

Proof. Fix $z \in bD$. For $|\zeta - z| < \varepsilon/2$, $\Phi = F^{\#}$, by (1.5). From (1.3) one obtains

$$d_\zeta \Phi(z, z) = d_\zeta F^{\#}(z, z) = \partial_\zeta r(z), \tag{3.23}$$

and therefore, at the point $\zeta = z$,

$$\begin{aligned} d_\zeta\, \text{Im}\,\Phi \wedge d_\zeta r &= \frac{1}{2i}(\partial_\zeta r - \bar{\partial}_\zeta r) \wedge (\partial_\zeta r + \bar{\partial}_\zeta r) \\ &= \frac{1}{i}\partial r \wedge \bar{\partial} r \neq 0. \end{aligned} \tag{3.24}$$

One can then find smooth real valued functions t_j, $3 \leq j \leq 2n$, with $t_j = 0$ for $\zeta = z$ and

$$d_\zeta r \wedge d_\zeta\, \text{Im}\,\Phi \wedge dt_3 \wedge \ldots \wedge dt_{2n} \neq 0$$

at $\zeta = z$. Lemma 3.4 for the fixed point z is now a consequence of the inverse function theorem. Since all the ingredients which enter into the above proof depend continuously on the point z, the Lemma will hold for points in a neighborhood of z, perhaps by choosing different constants. By compactness, finitely many such neighborhoods will cover bD and hence also $\{z: \delta_D(z) \leq a\}$ for sufficiently small $a > 0$. ∎

Remark. From (3.23) one also obtains $2\, d_\zeta \operatorname{Re} \Phi = d_\zeta r$ at $\zeta = z$; therefore

$$d_\zeta \operatorname{Re} \Phi \wedge d_\zeta \operatorname{Im} \Phi \neq 0.$$

So one can take Φ as the first coordinate of a complex valued C^1 coordinate system in a neighborhood of z. By the above, the Re Φ-axis is perpendicular to bD at $\zeta = z \in bD$, so that the Im Φ-axis will be perpendicular to the complex tangent space $T_z^{\mathbb{C}} bD$.

Because of (3.17) the proof of Proposition 3.3 is reduced to showing

$$(3.25) \qquad I(z) = \int_{bD} \frac{dS_\zeta}{|\Phi(\zeta, z)|^2 |\zeta - z|^{2n-3}} \lesssim \delta_D(z)^{-1/2} \qquad \text{for } z \in D.$$

Let $\gamma > 0$ be the smaller of the constants a and η in Lemma 3.4. For $z \in D$ we decompose $I(z) = I_1(z) + I_2(z)$, where

$$I_1(z) = \int_{bD \cap B(z, \gamma)} \Theta, \qquad I_2(z) = \int_{bD - B(z, \gamma)} \Theta,$$

and Θ denotes the integrand in (3.25). Obviously $I_2(z) \lesssim \gamma^{-(2n+1)} \cdot \operatorname{area}(bD)$ for all $z \in \bar{D}$. In order to estimate $I_1(z)$ we introduce the coordinate system $t = t(\zeta, z)$ given by Lemma 3.4. (Note that $I_1(z) \neq 0$ only if $\delta_D(z) < \gamma \leq a$.) We write $t = (t_1, t_2, t')$ with $t' \in \mathbb{R}^{2n-2}$, and $\delta = \delta_D(z)$. By (3.18) and Lemma 3.4 it follows that

$$(3.26) \qquad I_1(z) \lesssim \int_{0 < t_2 < 1; |t'| < 1} \frac{dt_2\, dt_3 \ldots dt_{2n}}{(t_2 + \delta + |t'|^2)^2 |t'|^{2n-3}}.$$

So the proof of (3.25), and therefore also of Proposition 3.3, will be complete once we prove

Lemma 3.5. *Let $J(\delta)$ denote the integral on the right side in* (3.26). *Then*

$$J(\delta) \lesssim \delta^{-1/2} \qquad \text{for } \delta > 0.$$

Proof. Integrating in t_2 one obtains

$$J(\delta) = \int_{|t'| < 1} \left[\frac{1}{\delta + |t'|^2} - \frac{1}{1 + \delta + |t'|^2} \right] \frac{dt'}{|t'|^{2n-3}}$$
$$\leq \int_{|t'| < 1} \frac{dt'}{(\delta + |t'|^2) |t'|^{2n-3}}.$$

Introduce polar coordinates in $t' \in \mathbb{R}^{2n-2}$ with $\rho = |t'|$. Then

$$J(\delta) \lesssim \int_0^1 \frac{d\rho}{\delta + \rho^2}$$
$$= \delta^{-1/2} \int_0^{\delta^{-1/2}} \frac{ds}{1 + s^2} \qquad (\text{substituting } \rho = \sqrt{\delta} s)$$
$$< \delta^{-1/2} \cdot \pi/2. \quad \blacksquare$$

3.4. Stability of the Estimate Under Perturbation

In many applications it is important to know how the estimate for the solution operator $\mathbf{S}_q^D$ for $\bar{\partial}$ depends on the domain D. We now show that the estimate is stable for small C^2 perturbations of bD. Let us make this notion precise. As usual, we assume that $D \subset\subset \mathbb{C}^n$ has a C^2 defining function $r = r_0$ defined on a neighborhood U of bD, and such that $dr \neq 0$ on U. For $\tau > 0$, define

$$\mathcal{N}^2(r_0, \tau) = \{r \in C^2(U, \mathbb{R}) \colon |r - r_0|_{2,U} < \tau\}.$$

For $r \in \mathcal{N}^2(r_0, \tau)$ we set

$$D^r = (D - U) \cup \{z \in U \colon r(z) < 0\}.$$

In particular, $D = D^{r_0}$. There is $\tau_0 = \tau_0(r_0)$, such that if $r \in \mathcal{N}^2(r_0, \tau_0)$, then D^r is a domain with C^2 boundary and defining function r. Also, if r_0 is strictly plurisubharmonic, U and τ_0 can be chosen so that any $r \in \mathcal{N}^2(r_0, \tau_0)$ is strictly plurisubharmonic.

Theorem 3.6. *Let $D \subset\subset \mathbb{C}^n$ be strictly pseudoconvex with C^2 boundary and with a strictly plurisubharmonic C^2 defining function r_0. Then there are constants $\tau_0 > 0$, and $C < \infty$, such that for any $r \in \mathcal{N}^2(r_0, \tau_0)$ and $1 \leq q \leq n$ there are solution operators $\mathbf{S}_q^r \colon C_{0,q}(\bar{D}^r) \to C_{0,q-1}(D^r)$ for $\bar{\partial}$ on D^r as in Theorem 2.7, such that*

$$|\mathbf{S}_q^r f|_{1/2, D^r} \leq C |f|_{\bar{D}^r}$$

for $f \in C_{0,q}(\bar{D}^r)$.

Proof. The proof involves checking how the constants involved in the construction of $\mathbf{S}_q$ in §1 and §2 and the estimates in §3 above change if r_0 is replaced by $r \in \mathcal{N}^2(r_0, \tau_0)$. Let us indicate the crucial steps.

The constants c and ε in (1.1) depend on the lowest eigenvalue of the Leviform of r and the modulus of continuity of the second order partial derivatives of r. So, for τ_1 sufficiently small, c and ε can be chosen independently of $r \in \mathcal{N}^2(r_0, \tau_1)$. If φ_{jk}, $1 \leq j, k \leq n$, are chosen with

$$\left| \varphi_{j,k} - \frac{\partial^2 r_0}{\partial \zeta_j \partial \zeta_k} \right|_{U'} \leq \frac{1}{2} c/2n^2,$$

then (1.3)–(1.8) will hold whenever $r \in \mathcal{N}^2(r_0, \tau_2)$, where $\tau_2 = \min(\tau_1, c/4n^2)$; moreover $D_\delta = D_\delta^{r_0}$ can be chosen independently of such r.

For $r \in \mathcal{N}^2(r_0, \tau_2)$ we thus obtain the generating form $L_r := L_{D^r} \in C_{1,0}^{1,\infty}(bD^r \times D^r)$, and, for $1 \leq q \leq n$, the Cauchy–Fantappiè form $\Omega_q(\hat{L}_r)$, the extension operator $\mathbf{E}_q^r$, and the parametrix $\mathbf{T}_q^r$. Since $\mathbf{E}_q^r \colon C_{0,q}(bD^r) \to C_{0,q}^\infty(\bar{D}_\delta)$, we can fix $0 < \delta' < \delta$ and $0 < \tau_3 \leq \tau_2$, such that $D^r \subset D_{\delta'}$ for $r \in \mathcal{N}^2(r_0, \tau_3)$. In the application of Theorem 2.5 to Theorem 2.7 we can then choose $K = \bar{D}_{\delta'}$ and $K \subset V_0 \subset\subset V \subset\subset D_\delta$, so that $\mathbf{T}_q^{V,V_0}$ is independent of $r \in \mathcal{N}^2(r_0, \tau_3)$. For such r we define

$$\mathbf{S}_q^{(r)} = \mathbf{T}_q^r + \mathbf{T}_q^{V, V_0} \circ \mathbf{E}_q^r.$$

Notice that there is $M < \infty$, such that $|\Omega_q(L_r)(\zeta, z)| \le M$ for all $r \in \mathcal{N}^2(r_0, \tau_3)$, $\zeta \in bD^r$ and $z \in \bar{D}_\delta$. Therefore

$$|\mathbf{E}_q^r f|_{D_\delta} \lesssim |f|_{bD^r}$$

uniformly in $r \in \mathcal{N}^2(r_0, \tau_3)$, and hence also

$$|\mathbf{T}_q^{V, V_0} \mathbf{E}_q^r f|_{\alpha, D^r} \le |\mathbf{T}_q^{V, V_0} \mathbf{E}_q^r f|_{\alpha, V_0} \lesssim |f|_{\bar{D}^r}$$

for $\alpha < 1$.

Clearly also the estimate

$$\left| \int_{D^r} f \wedge K_{q-1} \right|_{\alpha, D^r} \lesssim |f|_{D^r}$$

is uniform in r, so that one is left to show that the analogue of (3.8) for $\Omega_q(\hat{L}_r)$ is uniform in r.

First, notice that the estimate (3.1) in Lemma 3.1 is stable under small perturbations of D. This is elementary, though somewhat tedious, and we skip the details.

Next, by the above, the stability of (3.8) follows from the stability of (3.14). Here the main point is the choice of the coordinate system $t = t^{(r)}$ in Lemma 3.4. Certainly the estimate (3.18), with $\Phi = \Phi^r$, is uniform in r. Also, there is $\tau_4 \le \tau_3$, such that $d_\zeta(\operatorname{Im} \Phi^r) \wedge d_\zeta r \ne 0$ at $\zeta = z \in bD^r$ for $r \in \mathcal{N}(r_0, \tau_4)$. The functions $t_3, \dots t_{2n}$ can be chosen independently of r if τ_4 is sufficiently small. Finally the Jacobian matrix $J_{\mathbb{R}}(t^r)$ depends on the first order derivatives of r and Φ^r, the latter depending on second order derivatives of r. It follows that (3.22), and hence the rest of Lemma 3.4, including the choice of the constants a and η, is independent of $r \in \mathcal{N}(r_0, \tau_0)$ for some $\tau_0 \le \tau_4$. The estimate (3.26) for $I_1(z)$ is then uniform in $r \in \mathcal{N}(r_0, \tau_0)$, and so is the estimate for $I_2(z)$. ■

Exercises

E.3.1. For $m \in \mathbb{N}^+$, define $D^m \subset \mathbb{C}^2$ as in §2.5. Show that there is an integral operator

$$\mathbf{T}^m: C_{0,1}(\bar{D}^m) \to C(D^m)$$

such that

(i) $\bar{\partial}(\mathbf{T}^m f) = f$ if $\bar{\partial} f = 0$ on D^m;
(ii) $|\mathbf{T}^m f|_{1/2m, D^m} \le C |f|_{0,D}$ for some constant $C < \infty$ and all $f \in C_{0,1}(\bar{D}^m)$. (Hint: D^m is convex; construct the operator $\mathbf{T}_1^W$ associated to the canonical generating form $W = C^{(r_m)}$ (see IV.§3.4), and use E.1.1.).

E.3.2. Let $D \subset\subset \mathbb{C}^n$ have smooth boundary. A smooth complex valued vector field V on $\bar{D}$ is said to be **allowable** if for all $z \in bD$ one has $V_z \in T_z^{1,0} bD \oplus T_z^{0,1} bD$. Let D be strictly pseudoconvex and define Φ as in §1.1.

(i) Show that if V is an allowable vector field on $\bar{D}$, then $|V\Phi(\zeta, z)| = 0(|\zeta - z|)$.
(ii) Show that there is a vector field V on $\bar{D}$ with $V_z \in \mathbb{C}T_z bD$ for $z \in bD$, such that $(V\Phi)(z, z) \ne 0$ on bD.

(iii) Show that if $\mathbf{S}_q^D$ is the operator in Theorem 2.7 and if V is an admissible vector field on $\bar{D}$, then for every $\alpha < 1$ there is a constant $C_\alpha < \infty$ such that

$$|V(\mathbf{S}_q^D f)(z)| \leq C_\alpha |f|_0 \delta_D^{\frac{\alpha}{2}-1}(z) \qquad \text{for } z \in D.$$

Remark: This shows that the solution operator $\mathbf{S}_q^D$ for $\bar{\partial}$ is, essentially, "smoothing of order 1" in the *complex* tangential directions to bD, while Lemma 2.8 (with $m = 1$) shows that this is not true in *all* directions. This "nonisotropic" behavior was discovered by E.M. Stein [SteE 1].

E.3.3. Show that every sufficiently small "C^2 perturbation" of a bounded strictly pseudoconvex domain D with C^2 boundary is again a domain of the same type. (See §3.4 for the precise definition of C^2 perturbation of bD.)

Notes for Chapter V

The solution of the Cauchy–Riemann equations on a Stein domain D (i.e., the vanishing theorem $H_{\bar{\partial}}^q(D) = 0$ for $q \geq 1$) is a fundamental result in the classical theory of several complex variables developed in the early 1950s (see the Notes for Chapter VI). Combined with the known solutions of the Levi problem it implies the existence of solutions for $\bar{\partial}$ on pseudoconvex domains. The methods for solving $\bar{\partial}$ discussed in this chapter are much more recent.

The first *integral* solution operators for $\bar{\partial}$ on a *strictly* pseudoconvex domain D in $\mathbb{C}^n$ were constructed around 1969–70 by H. Grauert and I. Lieb [GrLi] and G.M. Henkin [Hen 2], by using a globally holomorphic generating form constructed shortly before by E. Ramirez [Ram] and G.M. Henkin [Hen 1], independently (see Chapter VII, §3). These results made use of the known classical methods for solving the Cauchy–Riemann equations. The direct and elementary construction presented in §1 and §2, which includes the solution of the Levi problem for D, and which does not require any a priori knowledge of solutions for $\bar{\partial}$, was discovered by R.M. Range [Ran 6]. A related construction was found by G.M. Henkin (see [HeLe]), whose ideas were originally published in somewhat different form in [AiYu]; Range's results were obtained independently of Henkin and [AiYu]. The generating form L_D was introduced by N. Kerzman and E.M. Stein [KeSt], who considered only the CF form $\Omega_0(L_D)$ of order 0 in the construction of a special Cauchy-type kernel (see Chapter VII, §1); it was applied to integral representations for $(0, q)$-forms in case $q \geq 1$ by R. Harvey and J. Polking [HaPo] and in [Ran 6].

The Levi problem was first solved in 1942 by K. Oka [Oka, VI] in $\mathbb{C}^2$, and in the early 1950s in arbitrary dimension by Oka [Oka, IX], H. Bremermann [Bre 1], and F. Norguet [Nor 1]. In 1958 H. Grauert [Gra] solved the Levi problem on complex manifolds by using the theory of coherent analytic sheaves and the finiteness theorem of L. Schwartz in Fréchet spaces. The basic *idea* of the solution of Levi's problem presented here is really the same as Grauert's, although the technical details are quite different. Similar solutions of the Levi problem by integral kernels were found by R. Harvey and J. Polking [HaPo], who used Fréchet space theory rather than the more elementary results for Banach spaces used here, and by Henkin and Leiterer [HeLe]. Theorem 2.2 was obtained first by H. Hefer [Hef] on Stein domains by using the solution of the Cousin I problem (or, equivalently, $H^q_{\bar\partial}(D) = 0$; see Chapter VI, §4).

L^∞ estimates for solutions of $\bar\partial u = f$, f a $\bar\partial$-closed $(0, q)$-form, were obtained by H. Grauert and I. Lieb [GrLi] and G.M. Henkin [Hen 2] in case $q = 1$, and by I. Lieb [Lie 2] in the general case $q \geq 1$. Related results were obtained by N. Øvrelid [Øvr 1]. Based on the work of Grauert and Lieb, N. Kerzman [Ker 1] then constructed a different solution operator for $\bar\partial$ on $(0, 1)$-forms and proved that it is bounded from L^∞ to Λ_α for any $\alpha < \frac{1}{2}$. The sharp estimate for $\alpha = \frac{1}{2}$ in case $q = 1$ was proved thereafter by G.M. Henkin and A.V. Romanov [HeRo], who used a modification of the solution operator in [Hen 2]. The $\frac{1}{2}$-estimate for arbitrary $q \geq 1$ was proved by R.M. Range and Y.T. Siu [RaSi]. In that paper integral solution operators for $\bar\partial$ were constructed on domains with *piecewise* strictly pseudoconvex boundary (see also P.L. Poljakov [Pol 2] for a similar result), and it was proved that these operators are bounded from L^∞ into Λ_α for any $\alpha < \frac{1}{2}$ in the general case, and into $\Lambda_{1/2}$ if the boundary is globally of class C^2 (see the last part of Theorem 3.9 in [RaSi]). The details of the estimations in §3 are based on [RaSi]; the importance of the local coordinate systems given in Lemma 3.4 was first noticed in [GrLi] and in [Hen 1]. Similar coordinate systems play an essential role in all estimates of integral kernels related to $\bar\partial$. Lemma 3.1, which is a straightforward generalization of a classical result of G.H. Hardy and J.E. Littlewood, was introduced in the present context in [HeRo]. The example in §2.5 for the case of a ball was suggested to Kerzman by E.M. Stein (see [Ker 1]); the generalization given here is in [Ran 4]. Further results regarding estimates for solutions of $\bar\partial$ will be discussed in Chapter VII, §5.

CHAPTER VI

Function Theory on Domains of Holomorphy in $\mathbb{C}^n$

In §1 of this chapter we first extend the fundamental vanishing theorem $H^q_{\bar\partial}(K) = 0$ for $q \geq 1$ on a Stein compactum K proved in Corollary V.2.6 to arbitrary *open* Stein domains (Theorem 1.4). The proof involves an approximation theorem for holomorphic functions on compact analytic polyhedra which is of independent interest, and which generalizes the classical Runge Approximation Theorem in the complex plane. We also discuss several variations of this approximation theorem. In particular we consider the Runge property for the exhaustion of a pseudoconvex domain D by *strictly* pseudoconvex domains which arises from the existence of a strictly plurisubharmonic exhaustion function on D. Together with the results of Chapter V this yields the solution of the Levi problem for arbitrary pseudoconvex domains. In §2 we apply these methods to solve the Cauchy–Riemann equations directly on a pseudoconvex domain D, i.e., we show that $H^q_{\bar\partial}(D) = 0$ for $q \geq 1$, and we prove that this property characterizes Stein domains. §3 deals with some topological properties of Stein domains D, for example, we show that if $D \subset \mathbb{C}^n$, then $H^r(D, \mathbb{C}) = 0$ for $r > n$. This section may be skipped without loss of continuity. Finally, in §4 and §5, the vanishing of $H^1_{\bar\partial}$ for Stein domains D is used to generalize to several variables the classical theorems of Mittag-Leffler and Weierstrass on the existence of global meromorphic functions with prescribed poles and zero sets on regions in the complex plane. §5 includes a detailed discussion of the new—strictly higher dimensional—phenomenon of a topological obstruction in the analog of the Weierstrass theorem.

When specialized to one variable, the results proved in this chapter include much of classical global function theory on arbitrary regions in the complex plane, but the proofs are quite different from those usually found in standard one variable texts. Keeping in mind the introductory nature of this book, our presentation is direct and elementary, in so far as sheaves and cohomology theory are not used; instead, we emphasize the concrete formulation in terms of the classical Cousin problems. This allows us to concentrate on the principal ideas without burdening the reader with additional technical baggage. On the other hand we believe that the reader should be made aware that these results are merely the starting point for the much richer fundamental theory of **cohomology of coherent analytic sheaves**,[1] which is indispensable for dealing with the higher dimensional phenomena of complex submanifolds of codimension > 1 and of singularities of analytic sets. In §6 we therefore present a brief introductory survey of this theory; our goal here is to help the reader to understand the importance of the relevant concepts and to gain some appreciation for these more advanced methods and results. Moreover, we hope that this discussion will provide motivation for further study.

§1. Approximation and Exhaustions

1.1. The Oka Approximation Theorem

The key ingredient for the various approximation theorems considered in §1 is an approximation theorem on compact analytic polyhedra (Theorem 1.1 below). The proof is based on Oka's fundamental idea to consider an embedding into higher dimensions in order to simplify the geometry ([Oka], I; see also the Notes for this chapter). We already used this idea in the construction of global holomorphic generating forms in Chapter V, §2.3. Compared to Oka's original proof (see also Hörmander [Hör 2], §2.7, for the $\bar{\partial}$-version of this proof), which dealt with the special case of polynomially convex sets, the proof given here is more direct and it gives immediately a more general result, as we already have available the solution of $\bar{\partial}$ on Stein compacta.

Theorem 1.1. *Let $K \subset \mathbb{C}^n$ be a Stein compactum and suppose $h_1, \ldots, h_l \in \mathcal{O}(K)$. Define*

$$K_l = \{z \in K: |h_j(z)| \leq 1, 1 \leq j \leq l\}.$$

Then K_l is Stein and $\mathcal{O}(K)$ is dense in $\mathcal{O}(K_l)$ in the supremum norm on K_l.

Proof. K_l is Stein by Lemma II.3.22. For the main part of the theorem we may assume that $l = 1$; the general case then follows by induction on l. So set $h_1 = h$ and choose an open neighborhood U such that $h \in \mathcal{O}(U)$. Define the

[1] In terms of this theory, the present chapter deals only with the vanishing of $H^q(D, \mathcal{O})$, $q \geq 1$, for the sheaf $\mathcal{O}$ of germs of holomorphic functions, and the immediate consequences thereof, while the general theory considers the cohomology groups $H^q(D, \mathcal{A})$ with coefficients in an arbitrary coherent analytic sheaf $\mathcal{A}$.

Oka map $\mu_h: U \to U \times \mathbb{C}$ by

(1.1) $$\mu_h(z) = (z, h(z)).$$

Then $K_1 = \mu_h^{-1}(K \times \bar{\Delta})$. We now prove

(1.2) *For every $f \in \mathcal{O}(K_1)$ there is $F \in \mathcal{O}(K \times \bar{\Delta})$ such that $F \circ \mu_h = f$ in a neighborhood of K_1.*

The proof of (1.2) is similar to the one of Proposition V.2.1. Assume that $f \in \mathcal{O}(V)$, where $V \subset U$ is a neighborhood of K_1. Choose $\chi \in C_0^\infty(V)$ such that $\chi \equiv 1$ on a neighborhood W of K_1. Since $\{(z, w) \in K \times \bar{\Delta}: h(z) - w = 0\} = \mu_h(K_1) \subset W \times \mathbb{C}$, the (0, 1)-form

$$\alpha = \frac{\bar{\partial}(\chi f)}{h(z) - w}$$

is smooth and $\bar{\partial}$-closed on a neighborhood of the Stein compactum $K \times \bar{\Delta}$. By Corollary V.2.6 there is $v \in C^\infty(K \times \bar{\Delta})$ with $\bar{\partial} v = \alpha$. Then $F = \chi f - (h(z) - w) \cdot v$ is in $\mathcal{O}(K \times \bar{\Delta})$, and F satisfies (1.2).

The proof of the Theorem is now immediate. Note that any $F \in \mathcal{O}(K \times \bar{\Delta})$ can be expanded in a Taylor series

$$F(z, w) = \sum_{\nu=0}^{\infty} a_\nu(z) w^\nu, \qquad a_\nu \in \mathcal{O}(K),\ \nu = 0, 1, 2, \ldots,$$

which converges uniformly on $K \times \bar{\Delta}$ (see Exercise E.I.1.9). By (1.2) one obtains

(1.3) $$f(z) = F(z, h(z)) = \sum_{\nu=0}^{\infty} a_\nu(z) [h(z)]^\nu,$$

with uniform convergence on K_1. The partial sums in (1.3) are in $\mathcal{O}(K)$, and we are done. ■

Corollary 1.2. *Let K be a Stein compactum and suppose $L \subset K$ is a compact subset with*

$$\hat{L}_{\mathcal{O}(K)} = L.$$

Then L is Stein and $\mathcal{O}(K)$ is dense in $\mathcal{O}(L)$ in the supremum norm over L.

Proof. Suppose $f \in \mathcal{O}(L)$ and choose an open neighborhood U of L, so that $f \in \mathcal{O}(U)$. By Proposition II.3.23, L is Stein and one can find a compact analytic polyhedron

$$K_l = \{z \in K: |h_j| \le 1, 1 \le j \le l\}$$

defined by functions $h_j \in \mathcal{O}(K)$, such that $L \subset K_l \subset U$. Since $f \in \mathcal{O}(K_l)$, Theorem 1.1 implies that f is the uniform limit on $K_l \supset L$ of functions in $\mathcal{O}(K)$. ■

1.2. The Solution of $\bar{\partial}$ on Stein Domains

The proof of the following Lemma involves a typical application of approximation techniques in order to extend analytic properties of Stein compacta to open Stein domains.

Lemma 1.3. *Let $D \subset \mathbb{C}^n$ be open and suppose there is a normal exhaustion $\{K_j, j = 1, 2, \dots\}$ of D with $H^q_{\bar{\partial}}(K_j) = 0$ for $q \geq 1$ and each j. Then $H^q_{\bar{\partial}}(D) = 0$ for $q \geq 2$. Moreover, if $\mathcal{O}(K_{j+1})|_{K_j}$ is dense in $\mathcal{O}(K_j)$ in the supremum norm on K_j for $j = 1, 2, \dots$, then also $H^1_{\bar{\partial}}(D) = 0$.*

PROOF. Suppose $q \geq 1$ and $f \in C^\infty_{0,q}(D)$ is $\bar{\partial}$-closed. By hypotheses, for each j there is $u_j \in C^\infty_{0,q-1}(K_j)$ such that $\bar{\partial} u_j = f$ near K_j.

Assume first that $q \geq 2$. Inductively, we will find $g_j \in C^\infty_{0,q-1}(K_j)$, $j = 1, 2, \dots$, such that

(1.4) $\bar{\partial} g_j = f$ in a neighborhood of K_j, and

(1.5) $g_j - g_{j-1} \equiv 0$ on K_{j-1} for $j \geq 2$.

Start with $g_1 = u_1$; suppose we already found $g_1, \dots, g_l$, such that (1.4) and (1.5) hold for $1 \leq j \leq l$. Then $\bar{\partial}(u_{l+1} - g_l) = 0$ near K_l, so there is $v_l \in C^\infty_{0,q-2}(K_l)$, such that $\bar{\partial} v_l = u_{l+1} - g_l$ near K_l. By multiplying v_l with a cutoff function χ_l which satisfies $\chi_l \equiv 1$ in a neighborhood of K_l, one extends $\chi_l \cdot v_l$ smoothly to a neighborhood of K_{l+1}, while retaining

$$\bar{\partial}(\chi_l v_l) = u_{l+1} - g_l \text{ near } K_l.$$

Therefore $g_{l+1} = u_{l+1} - \bar{\partial}(\chi_l v_l)$ will satisfy (1.4) and (1.5) with $j = l + 1$. It now follows that $g = \lim_{j\to\infty} g_j \in C^\infty_{0,q-1}(D)$ and $\bar{\partial} g = f$.

In case $q = 1$, we use a similar scheme, except we replace (1.5) by

$$(1.6) \qquad |g_j - g_{j-1}|_{0,K_{j-1}} \leq 2^{-j} \qquad \text{if } j \geq 2.$$

Again, set $g_1 = u_1$, and suppose $g_j \in C^\infty(K_j)$ has been found for $1 \leq j \leq l$, such that (1.4) and (1.6) hold. Since $\bar{\partial}(u_{l+1} - g_l) = 0$ near K_l, $u_{l+1} - g_l \in \mathcal{O}(K_l)$. By the additional hypotheses for $q = 1$, there is $h_{l+1} \in \mathcal{O}(K_{l+1})$ such that

$$|(u_{l+1} - g_l) - h_{l+1}|_{0,K_l} \leq 2^{-(l+1)}.$$

Therefore $g_{l+1} = u_{l+1} - h_{l+1} \in C^\infty(K_{l+1})$ satisfies (1.4) and (1.6). Now (1.6) implies that $g = \lim_{j\to\infty} g_j$ exists uniformly on K_l for each l; hence $g \in C(D)$. Furthermore, if $j \geq l + 1$, $g_j - g_{l+1} \in \mathcal{O}(K_{l+1})$, and $\lim_{j\to\infty}(g_j - g_{l+1}) = g - g_{l+1}$ uniformly on K_{l+1}. So $g - g_{l+1}$ is holomorphic on the interior of K_{l+1}; in particular, $g \in C^\infty(K_l)$ and $\bar{\partial} g = \bar{\partial} g_{l+1} = f$ on K_l. As this holds for each l, we are done. ∎

Theorem 1.4. *Let $D \subset \mathbb{C}^n$ be a Stein domain. Then*

$$H^q_{\bar{\partial}}(D) = 0 \qquad \text{for } q \geq 1.$$

Proof. By Lemma II.3.2 there is a normal exhaustion $\{K_j\}$ of D by Stein compacta, such that $(\hat{K}_j)_{\mathcal{O}(D)} = K_j$. In particular, $(\hat{K}_j)_{\mathcal{O}(K_{j+1})} = K_j$ for $j = 1, 2, \ldots$. Corollary V.2.6 and Corollary 1.2 show that the hypotheses of Lemma 1.3. are satisfied. ∎

1.2. Polynomial Convexity and the Theorem of Oka–Weil

We now apply Corollary 1.2 to a compact *polydisc* K in $\mathbb{C}^n$. If $\mathscr{P}$ denotes the algebra of holomorphic polynomials on $\mathbb{C}^n$, Theorem I.1.18 shows that $\mathscr{P}|_K$ is dense in $\mathcal{O}(K)$. Therefore, if $L \subset K$ and $\hat{L}_{\mathcal{O}(K)} = L$, it follows that $\mathscr{P}|_L$ is dense in $\mathcal{O}(L)$. Let us introduce the following terminology.

Definition. For a compact set $K \subset \mathbb{C}^n$, $\mathscr{P}(K)$ denotes the closure of $\mathscr{P}|_K$ in $C(K)$ in the supremum norm. The **polynomially convex hull $\hat{K}_{\mathscr{P}}$ of** K is defined by

$$\hat{K}_{\mathscr{P}} = \{z \in \mathbb{C}^n : |g(z)| \leq |g|_{0,K} \quad \text{for all } g \in \mathscr{P}\}.$$

K is called **polynomially convex** if $\hat{K}_{\mathscr{P}} = K$.

Of course, $\hat{K}_{\mathscr{P}}$ agrees with $\hat{K}_{\mathcal{O}(\mathbb{C}^n)}$, and, more generally, $\hat{K}_{\mathscr{P}} = \hat{K}_{\mathcal{O}(P)}$ for any polydisc P (open or compact) with $K \subset P$. So the remarks above can be reformulated as follows.

Theorem 1.5 (The Oka–Weil Theorem). *Let K be a compact polynomially convex set in $\mathbb{C}^n$. Then $\mathcal{O}(K) \subset \mathscr{P}(K)$.*

Theorem 1.5 extends the following classical Runge Approximation Theorem (see [Rud 1], Theorem 13.7) from one to several variables.

Theorem 1.6. *Let K be a compact subset of $\mathbb{C}$. Then $\mathcal{O}(K) \subset \mathscr{P}(K)$ if and only if $\mathbb{C}—K$ is connected.*

Indeed, for $K \subset \mathbb{C}$ the hypotheses in Theorems 1.5 and 1.6 are equivalent. More generally, one has the following result: the proof, which uses Theorem 1.6, is left as an exercise.

Lemma 1.7. *Let $K \subset \mathbb{C}$ be compact. Then $\hat{K}_{\mathscr{P}} = K \cup \{$bounded components of $\mathbb{C}—K\}$. In particular, K is polynomially convex if and only if $\mathbb{C}—K$ is connected.*

No such simple topological characterization of polynomial convexity exists in general; J. Wermer [Wer 2] constructed a compact set $K \subset \mathbb{C}^3$ which is biholomorphic to the closed unit polydisc $\bar{\Delta}^3$, and yet $\hat{K}_{\mathscr{P}} \neq K$.

In contrast to the situation in dimension one, the polynomial convexity of K is not necessary for $\mathcal{O}(K) \subset \mathscr{P}(K)$ in higher dimensions. This is related to the extension properties of holomorphic functions in *several* variables. For example, if $K = \{z \in \mathbb{C}^2 : |z| = 1\}$, then $\mathcal{O}(K) = \mathcal{O}(\overline{B(0, 1)}) \subset \mathcal{P}(K)$, but $\hat{K}_{\mathcal{P}} = \{z \in \mathbb{C}^2 : |z| \leq 1\} \neq K$. However, within the class of Stein compacta, polynomial convexity characterizes those sets on which polynomial approximation holds. (Notice that the example above is clearly not Stein.)

Theorem 1.8. *Let $K \subset \mathbb{C}^n$ be a Stein compactum and assume $\mathcal{O}(K) \subset \mathcal{P}(K)$. Then K is polynomially convex.*

PROOF. It is enough to show that $\hat{K}_{\mathcal{P}} \subset U$ for any open Stein neighborhood U of K.

First we show that for any such U

$$\hat{K}_{\mathcal{P}} \cap U = \hat{K}_{\mathcal{O}(U)}. \tag{1.7}$$

For $f \in \mathcal{P}(K)$ and $w \in \hat{K}_{\mathcal{P}}$ define

$$\hat{f}(w) = \lim_{\nu \to \infty} p_\nu(w), \tag{1.8}$$

where $p_\nu \in \mathcal{P}$ and $|p_\nu - f|_K \to 0$ as $\nu \to 0$. The limit in (1.8) is independent of the particular choice of the sequence $\{p_\nu\}$ which converges to f. Moreover,

$$|\hat{f}(w)| \leq |f|_K, \tag{1.9}$$

and $f \to \hat{f}(w)$ is a $\mathbb{C}$-algebra homomorphism $\mathcal{P}(K) \to \mathbb{C}$ (see Exercise E.1.5). Now fix $w \in \hat{K}_{\mathcal{P}} \cap U$. We will show that $\hat{f}(w) = f(w)$ for $f \in \mathcal{O}(U)$ if $\mathcal{O}(K) \subset \mathcal{P}(K)$. $Q = K \cup \{w\}$ is a Stein compactum contained in U. By Theorem V.2.2, any $f \in \mathcal{O}(U)$ can be written as

$$f - f(w) = \sum_{j=1}^{n} g_j(z_j - w_j), \tag{1.10}$$

with functions $g_j \in \mathcal{O}(Q)$. By hypothesis, f and $g_1, \ldots, g_n \in \mathcal{P}(K)$, so (1.10) implies

$$\hat{f}(w) - f(w) = \sum_{j=1}^{n} \hat{g}_j(w)(\hat{z}_j(w) - w_j) = 0.$$

By (1.9) we therefore obtain $|f(w)| = |\hat{f}(w)| \leq |f|_K$ for every $f \in \mathcal{O}(U)$; hence $w \in \hat{K}_{\mathcal{O}(U)}$, and (1.7) follows.

Since U is holomorphically convex, (1.7) implies $\hat{K}_{\mathcal{P}} \cap U \subset\subset U$. Therefore $\hat{K}_{\mathcal{P}} \cap U$ and $\hat{K}_{\mathcal{P}} - U$ are two disjoint compact sets. Define $g \in \mathcal{O}(\hat{K}_{\mathcal{P}})$ by $g \equiv 0$ in a neighborhood of $\hat{K}_{\mathcal{P}} \cap U$, and by $g \equiv 1$ in a neighborhood of $\hat{K}_{\mathcal{P}} - U$. By Theorem 1.5, $g \in \mathcal{P}(\hat{K}_{\mathcal{P}})$; so there is a polynomial p, such that $|p| < 1/2$ on $K \subset \hat{K}_{\mathcal{P}} \cap U$, while $|p| > 1/2$ on $\hat{K}_{\mathcal{P}} - U$. It follows that $\hat{K}_{\mathcal{P}} - U = \varnothing$, i.e. $\hat{K}_{\mathcal{P}} \subset U$. ∎

Polynomial approximation has been investigated intensively for a long time. One of the central questions has been to generalize the classical Weierstrass Theorem, which states $\mathcal{P}(K) = C(K)$ for $K = [0, 1]$, to curves in $\mathbb{C}^n$ or to more general sets. The notion of polynomial convexity arises naturally in this context: if a sequence of polynomials converges uniformly on $K \subset \mathbb{C}^n$, then it converges uniformly also on $\hat{K}_{\mathscr{P}}$. So functions in $\mathscr{P}(K)$ have a natural extension to continuous functions on $\hat{K}_{\mathscr{P}}$, and $\mathscr{P}(K) \simeq \mathscr{P}(\hat{K}_{\mathscr{P}})$. In a precise sense, $\hat{K}_{\mathscr{P}}$ is the largest compact set with this property: in the language of the Gelfand theory of commutative Banach algebras, $\hat{K}_{\mathscr{P}}$ is the "maximal ideal space" of $\mathscr{P}(K)$ (see Exercise E.1.5). So the structure of $\hat{K}_{\mathscr{P}}$ clearly will be relevant for polynomial approximation on K. For example, if K or $\hat{K}_{\mathscr{P}}$ contains an analytic set A of positive dimension, every $f \in \mathscr{P}(K)$ must be holomorphic on A, hence $\mathscr{P}(K) \neq C(K)$. In an influential paper, Wermer [Wer 1] proved that for a real analytic curve K in $\mathbb{C}^n$, either $\hat{K}_{\mathscr{P}} = K$ and $\mathscr{P}(K) = C(K)$, or else $\hat{K}_{\mathscr{P}} - K$ is a one-dimensional analytic set. This type of result has been generalized considerably; see T.W. Gamelin [Gam 2] for a survey and additional references.

1.4. Runge Domains and Runge Pairs

We now consider approximation by holomorphic functions on open sets.

Definition. An open set $D \subset \mathbb{C}^n$ is called a **Runge region** if the algebra of polynomials $\mathscr{P}$ is dense in $\mathcal{O}(D)$.

More generally, two open sets $D_1 \subset D_2$ are called a **Runge pair** (equivalently, one says that D_1 **is Runge in** D_2) if $\mathcal{O}(D_2)$ is dense in $\mathcal{O}(D_1)$.

Observe that D is Runge if and only if $(D, \mathbb{C}^n)$ is a Runge pair. Therefore the following result is a special case of Theorem 1.11 below. We state it separately because of the independent interest in polynomial approximation.

Theorem 1.9. *A Stein region D in $\mathbb{C}^n$ is Runge if and only if D is $\mathscr{P}$-convex, i.e., $\hat{K}_{\mathscr{P}} \cap D \subset\subset D$ for every compact set $K \subset D$.*

Remark. It is classical that an open set $D \subset \mathbb{C}$ is a Runge region if and only if D is simply connected (see Theorem 1.6).

We first prove a more general version of the Oka–Weil Theorem 1.5.

Theorem 1.10. *Let $D \subset \mathbb{C}^n$ be a Stein domain. Suppose $K \subset D$ is compact and $\hat{K}_{\mathcal{O}(D)} = K$. Then $\mathcal{O}(D)$ is dense in $\mathcal{O}(K)$.*

Remark. There is a converse to this result; see Exercise 1.8.

Proof. Notice that Theorem 1.5 followed directly from Corollary 1.2 by using the Taylor expansion. This tool is not available now, so the approximating functions must be obtained by an inductive procedure. Let $K_1 \subset K_2 \subset K_3 \ldots$ be a normal exhaustion of D with $K_1 = K$ and $(\hat{K}_j)_{\mathcal{O}(D)} = K_j$. Then each K_j is a Stein compactum, and $(\hat{K}_j)_{\mathcal{O}(K_{j+1})} = K_j$ for $j = 1, 2, \ldots$. By Corollary 1.2, $\mathcal{O}(K_{j+1})$ is dense in $\mathcal{O}(K_j)$ for each j. Therefore, given $f \in \mathcal{O}(K)$ and $\varepsilon > 0$, we can inductively choose functions $f_j \in \mathcal{O}(K_j)$, $j = 1, 2, \ldots$, with $f_1 = f$ and $|f_{j+1} - f_j| < \varepsilon \cdot 2^{-j}$ on K_j. It follows that for each l the sequence $\{f_j : j \geq l\}$ is Cauchy in $C(K_l)$; hence $F = \lim_{j\to\infty} f_j$ converges uniformly on each K_l and $F \in \mathcal{O}(D)$. Finally,

$$|F - f|_K \leq \sum_{j=1}^{\infty} |f_{j+1} - f_j|_K \leq \varepsilon \sum_{j=1}^{\infty} 2^{-j} = \varepsilon. \quad \blacksquare$$

Theorem 1.11. *The following statements are equivalent for two Stein regions* $D_1 \subset D_2$ *in* $\mathbb{C}^n$:

(i) $\mathcal{O}(D_2)$ *is dense in* $\mathcal{O}(D_1)$, *i.e.,* D_1 *is Runge in* D_2.
(ii) *For every compact set* $K \subset D_1$ *one has* $\hat{K}_{\mathcal{O}(D_2)} \cap D_1 = \hat{K}_{\mathcal{O}(D_1)}$.
(iii) D_1 *is* $\mathcal{O}(D_2)$ *convex, i.e.,* $\hat{K}_{\mathcal{O}(D_2)} \cap D_1 \subset\subset D_1$ *for every compact set* $K \subset D_1$.
(iv) *For every compact set* $K \subset D_1$ *one has* $\hat{K}_{\mathcal{O}(D_2)} \subset D_1$.
(v) *For every compact set* $K \subset D_1$ *one has* $\hat{K}_{\mathcal{O}(D_2)} = \hat{K}_{\mathcal{O}(D_1)}$.

Proof. It is obvious that (i) $\Rightarrow$ (ii) $\Rightarrow$ (iii) (note that $\hat{K}_{\mathcal{O}(D_1)} \subset\subset D_1$ since D_1 is Stein). Assuming (iii), notice that the function f defined by 0 on $K' = \hat{K}_{\mathcal{O}(D_2)} \cap D_1$ and by 1 on $K'' = \hat{K}_{\mathcal{O}(D_2)} - D_1$ is in $\mathcal{O}(\hat{K}_{\mathcal{O}(D_2)})$, since K' and K'' are disjoint and compact. By Theorem 1.10, there is $F \in \mathcal{O}(D_2)$ with $|F - f|_{\hat{K}_{\mathcal{O}(D_2)}} < 1/2$, i.e., $|F| < \frac{1}{2}$ on $K \subset K'$ and $|F| > \frac{1}{2}$ on K''; so $K'' = \emptyset$ and (iv) follows. (iv) implies (i) by applying Theorem 1.10 to D_2 and $(\hat{K}_j)_{\mathcal{O}(D_2)}$, $j = 1, 2, \ldots$, where $\{K_j\}$ is a normal exhaustion of D_1. Finally, it is obvious that (ii) and (iv) imply (v), and that (v) implies (ii). $\blacksquare$

The next theorem follows from Theorem 1.11 by a straightforward inductive procedure. The details of the proof are left to the reader.

Theorem 1.12. *Suppose* $D_1 \subset D_2 \subset D_3 \ldots$ *is an increasing sequence of Stein regions in* $\mathbb{C}^n$ *such that* D_j *is Runge in* D_{j+1} *for* $j = 1, 2, \ldots$. *Then* $D = \bigcup_{j=1}^{\infty} D_j$ *is Stein and* D_j *is Runge in* D *for each* j.

1.5. Holomorphic Peaking Functions

In order to solve the $\bar{\partial}$-equation and the Levi problem for an arbitrary pseudoconvex domain D, it is natural to consider an exhaustion of D by strictly pseudoconvex domains and to apply Lemma 1.3 or Theorem 1.12. Holomorphic peaking functions, which we now discuss, are a key ingredient in the proof of the required Runge property for such an exhaustion. An alternate (and perhaps more direct approach) could be based on the following theorem of H. Behnke and K. Stein [BeSt]: *If* $D_1 \subset D_2 \subset \ldots$ *is an increasing sequence of*

Stein domains in $\mathbb{C}^n$, *then* $D = \bigcup_{j=1}^{\infty} D_j$ *is Stein.* However, the main point of the proof of Behnke and Stein is a technical construction which replaces the sequence $\{D_j\}$ by a sequence $\{D_j'\}$ which does have the Runge property. Moreover, it is known that the analog of the Behnke–Stein Theorem does not hold in more general settings (see Fornaess [For 1]); so we prefer to discuss a method which does generalize. Also, peaking functions, as well as the Runge property discussed in the next section, are of independent interest. The Theorem of Behnke–Stein will be a simple consequence of the solution of the Levi problem.

Theorem 1.13. *Let* $K \subset \mathbb{C}^n$ *be a Stein compactum, and let* $\zeta \in K$ *be a strictly pseudoconvex boundary point, i.e., there are a neighborhood* U *of* ζ *and a* C^2 *strictly plurisubharmonic function* $r: U \to \mathbb{R}$, *such that* $r(\zeta) = 0$ *and*

$$K \cap U = \{z \in U: r(z) \le 0\}.$$

Then there is $h \in \mathcal{O}(K)$ *such that*

$$h(\zeta) = 1 \text{ and } |h(z)| < 1 \qquad \text{for } z \in K - \{\zeta\}. \tag{1.11}$$

Notice that we do not assume that $dr \neq 0$, so that $bK \cap U$ is not necessarily a real submanifold of U.

A function h which satisfies (1.11) is called a **peaking function for** ζ **on** K.

Proof. Let $u(z) = 1/\Phi(\zeta, z)$, where $\Phi(\zeta, z)$ is the function constructed in V.§1.1 from the Levi polynomial of r. We have $\operatorname{Re} u > 0$ on $K - \{\zeta\}$, and $\bar\partial u = 0$ for $|z - \zeta| < \varepsilon/2$. So $\bar\partial u$ extends as a smooth $\bar\partial$-closed (0, 1)-form to a neighborhood of K. By the hypothesis on K and Corollary V.2.6 there is $v \in C^\infty(K)$, such that $\bar\partial v = \bar\partial u$ in a neighborhood of K. Set $g = 1/(u - v + |v|_K)$. Then

$$\operatorname{Re} g = \frac{1}{\operatorname{Re}(u - v + |v|_K)} > 0 \qquad \text{on } K - \{\zeta\}, \tag{1.12}$$

and $g \in \mathcal{O}(K - \{\zeta\})$; moreover, for z near ζ, but $z \neq \zeta$,

$$g = \Phi(\zeta, \cdot)\frac{1}{1 + \Phi(\zeta, \cdot)(|v|_K - v)}. \tag{1.13}$$

Since $\bar\partial v = \bar\partial u = 0$ in a neighborhood of ζ, v, and, by (1.13), also g, is holomorphic at ζ. So $g \in \mathcal{O}(K)$ and $g(\zeta) = 0$. This and (1.12) show that the function $h = e^{-g}$ satisfies (1.11). ∎

Corollary 1.14. *Let* $D \subset\subset \mathbb{C}^n$ *be strictly pseudoconvex with* C^2 *boundary. Then for every* $\zeta \in bD$ *there is a peaking function* $h_\zeta \in \mathcal{O}(\bar D)$ *for* ζ *on* $\bar D$.

Proof. If $D = \{\varphi < 0\}$, where φ is strictly plurisubharmonic near bD, then $\bar D = \{\varphi \le 0\}$ is a Stein compactum, and $\zeta \in bD$ satisfies $\varphi(\zeta) = 0$. Now apply Theorem 1.13. ∎

1.6. The Runge Property for Strictly Plurisubharmonic Exhaustions

Proposition 1.15. *Let $D \subset \mathbb{C}^n$ be open and let $\varphi: D \to \mathbb{R}$ be a C^2 function such that for some real numbers $b^\# < c^\#$ φ is strictly plurisubharmonic on $\varphi^{-1}((b^\#, c^\#))$, and such that for each $b \in (b^\#, c^\#)$ the set $L_b = \{z \in D: \varphi(z) \leq b\}$ is compact. Then, for any $b, c \in \mathbb{R}$ with $b^\# < b < c < c^\#$, L_b is holomorphically convex in $D_c = \{z \in D: \varphi(z) < c\}$, i.e.,*

$$(\hat{L}_b)_{\mathcal{O}(D_c)} = L_b.$$

Proof. By Theorem V.1.5, D_c is holomorphically convex. Therefore $K = (\hat{L}_b)_{\mathcal{O}(D_c)} \subset D_c$ is compact. Let $\eta = \max\{\varphi(z): z \in K\}$; then $b \leq \eta < c$, and we must show $b = \eta$. L_η is a pseudoconvex compactum and hence Stein, by Corollary V.1.6.

Fix $\zeta \in K$, such that $\varphi(\zeta) = \eta$. Then ζ is a strictly pseudoconvex boundary point of L_η. Hence, by Theorem 1.13, there is a peaking function $h \in \mathcal{O}(L_\eta)$ for ζ. Suppose $b < \eta$; then $L_b \subset L_\eta - \{\zeta\}$, which implies $|h|_{L_b} < 1 = h(\zeta)$. Notice that $\hat{K}_{\mathcal{O}(D_c)} = K$ and $h \in \mathcal{O}(K)$, so that by Theorem 1.10 we can approximate h uniformly on K by functions in $\mathcal{O}(D_c)$. It follows that there is $f \in \mathcal{O}(D_c)$ with $|f|_{L_b} < |f(\zeta)|$. But this contradicts $\zeta \in K = (\hat{L}_b)_{\mathcal{O}(D_c)}$. So we must have $b = \eta$. ■

The main result of this section is now an easy consequence of Proposition 1.15.

Theorem 1.16. *Let φ and D satisfy the hypotheses of Proposition* 1.15. *Then, for any $b, c \in \mathbb{R}$ with $b^\# < b < c < c^\#$, the region D_b is Runge in D_c.*

Proof. Since both D_b and D_c are Stein, it is enough to verify condition (iv) in Theorem 1.11. Let K be a compact set in D_b. Choose $a \in \mathbb{R}$ such that $b^\# < a < b$ and $K \subset L_a$. By Proposition 1.15, $\hat{K}_{\mathcal{O}(D_c)} \subset (\hat{L}_a)_{\mathcal{O}(D_c)} = L_a$, so $\hat{K}_{\mathcal{O}(D_c)} \subset D_b$. ■

1.7. The Levi Problem for Pseudoconvex Domains

It is now very easy to extend Theorem V.1.5 to arbitrary pseudoconvex domains.

Theorem 1.17. *Let $D \subset \mathbb{C}^n$ be a pseudoconvex domain. Then D is Stein.*

Proof. By definition (cf. II.§2.10) there is a C^2 strictly plurisubharmonic exhaustion function $\varphi: D \to \mathbb{R}$. So each open set $D_c = \{z \in D: \varphi(z) < c\}$ is Stein and, by Theorem 1.16, D_c is Runge in $D_{c'}$ for $c < c'$. The conclusion now follows from Theorem 1.12. ■

In §2 we will give an alternate proof of Theorem 1.17 which is independent of Theorem 1.12.

1.8. Convexity with Respect to Plurisubharmonic Functions

Recall from Chapter II, §5.1 that a region $D \subset \mathbb{C}^n$ is convex with respect to the class $PS(D)$ of plurisubharmonic functions (PS-convex) if for each compact set $K \subset D$

$$\hat{K}_{PS(D)} = \{z \in D: u(z) \leq \sup_K u \quad \text{for all } u \in PS(D)\}$$

is relatively compact in D. We can now show that PS-convexity is really the same as holomorphic convexity in the following strong sense.

Theorem 1.18. *Let $D \subset \mathbb{C}^n$ be PS-convex and suppose that $K \subset D$ is compact. Then*

$$\hat{K}_{PS(D)} = \hat{K}_{\mathcal{O}(D)}.$$

Proof. Obviously $\hat{K}_{PS(D)} \subset \hat{K}_{\mathcal{O}(D)}$. To prove the reverse inclusion, suppose $w \in D - \hat{K}_{PS(D)}$; we will show that $w \notin \hat{K}_{\mathcal{O}(D)}$. We know that D is pseudoconvex, and by Theorem II.5.11 there is a C^∞ strictly plurisubharmonic exhaustion function φ for D such that $\varphi < 0$ on K and $a = \varphi(w) > 0$. By Theorem 1.13 there is $h \in \mathcal{O}(\{\varphi \leq a\})$ such that $|h| < 1$ on K and $h(w) = 1$. Choose $c > a$ such that $h \in \mathcal{O}(D_c)$. By Theorems 1.16 and 1.12 the region D_c is Runge in D. By approximating h by functions in $\mathcal{O}(D)$ we therefore can find $f \in \mathcal{O}(D)$ such that $|f|_K < |f(w)|$, that is, $w \notin \hat{K}_{\mathcal{O}(D)}$. ■

The following result is a useful generalization of the Runge property for a strictly plurisubharmonic exhaustion function proved in Theorem 1.16. Note that the hypothesis on D is of course satisfied for Stein domains.

Corollary 1.19. *Let $D \subset \mathbb{C}^n$ be PS-convex and suppose φ is an arbitrary plurisubharmonic function on D. For $c \in \mathbb{R}$ set $D_c = \{z \in D: \varphi(z) < c\}$. Then D_c is Runge in $D_{c'}$ for all $c < c'$.*

Proof. It is clear that each D_c is PS-convex and hence Stein by Theorem 1.18. Suppose $c < c'$ and let $K \subset D_c$ be compact. By Theorem 1.11 it suffices to show that $\hat{K}_{\mathcal{O}(D_{c'})} \subset D_c$. But this is obvious, since $\hat{K}_{\mathcal{O}(D_{c'})} = \hat{K}_{PS(D_{c'})}$ by Theorem 1.18, and the latter set is trivially contained in D_c. ■

Exercises

E.1.1. Suppose $K \subset \mathbb{C}^n$ is Stein and $h_1, \ldots, h_l \in \mathcal{O}(K)$. Define $\mu: K \to K \times \mathbb{C}^l$ by $\mu(z) = (z, h_1(z), \ldots, h_l(z))$, and let $\Delta = \{z \in \mathbb{C}: |z| < 1\}$. Show that for every $f \in \mathcal{O}(K')$, where $K' = \mu^{-1}(K \times \bar{\Delta}^l)$, there is $F \in \mathcal{O}(K \times \bar{\Delta}^l)$, such that $f = F \circ \mu$ on K.

E.1.2. Use Runge's Approximation Theorem to prove that a compact set $K \subset \mathbb{C}$ is polynomially convex if and only if $\mathbb{C} - K$ is connected.

E.1.3. Prove Lemma 1.7.

E.1.4. Let K be a compact set in $\mathbb{C}^n$ and let $\mathscr{P}(K) \subset C(K)$ be the closure of the holomorphic polynomials in $C(K)$ under uniform convergence on K.

(i) Show that $\mathscr{P}(K)$ is a closed $\mathbb{C}$-subalgebra of $C(K)$.
(ii) Show that every $\mathbb{C}$-algebra homomorphism $\Phi: \mathscr{P}(K) \to \mathbb{C}$ is continuous with norm 1.

E.1.5. Let K and $\mathscr{P}(K)$ be as in E.1.4.

(i) Show that every $z \in \hat{K}_{\mathscr{P}}$ defines a (unique) continuous homomorphism $\Phi_z: \mathscr{P}(K) \to \mathbb{C}$ with $\Phi_z(g) = g(z)$ for every holomorphic polynomial g.
(ii) Show that every continuous $\mathbb{C}$-algebra homomorphism $\Phi: \mathscr{P}(K) \to \mathbb{C}$ is of the form Φ_z for a unique $z \in \hat{K}_{\mathscr{P}}$. (Continuity is automatic, by E.1.4.)
(iii) Show that the algebras $\mathscr{P}(K)$ and $\mathscr{P}(\hat{K}_{\mathscr{P}})$ are isometrically isomorphic.
(iv) Show that $\hat{K}_{\mathscr{P}}$ is the largest compact set in $\mathbb{C}^n$ for which (i) and (iii) hold.

Remark. (i) and (ii) show that $\hat{K}_{\mathscr{P}}$ can be identified as the "maximal ideal space" of the Banach-algebra $\mathscr{P}(K)$, i.e., the space of all $\mathbb{C}$-algebra homomorphisms $\mathscr{P}(K) \to \mathbb{C}$.

E.1.6. Prove Theorem 1.12.

E.1.7. Let $D \subset \mathbb{C}^n$ be Stein and let φ be a continuous plurisubharmonic function on D. Suppose that $K = \{z \in D: \varphi(z) \leq 0\} \subset\subset D$. Prove that every $f \in \mathcal{O}(K)$ can be approximated uniformly on K by functions in $\mathcal{O}(D)$.

E.1.8. Suppose D is Stein and $K \subset D$ is a Stein compactum such that $\mathcal{O}(D)$ is dense in $\mathcal{O}(K)$. Show that $\hat{K}_{\mathcal{O}(D)} = K$. (Hint: Compare with Theorem 1.8.)

E.1.9. Suppose $D_1 \subset D_2$ is a Runge pair in $\mathbb{C}^n$. Show that $f \in \mathcal{O}^*(D_1)$ is the limit of a sequence $\{f_j\} \subset \mathcal{O}^*(D_2)$ in the topology of $\mathcal{O}(D_1)$ if and only if there is a sequence of continuous invertible functions on D_2 which converges to f compactly on D_1.

§2. $\bar{\partial}$-Cohomological Characterization of Stein Domains

2.1. The Solution of $\bar{\partial}$ on Pseudoconvex Domains

If $D \subset \mathbb{C}^n$ is pseudoconvex, we know, by Theorem 1.17, that D is Stein and hence, by Theorem 1.4, one can solve the $\bar{\partial}$-equation on D. Here we want to discuss an alternate proof, independent of Theorem 1.17.

Theorem 2.1. *Let $D \subset \mathbb{C}^n$ be pseudoconvex. Then*

$$H^q_{\bar{\partial}}(D) = 0 \qquad \textit{for } q \geq 1.$$

Proof. Let φ be a C^2 strictly plurisubharmonic exhaustion function for D. If $L_j = \{z \in D: \varphi(z) \leq j\}$, then $H^q_{\bar\partial}(L_j) = 0$ for $q \geq 1$ and $j = 1, 2, \ldots$ (see Chapter V, §2). Furthermore, by Proposition 1.15,

$$L_j \subset (\hat{L}_j)_{\mathcal{O}(L_{j+1})} \subset (\hat{L}_j)_{\mathcal{O}(D_{j+2})} = L_j$$

for $j = 1, 2, \ldots$; by Corollary 1.2, $\mathcal{O}(L_{j+1})$ is dense in $\mathcal{O}(L_j)$. The desired conclusion now follows by applying Lemma 1.3. ■

2.2. Extension of $\bar\partial$-Closed Forms

The following Lemma is a $\bar\partial$-cohomology version of Proposition V.2.1. Its proof is based on the same idea.

Lemma 2.2. *Let D be open in $\mathbb{C}^n$ and set $D_1 = \{z \in D: z_1 = 0\}$. Denote by ι the inclusion $D_1 \to D$. If $q \geq 0$ and $H^{q+1}_{\bar\partial}(D) = 0$, then for every $\bar\partial$-closed $C^\infty_{0,q}$ form f on D_1 there is $F \in C^\infty_{0,q}(D)$ with $\bar\partial F = 0$ and $\iota^* F = f$ on D_1.*

Proof. Denote by π the projection $\pi(z) = (0, z_2, \ldots z_n)$. D_1 and $D - \pi^{-1}(D_1)$ are two disjoint closed subsets of D, so there is $\chi \in C^\infty(D)$ such that $\chi \equiv 1$ in a neighborhood of D_1 and $\chi \equiv 0$ in a neighborhood of $D - \pi^{-1}(D_1)$. Then $g = \chi \cdot \pi^* f \in C^\infty_{0,q}(D)$, and since $\bar\partial g = \bar\partial \chi \wedge \pi^* f \equiv 0$ in a neighborhood of D_1, $z_1^{-1} \bar\partial g$ defines a smooth $\bar\partial$-closed $(0, q + 1)$-form on D. Since $H^{q+1}_{\bar\partial}(D) = 0$, there is $u \in C^\infty_{0,q}(D)$ with

$$\bar\partial u = z_1^{-1} \cdot \bar\partial g.$$

It follows that

$$F = \chi \cdot \pi^* f - z_1 u$$

is $\bar\partial$-closed on D and

$$\iota^* F = (\chi \circ \iota) \cdot \iota^* \pi^* f - 0 = (\pi \circ \iota)^* f = f. \quad ■$$

Corollary 2.3. *With D and D_1 as in Lemma 2.2, assume that $H^q_{\bar\partial}(D) = H^{q+1}_{\bar\partial}(D) = 0$ for some $q \geq 1$. Then $H^q_{\bar\partial}(D_1) = 0$.*

Proof. If $f \in C^\infty_{0,q}(D_1)$ and $\bar\partial f = 0$, choose F as in Lemma 2.2. Since $H^q_{\bar\partial}(D) = 0$, there is $u \in C^\infty_{0,q-1}(D)$ with $\bar\partial u = F$. Then $\iota^* u \in C^\infty_{0,q-1}(D_1)$ and

$$\bar\partial(\iota^* u) = \iota^*(\bar\partial u) = \iota^* F = f. \quad ■$$

2.3. The Characterization

We now prove the main result of this section.

Theorem 2.4. *The following statements are equivalent for a domain $D \subset \mathbb{C}^n$:*

(i) *D is pseudoconvex;*
(ii) *$H^q_{\bar\partial}(D) = 0$ for $q = 1, 2, \ldots, n$;*

(iii) *D is holomorphically convex, i.e., D is Stein;*
(iv) *D is locally Stein, i.e., each $z \in \bar{D}$ has a neighborhood U such that $U \cap D$ is Stein.*

PROOF. (i) ⇒ (ii) by Theorem 2.1, (iii) ⇒ (iv) is trivial, and (iv) implies that D is locally pseudoconvex, hence (i) holds. We now prove (ii) ⇒ (iii) by induction on the dimension n.

For $n = 1$ the implication is true since every $D \subset \mathbb{C}$ is Stein; so let $n > 1$ and assume the implication has been proved for regions in $\mathbb{C}^{n-1}$. Let H be a complex $(n-1)$-dimensional hyperplane in $\mathbb{C}^n$. After a linear change of coordinates we can assume that $H \cap D = D_1 = \{z \in D: z_1 = 0\}$. If D satisfies (ii), Corollary 2.3 implies $H^q_{\bar\partial}(D_1) = 0$ for $q = 1, \ldots, n-1$, and therefore, by inductive hypotheses, $D_1 \subset \mathbb{C}^{n-1}$ is Stein. So if $\zeta \in bD_1$, by Proposition II.3.4, there is $f \in \mathcal{O}(D_1)$ with $\lim \sup_{D_1 \ni z \to \zeta} |f(z)| = \infty$. We now apply Lemma 2.2 with $q = 0$ to obtain $F \in \mathcal{O}(D)$ with $F|_{D_1} = f$; hence F is unbounded at $\zeta \in bD$. Finally notice that the set of boundary points $\zeta \in bD$ which are boundary points of $H \cap D$ for some hyperplane H—though not necessarily equal to bD—is dense in bD. By Corollary II.3.20 it then follows that D is a domain of holomorphy and hence Stein. ∎

EXERCISES

E.2.1. Show: If $D \subset \mathbb{C}^n$ and $\dim_{\mathbb{C}} H^q_{\bar\partial}(D) < \infty$ for $q \geq 1$, then D is Stein. (Hint: Prove Lemma 2.2 under the above hypothesis; compare with the proof of Proposition V.2.1.)

E.2.2. Use Theorem 2.4 to prove the theorem of Behnke and Stein: If $D_1 \subset \ldots D_j \subset D_{j+1}$ is an increasing sequence of Stein domains in $\mathbb{C}^n$, then $D = \bigcup_{j=1}^{\infty} D_j$ is Stein.

E.2.3. Suppose $D \subset \mathbb{C}^n$ and $H^q_{\bar\partial}(D) = 0$ for some $q \geq 1$. Show that $H^{p,q}_{\bar\partial}(D) = 0$ for all $p \geq 0$.

E.2.4. Suppose $D \subset \mathbb{C}^n$ and that there are a neighborhood U of bD and $\varphi \in PS(U)$, such that $U \cap D = \{z \in U: \varphi(z) < 0\}$. Show that D is Stein.

E.2.5. Show that $H^n_{\bar\partial}(D) = 0$ for every open set $D \subset \mathbb{C}^n$.

§3. Topological Properties of Stein Domains

This section is mainly intended for readers with some knowledge of algebraic topology. It may be omitted without loss of continuity. On the other hand, even the less experienced reader may find it instructive to see how complex analytic properties of a region in $\mathbb{C}^n$ impose topological restrictions. A familiar example is the topological characterization of Runge domains in the complex plane (see Theorem 1.6).

3.1. De Rham Cohomology and Holomorphic Forms

For $r = 0, 1, 2, \ldots$, let $\Omega^r(D) = \{f \in C^\infty_{r,0}(D): \bar{\partial} f = 0\}$. Also, set $\Omega^{-1}(D) = \{0\}$. Differential forms in Ω^r have holomorphic coefficients, and therefore such forms are also called **holomorphic** ***r*-forms**. In particular, $\Omega^0(D) = \mathcal{O}(D)$. Notice that

$$df = \partial f \qquad \text{for } f \in \Omega^r.$$

The following fact is obvious:

(3.1) *A function $f \in C^\infty(D)$ is holomorphic if and only if $df \in \Omega^1(D)$.*

If interpreted suitably, this characterization of holomorphic 0-forms generalizes to d-cohomology classes of higher degree.

Lemma 3.1. *Suppose $H^q_{\bar{\partial}}(D) = 0$ for $1 \le q \le r$. Then $f \in C^\infty_r(D)$ is d-cohomologous to a holomorphic r-form if and only if $df \in \Omega^{r+1}(D)$.*

Proof. If $f - dg \in \Omega^r(D)$ for some $g \in C^\infty_{r-1}(D)$, then clearly $df \in \Omega^{r+1}$. To prove the reverse implication we must show

(3.2) *If $df \in \Omega^{r+1}(D)$, there is $g \in C^\infty_{r-1}(D)$ such that $f - dg \in \Omega^r(D)$.*

Every $f \in C^\infty_r$ can be written as

$$(3.3) \qquad f = \sum_{q=0}^{k} f_{r-q,q} \text{ for some } k \text{ with } 0 \le k \le r \qquad \text{where } f_{r-q,q} \in C^\infty_{r-q,q}(D).$$

We prove (3.2) by induction on the integer k appearing in (3.3).

Write f as in (3.3). Notice that $\bar{\partial} f_{r-k,k}$ is the only term of type $(r-k, k+1)$ in df; so, if $df \in \Omega^{r+1}$, we must have $\bar{\partial} f_{r-k,k} = 0$. If $k = 0$, this implies that $f \in \Omega^r$, i.e., (3.2) is true for such f. This proves the inductive beginning. If $k \ge 1$, we use $H^k_{\bar{\partial}}(D) = 0$ and Exercise E.2.3 to find $u \in C^\infty_{r-k,k-1}$ with $\bar{\partial} u = f_{r-k,k}$. Then

$$(3.4) \qquad f - du = \sum_{q=0}^{k-1} f_{r-q,q} - \partial u.$$

Since ∂u is of type $(r-k+1, k-1)$, (3.4) and the inductive hypothesis imply that (3.2) holds for $f - du$. So there is $g \in C^\infty_{r-1}(D)$ with $(f - du) - dg \in \Omega^r(D)$, i.e., $f - d(u + g) \in \Omega^r(D)$. This proves (3.2) for f. ∎

Since Lemma 3.1 applies, in particular, to Stein domains, it is now easy to see that the de Rham cohomology groups $H^r_d(D)$ of Stein domains can be calculated in terms of holomorphic forms.

Theorem 3.2. *Let $D \subset \mathbb{C}^n$ be a Stein domain. Then*

$$H^r_d(D) \simeq \{f \in \Omega^r(D): df = 0\}/d\Omega^{r-1}$$

for $r \ge 0$.

PROOF. The case $r = 0$ is trivial. If $r > 0$, consider the homomorphism

$$\alpha_r\colon \{f \in \Omega^r\colon df = 0\} \to H_d^r(D),$$

where $\alpha_r(f)$ is the d-cohomology class of f in $H_d^r(D)$. Lemma 3.1 implies that α_r is onto. Since $d\Omega^{r-1} \subset \ker \alpha_r$, the theorem will follow once we show $\ker \alpha_r \subset d\Omega^{r-1}$. Suppose $f \in \Omega^r$ is in $\ker \alpha_r$; then $f = du$ for some $u \in C_{r-1}^\infty(D)$. If $r - 1 = 0$, (3.1) implies that u is holomorphic; if $r - 1 > 0$, Lemma 3.1 applied to u shows that u may be replaced by a holomorphic $(r-1)$-form. In either case we see that $f \in d\Omega^{r-1}$. ∎

3.2. The Cohomology Groups of Stein Domains

By de Rham's Theorem, if M is a differentiable manifold, $H_d^r(M)$ is isomorphic to the cohomology group $H^r(M, \mathbb{C})$ for $r \geq 0$. So Theorem 3.2 implies the following holomorphic version of de Rham's Theorem.

Theorem 3.3. *Let $D \subset \mathbb{C}^n$ be a Stein domain. Then*

$$H^r(D, \mathbb{C}) \simeq \{f \in \Omega^r(D)\colon df = 0\}/d\Omega^{r-1}$$

for $r \geq 0$. In particular,

$$H^r(D, \mathbb{C}) = 0 \qquad \text{for } r > n.$$

Remark 3.4. It is a theorem in algebraic topology that

$$H^r(D, \mathbb{C}) = \operatorname{Hom}_{\mathbb{Z}}(H_r(D, \mathbb{Z}) \to \mathbb{C});$$

so Theorem 3.3 implies that for a Stein domain D, $H_r(D, \mathbb{Z})$ does not contain any free element if $r > n$ (such a cycle would define a nontrivial homomorphism $H_r(D, \mathbb{Z}) \to \mathbb{C}$). By using Morse Theory, one can even show that for a Stein domain $D \subset \mathbb{C}^n$

$$H_r(D, \mathbb{Z}) = H^r(D, \mathbb{Z}) = 0$$

if $r > n$ and that $H_n(D, \mathbb{Z})$ is free (see Andreotti–Fraenkel [AnFr] or Milnor [Mil], §7).

On the other hand, the cohomology groups $H^r(D, \mathbb{C})$ of Stein domains do not need to vanish for $r \leq n$.

Example 3.5. Let $D = \{z \in \mathbb{C}^n\colon 0 < |z_j| < 2, 1 \leq j \leq n\}$. D is Stein, since it is a product of planar domains. For $1 \leq r \leq n$ define the r-cycle γ_r in D by $\gamma_r(t_1, \ldots, t_r) = (e^{it_1}, \ldots, e^{it_r}, 1, \ldots, 1)$, $0 \leq t_j \leq 2\pi$; the holomorphic r-form

$$f_r = \frac{dz_1 \wedge \ldots \wedge dz_r}{z_1 \ldots z_r}$$

on D is d-closed. Since

(3.5) $$\int_{\gamma_r} f_r = (2\pi i)^r \neq 0,$$

f_r is not the differential of a holomorphic $(r-1)$-form on D. So, by Theorem 3.3, $H^r(D, \mathbb{C}) \neq 0$ for $1 \leq r \leq n$. Equation (3.5) also shows that γ_r is not a boundary in D, so that $H_r(D, \mathbb{Z}) \neq 0$ as well.

3.3. Runge Domains

A Runge domain $D \subset \mathbb{C}^1$ is necessarily simply connected and therefore $H^1(D, \mathbb{C}) = 0$. This result generalizes to higher dimensions as follows.

Theorem 3.6. *Let $D \subset \mathbb{C}^n$ be a Runge domain. Then*

$$H^r(D, \mathbb{C}) = 0 \qquad \text{for } r \geq n.$$

Proof. Since, in particular, D is Stein, by Theorem 3.3 it is enough to show that every holomorphic n-form f on D is the differential du of a holomorphic $(n-1)$-form on D. Suppose $f = h\, dz_1 \wedge \ldots \wedge dz_n \in \Omega^n(\mathbb{C}^n)$. Then

$$u(z_1, z_2, \ldots, z_n) = \left(\int_0^{z_1} h(\zeta, z_2, \ldots, z_n)\, d\zeta\right) dz_2 \wedge \ldots \wedge dz_n$$

is in $\Omega^{n-1}(\mathbb{C}^n)$ and $du = f$. So $\Omega^n(\mathbb{C}^n)|_D = d\Omega^{n-1}(\mathbb{C}^n)|_D \subset d\Omega^{n-1}(D)$. The Runge property of D implies that $\Omega^n(\mathbb{C}^n)|_D$, and hence $d\Omega^{n-1}(D)$, is dense in $\Omega^n(D)$ with respect to the natural topology on $\Omega^n(D)$. It is a general fact, valid for arbitrary differentiable manifolds M, that the exterior derivative $d\colon C_r^\infty(M) \to C_{r+1}^\infty(M)$ has closed range (see [Hör 2], §7.5). Therefore, $\Omega^n(D) =$ closure of $d\Omega^{n-1}(D) \subset dC_{n-1}^\infty(D)$. Lemma 3.1 then implies that $\Omega^n(D) = d\Omega^{n-1}(D)$. ∎

Exercises

E.3.1. Given $1 \leq k < n$, find a domain of holomorphy $D_k \subset \mathbb{C}^n$ such that $H^r(D_k, \mathbb{C}) \neq 0$ for $1 \leq r \leq k$, while $H^r(D_k, \mathbb{C}) = 0$ for $r > k$.

§4. Meromorphic Functions and the Additive Cousin Problem

Already in 1895 P. Cousin ("Sur les fonctions de n variables complexes," Acta Math. **19**(1985), 1–62) considered the generalization to several variables of the theorems of M.G. Mittag-Leffler and K. Weierstrass on the existence of meromorphic functions with prescribed poles and zeros. By using the Cauchy Integral Formula for product domains, Cousin was able to solve the "patching process" which is the major obstacle in several variables, and consequently he obtained the desired generalizations for such domains. Because of a lack of analogues of the Cauchy Integral, Cousin's method could not be extended to more general domains for a long time. Following H. Cartan ([Car], 471–473), these problems became known as Cousin's First and Second Problem (or simply Cousin I and II). In the next two sections we will show how the solution of $\bar{\partial}$ leads to the solution of these problems on domains of holomorphy.

4.1. Germs of Holomorphic Functions

A meromorphic function is an object which is locally the quotient of holomorphic functions. In order to make this concept precise it is convenient to introduce the language of germs of functions. In essence, the *germ* of a function contains the local information, while it disregards the domain of definition. Informally, this idea is used constantly. A typical example is the statement *the function f is holomorphic at the point* $z \in \mathbb{C}^n$.

We fix a point $a \in \mathbb{C}^n$. A **function element** (f, U) **at** a consists of a function $f: U \to \mathbb{C}$ defined on an open neighborhood U of a. Two function elements (f, U) and (g, V) at a are equivalent (at a) if there is an open neighborhood $W \subset U \cap V$ of a, such that $f|_W = g|_W$. The set of equivalence classes of function elements at a is called the set of **germs of (complex valued) functions at** a and is denoted by $\mathscr{F}_a$. The equivalence class of (f, U) at a is denoted by $\mathbf{f}_a$, and (f, U) is called a representative of the germ $\mathbf{f}_a$. A function element (f, U) defines germs $\mathbf{f}_z$ for each $z \in U$. Since $\mathscr{F}_z \cap \mathscr{F}_{z'} = \emptyset$ for $z \neq z'$, we have $\mathbf{f}_z \neq \mathbf{f}_{z'}$ for $z \neq z'$.

A germ $\mathbf{f}_a \in \mathscr{F}_a$ is continuous (of class C^k, $0 \leq k \leq \infty$, holomorphic) if $\mathbf{f}_a$ has a representative (f, U) with $f \in C(U)$ ($C^k(U)$, respectively $\mathcal{O}(U)$). We denote the corresponding sets of germs by $\mathscr{C}_a$, $\mathscr{C}_a^k$, and $\mathcal{O}_a$. If the dimension n of $\mathbb{C}^n$ matters, we will write ${}_n\mathcal{O}_a$ for $\mathcal{O}_a$. Obviously one has the following chain of strict inclusions (where $1 \leq k < \infty$)

$$\mathcal{O}_a \subset \mathscr{C}_a^\infty \subset \ldots \subset \mathscr{C}_a^{k+1} \subset \mathscr{C}_a^k \subset \ldots \subset \mathscr{C}_a^0 = \mathscr{C}_a \subset \mathscr{F}_a.$$

The main advantage of working with germs rather than function elements is that sets of germs, and especially the set ${}_n\mathcal{O}_a$ of holomorphic germs, have interesting algebraic properties which are useful for studying local properties of holomorphic functions and analytic sets. There is, of course, a natural way to define addition and multiplication of germs at a, which turns $\mathscr{F}_a$, $\mathscr{C}_a^k$, and $\mathcal{O}_a$ into commutative rings (in fact $\mathbb{C}$-algebras) with identity. A continuous germ $\mathbf{f}_a \in \mathscr{C}_a$ is invertible if and only if $f(a) \neq 0$. (The value $f(a)$ of the germ $\mathbf{f}_a$ is independent of the particular representative (f, U) of $\mathbf{f}_a$.) So $\mathscr{C}_a$(and similarly C_a^k and $\mathcal{O}_a$) has a unique maximal ideal $m_a = \{\mathbf{f}_a \in \mathscr{C}_a : f(a) = 0\}$.

The rings $\mathscr{C}_a^k$, $0 \leq k \leq \infty$, have many zero divisors (see Exercise E.4.2). Therefore the following result reflects a special property of holomorphic functions (see also Theorem I.1.20 for a global version).

Theorem 4.1. *The ring ${}_n\mathcal{O}_a$ of germs of holomorphic functions at $a \in \mathbb{C}^n$ is an integral domain.*

PROOF. We must show that if $\mathbf{f}_a \cdot \mathbf{g}_a = \mathbf{0}_a$ for two germs $\mathbf{f}_a, \mathbf{g}_a \in \mathcal{O}_a$, then at least one of them is zero. Suppose $\mathbf{f}_a \neq \mathbf{0}_a$. We can then choose representatives (f, U) and (g, U), where $U = P(a, r)$ is a sufficiently small polydisc, such that

$$f(z) \cdot g(z) = 0 \qquad \text{for } z \in U, \tag{4.1}$$

while there is $z_0 \in U$ with $f(z_0) \neq 0$. By continuity, $f(z) \neq 0$ for z in some neighborhood $W \subset U$ of z_0. By (4.1), $g(z) = 0$ for all $z \in W$. Hence, by the Identity Theorem I.1.19, $g(z) = 0$ for all $z \in U$, i.e., $\mathbf{g}_a = \mathbf{0}_a$. ∎

We leave it to the reader to supply the proof of the following concrete representation of $\mathcal{O}_a$.

Proposition 4.2. *The ring ${}_n\mathcal{O}_a$ is isomorphic to the ring of convergent power series in $(z_1 - a_1), \ldots, (z_n - a_n)$. Moreover, ${}_n\mathcal{O}_a$ is isomorphic to ${}_n\mathcal{O}_0$ for all $a \in \mathbb{C}^n$.*

We present without proof two more properties of $\mathcal{O}_a$ which are fundamental for the study of analytic sets. The proofs of these results make use of the Weierstrass Preparation Theorem. The interested reader may consult any of the standard references, for example [Nar 1] and Chapter 6 in [Hör 2].

Theorem 4.3. *$\mathcal{O}_a$ is a Noetherian unique factorization domain.*

Theorem 4.4. *If $\mathbf{f}_a$ and $\mathbf{g}_a \in \mathcal{O}_a$ are relatively prime, then there are representatives (f, U) and (g, U) such that $\mathbf{f}_z$ and $\mathbf{g}_z$ are relatively prime for all $z \in U$.*

Remark 4.5. In case of dimension $n = 1$, Theorems 4.3 and 4.4 are easy consequences of the fact that every nontrivial $\mathbf{f}_a \in {}_1\mathcal{O}_a$ has a unique representation

$$\mathbf{f}_a = (\mathbf{z} - \mathbf{a})_a^q \cdot \mathbf{u}_a$$

where $\mathbf{u}_a$ is a unit and q is an integer ≥ 0.

4.2. Meromorphic Functions

By Theorem 4.1, the ring ${}_n\mathcal{O}_a$ has no nontrivial zero divisors. Therefore ${}_n\mathcal{O}_a$ has a well defined quotient field, which we denote by ${}_n\mathcal{M}_a$ or simply $\mathcal{M}_a$. $\mathcal{M}_a$ is called the **field of germs of meromorphic functions at** a. Explicitly, a meromorphic germ $\mathbf{m}_a \in \mathcal{M}_a$ can be represented as a quotient

$$\mathbf{m}_a = \frac{\mathbf{f}_a}{\mathbf{g}_a},$$

where $\mathbf{f}_a, \mathbf{g}_a \in \mathcal{O}_a$, and $\mathbf{g}_a \neq \mathbf{0}_a$. We now define meromorphic functions by gluing together meromorphic germs in a special way.

Definition. Let $D \subset \mathbb{C}^n$ be open. A **meromorphic function** m on D is a mapping

$$m: D \to \bigcup_{z \in D} \mathscr{M}_z, \qquad z \to \mathbf{m}_z$$

with the following properties:

(i) $$\mathbf{m}_z \in \mathscr{M}_z \qquad \textit{for } z \in D;$$
(ii) for every $a \in D$ there are a connected neighborhood $U \subset D$ and holomorphic functions $f, g \in \mathcal{O}(U)$ with $g \not\equiv 0$, such that $\mathbf{m}_z = \mathbf{f}_z/\mathbf{g}_z$ for all $z \in U$.

The set of all meromorphic functions on D is denoted by $\mathscr{M}(D)$.

If $f, g \in \mathcal{O}(D)$, and $g \not\equiv 0$ on any connected component of D, then the quotient $m = f/g$ defines a meromorphic function $m \in \mathscr{M}(D)$ by setting $\mathbf{m}_z = \mathbf{f}_z/\mathbf{g}_z$. In particular, $\mathcal{O}(D) \subset \mathscr{M}(D)$. Conversely, if $m \in \mathscr{M}(D)$ with $\mathbf{m}_z \in \mathcal{O}_z$ for all $z \in D$, then m is locally represented by holomorphic function elements, and hence m defines a global holomorphic function $m \in \mathcal{O}(D)$.

It is obvious that $\mathscr{M}(D)$ carries the structure of a ring.

Proposition 4.6. *$\mathscr{M}(D)$ is a field if and only if D is connected.*

The proof is left as an exercise for the reader.

In case of dimension $n = 1$, meromorphic germs have a very simple representation: every $\mathbf{m}_a \in {}_1\mathscr{M}_a$ can be written uniquely as

$$\mathbf{m}_a = (\mathbf{z} - \mathbf{a})_a^q \cdot \mathbf{u}_a,$$

where $\mathbf{u}_a$ is a unit in $\mathcal{O}_a$ and $q \in \mathbb{Z}$. It follows that the singularities of a meromorphic function $m \in \mathscr{M}(D)$ are isolated if $D \subset \mathbb{C}^1$. Moreover, the only singularities are poles, that is, points at which the function has the limit ∞ in the Riemann sphere.

In case of two or more variables the singularities of meromorphic functions are more complicated. For example, if $m = z_1/z_2 \in \mathscr{M}(\mathbb{C}^2)$, every $a \in \mathbb{C} \cup \{\infty\}$ is a limit point of m at $(0, 0)$. Moreover, the singularities are never isolated.

Theorem 4.7. *Let $D \subset \mathbb{C}^n$ be open and suppose $m \in \mathscr{M}(D)$. Then the singular set $S(m) = \{z \in D: \mathbf{m}_z \notin \mathcal{O}_z\}$ is either empty or an analytic set of dimension $n - 1$.*

PROOF. Notice that the case $n = 1$ is trivial. For the general case, let $a \in D$ and write $\mathbf{m}_a = \mathbf{f}_a/\mathbf{g}_a$, where, by Theorems 4.3 and 4.4 we can assume that $f, g \in \mathcal{O}(U)$, and $\mathbf{f}_z$ and $\mathbf{g}_z$ are relatively prime for all $z \in U$. This implies that for $z \in U$, $\mathbf{m}_z \notin \mathcal{O}_z$ precisely if $g(z) = 0$, hence

$$S(m) \cap U = \{z \in U: g(z) = 0\}.$$

So $S(m)$ is locally the zero set of a single nontrivial holomorphic function; hence $S(m)$ is an analytic set, and, if $S(m) \neq \varnothing$, $\dim S(m) = n - 1$ by Theorem I.3.12. ∎

4.3. The Mittag-Leffler Problem and the Additive Cousin Problem

We now discuss the construction of global meromorphic functions with prescribed local singularities. Let us first recall the classical one variable Theorem of Mittag-Leffler.

Let $\{a_j: j = 1, 2, \ldots\}$ be a locally finite subset of the region $D \subset \mathbb{C}^1$. Given finite order principal parts $p_j = \sum_{\nu=1}^{n_j} c_\nu^j (z - a_j)^{-\nu}$ at each point a_j, there is $m \in \mathscr{M}(D)$ which is holomorphic on D except at the points a_j, and such that $m - p_j$ is holomorphic at a_j for $j = 1, 2, \ldots$.

In several variables one must reformulate the problem in order to ensure that the singularities of the given principal parts match up—in $\mathbb{C}^1$ they always do, since the singularities are isolated points. We now state the classical formulation of P. Cousin's First Problem.

The additive Cousin problem. *Let D be a region in $\mathbb{C}^n$ and let $\{U_j, j \in J\}$ be an open covering of D. Suppose the meromorphic functions $m_j \in \mathscr{M}(U_j)$ satisfy*

$$m_j - m_i \in \mathcal{O}(U_i \cap U_j) \qquad \text{for all } i, j \in J; \tag{4.2}$$

find $m \in \mathscr{M}(D)$ such that

$$m - m_j \in \mathcal{O}(U_j) \qquad \text{for each } j \in J. \tag{4.3}$$

A collection of functions $\{m_j \in \mathscr{M}(U_j), j \in J\}$ which satisfies (4.2) is called a **Cousin I distribution** on D.

Clearly the Mittag-Leffler Theorem is equivalent to the solution of the additive Cousin problem for $D \subset \mathbb{C}^1$. We will solve the additive Cousin problem by reducing it to the following theorem which involves only holomorphic functions. This sort of reduction was already used by Cousin in his solution of the additive problem for product domains in $\mathbb{C}^n$.

Theorem 4.8. *Let D be a region in $\mathbb{C}^n$ such that $H^1_{\bar\partial}(D) = 0$. Let $\{U_j: j \in J\}$ be an open covering of D. If the functions $g_{ij} \in \mathcal{O}(U_i \cap U_j)$, $i, j \in J$, satisfy*

$$g_{ij} = -g_{ji}; \qquad g_{ij} + g_{jk} + g_{ki} = 0 \text{ whenever } U_i \cap U_j \cap U_k \neq \varnothing, \tag{4.4}$$

then there are functions $g_j \in \mathcal{O}(U_j)$ such that

$$g_{ij} = g_j - g_i \qquad \text{for all } i, j \in J \text{ with } U_i \cap U_j \neq \varnothing. \tag{4.5}$$

Before proving Theorem 4.8, we give its most important application, first proved in 1937 by K. Oka ([Oka],II). Notice that it includes the solution of the Mittag-Leffler problem.

Theorem 4.9. *Let $D \subset \mathbb{C}^n$ be Stein (it is enough to assume $H^1_{\bar\partial}(D) = 0$). Then every additive Cousin problem on D has a solution.*

Proof of Theorem 4.9. Suppose the functions $m_j \in \mathcal{M}(U_j)$ satisfy (4.2). It follows that the functions $g_{ij} = m_j - m_i$ satisfy (4.4) in Theorem 4.8. Let $g_j \in \mathcal{O}(U_j)$, $j \in J$, be the functions given by Theorem 4.8. By (4.5), $g_j - g_i = g_{ij} = m_j - m_i$ on $U_i \cap U_j$, or, equivalently,

$$m_j - g_j = m_i - g_i \qquad \text{on } U_i \cap U_j. \tag{4.6}$$

So there is a unique $m \in \mathcal{M}(D)$ with $\mathbf{m}_z = (\mathbf{m}_j - \mathbf{g}_j)_z$ for $z \in U_j$. Clearly $m - m_j = -g_j \in \mathcal{O}(U_j)$ for each $j \in J$. ∎

Proof of Theorem 4.8. The proof involves two steps: first we find functions $v_j \in C^\infty(U_j)$, $j \in J$, which satisfy (4.5) (this works for *any* region D); then we use $H^1_{\bar\partial}D) = 0$ to change the v_j's into holomorphic functions.

Step 1. Let $\{V_\alpha, \alpha \in I\}$ be a locally finite refinement of the given covering $\{U_j, j \in J\}$, with refinement map $\tau: I \to J$, so that $V_\alpha \subset U_{\tau(\alpha)}$. Choose a partition of unity $\{\chi_\alpha \in C_0^\infty(V_\alpha), \alpha \in I\}$. For any $\alpha \in I$ and $j \in J$ define

$$(\chi_\alpha \cdot g_{\tau(\alpha)j})(z) = \begin{cases} \chi_\alpha(z) g_{\tau(\alpha)j}(z) & \text{if } z \in V_\alpha \cap U_j; \\ 0 & \text{if } z \in U_j - V_\alpha; \end{cases}$$

then $\chi_\alpha \cdot g_{\tau(\alpha)j} \in C^\infty(U_j)$. Since $\{V_\alpha\}$ is locally finite, the sum

$$v_j = \sum_{\alpha \in I} \chi_\alpha g_{\tau(\alpha)j} \tag{4.7}$$

defines a C^∞ function on U_j. On $U_i \cap U_j$ we obtain, using (4.4),

$$\begin{aligned} v_j - v_i &= \sum_{\alpha \in I} \chi_\alpha (g_{\tau(\alpha)j} - g_{\tau(\alpha)i}) \\ &= \sum_{\alpha \in I} \chi_\alpha (g_{ij}) = g_{ij}, \end{aligned} \tag{4.8}$$

since $\sum \chi_\alpha \equiv 1$.

Step 2. The functions v_j found above are not holomorphic, but by (4.8) and $\bar\partial g_{ij} = 0$,

$$\bar\partial v_i = \bar\partial v_j \text{ on } U_i \cap U_j \qquad \text{for all } i, j \in J. \tag{4.9}$$

So the locally defined $C^\infty_{0,1}$ forms $\bar\partial v_j$ patch together to a global form $f \in C^\infty_{0,1}(D)$ with $f = \bar\partial v_j$ on U_j. Clearly $\bar\partial f = 0$ on D. Since $H^1_{\bar\partial}(D) = 0$, there is $u \in C^\infty(D)$ with $\bar\partial u = f$. Then $g_j = v_j - u$ satisfies $\bar\partial g_j = 0$, so $g_j \in \mathcal{O}(U_j)$, and, by (4.8), on $U_i \cap U_j$ one has

$$g_j - g_i = (v_j - u) - (v_i - u) = v_j - v_i = g_{ij}. \quad ∎$$

4.4. Examples

We have seen that $H^1_{\bar\partial}(D) = 0$ is a sufficient condition for the solution of the additive Cousin problem on D. A domain with this property need not be Stein. For example, $H^1_{\bar\partial}(\mathbb{C}^n - \{0\}) = 0$ for $n \geq 3$ (see Exercise E.4.7), and, by Hartogs'

Theorem, $\mathbb{C}^n - \{0\}$ is not a domain of holomorphy for $n \geq 2$. On the other hand, contrary to the situation in one variable, the additive Cousin problem cannot be solved on arbitrary domains in higher dimensions. We will verify this for $D = \mathbb{C}^2 - \{0\}$. Since D is not Stein, we know, by Theorem 2.4 and E.2.5, that $H^1_{\bar\partial}(D) \neq 0$.

Consider the covering $\{U_1, U_2\}$ of $\mathbb{C}^2 - \{0\}$ given by $U_j = \{(z_1, z_2) \in \mathbb{C}^2: z_j \neq 0\}, j = 1, 2$. We define $m_j \in \mathscr{M}(U_j)$ by $m_1 = 1/z_1 z_2$ and $m_2 = 0$. Then

$$m_1 - m_2 = \frac{1}{z_1 z_2} \in \mathcal{O}(U_1 \cap U_2),$$

so the compatibility condition (4.2) is satisfied. Suppose $m \in \mathscr{M}(\mathbb{C}^2 - 0)$ satisfies $m - m_j \in \mathcal{O}(U_j)$ for $j = 1, 2$. It follows that $m \in \mathcal{O}(U_2)$. By Theorem II.1.5, m can be represented by a Laurent series

$$m = \sum a_{\nu_1 \nu_2} z_1^{\nu_1} z_2^{\nu_2} \qquad \text{on } U_2$$

with $a_{\nu_1 \nu_2} = 0$ whenever $\nu_1 < 0$. On the other hand,

$$m - m_1 = \sum_{\nu_1 \geq 0} a_{\nu_1 \nu_2} z_1^{\nu_1} z_2^{\nu_2} - \frac{1}{z_1 z_2} \tag{4.10}$$

would have to be holomorphic on U_1. This is not possible since the term $-(z_1 z_2)^{-1}$ on the right side of (4.10) is not cancelled by any other term in the sum. So the additive Cousin problem for the Cousin I distribution $\{m_1, m_2\}$ on $\mathbb{C}^2 - \{0\}$ has no solution.

Exercises

E.4.1. (i) Show that for each $k = 0, 1, \ldots, \infty$ a germ $\mathbf{f}_a \in \mathscr{C}^k_a$ at $a \in \mathbb{R}^n$ is invertible if and only if $f(a) \neq 0$.
(ii) Show that $\mathscr{C}^k_a$ has a unique maximal ideal.
(iii) Show that (i) and (ii) hold also for the ring ${}_n\mathcal{O}_a$, $a \in \mathbb{C}^n$.
(iv) Show that (i) and (ii) are false for the ring $\mathscr{F}_a$, for $a \in \mathbb{R}^n$ with $n \geq 1$.

E.4.2. Show that $\mathscr{C}^k_a$, $a \in \mathbb{R}^n$, has nontrivial zero divisors.

E.4.3. (i) Let $D = \{z \in \mathbb{C}: 0 < |z| < 1\}$. Show that a function $f \in \mathcal{O}(D)$ is meromorphic at 0 according to the definition in §4.2 if and only if f has a pole or removable singularity at 0.
(ii) Generalize (i) to $\mathbb{C}^n$ as follows: Let $D \subset \mathbb{C}^n$ be a Reinhardt domain with center 0, such that $0 \in bD$. Show that a Laurent series $\sum_{\nu \in \mathbb{Z}^n} a_\nu z^\nu$ which converges on D defines a meromorphic function at 0 if and only if there are only finitely many multi-indices $\nu \in \mathbb{Z}^n$ with $\nu \notin \mathbb{N}^n$ and $a_\nu \neq 0$.

E.4.4. Prove Proposition 4.6.

E.4.5. Show that if the conclusion of Theorem 4.8 holds on the region $D \subset \mathbb{C}^n$, then $H^1_{\bar\partial}(D) = 0$. (Hint: Use $H^1_{\bar\partial}(\bar{B}) = 0$ for every ball $B \subset\subset D$.)

E.4.6. (i) For $j = 1, 2, 3$ set $U_j = \{z \in \mathbb{C}^3: z_j \neq 0\}$. Use Laurent series to show that every Cousin I distribution $\{m_j \in \mathscr{M}(U_j), j = 1, 2, 3\}$ for the cover $\{U_1, U_2, U_3\}$ of $\mathbb{C}^3 - \{0\}$ has a solution. (Hint: Prove a special case of Theorem 4.8.)
(ii) Use (i) to show that $H^1_{\bar\partial}(\mathbb{C}^3 - \{0\}) = 0$. (Hint: U_j is Stein and see E.4.5.)

E.4.7. Generalize the methods used in E.4.6 to show that

(i) $H^q_{\bar\partial}(\mathbb{C}^n - \{0\}) = 0$ for $q \neq 0, n-1$,
(ii) $\dim_{\mathbb{C}} H^{n-1}_{\bar\partial}(\mathbb{C}^n - \{0\}) = \infty$.

E.4.8. Let $m \in \mathscr{M}(D)$ and let $S(m) \subset D$ be its singular set (cf. Theorem 4.7). A point $a \in S(m)$ is called a **pole** for m if

$$\lim_{\substack{z \to a \\ z \notin S(m)}} m(z) = \infty,$$

and a **point of indeterminacy** for m if m has no limit at a in the closed Riemann sphere. Show that if $\mathbf{m}_a = \mathbf{f}_a/\mathbf{g}_a$, where $\mathbf{f}_a, \mathbf{g}_a \in \mathcal{O}_a$, and if a is a point of indeterminacy, then $f(a) = g(a) = 0$. (The converse is true also, provided $\mathbf{f}_a$ and $\mathbf{g}_a$ are relatively prime, but is much deeper.)

§5. Holomorphic Functions with Prescribed Zeroes

5.1. The Multiplicative Cousin Problem

We now turn to the problem of finding global holomorphic functions with prescribed zeroes. Since the zeroes of holomorphic functions of two or more variables are never isolated, prescribing the local data for zero sets will involve compatibility conditions analogous to those which led us from the Mittag-Leffler problem to the formulation of the additive Cousin problem. We denote by $\mathcal{O}^*(U)$ the set of invertible functions $f \in \mathcal{O}(U)$; $f \in \mathcal{O}^*(U)$ if and only if $f \in \mathcal{O}(U)$ and $f(z) \neq 0$ for all $z \in U$. Similarly, $C^*(U)$ and $\mathscr{M}^*(U)$ denote the sets of invertible elements in $C(U)$, respectively $\mathscr{M}(U)$. Notice that $f \in \mathscr{M}^*(U)$ if and only if $f \in \mathscr{M}(U)$ and f is not identically zero on any component of U.

Problem Z. *Given an open covering $\{U_j, j \in J\}$ of a region $D \subset \mathbb{C}^n$, and holomorphic functions $f_j \in \mathcal{O}(U_j)$, not identically zero on any component of U_j, which satisfy*

$$f_j f_i^{-1} \in \mathcal{O}^*(U_i \cap U_j) \tag{5.1}$$

whenever $U_i \cap U_j \neq \varnothing$, find a global holomorphic function $f \in \mathcal{O}(D)$ such that $f f_j^{-1} \in \mathcal{O}^*(U_j)$ for all $j \in J$.

The classical Weierstrass Theorem in one variable is a special case of the solution of the above problem for $D \subset \mathbb{C}^1$. In fact, suppose $D \subset \mathbb{C}^1$, $\{a_j: j = 1, 2, \ldots\} \subset D$ is locally finite in D, and n_j is some positive integer for each $j = 1, 2, \ldots$. Choose neighborhoods U_j of a_j, such that $U_i \cap U_j = \varnothing$ for $i, j \geq 1$ and define $f_j = (z - a_j)^{n_j} \in \mathcal{O}(U_j)$. Let $U_0 = D - \{a_j: j = 1, 2, \ldots\}$ and set $f_0 \equiv 1$ on U_0. Clearly the collection of functions $f_j \in \mathcal{O}(U_j)$, $j = 0, 1, 2, \ldots$ so defined satisfies (5.1). The solution f of Problem Z for this data is a holomorphic function on D which has a zero of order n_j at a_j for $j = 1, 2, \ldots$, and no other zeroes on D.

More generally, in his classical 1895 paper, P. Cousin considered the problem of finding *meromorphic* functions with locally prescribed zeroes and poles. This is now referred to as the multiplicative Cousin problem or Cousin II problem.

The multiplicative Cousin problem. *Let* $\{U_j, j \in J\}$ *be an open covering of the region* $D \subset \mathbb{C}^n$ *and suppose the functions* $m_j \in \mathscr{M}^*(U_j), j \in J$, *satisfy*

$$m_j m_i^{-1} \in \mathscr{O}^*(U_i \cap U_j) \tag{5.2}$$

whenever $U_i \cap U_j \neq \varnothing$. *Find a global meromorphic function* $m \in \mathscr{M}(D)$, *such that*

$$m m_j^{-1} \in \mathscr{O}^*(U_j) \tag{5.3}$$

for all $j \in J$.

A collection of functions $\{m_j \in \mathscr{M}^*(U_j)\}$ which satisfies (5.2) is called a **Cousin II distribution** on D. Notice that a solution m of a *holomorphic* Cousin II distribution (i.e., $m_j \in \mathscr{O}(U_j)$) is necessarily holomorphic by (5.3). So the solution of the multiplicative Cousin problem includes the solution of Problem Z.

Formally, the multiplicative Cousin problem is the exact multiplicative analogon of the additive Cousin problem. However, there is a profound difference which shows up only in dimension ≥ 2. In his 1895 paper, P. Cousin had solved both the additive and the multiplicative problems on products of planar domains, but he must have overlooked the fact that his argument in the multiplicative case required all domains but one to be simply connected. This was pointed out in 1917 by T.H. Gronwall ("On expressibility of uniform functions of several complex variables as quotient of two functions of entire character," *Trans. Amer. Math. Soc.* **18**(1917), 50–64). Gronwall gave an example of a product of two (multiply) connected planar domains on which the multiplicative problem is not always solvable. So, in contrast to what we have seen in §4 for the additive problem, the multiplicative Cousin problem is not universally solvable on Stein domains in two or more variables.

The situation remained obscure for a long time. In 1934 it was still unknown whether the multiplicative problem could be solved on every simply connected domain of holomorphy (see H. Behnke and P. Thullen [BeTh], p. 102). Finally, in 1953, J.P. Serre ([Ser 1], p. 63) produced a counterexample in this case as well. In the meantime though, K. Oka ([Oka], III) had proved in 1939 a remarkable theorem which showed that on Stein domains D the obstruction to the solution of the multiplicative problem is a topological one—obviously not the fundamental group $\pi_1(D)$, since by the general Weierstrass Theorem the problem is solvable on *any* region $D \subset \mathbb{C}^1$. This topological obstruction was investigated further by K. Stein, first in 1941 [SteK 1], and then 10 years later in his pioneering paper [SteK 2], in which he introduced the class of "Stein manifolds" and gave a homological characterization of the obstruction. In

1953, J.P. Serre [Ser 1] gave these results their definitive form in terms of the cohomology group $H^2(D, \mathbb{Z})$—notice that $H^2(D, \mathbb{Z}) = 0$ for all regions D in the complex plane (cf. Remark 3.4).

In §5.3 we will prove Oka's Theorem using a simple variant of the original formulation of the topological obstruction, and in §5.4 we will present an elementary proof for the vanishing of this obstruction in some special cases, including all regions in $\mathbb{C}^1$. In §6.4 we will use the language of sheaf cohomology in order to translate Oka's topological condition into the cohomological one referred to above.

5.2. Oka's Counterexample

We first discuss an example of Oka which shows that Problem Z does not always have a solution, even for Stein domains.

Define the domain $D \subset \mathbb{C}^2$ by

$$D = \{z \in \mathbb{C}^2 \colon 3/4 < |z_j| < 5/4, j = 1, 2,\}.$$

D is a product domain and hence Stein. Let

$$A = \{z \in D \colon z_2 - z_1 + 1 = 0\}.$$

Notice that $A \cap \{z \in D \colon \operatorname{Im} z_1 = 0\} = \varnothing$, so that $Y^{\pm} = A \cap \{z \in D \colon \operatorname{Im} z_1 \gtrless 0\}$ are disjoint closed subvarieties of D with $A = Y^+ \cup Y^-$. We will show that the Cousin II distribution given by

$$\begin{aligned} f_1 &= z_2 - z_1 + 1 \qquad && \text{on } U_1 = D - Y^- \\ f_2 &= 1 && \text{on } U_2 = D - Y^+ \end{aligned} \tag{5.4}$$

has no solution on D (notice that $f_2 f_1^{-1} \in \mathcal{O}^*(U_1 \cap U_2)$). This will be a simple consequence of properties of the winding number of closed curves in $\mathbb{C}^1$, which we briefly recall (for more details, see [Ahl], 4.2.1). Given a C^1 function $\varphi \colon [0, 2\pi] \to \mathbb{C} - \{0\}$ with $\varphi(0) = \varphi(2\pi)$, the winding number (around 0)—or "change in argument"—of φ is defined by

$$N(\varphi) = \frac{1}{2\pi i} \int_0^{2\pi} \frac{\varphi'(\theta)}{\varphi(\theta)}\, d\theta.$$

$N(\varphi)$ is always an integer. In fact, consider

$$h(s) = \varphi(s) \exp\left[-\int_0^s \frac{\varphi'(\theta)}{\varphi(\theta)}\, d\theta \right];$$

then $h'(s) = 0$ on $(0, 2\pi)$, hence $h(2\pi) = h(0) = \varphi(0)$. Since

$$h(2\pi) = \varphi(2\pi) \exp(-2\pi i N(\varphi)) = \varphi(0) \exp(-2\pi i N(\varphi)),$$

it follows that $\exp(-2\pi i N(\varphi)) = 1$.

If $\{\varphi_t \colon t \in I\}$ is a continuous family of closed C^1 curves in $\mathbb{C} - \{0\}$ on the interval I (i.e., $\varphi_t(\theta)$ and $\varphi_t'(\theta)$ are continuous on $[0, 2\pi] \times I$), then $N(\varphi_t)$ is continuous in $t \in I$, and hence, since integer valued, constant in t. Furthermore, if $\varphi(\theta) = g(e^{i\theta})$, where g is C^1 and $\neq 0$ on $\{z \in \mathbb{C}: |z| \leq 1\}$, then $N(g(e^{i\theta})) = 0$.

Returning to our example, let us first consider the curves $f_1(\zeta, e^{i\theta}) = e^{i\theta} - \zeta + 1$ for $\zeta = 1$ and $\zeta = -1$. Since $f_1(1, e^{i\theta}) = e^{i\theta}$ and $f_1(-1, e^{i\theta}) = e^{i\theta} + 2$, it follows that

$$N(f_1(-1, e^{i\theta})) = 0 \neq 1 = N(f_1(1, e^{i\theta})). \tag{5.5}$$

Assuming that the Cousin II distribution (5.4) has a solution on D, we will obtain a contradiction to (5.5).

In fact, suppose there is $f \in \mathcal{O}(D)$ such that $f = f/f_2 \in \mathcal{O}^*(U_2)$, and $h = f/f_1 \in \mathcal{O}^*(U_1)$. By considering the continuous family of closed curves $\varphi_t(\theta) = f(e^{it}, e^{i\theta})$ for $0 \leq \theta \leq 2\pi$ and $-\pi \leq t \leq 0$ one obtains

$$N(f(-1, e^{i\theta})) = N(\varphi_{-\pi}) = N(\varphi_0) = N(f(1, e^{i\theta})), \tag{5.6}$$

since $f \neq 0$ on U_2. Similarly, since $h(e^{it}, e^{i\theta}) \neq 0$ for $0 \leq \theta \leq 2\pi$ and $0 \leq t \leq \pi$, it follows that

$$N(h(-1, e^{i\theta})) = N(h(1, e^{i\theta})). \tag{5.7}$$

From $f = h \cdot f_1$ on U_1, it follows that

$$N(f(\zeta, e^{i\theta}) = N(h(\zeta, e^{i\theta})) + N(f_1(\zeta, e^{i\theta}))$$

for $\zeta = 1$ and $\zeta = -1$. Therefore, by (5.6) and (5.7),

$$N(f_1(-1, 0^{i\theta})) = N(f_1(1, e^{i\theta})),$$

which contradicts (5.5). ∎

Remark. The above argument does not use at all the fact that the solution f is *holomorphic*; clearly it is sufficient to assume $f \in C^1(D)$ and that $f/f_1 \in C^1(U_1)$ is not zero on U_1. The winding number can also be defined for curves which are only continuous, so that the Cousin II distribution (5.4) does not even have a continuous solution. In fact, this will follow immediately from the non-existence of a holomorphic solution and Oka's Theorem (Theorem 5.2 below).

5.3. Reduction to the Additive Problem

It is natural to attempt to solve the multiplicative Cousin problem by taking logarithms and solving the resulting additive problem, as in §4. Obviously one cannot define $\log m_j$ for $m_j \in \mathcal{M}^*(U_j)$, even if U_j is simply connected, unless $m_j \in \mathcal{O}^*(U_j)$. So, the proper place to start is with the functions $g_{ij} = m_j m_i^{-1}$, which, according to (5.2), are in $\mathcal{O}^*(U_i \cap U_j)$. Assuming $U_i \cap U_j$ is simply connected, in order to apply Theorem 4.8 to the functions $\{\log g_{ij}\}$ one must choose the branches of the logarithms so that

$$c_{ijk} = \log g_{ij} + \log g_{jk} + \log g_{ki} = 0 \text{ on } U_i \cap U_j \cap U_k. \tag{5.8}$$

Since $g_{ij} \cdot g_{jk} \cdot g_{ki} = 1$, one sees that, in general, the best one can say is that $c_{ijk} = n_{ijk} \cdot 2\pi\sqrt{-1}$ for some integer n_{ijk}. How to achieve $n_{ijk} = 0$ appears to be a major difficulty. In fact, in general this cannot be done unless D satisfies a suitable topological condition. For the time being we bypass this obstacle by introducing a seemingly more elementary condition.

Theorem 5.1. *Let $D \subset \mathbb{C}^n$ satisfy $H^1_{\bar\partial}(D) = 0$, and suppose $\mathcal{U} = \{U_j : j \in J\}$ is an open covering of D. If $g_{ij} \in \mathcal{O}^*(U_i \cap U_j)$, $i, j \in J$, satisfy*

$$(5.9) \qquad g_{ij} = g_{ji}^{-1}; \qquad g_{ij} \cdot g_{jk} \cdot g_{ki} = 1$$

whenever $U_i \cap U_j \cap U_k \neq \emptyset$, then there are holomorphic functions $g_j \in \mathcal{O}^*(U_j)$ which satisfy

$$(5.10) \qquad g_{ij} = g_j \cdot g_i^{-1} \text{ on } U_i \cap U_j \qquad \text{for all } i, j \in J$$

if and only if there are nonvanishing continuous functions $g_j \in C^*(U_j)$ which satisfy (5.10).

Except for the hypothesis of the existence of a continuous solution, Theorem 5.1 is the precise multiplicative analog of Theorem 4.8. Before proving Theorem 5.1 we want to discuss one of its main applications.

Theorem 5.2. ([Oka], III). *Let D be a Stein domain in $\mathbb{C}^n$ (it is enough to assume $H^1_{\bar\partial}(D) = 0$). Then Problem Z has a holomorphic solution if and only if there is a continuous solution, i.e., there is $c \in C(D)$, such that $c \cdot f_j^{-1} \in C^*(U_j)$ for each $j \in J$.*

Historically, Theorem 5.2 was the first example of what is now known as the "Oka principle": On Stein domains, problems which have local holomorphic solutions will have a *global* holomorphic solution provided there is a continuous global solution. Stated differently, on Stein domains the obstruction for passing from local to global analytic statements is purely topological. Another famous example of the Oka principle was found by H. Grauert in 1956; we state only one part of it. *On Stein domains a holomorphic vector bundle is holomorphically trivial if and only if it is topologically trivial.* Grauert's results are much more general; the interested reader should consult the exposition of H. Cartan in [Car], p. 752–776.

PROOF OF THEOREM 5.2 (assuming Theorem 5.1). Only one implication is nontrivial. Let $f_j \in \mathcal{O}(U_j)$ be given, such that $g_{ij} = f_j f_i^{-1} \in \mathcal{O}^*(U_i \cap U_j)$ (cf. (5.1)). If $c \in C(D)$ is a continuous solution, we define $c_j = c/f_j \in C^*(U_j)$. Then $c_j c_i^{-1} = g_{ij}$ on $U_i \cap U_j$. So the hypotheses of Theorem 5.1 are satisfied, and we conclude that there are functions $g_j \in \mathcal{O}^*(U_j)$ with $g_j g_i^{-1} = g_{ij}$. It follows that

$$(5.11) \qquad f_j g_j^{-1} = f_i g_i^{-1} \text{ on } U_i \cap U_j \qquad \text{for all } i, j \in J.$$

Therefore we can define $f \in \mathcal{O}(D)$ by setting $f = f_j g_j^{-1}$ on U_j. Since $f/f_j = g_j^{-1} \in \mathcal{O}^*(U_j)$, f is the desired holomorphic solution. ∎

Proof of Theorem 5.1. Again, only one implication is nontrivial. We first assume that the open sets U_j are simply connected for all $j \in J$. If $c_j \in C^*(U_j)$ satisfies $c_j/c_i = g_{ij}$ on $U_i \cap U_j$, we define $h_j = \log c_j$ on U_j by choosing some branch of the logarithm. Set

$$f_{ij} = h_j - h_i \text{ whenever } U_i \cap U_j \neq \emptyset; \tag{5.12}$$

then

$$\exp f_{ij} = \exp h_j/\exp h_i = c_j/c_i = g_{ij} \in \mathcal{O}^*(U_i \cap U_j), \tag{5.13}$$

so $f_{ij} \in \mathcal{O}(U_i \cap U_j)$. Equation (5.12) implies that the conditions (4.4) in Theorem 4.8 are satisfied (notice how the difficulty of choosing the right branches for $\log g_{ij}$ has disappeared), so we obtain functions $f_j \in \mathcal{O}(U_j)$ for $j \in J$, such that $f_j - f_i = f_{ij}$. It follows that $g_j = \exp f_j \in \mathcal{O}^*(U_j)$, and, by (5.13),

$$g_j/g_i = \exp f_j/\exp f_i = \exp f_{ij} = g_{ij} \text{ on } U_i \cap U_j.$$

We now consider the case of an arbitrary covering $\{U_j, j \in J\}$. Choose a locally finite refinement $\{V_\alpha, \alpha \in I\}$ of $\{U_j\}$ by simply connected open sets V_α (for example, V_α could be a ball). Let $\tau: I \to J$ be the refinement map. For $V_\alpha \cap V_\beta \neq \emptyset$, define $g'_{\alpha\beta} = g_{\tau(\alpha)\tau(\beta)}|_{V_\alpha \cap V_\beta}$; then $\{g'_{\alpha\beta}\}$ satisfies (5.9) with respect to the covering $\{V_\alpha\}$. If $\{c_j\}$ is a continuous solution for the original problem and $V_\alpha \cap V_\beta \neq \emptyset$, we have $g'_{\alpha\beta} = c'_\beta/c'_\alpha$, with $c'_\alpha = c_{\tau(\alpha)}$. So, by the first part of the proof there are holomorphic functions $g'_\alpha \in \mathcal{O}^*(V_\alpha)$, $\alpha \in I$, which satisfy

$$g'_{\alpha\beta} = g'_\beta/g'_\alpha \text{ on } V_\alpha \cap V_\beta. \tag{5.14}$$

Suppose $V_\alpha \cap V_\beta \cap U_j \neq \emptyset$; since $V_\alpha \cap V_\beta \cap U_j \subset U_{\tau(\alpha)} \cap U_{\tau(\beta)} \cap U_j$, (5.9) implies

$$g_{\tau(\alpha)\tau(\beta)} = g_{j\tau(\beta)}/g_{j\tau(\alpha)}.$$

Together with (5.14) this implies

$$g'_\beta g_{\tau(\beta)j} = g'_\alpha g_{\tau(\alpha)j}$$

on $V_\alpha \cap V_\beta \cap U_j$, so we can define $g_j \in \mathcal{O}^*(U_j)$ by setting $g_j = g'_\alpha g_{\tau(\alpha)j}$ on $V_\alpha \cap U_j$. Finally, on $V_\alpha \cap U_i \cap U_j$,

$$\begin{aligned} g_j/g_i &= g'_\alpha g_{\tau(\alpha)j}/g'_\alpha g_{\tau(\alpha)i} = g_{i\tau(\alpha)} \cdot g_{\tau(\alpha)j} \\ &= g_{ij}, \end{aligned}$$

where we have used again (5.9); this proves (5.10). ∎

5.4. The Topological Obstruction

In order to apply Theorem 5.1 or 5.2 one needs to find domains on which the respective problems have continuous solutions. We now single out the relevant topological condition and verify it for certain regions in $\mathbb{C}^n$.

Definition. A topological space X has the **Oka property** if for any open covering $\{U_j: j \in J\}$ of X and functions $c_{ij} \in C^*(U_i \cap U_j)$ which satisfy

$$(5.15) \qquad c_{ij} = c_{ji}^{-1}; \qquad c_{ij} \cdot c_{jk} \cdot c_{ki} = 1 \text{ on } U_i \cap U_j \cap U_k,$$

there are functions $c_j \in C^*(U_j)$ such that

$$(5.16) \qquad c_j/c_i = c_{ij} \qquad \text{for all } i, j \in J \text{ with } U_i \cap U_j \neq \varnothing.$$

We can now state and prove a more general version of Theorem 5.2.

Theorem 5.3. *Suppose $D \subset \mathbb{C}^n$ has the Oka property and $H^1_{\bar\partial}(D) = 0$. Then every multiplicative Cousin problem on D has a solution.*

Proof. Given an open covering $\{U_j, j \in J\}$ of D and functions $m_j \in \mathcal{M}^*(U_j)$ with $g_{ij} = m_j m_i^{-1} \in \mathcal{O}^*(U_i \cap U_j)$, the Oka property implies the existence of functions $c_j \in C^*(U_j)$ with $c_j c_i^{-1} = g_{ij}$. So Theorem 5.1 applies, and one can proceed as in the proof of Theorem 5.2. ∎

In order to apply Theorem 5.3 one needs some simple sufficient conditions for the Oka property. Notice that the Oka example in §5.2 (a product of two annuli) does not have the Oka property, otherwise Theorem 5.3 would apply. So the following result is optimal for product domains in $\mathbb{C}^n$.

Theorem 5.4. *Let D_ν, $1 \le \nu \le n$, be open subsets of $\mathbb{C}^1$ and suppose that all but one of them are simply connected. Then $D = D_1 \times D_2 \times \cdots \times D_n \subset \mathbb{C}^n$ has the Oka property. In particular, every region $D \subset \mathbb{C}^1$ has the Oka property.*

Remark 5.5. It is shown in §6.4 that for paracompact manifolds X, and hence, in particular, for regions in $\mathbb{C}^n$, the Oka property for X is equivalent to the vanishing of the Čech cohomology group $\check{H}^2(X, \mathbb{Z})$; for such X the latter group agrees with the singular cohomology group $H^2(X, \mathbb{Z})$. So Theorem 5.4 would follow from standard results in algebraic topology. The proof presented here, essentially due to Oka ([Oka], III), is more elementary. More importantly, the method of proof is of fundamental importance for proving vanishing theorems for sheaf cohomology groups. Since we want to keep the proof of Theorem 5.4 independent of the results outlined in §6, the language of sheaf cohomology will not be used explicitly at this point.

For the proof of Theorem 5.4 we fix an open covering $\{U_j, j \in J\}$ of D and functions $c_{ij} \in C^*(U_i \cap U_j)$ which satisfy (5.15). We say that a set $K \subset D$ has *property S* (with respect to $\{c_{ij}\}$) if there is a solution of (5.16) on some open neighborhood $W \subset D$ of K, i.e., if there are functions $c_j \in C^*(U_j \cap W)$, $j \in J$, such that $c_j/c_i = c_{ij}$ on $U_i \cap U_j \cap W$. By a sequence of Lemmas we first prove property S for certain compact subsets K of D. By an exhaustion procedure we will then obtain property S for D itself, completing the proof of the Theorem.

The key step of the proof is contained in the following "patching" lemma.

Lemma 5.6. *Let $K = K_1 \times \cdots \times K_n \subset D$, where $K_1 \subset \mathbb{C}$ is a compact set whose boundary consists of finitely many compact intervals parallel to the coordinate axis, and $K_\nu \subset \mathbb{C}$ is compact and simply connected for $\nu \geq 2$. Let $\lambda \in \mathbb{R}$. If $K' = \{z \in K: \operatorname{Re} z_1 \leq \lambda\}$ and $K'' = \{z \in K: \operatorname{Re} z_1 \geq \lambda\}$ have property S, so does $K = K' \cup K''$.*

Proof of Lemma 5.6. Let W_μ, $\mu = 1, 2$, be neighborhoods of K' and K'' on which we have solutions $c_j^{(\mu)} \in C^*(W_\mu \cap U_j)$ of (5.16). Since $K' \cap K'' = \{z_1 \in K_1: \operatorname{Re} z_1 = \lambda\} \times K_2 \times \cdots \times K_n$ is a finite union of disjoint, compact, simply connected sets, we may choose W_1 and W_2 so that $W_{12} = W_1 \cap W_2$ is a simply connected (though not necessarily connected) neighborhood of $K' \cap K''$. If $V = W_{12} \cap U_j \cap U_i \neq \varnothing$, one has

$$c_j^{(1)}/c_i^{(1)} = c_{ij} = c_j^{(2)}/c_i^{(2)} \text{ on } V, \tag{5.17}$$

and hence

$$c_j^{(2)}/c_j^{(1)} = c_i^{(2)}/c_i^{(1)} \text{ on } V.$$

It follows that there is a function $F \in C^*(W_{12})$ with $F = c_j^{(2)}/c_j^{(1)}$ on $W_{12} \cap U_j$ for each $j \in J$. Choose a branch of $\log F$ on W_{12}, and write $\log F = u_2 - u_1$, where $u_\mu \in C(W_\mu)$, $\mu = 1, 2$. Then $F_\mu = \exp u_\mu \in C^*(W_\mu)$, and

$$F = F_2/F_1 \text{ on } W_{12}. \tag{5.18}$$

We now use (5.18) to patch together the two solutions $\{c_j^{(\mu)}\}$, $\mu = 1, 2$, as follows. If $j \in J$ with $U_j \cap W_{12} \neq \varnothing$, (5.18) implies $c_j^{(2)}/F_2 = c_j^{(1)}/F_1$ on $U_j \cap W_{12}$. So, if $W = W_1 \cup W_2$, there is $c_j \in C^*(U_j \cap W)$ with $c_j = c_j^{(\mu)}/F_\mu$ on $U_j \cap W_\mu$, $\mu = 1, 2$. W is a neighborhood of K, and $\{c_j\}$ is a solution of (5.16) on W. In fact, if $U_i \cap U_j \neq \varnothing$, the equation

$$c_j/c_i = (c_j^{(\mu)}/F_\mu)/(c_i^{(\mu)}/F_\mu) = c_j^{(\mu)}/c_i^{(\mu)} = c_{ij}$$

holds on $U_i \cap U_j \cap W_\mu$ for $\mu = 1, 2$, and hence $c_j/c_i = c_{ij}$ on $U_i \cap U_j \cap W$. ∎

Lemma 5.7. *Let K be as in Lemma 5.6, and let h denote either one of the functions $\operatorname{Re} z_1$, $\operatorname{Im} z_1$. Suppose $K(\lambda) = \{z \in K: h(z) = \lambda\}$ has property S for each $\lambda \in \mathbb{R}$. Then K has property S.*

Proof. Since the geometric assumptions on K_1 in Lemma 5.6 are not changed if K_1 is multiplied by $\sqrt{-1}$, it is enough to prove the case $h(z) = \operatorname{Re} z_1$. Let m be the supremum of the set of those $\lambda \in \mathbb{R}$ for which $\{z \in K: \operatorname{Re} z_1 \leq \lambda\}$ has property S. The Lemma is proved if we show $m = \infty$. Suppose that $m < \infty$. By hypothesis, $K(m)$ has property S, so there is a solution of (5.16) on an open neighborhood W of $K(m)$. Choose $\varepsilon > 0$ such that

$$K'' = \{z \in K: m - \varepsilon \leq \operatorname{Re} z_1 \leq m + \varepsilon\} \subset W.$$

Then K'' and $K' = \{z \in K: \operatorname{Re} z_1 \leq m - \varepsilon\}$ have property S, and, by Lemma 5.6, so does $K' \cup K'' = \{z \in K: \operatorname{Re} z_1 \leq m + \varepsilon\}$. This contradicts the supremum property of m. ∎

Lemma 5.8. *Let $K = K_1 \times \cdots \times K_n \subset D$, where each $K_\nu \subset \mathbb{C}$ is a compact set whose boundary consists of finitely many compact intervals parallel to the coordinate axis. Suppose that all but one of the K_ν's are simply connected. Then K has property S.*

Proof. Without loss of generality assume that K_ν is simply connected for $1 \leq \nu \leq n - 1$. We will prove by induction on l that for $a \in K$ the set

$$L_l(a) = K_1 \times \cdots \times K_l \times \{a_{l+1}\} \times \cdots \times \{a_n\}$$

has property S. The case $l = n$ will prove the Lemma. Now, for $l = 0$, $L_0(a)$ is just the single point $\{a\}$, and property S is trivially satisfied in this case: just choose $j \in J$, such that $a \in U_j$, and define $c_j = 1$, $c_i = c_{ji}$ for $i \neq j$ with $U_i \cap U_j \neq \varnothing$. Then $\{c_k\}$ defines a solution of (5.16) on the neighborhood $W = U_j$ of $L_0(a)$.

Suppose we proved property S for all $L_l(a)$, $a \in K$, where $0 \leq l < n$. Set $Q = L_{l+1}(a)$, and define, for $\xi, \eta \in \mathbb{R}$, $Q(\eta) = \{z \in Q: \operatorname{Im} z_{l+1} = \eta\}$ and $Q(\eta)(\xi) = \{z \in Q(\eta): \operatorname{Re} z_{l+1} = \xi\}$. Then $Q(\eta)(\xi)$ is either empty or equal to $K_1 \times \cdots \times K_l \times \{\xi + i\eta\} \times \{a_{l+2}\} \times \cdots \times \{a_n\}$, and hence has property S, by inductive assumption. An application of Lemma 5.7 with $K = Q(\eta)$ and $h = \operatorname{Re} z_{l+1}$, combined with a permutation of the coordinates, shows that $Q(\eta)$ has property S for each η. Hence, again by Lemma 5.7, now applied to $K = Q$ and $h = \operatorname{Im} z_{l+1}$, $Q = L_{l+1}(a)$ has property S. ∎

We now complete the proof of Theorem 5.4. Without loss of generality we assume that $D = D_1 \times \cdots \times D_n$ is connected and that $D_2, \ldots, D_n$ are simply connected. Choose normal exhaustions $\{D_\nu^{(\kappa)}: \kappa = 1, 2, \ldots\}$ of D_ν for $1 \leq \nu \leq n$, with the following properties:

(5.19) $K^{(\kappa)} = \bar{D}_1^{(\kappa)} \times \cdots \times \bar{D}_n^{(\kappa)}$ *satisfies the hypothesis of* Lemma 5.8, *and hence has property S*;

(5.20) *each function $f \in C^*(K^{(\kappa)})$ has a continuous extension $\tilde{f} \in C^*(K^{(\kappa+1)})$.*

Statement (5.19) is straightforward, and the proof of (5.20) is indicated in Exercise E.5.2.

By induction on κ we will construct solutions $\{c_j^{(\kappa)}\}$ of (5.16) on $K^{(\kappa)}$, such that

$$c_j^{(\kappa+1)} = c_j^{(\kappa)} \text{ on } K^{(\kappa)} \cap U_j. \tag{5.21}$$

For $\kappa = 1$ we choose any solution $\{c_j^{(1)}\}$ on $K^{(1)}$. Suppose we already constructed $\{c_j^{(\nu)}\}$, $1 \leq \nu \leq \kappa$, such that (5.21) holds with κ replaced by $\nu = 1, \ldots, \kappa - 1$. By (5.19) there are functions $b_j \in C^*(U_j \cap K^{(\kappa+1)})$, $j \in J$, which satisfy $b_j/b_i = c_{ij}$ on $K^{(\kappa+1)} \cap U_i \cap U_j$. This implies $b_j/c_j^{(\kappa)} = b_i/c_i^{(\kappa)}$ on $K^{(\kappa)} \cap U_i \cap U_j$.

Let $f \in C^*(K^{(\kappa)})$ be defined by $f = b_j/c_j^{(\kappa)}$ on $K^{(\kappa)} \cap U_j$; by (5.20) there is an extension $\tilde{f} \in C^*(K^{(\kappa+1)})$ of f. Now define $c_j^{(\kappa+1)} = b_j/\tilde{f} \in C^*(K^{(\kappa+1)} \cap U_j)$ for each $j \in J$. Then (5.21) holds, and

$$c_j^{(\kappa+1)}/c_i^{(\kappa+1)} = b_j/b_i = c_{ij} \text{ on } K^{(\kappa+1)} \cap U_i \cap U_j, \tag{5.22}$$

so $\{c_j^{(\kappa+1)}\}$ is a solution on $K^{(\kappa+1)}$.

It now follows from (5.21) that $c_j = \lim_{\kappa\to\infty} c_j^{(\kappa)}$ defines a function in $C^*(U_j)$ for each $j \in J$, and, by (5.22), $c_j/c_i = c_{ij}$ on $U_i \cap U_j$. So D has property S and Theorem 5.4 is proved. ■

Remark. As is well known, in case of *one* complex variable, the solution of the multiplicative Cousin problem (i.e., the general Weierstrass Theorem) can be obtained more directly. Since the zeroes and poles of meromorphic functions in $\mathbb{C}^1$ are isolated points, it is enough to consider special coverings $\{U_j\}$, and one may assume that $m_j \in \mathcal{M}^*(U_j)$ is rational. A solution on a *compact* set K is simply given by a *finite* product of the m_j's. The main difficulty then involves an exhaustion and approximation process. In contrast, the main difficulty in *several* variables appears already when looking for solutions on compact sets. The given *local* distribution cannot be trivially replaced by a global distribution, as in one variable; rather, the local data have to be patched together carefully.

5.5. Divisors

We first review the classical notion of a divisor in one complex variable.

The **divisor** div m of an invertible meromorphic function $m \in \mathcal{M}^*(D)$, $D \subset \mathbb{C}^1$, is the map div m: $D \to \mathbb{Z}$ given by div $m(a) = q_a$ = the order of vanishing of m at a; that is, q_a is the unique integer such that m can be written $m(z) = (z-a)^{q_a}u(z)$, where u is holomorphic at a and $u(a) \neq 0$. Notice that $\{a \in D: \text{div } m(a) \neq 0\}$ is a discrete sequence $\{a_\nu: \nu = 1, 2, \ldots\}$ in D, and that div m is completely determined by the formal series $\sum_\nu q_\nu \cdot a_\nu$, where $q_\nu = \text{div } m(a_\nu)$.

More generally, any map $\delta: D \to \mathbb{Z}$ with $\{a \in D: \delta(a) \neq 0\}$ locally finite in D is called a divisor on D. The set $\mathcal{D}(D)$ of divisors on D is closed under addition (of $\mathbb{Z}$-valued functions), so $\mathcal{D}(D)$ is in fact a commutative group. It is easy to see that

$$\text{div}(f \cdot g) = \text{div } f + \text{div } g \qquad \text{for } f, g \in \mathcal{M}^*(D).$$

A divisor $\delta \in \mathcal{D}(D)$ is called **principal** if $\delta = \text{div } m$ for some $m \in \mathcal{M}^*(D)$. δ is called **positive** if $\delta \geq 0$ on D. Notice that for $m \in \mathcal{M}^*(D)$, div $m \geq 0$ if and only if $m \in \mathcal{O}(D)$.

A fundamental problem in classical complex analysis is the characterization of principal divisors and the description of the **divisor class group** $\mathcal{D}(D)/\text{div } \mathcal{M}^*(D)$. The solution of the multiplicative Cousin problem for arbitrary $D \subset \mathbb{C}^1$ implies that every divisor on D is principal, and the problem is settled.

The situation is, of course, completely different if D is replaced by a compact Riemann surface; it is only in this setting that nonprincipal divisors appear in one complex variable.

As seen by the Oka example, the analogous problems become much more complicated, even for Stein domains in $\mathbb{C}^n$, as soon as $n \geq 2$. Divisors which are not principal appear already in the noncompact case. Geometrically, a divisor of a meromorphic function $m \in \mathcal{M}^*(D)$ describes the zero and singular sets of m, with certain multiplicities attached to the "branches" of these sets. In case $n = 1$, the "branches" are simply isolated points; in general, the "branches" turn out to be analytic sets of dimension $n - 1$ (see Theorem I.3.12 and Theorem 4.7) which are *irreducible* in the following sense: *an analytic set $A \subset D$ is irreducible if A cannot be decomposed as $A = A_1 \cup A_2$, where A_1, A_2 are analytic sets and $A_i \neq A$ for $i = 1, 2$.* One is thus led to define a divisor on $D \subset \mathbb{C}^n$ as a map

$$\delta: \{A \subset D: A \text{ irreducible analytic set, } \dim A = n - 1\} \to \mathbb{Z}$$

with the property that $\{A: \delta(A) \neq 0\}$ is locally finite. Such a map is completely described by the formal series $\sum_\nu q_\nu A_\nu$, where $\delta(A_\nu) = q_\nu$ and $\delta(A) = 0$ for $A \neq A_\nu$, $\nu = 1, 2, \ldots$. However, in order to precisely relate this geometric representation of a divisor to the analytic problems discussed earlier in this paragraph, one needs a fairly detailed description of the local structure of analytic sets. Therefore we will now represent divisors in a different way, which is more suitable for our present needs.

Denote by ${}_n\mathcal{O}_a^*$ and ${}_n\mathcal{M}_a^*$ the (multiplicative) groups of invertible elements in ${}_n\mathcal{O}_a$ and in ${}_n\mathcal{M}_a$, respectively. Clearly $\mathcal{O}_a^*$ is a subgroup of $\mathcal{M}_a^*$. Since the local structure of the zero and singular set of $\mathbf{m} \in \mathcal{M}_a^*$ is not affected if $\mathbf{m}$ is replaced by $\mathbf{m} \cdot \mathbf{u}$, where $\mathbf{u} \in \mathcal{O}_a^*$, the relevant information is captured by the equivalence class $[\mathbf{m}]$ of $\mathbf{m}$ in the quotient group ${}_n\mathcal{M}_a^*/{}_n\mathcal{O}_a^*$.

Definition. A **divisor on the region** $D \subset \mathbb{C}^n$ is a map

$$\delta: D \to \bigcup_{a \in D} {}_n\mathcal{M}_a^*/{}_n\mathcal{O}_a^*$$

with $\delta(a) \in {}_n\mathcal{M}_a^*/{}_n\mathcal{O}_a^*$, which has the following property:

(5.23) *for each $a \in D$ there are a neighborhood $U \subset D$ of a and $m \in \mathcal{M}^*(U)$, such that $\delta(z) = [\mathbf{m}_z]$ for $z \in U$.*

The set of all divisors on D is denoted by $\mathcal{D}(D)$. Every $m \in \mathcal{M}^*(D)$ defines a divisor $\operatorname{div} m \in \mathcal{D}(D)$ by setting $\operatorname{div} m(a) = [\mathbf{m}_a]$ for $a \in D$. Such divisors are called **principal**. A divisor $\delta \in \mathcal{D}(D)$ is **positive** if $\delta(a) \in \mathcal{O}_a/\mathcal{O}_a^*$ for all $a \in D$. Clearly $\mathcal{D}(D)$ inherits the structure of a (multiplicative) commutative group; the map $m \to \operatorname{div} m$ defines a homomorphism div: $\mathcal{M}^*(D) \to \mathcal{D}(D)$.

Notice that in case $n = 1$ the group ${}_1\mathcal{M}_a^*/{}_1\mathcal{O}_a^*$ is particularly simple: the map which assigns to $\mathbf{m} \in {}_1\mathcal{M}_a^*$ its order of vanishing at a defines an isomorphism ${}_1\mathcal{M}_a^*/{}_1\mathcal{O}_a^* \to \mathbb{Z}$. Modulo this canonical isomorphism, the concept of divisor just introduced is, for $n = 1$, identical to the classical one discussed above.

It follows from (5.23) that a divisor on D defines a Cousin II distribution on D (in many different ways). Conversely, any Cousin II distribution defines a unique divisor. It is then obvious that the multiplicative Cousin problem can be reformulated as follows: *When is a divisor $\delta \in \mathscr{D}(D)$ principal?* Similarly, we can reformulate Theorem 5.3:

Theorem 5.3′. *Suppose $D \subset \mathbb{C}^n$ has the Oka property and $H^1_{\bar\partial}(D) = 0$. Then every divisor on D is principal.*

Since not every Stein domain has the Oka property—unless $n = 1$—Theorem 5.3′ does not suffice for characterizing principal divisors. According to the Oka principle, such a characterization should be purely topological, at least on Stein domains. In §6.4 we will describe, using the language of sheaf cohomology, a fundamental topological invariant of a divisor, the so-called **Chern class**, which completely settles the problem: *A divisor on a Stein domain is principal if and only if its Chern class is zero.*

5.6. Quotient Representation of Meromorphic Functions

By definition, a meromorphic function is *locally* the quotient of two (locally defined) holomorphic functions. It is classical that every $m \in \mathcal{M}(D)$, $D \subset \mathbb{C}^1$, has a *global* quotient representation $m = f/g$, with $f, g \in \mathcal{O}(D)$ (this will be a special case of Theorem 5.10 below). Whether this remains true if $D \subset \mathbb{C}^n$, $n \geq 2$, is a deep question which has motivated much of the early work in several complex variables. H. Poincaré ("Sur les fonctions de deux variables," Acta Math. **2**(1883), 97–113) gave an affirmative answer for $D = \mathbb{C}^2$. We now discuss this problem—called the Poincaré Problem, following H. Cartan ([Car], p. 471)—on Stein domains.

The main additional information needed is the following quotient representation for divisors; notice that this result is trivial in one variable.

Lemma 5.9. *Let $D \subset \mathbb{C}^n$. Every divisor δ on D is the quotient $\delta = \delta^+/\delta^-$ of two positive divisors δ^+, δ^- on D, such that $\delta^+(z)$ and $\delta^-(z)$ are relatively prime for $z \in D$. (This property is defined invariantly in $\mathcal{O}_z/\mathcal{O}_z^*$ in terms of representative germs in $\mathcal{O}_z$.)*

Proof. We will need some of the algebraic properties of ${}_n\mathcal{O}_a$ stated in §4.1. Let $\delta \in \mathscr{D}(D)$. By (5.23), for each $a \in D$ there are a neighborhood U_a and holomorphic functions $f^{(a)}$, $g^{(a)} \in \mathcal{O}(U_a)$, such that $\delta(z) = [\mathbf{f}_z^{(a)}/\mathbf{g}_z^{(a)}]$ for $z \in U_a$. Moreover, by Theorems 4.3 and 4.4, we can assume—after shrinking U_a—that

(5.24) $\qquad \mathbf{f}_z^{(a)}$ *and* $\mathbf{g}_z^{(a)}$ *are relatively prime for all* $z \in U_a$.

If $z \in U_a \cap U_b$, $\mathbf{f}_z^{(a)}/\mathbf{g}_z^{(a)}$ and $\mathbf{f}_z^{(b)}/\mathbf{g}_z^{(b)}$ define the same equivalence class in $\mathcal{M}_z^*/\mathcal{O}_z^*$, hence there is $\mathbf{u} \in \mathcal{O}_z^*$, such that $\mathbf{f}_z^{(a)} \cdot \mathbf{g}_z^{(b)} = \mathbf{u} \cdot \mathbf{f}_z^{(b)} \cdot \mathbf{g}_z^{(a)}$ in $\mathcal{O}_z$. By (5.24) and Theorem 4.4, there are units $\mathbf{v}_1, \mathbf{v}_2 \in \mathcal{O}_z^*$, such that $\mathbf{f}_z^{(a)} = \mathbf{v}_1 \cdot \mathbf{f}_z^{(b)}$ and $\mathbf{g}_z^{(a)} = \mathbf{v}_2 \cdot \mathbf{g}_z^{(b)}$. Hence $[\mathbf{f}_z^{(a)}] = [\mathbf{f}_z^{(b)}]$ for all $z \in U_a \cap U_b$. This implies that the map δ^+ on D given by $\delta^+(a) = [\mathbf{f}_a^{(a)}]$ for $a \in D$ satisfies (5.23), and therefore is a divisor. Similarly, $\delta^-(a) = [\mathbf{g}_a^{(a)}]$ defines $\delta^- \in \mathcal{D}(D)$. By construction, δ^+ and δ^- are positive, and $\delta = \delta^+/\delta^-$. Moreover, by (5.24), $\delta^+(z)$ and $\delta^-(z)$ are relatively prime for all $z \in D$. ∎

We now can solve the Poincaré Problem on Stein domains in the following strong form.

Theorem 5.10. *Suppose* $D \subset \mathbb{C}^n$ *has the Oka property and* $H^1_{\bar\partial}(D) = 0$. *If* $m \in \mathcal{M}^*(D)$, *then there are* F, $G \in \mathcal{O}(D)$ *such that* $m = F/G$ *and* $\mathbf{F}_z$ *and* $\mathbf{G}_z$ *are relatively prime at every* $z \in D$.

PROOF. By Lemma 5.9 we can write $\operatorname{div} m = \delta^+/\delta^-$, where δ^+ and δ^- are positive relatively prime divisors. By Theorem 5.3′, there exist f, $G \in \mathcal{O}(D)$, such that $\delta^+ = \operatorname{div} f$ and $\delta^- = \operatorname{div} G$. It follows that $\mathbf{f}_z$ and $\mathbf{G}_z$ are relatively prime at every $z \in D$, and since $\operatorname{div} m = \operatorname{div}(f/G)$, $m \cdot (f/G)^{-1} = h \in \mathcal{O}^*(D)$. Therefore, if $F = f \cdot h$, one obtains $m = F/G$, with F, G as required. ∎

Remark 5.11. By using Cartan's Theorem A for coherent analytic sheaves (see §6.8) one can prove a weaker version of Theorem 5.10 on arbitrary Stein domains D: *every meromorphic function* $m \in \mathcal{M}^*(D)$ *is the quotient* F/G *of two global holomorphic functions on* D. As seen by the example below, in general it will not be possible to find a quotient representation $m = F/G$ with $\mathbf{F}_z$ and $\mathbf{G}_z$ relatively prime for all $z \in D$.

Example 5.12. On the domain $D \subset \mathbb{C}^2$ of the Oka example discussed in §5.2, we introduce the following Cousin I distribution (the notation is as in §5.2):

$$m_1 = 1/f_1 \in \mathcal{M}(U_1), \qquad m_2 = 1 \in \mathcal{M}(U_2).$$

Notice that $m_1 - m_2 \in \mathcal{O}(U_1 \cap U_2)$. Since D is Stein, by Theorem 4.9 there is $m \in \mathcal{M}(D)$, such that $m - m_i \in \mathcal{O}(U_i)$, $i = 1, 2$. Clearly $m \not\equiv 0$ on D, so $m \in \mathcal{M}^*(D)$. Suppose $m = F/G$, with F, $G \in \mathcal{O}(D)$, and $\mathbf{F}_z$, $\mathbf{G}_z$ relatively prime at every $z \in D$. It follows that $\operatorname{div}(F/G)(z) = \operatorname{div} m(z) = \operatorname{div}(1/f_1)(z)$, and hence $\operatorname{div} G(z) = \operatorname{div} f_1(z)$ for $z \in Y^+$. On the other hand, m is holomorphic on $U_2 = D - Y^+$, and therefore $\mathbf{F}_z/\mathbf{G}_z \in \mathcal{O}_z$ for $z \in U_2$, which implies $\mathbf{G}_z \in \mathcal{O}_z^*$. Thus $G \in \mathcal{O}(D)$ satisfies $G/f_1 \in \mathcal{O}^*(D - Y^-)$ and $G/1 \in \mathcal{O}^*(D - Y^+)$, and we would have found a solution on D for the multiplicative Cousin distribution given in §5.2; but we saw that this is impossible.

EXERCISES

E.5.1. Let $D \subset \mathbb{C}$ be open. Show that there is a normal exhaustion $\{K_j, j = 1, 2, \ldots\}$ of D by compact sets such that the following hold for each j:

(i) each component of $\mathbb{C} - K_j$ contains a component of $\mathbb{C} - D$;
(ii) if $f \in C(K_j)$ is nowhere 0 on K_j, then f has a continuous extension $\tilde{f}$ to K_{j+1} without zeroes on K_{j+1}.

(Hint: For (ii), note that a nonzero continuous function on $\{z: 1 \leq |z| \leq 2\}$ has a nonzero continuous extension to $\mathbb{C} - \{0\}$ (for example, the extension can be chosen constant along radial directions), and modify this idea to handle a suitable exhaustion.)

E.5.2. Use E.5.1 to show that the product domain D in Theorem 5.4 has a normal exhaustion $\{K^{(\kappa)}, \kappa = 1, 2, \ldots\}$ which satisfies (5.19) and (5.20) in the text.

E.5.3. Let $\mathscr{U} = \{U_j : j \in J\}$ be an open cover of $D \subset \mathbb{C}^n$ such that $U_i \cap U_j$ is simply connected for all $i, j \in J$, and let $\{g_{ij} \in \mathcal{O}^*(U_i \cap U_j)\}$ be a Cousin II distribution. As in (5.8), define integers n_{ijk} for $U_i \cap U_j \cap U_k \neq \varnothing$ by

$$n_{ijk} 2\pi\sqrt{-1} = \log g_{ij} + \log g_{jk} + \log g_{ki}$$

for some choice of the branches of the logarithms. Show that if $\tilde{n}_{ijk} \in \mathbb{Z}$ are defined as above with respect to some other choice of branches, then for each pair $i, j \in J$ with $U_i \cap U_j \neq \varnothing$ there is $m_{ij} \in \mathbb{Z}$, so that

$$n_{ijk} - \tilde{n}_{ijk} = m_{jk} - m_{ik} + m_{ij}$$

whenever $U_i \cap U_j \cap U_k \neq \varnothing$.

(In the language of Čech cohomology (cf. §6.2) this shows that the cohomology class of $\{n_{ijk}\}$ in $H^2(\mathscr{U}, \mathbb{Z})$ is determined uniquely by $\{g_{ij}\}$.)

E.5.4. Use Lemma 5.9. to show that the solution of Problem Z for a region D in $\mathbb{C}^n$ implies the solution of the Cousin II Problem on D.

§6. Preview: Cohomology of Coherent Analytic Sheaves

The reader will certainly have noticed the many formal similarities in the formulation and proof of the Cousin problems and of the Oka property. *Cohomology theory with coefficients in a sheaf of Abelian groups*—a generalization of ordinary cohomology with coefficients in a (constant) group—is the natural abstract tool for working efficiently with such formal structures.

Sheaves were introduced in algebraic topology in 1946 by J. Leray (C.R. Acad. Sci. Paris **222** (1946), 1366–1368), but they had already appeared implicitly in complex analysis in a 1944 paper of H. Cartan ([Car], 565–613) and, independently, though somewhat later, in the work of K. Oka ([Oka], VII). The roots are, of course, much older: connected sets of "germs"—which are the basic building blocks of sheaves—are already present in K. Weierstrass' concept of "analytic configuration" generated by analytic continuation of a holomorphic function element.

H. Cartan immediately realized the great importance of sheaves and of the corresponding cohomology theory, and already in his 1949 paper ([Car], 618–653) he systematically introduced and used sheaves in complex analysis and formalized the fundamental concept of *coherent analytic sheaf*, which during the 1950s and 60s evolved into one of the principal objects of study in modern complex analysis. To keep matters in perpective, it should be emphasized that sheaf theory is, in essence, a very general formal tool for passing from local to global information, but that the amazing success and wide applications it has found in complex analysis ultimately rest on very deep analytic-algebraic results which state that certain "natural" sheaves in complex analysis are *coherent* analytic sheaves.

In the present paragraph we give a brief introduction to sheaf cohomology, reformulate the Cousin problems and their solution in this abstract language, and—most importantly—we discuss the concept of coherent analytic sheaf, culminating in the formulation of Cartan's famous Theorems A and B. This section is strictly expository—no proofs are given, except for some simple statements which might help in understanding the basics. The reader is strongly encouraged to read the classic surveys of H. Cartan ([Car], 669–683) and J.P. Serre [Ser 1] presented at the 1953 "Colloque sur les functions de plusieurs variables" in Bruxelles. Those wishing to study these topics more in depth are referred to the 1951–52 Cartan Seminar [Car 1], to the lecture notes of B. Malgrange [Mal] and R. Narasimhan [Nar 1], to the books of L. Hörmander [Hör 2] and of L. Kaup and B. Kaup [KaKa], and—for a complete and authoritative account representing the "latest state-of-the-art"—to the work of H. Grauert and R. Remmert [GrRe 1, 2], the two major contributors to this theory since H. Cartan and K. Oka.

6.1. Sheaves

As an introduction to the abstract concept of a sheaf we begin with the examples of foremost interest in complex analysis. In §4.1 we had already introduced the ring ${}_n\mathcal{O}_a$ of germs of holomorphic functions at the point $a \in \mathbb{C}^n$. We now consider the collection

$$\mathcal{O} = \mathcal{O}_{\mathbb{C}^n} := \bigcup_{a \in \mathbb{C}^n} {}_n\mathcal{O}_a$$

of *all* holomorphic germs in $\mathbb{C}^n$ and equip this set with a special topology which reflects the intimate relationship between the germs defined at different points by the same function. A basis for the open sets of this topology—the so-called **sheaf topology**—is given by all subsets $U_f \subset \mathcal{O}$ of the form

$$U_f = \{p \in \mathcal{O} : p = \mathbf{f}_z \text{ for some } z \in U\}, \tag{6.1}$$

where $U \subset \mathbb{C}^n$ is open and $f \in \mathcal{O}(U)$. Topologized in this way, $\mathcal{O}_{\mathbb{C}^n}$ is called the **sheaf of germs of holomorphic functions on** $\mathbb{C}^n$, or simply the **structure sheaf of** $\mathbb{C}^n$. There is a natural projection $\pi\colon \mathcal{O}_{\mathbb{C}^n} \to \mathbb{C}^n$ defined by $\pi(\mathbf{f}_a) = a$ for $\mathbf{f}_a \in \mathcal{O}_a$. It readily follows from the definition of the topology of $\mathcal{O}$ that π is a local homeomorphism, i.e., each point $p \in \mathcal{O}$ has a neighborhood W (which may be chosen of the form U_f) so that $\pi|_W\colon W \to \pi(W)$ is a homeomorphism.

By an analogous construction one obtains for each $k \in \mathbb{N}$ or $k = \infty$, the **sheaf** $\mathscr{C}^k_{\mathbb{C}^n}$ **of germs of** C^k **functions on** $\mathbb{C}^n$. Clearly these concepts can be generalized to arbitrary complex manifolds in place of $\mathbb{C}^n$, or, in case of the sheaves $\mathscr{C}^k$, to C^k manifolds.

Similarly, in the collection

$$\mathscr{M} = \mathscr{M}_{\mathbb{C}^n} := \bigcup_{a \in \mathbb{C}^n} \mathscr{M}_a$$

of all germs of meromorphic functions (see §4.2) one takes as a basis for the open sets of the sheaf topology of $\mathscr{M}$ the sets

$$U_{f/g} = \{p \in \mathscr{M}\colon p = \mathbf{f}_z/\mathbf{g}_z \textit{ for some } z \in U\}, \tag{6.2}$$

where $U \subset \mathbb{C}^n$ is open, and $f, g \in \mathcal{O}(U)$, with $\mathbf{g}_z \neq 0$ for all $z \in U$. Again, the natural projection $\pi\colon \mathscr{M}_{\mathbb{C}^n} \to \mathbb{C}^n$ is then a local homeomorphism, and the resulting space $\mathscr{M}$ is called the **sheaf of germs of meromorphic functions**.

The relevant topological features of these examples are formalized as follows.[1]

Definition. Let X be a topological space. A **sheaf** $\mathscr{S}$ **over** X is a topological space $\mathscr{S}$ together with a surjective local homeomorphism $\pi\colon \mathscr{S} \to X$.

If $\mathscr{S}$ is a sheaf over X, the projection π is an open map, and each **stalk** $\mathscr{S}_x := \pi^{-1}(x)$ **over** $x \in X$ is a closed, discrete subset of $\mathscr{S}$. A **subsheaf of** $\mathscr{S}$ is a subset $\mathscr{S}' \subset \mathscr{S}$ such that $\pi|_{\mathscr{S}'}\colon \mathscr{S}' \to X$ is a sheaf. Clearly $\mathscr{S}'$ is a subsheaf if and only if $\mathscr{S}'$ is open in $\mathscr{S}$ and $\pi(\mathscr{S}') = X$. Given a subspace Y of X, the **restriction of** $\mathscr{S}$ **to** Y, denoted by $\mathscr{S}_Y$, is defined by $\mathscr{S}_Y = \pi^{-1}(Y)$, with projection $\pi_Y = \pi|_{\mathscr{S}_Y}\colon \mathscr{S}_Y \to Y$; $\mathscr{S}_Y$ is a sheaf over Y. If $\pi_i\colon \mathscr{S}_i \to X$, $i = 1, 2$, are two sheaves over X, one defines

$$\mathscr{S}_1 \oplus \mathscr{S}_2 = \{(p_1, p_2) \in \mathscr{S}_1 \times \mathscr{S}_2\colon \pi_1(p_1) = \pi_2(p_2)\},$$

equipped with the relative topology of $\mathscr{S}_1 \times \mathscr{S}_2$. One checks that $\pi\colon \mathscr{S}_1 \oplus \mathscr{S}_2 \to X$ defined by $\pi(p_1, p_2) = \pi_1(p_1)(=\pi_2(p_2))$ is a local homeomorphism, and hence $\mathscr{S}_1 \oplus \mathscr{S}_2$ is a sheaf over X, called the **direct sum** or **Whitney sum** of $\mathscr{S}_1$ and $\mathscr{S}_2$. The direct sum of finitely many sheaves is defined analogously.

Of fundamental importance in sheaf theory is the following concept: a continuous map $s\colon Y \to \mathscr{S}$ from a subset Y of X into the sheaf $\mathscr{S}$ over X is

[1] In the literature one often finds an equivalent, more algebraic definition of sheaves involving the concept of "pre-sheaf". A sheaf is a presheaf which satisfies some additional special conditions.

called a **section of** $\mathscr{S}$ **over** Y if $\pi \circ s(x) = x$ for all $x \in Y$. The collection of all sections of $\mathscr{S}$ over Y is denoted by $\Gamma(Y, \mathscr{S})$.

Lemma 6.1. *Let $D \subset \mathbb{C}^n$ be open. Given $f \in \mathcal{O}(D)$, define $s_f: D \to \mathcal{O}_{\mathbb{C}^n}$ by $s_f(a) = \mathbf{f}_a$ for $a \in D$. Then s_f is continuous, and hence a section of $\mathcal{O}$. The map $\mathcal{O}(D) \to \Gamma(D, \mathcal{O})$ so defined is one-to-one and onto.*

The proof is an elementary consequence of the definitions and is left to the reader. In the following we shall not distinguish between holomorphic functions and sections of the sheaf $\mathcal{O}$.

Corresponding results hold for the sheaves $\mathscr{C}_X^k$ over a differentiable manifold X. The reader should also check that a meromorphic function m on $D \subset C^n$ according to the definition given in §4.2 is precisely a section over D in the sheaf of germs $\mathscr{M}$ of meromorphic functions, as defined here.

The following property of sections in a sheaf is also easy to verify.

Lemma 6.2. *If $s_1, s_2 \in \Gamma(Y, \mathscr{S})$ are sections with $s_1(x_0) = s_2(x_0)$ for some $x_0 \in Y$, then $s_1(x) = s_2(x)$ for all x in a neighborhood of x_0 in Y.*

Corollary 6.3. *If the sheaf $\mathscr{S}$ is a Hausdorff space, then two sections s_1 and s_2 of $\mathscr{S}$ over Y which agree at one point $x_0 \in Y$ agree on the connected component of Y which contains x_0.*

Proof. The set $\Omega = \{x \in Y: s_1(x) = s_2(x)\}$ contains x_0 and is open, by Lemma 6.2; if $\mathscr{S}$ is Hausdorff, Ω is also closed. ■

This "*Identity Theorem for sections*" suggests that Hausdorff sheaves are quite exceptional; we leave it to the reader to verify that the sheaves $\mathscr{C}_{\mathbb{R}^n}^k$ are *not* Hausdorff for any k, while the sheaf $\mathcal{O}_{\mathbb{C}^n}$ *is* Hausdorff. The proof of the latter fact uses the Identity Theorem for holomorphic functions, so Corollary 6.3 does not give a *new* proof of that theorem.

In most applications the stalks $\mathscr{S}_x$ of the sheaf $\mathscr{S}$ carry some natural algebraic structure which depends continuously on x, in the sense that the corresponding operations induced on sections by pointwise action lead again to sections. Let us consider more precisely one of the most common situations. A sheaf $\mathscr{S}$ over X is called a **sheaf of Abelian groups over** X if each $\mathscr{S}_x$, $x \in X$ carries the structure of an Abelian group, so that if $Y \subset X$ and $s_1, s_2 \in \Gamma(Y, \mathscr{S})$, then $s_1 - s_2: Y \to \mathscr{S}$ defined by $(s_1 - s_2)(x) = s_1(x) - s_2(x) \in \mathscr{S}_x$ for $x \in Y$ is continuous on Y, and hence is a section. It follows that if $\mathscr{S}$ is a sheaf of Abelian groups, then $\Gamma(Y, \mathscr{S})$ inherits the structure of an Abelian group. In particular, there is an element $O \in \Gamma(X, \mathscr{S})$—called the **zero-section**—defined by $O(x) = 0_x$, where 0_x is the neutral element in the group $\mathscr{S}_x$. Sheaves with other algebraic structures are defined analogously. The sheaves $\mathscr{C}_X^k$ and $\mathcal{O}_{\mathbb{C}^n}$ are examples of **sheaves of commutative rings with identity** (in particular, they are sheaves of Abelian groups). Moreover, $\mathscr{C}_{\mathbb{C}^n}^k$ and $\mathcal{O}_{\mathbb{C}^n}^l = \mathcal{O}_{\mathbb{C}^n} \oplus \ldots \oplus \mathcal{O}_{\mathbb{C}^n}$ (l direct summands) carry a natural structure of a **sheaf of $\mathcal{O}$-modules**: the stalks over each point $a \in \mathbb{C}^n$ are $\mathcal{O}_a$-modules, and the natural continuity conditions at the level of sections are satisfied. (See Exercises E.6.5 and E.6.6 for the precise formulations.)

We briefly discuss a number of routine definitions and facts which will occur in the following sections.

If $\mathcal{S}$ is a sheaf of Abelian groups over X, a subsheaf $\mathcal{S}' \subset \mathcal{S}$ will typically be required to be a **sheaf of subgroups**, i.e., $\mathcal{S}'_x$ is a subgroup of $\mathcal{S}_x$ for each $x \in X$. If $\mathcal{S}'$ and $\mathcal{S}$ are sheaves of Abelian groups over X, a continuous map φ: $\mathcal{S}' \to \mathcal{S}$ is called a **sheaf homomorphism** or simply a **homomorphism** if φ preserves stalks (i.e., $\varphi(\mathcal{S}'_x) \subset \mathcal{S}_x$ for each $x \in X$) and if the restriction φ_x: $\mathcal{S}'_x \to \mathcal{S}_x$ of φ to $\mathcal{S}'_x$ is a (group) homomorphism for each x. If φ: $\mathcal{S}' \to \mathcal{S}$ is a sheaf homomorphism, then the **kernel of** φ and the **image of** φ, defined by

$$\mathcal{K}er\ \varphi = \bigcup_{x \in X} \ker \varphi_x \quad \text{and} \quad \mathcal{I}m\ \varphi = \bigcup_{x \in X} \operatorname{Im} \varphi_{x'}$$

are sheaves of subgroups of $\mathcal{S}'$ and $\mathcal{S}$, respectively. Every sheaf homomorphism φ: $\mathcal{S}' \to \mathcal{S}$ gives rise to an **exact sequence**[1]

$$(6.3) \qquad O \to \mathcal{K}er\ \varphi \xrightarrow{\iota} \mathcal{S}' \xrightarrow{\varphi} \mathcal{I}m\ \varphi \to O,$$

where O is the trivial zero sheaf, whose stalk O_x consists only of the neutral element for each x, and ι is the inclusion.

If $\mathcal{S}' \subset \mathcal{S}$ is a sheaf of subgroups of the sheaf of Abelian groups $\mathcal{S}$, one defines the **quotient sheaf** $\mathcal{S}/\mathcal{S}'$ as the union of all the quotient groups $\mathcal{S}_x/\mathcal{S}'_x$ for $x \in X$, equipped with the quotient topology, i.e., the finest topology which makes the stalkwise defined quotient map q: $\mathcal{S} \to \mathcal{S}/\mathcal{S}'$ continuous. One checks that $\mathcal{S}/\mathcal{S}'$ is a sheaf of Abelian groups. Obviously q is then a sheaf homomorphism; the sequence (6.3) corresponding to q reads

$$(6.4) \qquad O \to \mathcal{S}' \to \mathcal{S} \xrightarrow{q} \mathcal{S}/\mathcal{S}' \to O.$$

6.2. Čech Cohomology with Coefficients in a Sheaf

In this section we give a brief outline of the basics of abstract sheaf cohomology in terms of Čech cohomology. Applications to complex analysis will be discussed in the following sections. We assume throughout this section that X is a paracompact Hausdorff space.

Let $\mathcal{S}$ be a sheaf of Abelian groups over X, and let $\mathcal{U} = \{U_j : j \in J\}$ be an open cover of X. A **q-cochain** f **for** $\mathcal{U}$ **with coefficients in** $\mathcal{S}$, where q is an

[1] Exactness of a sequence of sheaf homomorphisms $\cdots \to \mathcal{S}_{i-1} \xrightarrow{\varphi_{i-1}} \mathcal{S}_i \xrightarrow{\varphi_i} \mathcal{S}_{i+1} \xrightarrow{\varphi_{i+1}} \cdots$ is defined as in algebra: the sequence is exact at $\mathcal{S}_i$ if $\mathcal{I}m\ \varphi_{i-1} = \mathcal{K}er\ \varphi_i$; the sequence is exact if it is exact at every place.

integer ≥ 0, is a map f which assigns to each $(q + 1)$-tuple $(j_0, j_1, \ldots, j_q) \in J^{q+1}$ with

$$U(j_0, \ldots, j_q) := U_{j_0} \cap \ldots \cap U_{j_q} \neq \emptyset$$

a section

$$f(j_0, \ldots, j_q) \in \Gamma(U(j_0, \ldots, j_q), \mathscr{S}).$$

The set of all q-cochains is an Abelian group which is denoted by $C^q(\mathscr{U}, \mathscr{S})$.

The **coboundary map** δ_q: $C^q(\mathscr{U}, \mathscr{S}) \to C^{q+1}(\mathscr{U}, \mathscr{S})$ is defined by

$$(\delta_q f)(j_0 \ldots j_{q+1}) = \sum_{k=0}^{q+1} (-1)^k f(j_0 \ldots \hat{j}_k \ldots j_{q+1})|_{U(j_0 \ldots j_k \ldots j_{q+1})},$$

where $\hat{j}_k$ denotes omission of that index; δ_q is a homomorphism, and it is straightforward to verify that

$$\delta_{q+1} \circ \delta_q = 0 \qquad \text{for any } q \geq 0. \tag{6.5}$$

The kernel of δ_q is called the group $Z^q(\mathscr{U}, \mathscr{S})$ of **q-cocycles**, and, for $q \geq 1$, the image of δ_{q-1} in C^q is called the group $B^q(\mathscr{U}, \mathscr{S})$ of **q-coboundaries**. For $q = 0$ one sets $B^0(\mathscr{U}, \mathscr{S}) = 0$. By (6.5), one has $B^q \subset Z^q$ for all $q \geq 0$. The quotient group

$$H^q(\mathscr{U}, \mathscr{S}) := Z^q(\mathscr{U}, \mathscr{S})/B^q(\mathscr{U}, \mathscr{S})$$

is called the **q-th Čech cohomology group of $\mathscr{U}$ with coefficients in $\mathscr{S}$**. The cohomology class of $f \in Z^q(\mathscr{U}, \mathscr{S})$ in $H^q(\mathscr{U}, \mathscr{S})$ is denoted by $[f]$.

In order to eliminate the dependence on the covering $\mathscr{U}$, one now takes the *direct limit* of $H^q(\mathscr{U}, \mathscr{S})$ over all open coverings. This process is similar to the one used in the definition of germs of functions. In somewhat more detail, suppose $\mathscr{V} = \{V_i: i \in I\}$ is a refinement of $\mathscr{U}$ with refining map $\tau: I \to J$, so that $V_i \subset U_{\tau(i)}$ for all $i \in I$. τ induces homomorphisms

$$\tau_q^*: C^q(\mathscr{U}, \mathscr{S}) \to C^q(\mathscr{V}, \mathscr{S}), \qquad q \geq 0,$$

defined by setting

$$(\tau_q^* f)(i_0, \ldots, i_q) := f(\tau(i_0), \ldots, \tau(i_q))|_{V(i_0, \ldots, i_q)}$$

for $f \in C^q(\mathscr{U}, \mathscr{S})$. Since $\tau_{q+1}^* \circ \delta_q = \delta_q \circ \tau_q^*$, τ_q^* induces a homomorphism

$$\rho_q^{\mathscr{U}\mathscr{V}}: H^q(\mathscr{U}, \mathscr{S}) \to H^q(\mathscr{V}, \mathscr{S}),$$

which turns out to depend only on the coverings and not on the particular refining map τ. Two cohomology classes $[f] \in H^q(\mathscr{U}, \mathscr{S})$ and $[g] \in H^q(\mathscr{U}', \mathscr{S})$ are said to be equivalent if there is a common refinement $\mathscr{V}$ of $\mathscr{U}$ and $\mathscr{U}'$, such that $\rho_q^{\mathscr{U}\mathscr{V}}([f]) = \rho_q^{\mathscr{U}'\mathscr{V}}([g])$. The direct limit

$$H^q(X, \mathscr{S}) = \varinjlim_{\mathscr{U}} H^q(\mathscr{U}, \mathscr{S})$$

is, by definition, the set of all equivalence classes in the disjoint union

$\bigcup_{\mathscr{U}} H^q(\mathscr{U}, \mathscr{S})$ over all open covers $\mathscr{U}$ of X. $H^q(X, \mathscr{S})$ inherits the structure of an Abelian group in the obvious way. $H^q(X, \mathscr{S})$ is called the **q-th Čech cohomology group of X with coefficients in $\mathscr{S}$**.

For each covering $\mathscr{U}$ there are natural homomorphisms $H^q(\mathscr{U}, \mathscr{S}) \to H^q(X, \mathscr{S})$ which, in general, are neither one-to-one nor onto. The following result is elementary.

Lemma 6.4. *For any open cover $\mathscr{U}$ of X and sheaf $\mathscr{S}$ of Abelian groups one has*

(i) $$H^0(X, \mathscr{S}) = H^0(\mathscr{U}, \mathscr{S}) = Z^0(\mathscr{U}, \mathscr{S}) = \Gamma(X, \mathscr{S}),$$

and

(ii) *if $\mathscr{V}$ is a refinement of $\mathscr{U}$, then*

$$H^1(\mathscr{U}, \mathscr{S}) \to H^1(\mathscr{V}, \mathscr{S})$$

is injective, and hence $H^1(\mathscr{U}, \mathscr{S}) \to H^1(X, \mathscr{S})$ is also injective.

Notice that (i) is an immediate consequence of the definitions. The proof of (ii) is an abstract reformulation of the second part of the proof of Theorem 5.1.

Much deeper is **Leray's Theorem on Cohomology**.

Theorem 6.5. *Let X be paracompact, and assume that the open cover $\mathscr{U}$ is acyclic for the sheaf $\mathscr{S}$ of Abelian groups (this means that $H^q(U_{i_0} \cap \ldots \cap U_{i_l}, \mathscr{S}) = 0$ for all $q \geq 1$, $l \geq 0$, and $U_{i_\nu} \in \mathscr{U}$). Then*

$$H^q(\mathscr{U}, \mathscr{S}) \to H^q(X, \mathscr{S})$$

is an isomorphism for all $q \geq 0$.

Given a sheaf homomorphism $\varphi: \mathscr{S}' \to \mathscr{S}$ and an open cover $\mathscr{U}$ of X, it is easy to see that φ induces homomorphisms $C^q(\mathscr{U}, \mathscr{S}') \to C^q(\mathscr{U}, \mathscr{S})$, $q \geq 0$, by sending $f \in C^q(\mathscr{U}, \mathscr{S}')$ to the composition $\varphi \circ f \in C^q(\mathscr{U}, \mathscr{S}')$. One shows by routine arguments that these homomorphisms induce homomorphisms $\varphi^q_{\mathscr{U}}: H^q(\mathscr{U}, \mathscr{S}') \to H^q(\mathscr{U}, \mathscr{S})$ and

$$\varphi^q: H^q(X, \mathscr{S}') \to H^q(X, \mathscr{S}), \qquad q \geq 0,$$

at the cohomology level. The usual "functorial" properties are satisfied: *if id: $\mathscr{S} \to \mathscr{S}$ is the identity, so is id^q for $q \geq 0$, and if $\varphi: \mathscr{S}' \to \mathscr{S}$, $\psi: \mathscr{S} \to \mathscr{S}''$ are homomorphisms, then $(\psi \circ \varphi)^q = \psi^q \circ \varphi^q$ for $q \geq 0$.*

One of the most useful relationships between cohomology groups is expressed by the **long exact cohomology sequence**, as follows.

Theorem 6.6. *Suppose*

(6.7) $$0 \to \mathscr{S}' \overset{\varphi}{\to} \mathscr{S} \overset{\psi}{\to} \mathscr{S}'' \to 0$$

is an exact sequence of sheaf homomorphisms over X. Then there exist connect-

ing homomorphisms

$$\delta^q: H^q(X, \mathscr{S}'') \to H^{q+1}(X, \mathscr{S}') \qquad \text{for } q = 0, 1, \ldots,$$

such that the sequence of groups

$$0 \to H^0(X, \mathscr{S}') \xrightarrow{\varphi^0} H^0(X, \mathscr{S}) \xrightarrow{\psi^0} H^0(X, \mathscr{S}'') \xrightarrow{\delta^0} H^1(X, \mathscr{S}') \xrightarrow{\varphi^1} \cdots$$
$$\xrightarrow{\varphi^q} H^q(X, \mathscr{S}) \xrightarrow{\psi^q} H^q(X, \mathscr{S}'') \xrightarrow{\delta^q} H^{q+1}(X, \mathscr{S}') \to \cdots$$

is exact.

The following immediate consequence is one of the most important applications of Theorem 6.6.

Corollary 6.7. *If in the exact sequence* (6.7) *one has* $H^1(X, \mathscr{S}') = 0$, *then*

$$\psi^0: \Gamma(X, \mathscr{S}) \to \Gamma(X, \mathscr{S}'')$$

is surjective.

PROOF. The relevant portion of the cohomology sequence reads

$$\cdots \to H^0(X, \mathscr{S}) \xrightarrow{\psi^0} H^0(X, \mathscr{S}'') \xrightarrow{\delta^0} 0 \to \cdots;$$

exactness at $H^0(X, \mathscr{S}'')$ implies the desired result. ∎.

Because of the importance of Corollary 6.7 in the discussion later on, we give a complete proof of it, independent of Theorem 6.6.

DIRECT PROOF OF COROLLARY 6.7. Let $s'' \in \Gamma(X, \mathscr{S}'')$. Since ψ is surjective, for each $x \in X$ there are a neighborhood U_x of x and a section $\hat{s}_x \in \Gamma(U_x, \mathscr{S})$ with $\psi \circ \hat{s}_x = s''$ on U_x (use Exercise E.6.2 and Lemma 6.2). Let $\mathscr{U} = \{U_x : x \in X\}$. Then $\hat{s} = \{\hat{s}_x\}$ defines a 0-cochain in $C^0(\mathscr{U}, \mathscr{S})$. In general, $\delta_0 \hat{s} \neq 0$, so $\hat{s}$ is not a global section of $\mathscr{S}$. One therefore tries to modify $\hat{s}$ by subtracting a 0-cochain of the form $\varphi \circ g$, where $g \in C^0(\mathscr{U}, \mathscr{S}')$ satisfies

$$\delta_0(\varphi \circ g) = \delta_0 \hat{s}. \tag{6.8}$$

It then follows that $s := \hat{s} - \varphi \circ g$ is in $Z^0(\mathscr{U}, \mathscr{S}) = \Gamma(X, \mathscr{S})$, and $\psi \circ s = \psi \circ \hat{s} - \psi \circ \varphi \circ g = s'' - 0 = s''$, i.e., $\psi^0(s) = s''$. In order to find g as above, notice that since $\psi \circ (\delta_0 \hat{s}) = \delta_0(\psi \circ \hat{s}) = \delta_0 s'' = 0$, we have $\delta_0 \hat{s} \in B^1(\mathscr{U}, \mathscr{K}er\, \psi)$. By using the exactness of (6.7) at $\mathscr{S}$, after passing to a suitable refinement $\mathscr{V} = \{V_x : x \in X\}$, $V_x \subset U_x$, one can lift $\delta_0 \hat{s}$ to a 1-cochain $s' \in C^1(\mathscr{V}, \mathscr{S}')$ with

$$\varphi \circ s' = \delta_0 \hat{s}. \tag{6.9}$$

Since $\varphi \circ (\delta_1 s') = \delta_1(\varphi \circ s') = \delta_1(\delta_0 \hat{s}) = 0$, and since φ is injective, one has $\delta_1 s' = 0$, i.e., $s' \in Z^1(\mathscr{V}, \mathscr{S}')$. By the hypothesis $H^1(X, \mathscr{S}') = 0$, after perhaps refining the cover $\mathscr{V}$ further, we may assume that $s' \in B^1(\mathscr{V}, \mathscr{S}')$, i.e., $s' = \delta_0 g$ for some $g \in C^0(\mathscr{V}, \mathscr{S}')$. By (6.9), $\delta_0 \hat{s} = \varphi \circ s' = \varphi \circ (\delta_0 g) = \delta_0(\varphi \circ g)$, so (6.8) holds, and we are done. ∎.

The complete proof of Theorem 6.6 uses basically the same ideas and techniques as the proof of the special case given above, but it is, of course, more tedious and complex. The connecting homomorphism δ^q is given by $(\varphi^{q+1})^{-1} \circ \delta_q \circ (\psi^q)^{-1}$, and part of the proof involves showing that this is well defined.

Remark 6.8. A few comments are in order on the relationship between sheaf cohomology and ordinary cohomology theory with coefficients in an Abelian group G, as defined for example in algebraic topology by the Eilenberg–Steenrod axioms. For more details, the reader should consult [God] or [Spa]. First of all, sheaf cohomology theory can be defined on arbitrary topological spaces by means of "flabby resolutions," and the resulting groups are uniquely determined by a system of natural axioms which include, among others, the existence of the long exact cohomology sequence as given in Theorem 6.6. The Čech cohomology groups constructed here satisfy these "*Axioms of Sheaf Cohomology*" provided the underlying space X is paracompact—we remind the reader that this includes all differentiable manifolds countable at infinity, and in particular open sets D in $\mathbb{C}^n$—so that for such spaces, the Čech groups $H^q(X, \mathscr{S})$ constructed here are *the* cohomology groups of X with coefficients in the sheaf $\mathscr{S}$. Depending on X and $\mathscr{S}$, there are often other useful ways for computing the groups $H^q(X, \mathscr{S})$ than by means of open covers. For example, it is of fundamental importance for complex analysis that the groups $H^q(D, \mathcal{O})$ can be computed via the $\bar{\partial}$-complex (see Theorem 6.9 below).

Now let G be an Abelian group. The space $\mathscr{G} = X \times G$, where G carries the discrete topology, and the projection $\pi\colon \mathscr{G} \to X$ onto the first factor define a sheaf of Abelian groups over X, called the **constant G-sheaf over X**. The construction of the (sheaf)-cohomology groups $H^q(X, \mathscr{G})$ given above coincides for the *constant* sheaf $\mathscr{G}$ with the classical construction of the Čech cohomology groups $\check{H}^q(X, G)$ with coefficients in G. For paracompact spaces the Čech cohomology theory with coefficients in G is known to satisfy the Eilenberg–Steenrod axioms. Thus, by the uniqueness property of those axioms, it follows that

$$H^q(X, \mathscr{G}) = \check{H}^q(X, G) = H^q(X, G) \tag{6.10}$$

if X is paracompact, where the group on the right is the cohomology group of X with coefficients in G, as defined, for example, by means of the singular simplicial theory.

We shall use (6.10) in §6.4 in order to identify $H^q(D, \mathbb{Z})$ for D in $\mathbb{C}^n$ in terms of the Čech cohomology groups with coefficients in the constant sheaf $\mathbb{Z}$ over D.

6.3. Cohomological Formulation of the Cousin Problems

We now return to complex analysis and the Cousin problems on a domain D in $\mathbb{C}^n$.[1] The sort of relations involved in Theorem 4.8 can be conveniently

[1] All results in this and the following sections remain true if D is replaced by a complex manifold M.

formulated in terms of sheaf cohomology. Let $\mathscr{U}$ be an open cover of D and consider a 1-cochain $g \in C^1(\mathscr{U}, \mathscr{O})$. We write g_{ij} for $g(i,j) \in \Gamma(U_i \cap U_j, \mathscr{O}) = \mathscr{O}(U_i \cap U_j)$, and $g = \{g_{ij}\}$. Suppose $g \in Z^1(\mathscr{U}, \mathscr{O})$, i.e.,

$$(\delta_1 g)(i, j, k) = g_{jk} - g_{ik} + g_{ij} = 0$$

for all i, j, k. Taking $i = j = k$ gives $g_{ii} = 0$, and taking $k = i$ then gives $g_{ji} + g_{ij} = 0$. It is now obvious that

(6.11) $\{g_{ij}\} \in Z^1(\mathscr{U}, \mathscr{O})$ *if and only if* (4.4) *holds*, and

(6.12) $\{g_{ij}\} \in B^1(\mathscr{U}, \mathscr{O})$ *if and only if* (4.5) *holds.*

Thus Theorem 4.8 can be restated as follows:

(6.13) *If* $H^1_{\bar\partial}(D) = 0$, *then* $H^1(\mathscr{U}, \mathscr{O}) = 0$ *for any open cover* $\mathscr{U}$ *of* D.

The converse of (6.13) holds as well (see Exercise E.4.5). Since, by Lemma 6.4. (ii) the condition on the right in (6.13) is equivalent to $H^1(D, \mathscr{O}) = 0$, one therefore obtains

(6.14) $H^1_{\bar\partial}(D) = 0$ *if and only if* $H^1(D, \mathscr{O}) = 0$.

The above statement is a special case of the following much more general result, which is known as **Dolbeault's Isomorphism.**

Theorem 6.9 ([Dol]). *For every* $q \geq 0$ *the groups* $H^q_{\bar\partial}(D)$ *and* $H^q(D, \mathscr{O})$ *are naturally isomorphic.*

Corollary 6.10. *If* D *is Stein, then*

$$H^q(D, \mathscr{O}) = 0 \qquad \textit{for } q \geq 1.$$

Proof. Use Theorem 6.9 and Theorem 1.4. ∎

Corollary 6.11. *If* $\mathscr{U}$ *is a covering of* D *by Stein domains, then*

$$H^q(\mathscr{U}, \mathscr{O}) = H^q(D, \mathscr{O}) = H^q_{\bar\partial}(D) \qquad \textit{for } q \geq 0.$$

Proof. Since finite intersections of Stein domains are Stein, Corollary 6.10 implies that $H^q(U_{i_0} \cap \ldots \cap U_{i_l}, \mathscr{O}) = 0$ for $q \geq 1$ and $l \geq 0$. Thus $\mathscr{U}$ is acyclic for $\mathscr{O}$, and the desired result follows from Leray's Theorem 6.5. ∎

Returning to the Cousin problem, Theorem 4.9 now says that the additive Cousin problem is solvable on D whenever $H^1(D, \mathscr{O}) = 0$. In terms of the machinery of sheaf cohomology, the "abstract" proof of Theorem 4.9 goes as follows. It is readily seen that in the present terminology a Cousin I distribution $\{m_j \in \mathscr{M}(U_j)\}$ on D (cf. (4.2)) defines a global section s over D in the quotient sheaf $\mathscr{M}/\mathscr{O}$, and that the solution of the Cousin problem boils down to lifting s to a *global* section $m \in \Gamma(D, \mathscr{M})$, i.e., showing that $\Gamma(D, \mathscr{M}) \to \Gamma(D, \mathscr{M}/\mathscr{O})$ is surjective. Given that $H^1(D, \mathscr{O}) = 0$, this follows from Corollary

6.7 applied to the exact sequence

$$0 \to \mathcal{O} \xrightarrow{\iota} \mathcal{M} \xrightarrow{q} \mathcal{M}/\mathcal{O} \to 0.$$

The formulation of the multiplicative Cousin problem is completely analogous. The sets $\mathcal{O}^* \subset \mathcal{O}$ and $\mathcal{M}^* \subset \mathcal{M}$ of invertible elements are sheaves of Abelian groups under multiplication. The quotient sheaf $\mathcal{M}^*/\mathcal{O}^*$ is called the **sheaf of germs of divisors**. The reader should convince himself that a divisor on D, as defined in §5.5, is precisely a section in $\mathcal{M}^*/\mathcal{O}^*$ over D. Since a Cousin II distribution $\{m_j \in \mathcal{M}^*(U_j)\}$ on D defines a divisor, i.e., a section $s^* \in \Gamma(D, \mathcal{M}^*/\mathcal{O}^*)$, the solution of the Cousin II problem is reduced to deciding whether the induced map $\Gamma(D, \mathcal{M}^*) \to \Gamma(D, \mathcal{M}^*/\mathcal{O}^*)$ is surjective. In analogy to the additive Cousin problem, this will be the case if $H^1(D, \mathcal{O}^*) = 0$; just apply Corollary 6.7 to the exact sequence

$$0 \to \mathcal{O}^* \to \mathcal{M}^* \to \mathcal{M}^*/\mathcal{O}^* \to 0. \tag{6.15}$$

6.4. The Oka Property and Chern Classes

We now analyze the condition $H^1(D, \mathcal{O}^*) = 0$ further. An examination of the reduction to the additive problem suggested in §5.3 leads one to consider the sheaf homomorphism $E: \mathcal{O} \to \mathcal{O}^*$ defined by $E(\mathbf{f}_z) = [\exp(2\pi i f)]_z$, whose kernel is the constant sheaf $\mathbb{Z}$. Notice that E is surjective. Application of Theorem 6.6 to the exact sequence

$$0 \to \mathbb{Z} \xrightarrow{\iota} \mathcal{O} \xrightarrow{E} \mathcal{O}^* \to 0 \tag{6.16}$$

gives the following portion of the exact cohomology sequence:

$$\cdots \to H^1(D, \mathcal{O}) \xrightarrow{E^1} H^1(D, \mathcal{O}^*) \xrightarrow{\delta^1} H^2(D, \mathbb{Z}) \to H^2(D, \mathcal{O}) \to \cdots \tag{6.17}$$

By Corollary 6.10 we therefore obtain

Lemma 6.12. *If D is Stein, then*

$$\delta^1: H^1(D, \mathcal{O}^*) \to H^2(D, \mathbb{Z})$$

is an isomorphism.

Corollary 6.13. *If D* is Stein and $H^2(D, \mathbb{Z}) = 0$, *then the multiplicative Cousin problem is universally solvable on D.*

Next we show that the condition $H^2(D, \mathbb{Z}) = 0$ is equivalent to the Oka property for D as formulated in §5.4. We consider the sheaf $\mathscr{C}$ of germs of continuous functions on a paracompact space X and the subsheaf $\mathscr{C}^* \subset \mathscr{C}$ consisting of invertible germs. Notice that $\mathscr{C}^*$ is a sheaf of Abelian groups under multiplication. Clearly the Oka property states that $H^1(\mathscr{U}, \mathscr{C}^*) = 0$ for every covering $\mathscr{U}$ of X. By Lemma 6.4. (ii) it follows that

(6.18) *X satisfies the Oka property if and only if $H^1(X, \mathscr{C}^*) = 0$.*

We will need the following elementary fact.

Lemma 6.14. *If X is paracompact, then $H^q(X, \mathscr{C}) = 0$ for $q \geq 1$.*

The proof of Lemma 6.14 uses a construction based on the existence of partitions of unity analogous to the proof of Step 1 in Theorem 4.8, which, in the present language, states that $H^1(D, \mathscr{C}^\infty) = 0$. The same method of proof gives the following for any $k = 1, 2, \ldots, \infty$: *if X is a C^k manifold, then $H^q(X, \mathscr{C}^k) = 0$ for $q \geq 1$.* The details are left to the reader.

The continuous analogue of (6.16),

$$O \to \mathbb{Z} \xrightarrow{\iota} \mathscr{C} \xrightarrow{E} \mathscr{C}^* \to O,$$

leads to the exact sequence

$$\cdots \to H^1(X, \mathscr{C}) \xrightarrow{E^1} H^1(X, \mathscr{C}^*) \xrightarrow{\delta_c^1} H^2(X, \mathbb{Z}) \to H^2(X, \mathbb{C}) \to \cdots.$$

By Lemma 6.14 we therefore obtain

Lemma 6.15. *The map*

$$\delta_c^1 \colon H^1(X, \mathscr{C}^*) \to H^2(X, \mathbb{Z})$$

is an isomorphism, and hence X has the Oka property if and only if $H^2(X, \mathbb{Z}) = 0$.

Corollary 6.16. *If D is Stein, the inclusion $\mathscr{O}^* \hookrightarrow \mathscr{C}^*$ induces an isomorphism*

(6.19) $$H^1(D, \mathscr{O}^*) \simeq H^1(D, \mathscr{C}^*).$$

The fact that the two groups are isomorphic is clear from Lemmas 6.12 and 6.15. To see that the isomorphism is induced by the inclusion map requires some "diagram chasing" which we leave to the reader.

The statement in Corollary 6.16 is the cohomological formulation of the Oka principle for the multiplicative Cousin problem (cf. the remarks after Theorem 5.2).

Finally, we analyze the characterization of principal divisors in case $H^1(D, \mathscr{O}^*) \neq 0$. Consider the segment

(6.20) $$\cdots \to \Gamma(D, \mathscr{M}^*) \xrightarrow{q} \Gamma(D, \mathscr{M}^*/\mathscr{O}^*) \xrightarrow{\alpha} H^1(D, \mathscr{O}^*) \to \cdots$$

of the exact cohomology sequence associated to the exact sequence (6.15). A divisor $\mathscr{d} \in \Gamma(D, \mathscr{M}^*/\mathscr{O}^*)$ is principal if and only if $\mathscr{d}$ is in the image of the map q, and by exactness of (6.20), this is equivalent to $\mathscr{d} \in \ker \alpha$. The homomorphism $\tau = \delta^1 \circ \alpha$,

$$\tau: \Gamma(D, \mathscr{M}^*/\mathscr{O}^*) \to H^2(D, \mathbb{Z}),$$

where δ^1 is the map in (6.17), is called the **Chern map**, and the image $\tau(\mathscr{d}) \in H^2(D, \mathbb{Z})$ is called the **Chern class of the divisor** $\mathscr{d}$. The above arguments, together with Lemma 6.12, imply the following result which we had announced at the end of §5.5.

Theorem 6.17. *If D is Stein, then a divisor on D is principal if and only if its Chern class is* 0.

At this point the reader should turn back to the beginning of §5.3 where, essentially, we had explicitly constructed the Chern class of the divisor $\mathscr{d}$ associated to the Cousin II distribution $\{m_j \in \mathscr{M}^*(U_j)\}$. In fact, the collection $\{n_{ijk}\}$ of integers obtained there is easily seen to define a 2-cocycle in $Z^2(\mathscr{U}, \mathbb{Z})$, whose cohomology class in $H^2(D, \mathbb{Z})$ is the Chern class of $\mathscr{d}$.

6.5. Two Fundamental Problems

After having seen sheaves in action on familiar grounds we are now in a position to apply these tools to much more general settings and to explain H. Cartan's fundamental theorems on sheaf cohomology on Stein domains.

Let us first mention two classical problems in the theory of analytic sets which have stimulated much of the pioneering work of Cartan and Oka. We fix an analytic set A in the domain D in $\mathbb{C}^n$.

PROBLEM 1 (Global definition of analytic sets). Can A be defined by global holomorphic functions, i.e., is there a subset $\mathscr{F} \subset \mathscr{O}(D)$ such that

$$A = \{z \in D: f(z) = 0 \text{ for all } f \in \mathscr{F}\}?$$

Clearly this is equivalent to

PROBLEM 1′. Given $a \in D - A$, is there $f \in \mathscr{O}(D)$ such that $f(z) = 0$ for $z \in A$ but $f(a) \neq 0$?

To formulate the other problem we need the concept of a **holomorphic function on the analytic set** A, which is defined as follows: a continuous function $f: A \to \mathbb{C}$ is said to be holomorphic on A if f is *locally* the restriction of holomorphic functions in the ambient space, i.e., if for every $a \in A$ there are a neighborhood U_a of a in $\mathbb{C}^n$ and a function $h_a \in \mathscr{O}(U_a)$ such that $f(z) = h_a(z)$ for $z \in A \cap U_a$. This definition is equivalent to the one given in I.§2.6 in case A is a complex submanifold (cf. Theorem I.2.10).

PROBLEM 2 (Global extension of holomorphic functions). Does every holomorphic function f on A have an extension to a holomorphic function F on D, i.e., is there $F \in \mathcal{O}(D)$ with $F(z) = f(z)$ for $z \in A$?

Both these problems involve passing from local information to global information. In order to formulate them in the language of sheaves we introduce the **ideal sheaf** $\mathcal{I} = \mathcal{I}(A)$ **of** A on D as follows: if $z \notin A$, we set $\mathcal{I}_z = {}_n\mathcal{O}_z$, and for $z \in A$ we set $\mathcal{I}_z = \{\mathbf{f}_z \in {}_n\mathcal{O}_z : f|_A = 0\}$. One easily checks that $\mathcal{I}(A)$ is a *subsheaf of ideals of* $\mathcal{O}_D$; that is, a subsheaf of $\mathcal{O}_D$ which is a sheaf of $\mathcal{O}_D$-modules. Problems 1 and 1′ are obviously a special case of the following more general

PROBLEM A. Is each stalk $\mathcal{I}_z(A)$, $z \in D$, generated as $\mathcal{O}_z$-module by the images in $\mathcal{I}_z$ of global sections of $\mathcal{I}(A)$ over D?

For Problem 2, notice that after natural identifications a holomorphic function on A is precisely a section over A in the sheaf $\mathcal{O}/\mathcal{I}(A)$, and that each section $f \in \Gamma(A, \mathcal{O}/\mathcal{I}(A))$ has a unique extension $\hat{f} \in \Gamma(D, \mathcal{O}/\mathcal{I})$ given by $\hat{f}(z) = 0$ for $z \notin A$. Problem 2 is thus equivalent to asking whether the map $\Gamma(D, \mathcal{O}) \to \Gamma(D, \mathcal{O}/\mathcal{I})$ induced by the quotient map $q: \mathcal{O} \to \mathcal{O}/\mathcal{I}$ is surjective. Hence, by Corollary 6.7, in order to give an affirmative answer for Problem 2 it suffices to do so for

PROBLEM B. Is $H^1(D, \mathcal{I}(A)) = 0$?

It is easy to see that Problem 2 and Problem B are in fact equivalent whenever $H^1(D, \mathcal{O}) = 0$.

If $A = \varnothing$, then $\mathcal{I}(A) = \mathcal{O}_D$; so Problem A is trivial in this case, while the answer to Problem B is positive for a Stein domain D (Corollary 6.10). The results we have proved so far allow us to solve Problem B in another special case. (Observe that Problem A is still trivial in this case.)

Theorem 6.18. *Suppose D is Stein and that $A = Z(f) = \{z \in D: f(z) = 0\}$, where $f \in \mathcal{O}(D)$ satisfies $df \neq 0$ on A. Then $H^1(D, \mathcal{I}(A)) = 0$.*

PROOF. The hypotheses imply that A is a complex submanifold of dimension $n - 1$ and that $\mathcal{I}(A) = f\mathcal{O}_D$. Since $\mathcal{O}_a$ is an integral domain, multiplication by f defines a sheaf isomorphism $\varphi_f: \mathcal{O}_D \xrightarrow{\sim} f\mathcal{O}_D = \mathcal{I}(A)$, and hence, since D is Stein, $H^1(D, \mathcal{I}(A)) \simeq H^1(D, \mathcal{O}) = 0$. ∎

6.6. Coherent Analytic Sheaves

The simple method used in proof of Theorem 6.18 to reduce Problem B to the structure sheaf $\mathcal{O}$ does not work in case A has codimension ≥ 2, even if A is nonsingular. The major difficulty involves finding a good description of the ideal sheaf $\mathscr{I}(A)$ in terms of the sheaf $\mathcal{O}$. Thinking in analogy to the Cousin problems, it seems reasonable to begin the study of ideal sheaves at the local level and to use the machinery of sheaves to pass to global results. Moreover, application of algebraic methods immediately leads one to consider not just ideals but sheaves of $\mathcal{O}$-submodules of $\mathcal{O}^l$ for $l \geq 1$, or more general abstract sheaves of $\mathcal{O}$-modules; these are the so-called **analytic sheaves.**

Correspondingly, all homomorphisms between analytic sheaves will be assumed to be **analytic homomorphisms**, i.e., $\mathcal{O}$-module homomorphisms. But aside from this formal generalization of the objects to be studied there is a much more fundamental obstacle even at the local level, namely algebraic information—as given, for example, in Theorems 4.3 and 4.4—is available at first only at the pointwise level. Let us quote H. Cartan: "... avant de pouvoir faire le passage du local au global, il faut approfondir les propriétés locales, c'est-à-dire voir comment les propriétés ponctuelles s'organisent localment" ([Car], p. 619). Of fundamental importance for studying this "local organization" is the notion of *coherence*, first introduced by H. Cartan in 1944 ([Car], p. 572). Already at that time—notice that the formal concept of a sheaf was not yet known—Cartan formulated, but could not yet prove, the principal coherence problems in complex analysis, fully recognizing their far-reaching consequences. The same problems, though in somewhat different language, were studied independently a few years later by K. Oka ([Oka], VII), who was not aware of Cartan's 1944 paper.[1] (See Cartan's comments in [Oka], p. 107, and the footnote in [Oka], p. 110.) The notion of *coherent analytic sheaf* was finally firmly established in Cartan's 1949 paper ([Car], 618–653).

To explain this concept in more detail, let us consider first an ideal sheaf $\mathscr{I} \subset \mathcal{O}$, not necessarily the ideal sheaf of an analytic set. Since the ring $\mathcal{O}_a$ is Noetherian (Theorem 4.3), the ideal $\mathscr{I}_a \subset \mathcal{O}_a$ is finitely generated, i.e., there are germs $\mathbf{f}_a^{(1)}, \ldots, \mathbf{f}_a^{(l)}$ which generate $\mathscr{I}_a$ over $\mathcal{O}_a$. We may, of course, assume that the generators $\mathbf{f}_a^{(\nu)}$ have representatives $f^{(\nu)} \in \Gamma(U, \mathscr{I}) \subset \mathcal{O}(U)$, $\nu = 1, \ldots, l$, for some neighborhood U of a, but it is not all clear whether $\mathbf{f}_z^{(1)}, \ldots, \mathbf{f}_z^{(l)}$ will generate $\mathscr{I}_z$ over $\mathcal{O}_z$ for all $z \neq a$ near a. In general, this is indeed false: consider for an open set $D \neq \mathbb{C}^n$ the trivial extension $\hat{\mathcal{O}}_D$ of $\mathcal{O}_D$ to $\mathbb{C}^n$ defined by $\hat{\mathcal{O}}_D|_D = \mathcal{O}_D$ and $\hat{\mathcal{O}}_D|_{\mathbb{C}^n - D} = 0$ (this *is* an ideal sheaf!) and take $a \in bD$. One therefore introduces the following condition: an analytic sheaf $\mathscr{A}$ (i.e., an $\mathcal{O}$-module sheaf) over D is said to be of **finite type** if for every $a \in D$ there are an open neighborhood U of a and finitely many sections $s_1, \ldots, s_l \in \Gamma(U, \mathscr{A})$ such that

(6.21) *the $\mathcal{O}_z$-module $\mathscr{A}_z$ is generated by $s_1(z), \ldots, s_l(z)$ for all $z \in U$.*

Notice that (6.21) is equivalent to the following:

(6.22) *The homomorphism $\varphi\colon \mathcal{O}_U^l \to \mathscr{A}_U$ defined by $\varphi_z(b_1, \ldots, b_l) = \sum_{j=1}^{l} b_j s_j(z)$ for $(b_1, \ldots, b_l) \in \mathcal{O}_z^l$ and $z \in U$ is surjective.*

[1] The turmoil of the war years made scientific contacts very difficult, if not impossible, at that time.

Examples. It is obvious that the structure sheaf $\mathcal{O}$ is of finite type and that, given an analytic sheaf $\mathscr{A}$ and sections $s_1, \ldots, s_l \in \Gamma(U, \mathscr{A})$, the submodule sheaf $\mathscr{I}m\ \varphi$ of $\mathscr{A}$, where φ is defined as in (6.22), is of finite type. Quotient sheaves of sheaves of finite type are again of finite type, but $\mathcal{O}$-submodules of a sheaf of finite type are not necessarily of finite type. For example, the ideal sheaf $\hat{\mathcal{O}}_D \subset \mathcal{O}$ defined above is *not* of finite type.

The following deep result is of fundamental importance for the solution of Problems A and B.

Theorem 6.19. *The ideal sheaf $\mathscr{I}(A)$ of an analytic set $A \subset D$ is of finite type.*

The first proof of Theorem 6.19 was published in 1950 by H. Cartan ([Car], p. 631), but it seems that Oka knew a proof already in 1948 (see Cartan's comments in [Oka], p. 107). Oka published a proof in 1951 ([Oka], VIII).

Let us now consider an analytic sheaf $\mathscr{A}$ of finite type. In order to continue with the local description of $\mathscr{A}$, one would like that the kernel of the homomorphism $\varphi: \mathcal{O}_U^l \to \mathscr{A}_U$ in (6.22) is of finite type as well. In case $\mathscr{A}$ is an $\mathcal{O}$-submodule of $\mathcal{O}$ or $\mathcal{O}^p$ for some $p \geq 1$, this is always true as a consequence of **Oka's Coherence Theorem** (Theorem 6.20 below). This result is even more fundamental than Theorem 6.19. First we formulate the relevant abstract concept of coherence.

Definition. An analytic sheaf $\mathscr{A}$ over D is said to be **coherent** if $\mathscr{A}$ is of finite type and if for every homomorphism $\varphi: \mathcal{O}_U^l \to \mathscr{A}_U$ defined by sections $s_1, \ldots, s_l \in \Gamma(U, \mathscr{A})$ as in (6.22), the **sheaf of relations**

$$\mathscr{R}el_U(s_1, \ldots, s_l) := \mathscr{K}er\ \varphi = \bigcup_{z \in U} \left\{(b_1, \ldots, b_l) \in \mathcal{O}_z^l : \sum_{j=1}^{l} b_j s_j(z) = 0\right\}$$

is of finite type over U.

We note that coherence is a *local* property.

Theorem 6.20. ([Oka], VII). *The sheaf $\mathcal{O}_{\mathbb{C}^n}$ of germs of holomorphic functions on $\mathbb{C}^n$ is coherent.*

One of the consequences of Oka's Coherence Theorem is that the class of coherent analytic sheaves contains many interesting examples and that it has a rich formal theory. In particular this class is closed under most standard algebraic constructions. For example, direct sums and quotients of coherent analytic sheaves are again coherent (hence $\mathcal{O}^l$ is coherent for any $l \geq 1$), and the kernel and image of an analytic homomorphism between coherent analytic sheaves are coherent.

Theorem 6.19 and 6.20 obviously imply

Theorem 6.19′. *The ideal sheaf $\mathscr{I}(A)$ of an analytic set $A \subset D$ is coherent.*

We conclude this section by formulating a local representation of a coherent analytic sheaf in terms of the structure sheaf $\mathcal{O}$.

Lemma 6.21. *Suppose $\mathscr{A}$ is a coherent analytic sheaf over D. Then for every point $a \in D$ there are a neighborhood U of a, positive integers l_1, l_2, and a homomorphism $\psi\colon \mathcal{O}_U^{l_1} \to \mathcal{O}_U^{l_2}$ so that*

$$\mathscr{A}_U \simeq \mathcal{O}_U^{l_2}/\mathscr{I}m\,\psi. \tag{6.23}$$

Proof. Since $\mathscr{A}$ is of finite type, given $a \in D$ there is a surjective homomorphism $\varphi\colon \mathcal{O}_U^{l_2} \to \mathscr{A}_U$ for some neighborhood U of a. By the definition of coherence, $\mathscr{K}er\,\varphi$ is of finite type, so, after shrinking U, there is a homomorphism $\psi\colon \mathcal{O}_U^{l_1} \to \mathcal{O}_U^{l_2}$ with $\mathscr{I}m\,\psi = \mathscr{K}er\,\varphi$, and (6.23) follows. ■

Remark. The converse of Lemma 6.21 is true as well as a consequence of the formal theory of coherent analytic sheaves and Oka's Coherence Theorem.

6.7. Cartan's Theorems A and B

Our initial goal was the solution of Problems A and B in §6.5 for ideal sheaves $\mathscr{I}(A)$ of an analytic set. In the preceding section we were led to consider coherent analytic sheaves which are, as we saw, those sheaves which have a good *local* representation in terms of the structure sheaf $\mathcal{O}$ (cf. Lemma 6.21). The final major problem left is to patch the local information together to obtain *global* results. In essence, this step involves a very sophisticated generalization of the patching process involved in the Cousin II problem, as seen perhaps most clearly in the proof of the Oka property given in §5.4. The origins of this patching process can thus be traced to P. Cousin and to Oka's 1939 paper ([Oka], III), but it was the deep analytic work of H. Cartan—in a series of papers beginning in 1940 ([Car], 539–653)—and the solution of the fundamental coherence problems by Cartan and Oka which made it possible to pass from the scalar valued case—as it appears in the Cousin II problem—to the general matrix valued case which is needed to patch together the sheaf homomorphisms appearing in the local representations in Lemma 6.21. Cartan's work culminated in the proof of the following fundamental theorem first presented in the 1951–52 Cartan Seminar [Car 1] (see also [Car], 669–683). Note that the results apply, in particular, to Stein domains in $\mathbb{C}^n$, and that by Theorem 6.19′ they include the solution of Problems A and B formulated in §6.5.

Cartan's Fundamental Theorem. *Let M be a Stein manifold and let $\mathscr{A}$ be a coherent analytic sheaf on M. Then*

(A) *Each stalk* $\mathscr{A}_z$, $z \in M$, *is generated over* $\mathcal{O}_z$ *by global sections in* $\Gamma(M, \mathscr{A})$.
(B) $H^q(M, \mathscr{A}) = 0$ *for* $q \geq 1$.

Remark 6.22. In all the applications we had considered earlier we had seen the fundamental role of the vanishing of the *first* cohomology group $H^1(D, \mathscr{A})$ (recall Corollary 6.7 for the underlying abstract principle). In fact, the vanishing of H^1 is the essential property: it is not hard to show that a complex manifold M which satisfies $H^1(M, \mathscr{I}) = 0$ for every coherent ideal sheaf $\mathscr{I} \subset \mathcal{O}_M$ is necessarily Stein (see Exercise E.6.13 for the case of domains in $\mathbb{C}^n$), so that Cartan's Theorem holds for M in full generality. In particular, Theorem A is a consequence of Theorem B.

6.8. Some Applications

It is not possible to present here an adequate picture of the numerous applications of Cartan's Theorem. But in order to give the reader at least a glimpse of the power and elegance of these methods, we conclude this paragraph by briefly discussing three applications which deal with topics we had encountered earlier in special cases. Again, we limit the statements to domains in $\mathbb{C}^n$, though they are of course valid in much more general settings.

We first present the solution of the weak version of the Poincaré problem (cf. §5.6, in particular, Remark 5.11).

Theorem 6.23. *Let D be Stein. Then every meromorphic function* $m \in \mathscr{M}(D)$ *is the quotient of two global holomorphic functions on D.*

PROOF. Without loss of generality we may assume that $m \in \mathscr{M}^*(D)$. Define the sheaf homomorphism $\mu: \mathscr{M}_D \to \mathscr{M}_D$ by $\mu(\mathbf{s}_z) = \mathbf{m}_z\mathbf{s}_z$ for $\mathbf{s}_z \in \mathscr{M}_z$. The image $\mu(\mathcal{O}_D) = m\mathcal{O}_D$ is an analytic sheaf isomorphic to $\mathcal{O}_D$, and hence is coherent. The formal theory implies that the intersection of two coherent subsheaves of a given sheaf is coherent, so $\mathscr{S} = \mathcal{O}_D \cap m\mathcal{O}_D$ is coherent. By Theorem A there exists a nontrivial global section $f \in \Gamma(D, \mathscr{S})$. Thus $f \in \mathcal{O}(D)$, and there exists $g \in \mathcal{O}(D)$ with $f = mg$; so $m = f/g$. ∎

Next we analyze the image of the Chern map in $H^2(D, \mathbb{Z})$ (cf. §6.4). We have seen in Corollary 6.13 that for a Stein domain D the condition $H^2(D, \mathbb{Z}) = 0$ is sufficient for the universal solvability of the multiplicative Cousin problem. At that time we did not address the question of whether this condition is also necessary. The following result, together with the characterization of principal divisors given by Theorem 6.17, shows that this is indeed the case.

Theorem 6.24. *Let D be Stein. Then the Chern map*

$$\tau: \Gamma(D, \mathscr{M}^*/\mathcal{O}^*) \to H^2(D, \mathbb{Z})$$

is surjective. Moreover, already its restriction $\tau^+: \Gamma(D, \mathcal{O}/\mathcal{O}^*) \to H^2(D, \mathbb{Z})$ *to positive divisors is surjective.*

Proof. Recall that $\tau = \delta^1 \circ \alpha$. Since $\delta^1: H^1(D, \mathcal{O}^*) \to H^2(D, \mathbb{Z})$ is an isomorphism, it is enough to show that the homomorphism $\alpha: \Gamma(D, \mathcal{M}^*/\mathcal{O}^*) \to H^1(D, \mathcal{O}^*)$ in (6.20), respectively its restriction to $\Gamma(D, \mathcal{O}/\mathcal{O}^*)$, is surjective. In order to see this, let $[g] \in H^1(D, \mathcal{O}^*)$ be represented by a cocycle $g = \{g_{ij}\} \in Z^1(\mathcal{U}, \mathcal{O}^*)$ for some locally finite cover $\mathcal{U} = \{U_j: j \in J\}$ of D. We will show that there is $s = \{s_j\} \in C^0(\mathcal{U}, \mathcal{O} \cap \mathcal{M}^*)$, such that

$$s_j s_i^{-1} = g_{ij} \in \mathcal{O}^*(U_i \cap U_j). \tag{6.24}$$

(Notice that in general s will *not* be in $C^0(\mathcal{U}, \mathcal{O}^*)$ unless $H^1(D, \mathcal{O}^*) = 0$, in which case the theorem is trivially true.) The relation (6.24) implies that $\{s_j \in \mathcal{O}(U_j), j \in J\}$ is a holomorphic Cousin II distribution whose associated divisor $\mathscr{d}(s)$—which is clearly positive—is mapped by α onto $[g]$.

We shall find $s \in C^0(\mathcal{U}, \mathcal{O} \cap \mathcal{M}^*)$ satisfying (6.24) as a global section of a special coherent analytic sheaf $\mathcal{S}$, which we now define.[1] Consider the *disjoint* union $\mathcal{X} = \bigcup_{j \in J} \mathcal{O}_{U_j}$ and introduce an equivalence relation $\overset{g}{\sim}$ in $\mathcal{X}$ by defining

$$a_{jz} \overset{g}{\sim} a_{iw} \text{ if and only if } z = w \in U_i \cap U_j \text{ and } a_{jz} = a_{iz} g_{ij}(z). \tag{6.25}$$

It follows from the cocycle condition for $\{g_{ij}\}$ that (6.25) is indeed an *equivalence* relation. It is easily seen that the quotient space $\mathcal{S}$ of $\mathcal{X}$ defined by $\overset{g}{\sim}$ carries the structure of an analytic sheaf over D which is *locally* isomorphic to $\mathcal{O}$. Hence $\mathcal{S}$ is coherent, and by Theorem A there exists a nontrivial global section $s \in \Gamma(D, \mathcal{S})$. For each $j \in J$ the sheaf $\mathcal{S}_{U_j}$ has a natural identification with $\mathcal{O}_{U_j}$. Correspondingly, $s|_{U_j}$ can be represented by $s_j \in \Gamma(D, \mathcal{O}_{U_j}) = \mathcal{O}(U_j)$, and if $z \in U_i \cap U_j$ one must have $s_i(z) \overset{g}{\sim} s_j(z)$. The desired conclusion (6.24) then follows from (6.25). ∎

Finally we consider the following general situation which occurs in numerous applications.

Theorem 6.25. *Let $\mathcal{A}$ be a coherent analytic sheaf over the Stein domain D. Let $s_1, \ldots, s_l \in \Gamma(D, \mathcal{A})$ be global sections which generate the $\mathcal{O}$-submodule sheaf $\mathcal{S} \subset \mathcal{A}$,* i.e., *$s_1(z), \ldots, s_l(z)$ generate $\mathcal{S}_z$ over $\mathcal{O}_z$ for each point $z \in D$. Then $s_1, \ldots, s_l$ generate the $\mathcal{O}(D)$-*module *$\Gamma(D, \mathcal{S})$ over $\mathcal{O}(D)$.*

Proof. Consider the homomorphism $\varphi: \mathcal{O}_D^l \to \mathcal{A}$ defined by $s_1, \ldots, s_l$ as in (6.22). Then $\mathcal{S} = \mathcal{I}m\, \varphi$ is coherent, and $\mathcal{K}er\, \varphi$ is coherent as well, by the definition of coherence for $\mathcal{A}$. So $H^1(D, \mathcal{K}er\, \varphi) = 0$ by Theorem B, and application of Corollary 6.7 to the exact sequence

$$0 \to \mathcal{K}er\, \varphi \to \mathcal{O}_D^l \overset{\varphi}{\to} \mathcal{S} \to 0$$

[1] The reader familiar with line bundles will of course have recognized that we are looking for a global holomorphic section s in the holomorphic line bundle defined by the 1-cocycle $\{g_{ij}\}$.

implies that $\varphi^0: \Gamma(D, \mathcal{O}^l) \to \Gamma(D, \mathcal{S})$ is surjective, i.e., given $s \in \Gamma(D, \mathcal{S})$, there is $(f_1, \ldots, f_l) \in \Gamma(D, \mathcal{O}^l) = l$-fold direct sum of $\mathcal{O}(D)$, so that $\varphi^0(f_1, \ldots, f_l) = \sum_{j=1}^{l} f_j s_j = s$. ∎

We conclude with the following concrete application of Theorem 6.25 which generalizes Hefer's Theorem V.2.2.

Corollary 6.26. *Let $D \subset \mathbb{C}^n$ be Stein and let $A = \{(z, w) \in D \times D: z = w\}$. Then for every $F \in \mathcal{O}(D \times D)$ which vanishes on A there are functions $G_j \in \mathcal{O}(D \times D)$, $j = 1, \ldots, n$, such that*

$$F = \sum_{j=1}^{n} (z_j - w_j) G_j \qquad \text{on } D \times D.$$

Proof. Consider the sections $s_j \in \Gamma(D \times D, \mathcal{O}_{D \times D})$, $1 \le j \le n$, defined by $s_j(z, w) = z_j - w_j$. It is easily seen that $s_1, \ldots, s_n$ generate the ideal sheaf $\mathcal{I}(A)$ of A, and since a function $F \in \mathcal{O}(D \times D)$ which vanishes on A is precisely a section in $\Gamma(D \times D, \mathcal{I}(A))$, the desired conclusion follows from Theorem 6.25 applied to $\mathcal{S} = \mathcal{I}(A) \subset \mathcal{O}_{D \times D}$ over the Stein domain $D \times D$. ∎

Exercises

E.6.1. Let $\mathcal{S}$ be a sheaf of Abelian groups over X and let $\mathcal{S}' \subset \mathcal{S}$ be a subsheaf of groups. Show that the quotient space $\mathcal{S}/\mathcal{S}'$ carries a natural structure of a sheaf of Abelian groups over X.

E.6.2. Let $\mathcal{S}$ be a sheaf over X. Show: if $s_x \in \mathcal{S}_x$, there is a neighborhood U of x and $s \in \Gamma(U, \mathcal{S})$, such that $s(x) = s_x$.

E.6.3. Prove Lemma 6.2 and Corollary 6.3.

E.6.4. Show that $\mathcal{O}_{\mathbb{C}^n}$ is a Hausdorff space.

E.6.5. A sheaf $\mathcal{R}$ over X is called a **sheaf of commutative rings with identity** if each stalk $\mathcal{R}_x$, $x \in X$, is a commutative ring with identity 1_x such that

(i) the maps $\varphi, \psi: \mathcal{R} \oplus \mathcal{R} \to \mathcal{R}$ defined by $\varphi(a_x, b_x) = a_x - b_x$ and $\psi(a_x, b_x) = a_x b_x$, respectively, are continuous, and
(ii) $x \to 1_x$ is continuous, i.e., a section.

Show that $\mathcal{O}_{\mathbb{C}^n}$ and $\mathscr{C}_X^k$, where X is a C^k manifold, are sheaves of commutative rings with identity, the operations being defined in the obvious way.

E.6.6. Let $\mathcal{R}$ be a sheaf of commutative rings with identity over X (cf. E.6.5.). A sheaf $\mathcal{S}$ of Abelian groups over X is called an **$\mathcal{R}$-module sheaf** if each stalk $\mathcal{S}_x$ carries the structure of an $\mathcal{R}_x$-module, so that the map $\mathcal{R} \oplus \mathcal{S} \to \mathcal{S}$ given by $(a_x, s_x) \mapsto a_x s_x$ is continuous. Show that $\mathcal{O}_{\mathbb{C}^n}^l$, $l \ge 1$, is an $\mathcal{O}$-module sheaf.

E.6.7. Let $D \subset \mathbb{C}^n$ be open and define the sheaf $\mathcal{S}$ by

$$\mathcal{S}_a = \begin{cases} \mathcal{O}_a & \text{for } a \in D \\ 0 & \text{for } a \notin D. \end{cases}$$

(i) Show that $\mathcal{S}$ is a sheaf of commutative rings with identity over $\mathbb{C}^n$ (cf. E.6.5).
(ii) Show: $1_a = 0_a$ if and only if $a \notin D$.

E.6.8. For a sheaf $\mathcal{S}$ of Abelian groups over X the support of $\mathcal{S}$ is the set

$$\operatorname{supp} \mathcal{S} = \{x \in X : \mathcal{S}_x \neq \{0_x\}\}.$$

(i) Show that if $D \subset \mathbb{C}^n$ and if $\mathcal{S} \subset \mathcal{O}_D$ is an ideal sheaf of finite type, then $\operatorname{supp} \mathcal{S}$ is an analytic set in D. (The same is true for arbitrary coherent analytic sheaves; this uses the formal theory of such sheaves.)
(ii) If $\mathcal{I}(A)$ is the ideal sheaf of an analytic set $A \subset D$, show that $\operatorname{supp} \mathcal{O}_D/\mathcal{I}(A) = A$.

E.6.9. Let $\varphi: \mathcal{S}' \to \mathcal{S}$ be a sheaf homomorphism between sheaves of Abelian groups. Show that $\mathcal{K}er\ \varphi$ and $\mathcal{I}m\ \varphi$ are subsheaves of $\mathcal{S}'$ and $\mathcal{S}$, respectively.

E.6.10. Prove the Dolbeault Isomorphism (Theorem 6.9) in case $q = 1$. (Hint: Refine the arguments used in the proof of Theorem 4.8.)

E.6.11. Show that if $D \subset \mathbb{C}^n$ is Stein and $H^2(D, \mathbb{Z}) = 0$, then the hypotheses of Theorem 6.18 are satisfied for every $(n-1)$-dimensional complex submanifold M of D.

E.6.12. Let M be a complex submanifold of the region D in $\mathbb{C}^n$. Show that the ideal sheaf $\mathcal{I}(M)$ of M is of finite type.

E.6.13 Let D be open in $\mathbb{C}^n$. Show that D is Stein if $H^1(D, \mathcal{I}) = 0$ for every ideal sheaf $\mathcal{I} \subset \mathcal{O}_D$ with $\operatorname{supp} \mathcal{I}$ (cf. E.6.8) a discrete set in D. (Hint: Use Proposition II.3.4 and the solution of Problem 2 in §6.5 to show that D is holomorphically convex.)

Notes for Chapter VI

Most of the results in this chapter are due to K. Oka, although in many places the presentation differs considerably from the original work. K. Oka was inspired to his fundamental investigations by the 1934 Ergebnisbericht of H. Behnke and P. Thullen [BeTh], who had summarized the state of the theory up to that time and had singled out several unresolved basic problems in global function theory. When reading Oka's work, one is still awed by the far-reaching vision Oka had in 1936, when he set as his goal the solution of these long outstanding problems. In the introduction to his first memoir, after a very brief reference to the unsolved problems, Oka simply states: "The present memoir and those which will follow are meant to treat these problems" ([Oka], p. 1). And in the next sentence, "Now I have noticed that one can sometimes reduce the difficulty of these problems by raising suitably the dimension of the spaces in which one works," he presents a key idea which has been extremely fruitful up to the present time. For example, notice that the fundamental construction of a global holomorphic generating form in a neighborhood of a holomorphically convex compactum given in Chapter V, §2.3, makes essential use of this idea, as does the proof of the Oka Approximation Theorem (Theorem 1.1) given in this chapter.

The proof of the Oka Approximation Theorem is now very simple, as we have available at the outset the solution of $\bar{\partial}$ on Stein compacta, while in Oka's original approach this result—formulated as the solution of the Cousin I problem—had to be deduced from the polydisc case (known since 1895, thanks to P. Cousin) by an ingenious inductive procedure. Also in Oka's first memoir only the simpler case of polynomially convex sets was handled. The approximation theorem corresponding to this special case—stated here as Theorem 1.5—was proved first by A. Weil [Wei] in 1935, who had obtained it as a consequence of a generalization of the Cauchy Integral formula, in the spirit of the classical proof of the Runge Approximation Theorem in the plane. Weil and Oka were aware that Weil's integral formula proof would hold in much greater generality if a certain decomposition for holomorphic functions—now known as Hefer's Theorem (cf. Theorem V.2.2)—would hold. But Oka's method of passing to higher dimensions was more direct and powerful, and already shortly thereafter Oka solved the First Cousin Problem on arbitrary domains of holomorphy D in $\mathbb{C}^n$ [Oka, II], and without explicitly stating so, he obtained, by the obvious extension of his earlier methods, the general case of Theorem 1.1.

The solution of the Cauchy–Riemann equations on Stein domains (Theorem 1.4) is due to Oka in case $q = 1$ (formulated as the solution of the Cousin I problem). The general case was obtained first in 1953 as a consequence of Cartan's Theorem B and of the Dolbeault Isomorphism (see [Dol]).

The notion of Runge pair (§1.4), which became standard during the 1950s, was introduced in [Beh], where Theorem 1.11 was proved. Theorem 1.12 is due to K. Stein [SteK 3]. The Runge property for strictly plurisubharmonic exhaustions (Theorem 1.16) is due to Oka ([Oka], IX). It has undergone considerable generalizations since then (cf., for example, Corollary 1.19, [DoGr] and [Nar 1]). The original proof made use of the fact that for each $P \in bD_c$ there is a function $f \in \mathcal{O}(\bar{D}_c)$ with $f(P) = 0$ and $f(z) \neq 0$ for $z \in \bar{D}_c - \{P\}$. The simpler argument based on holomorphic peaking functions (§1.5) has been used since H. Rossi's proof of the Local Maximum Modulus Principle [Rss] (cf. [GuRo], Chapter IXC). As already mentioned in Chapter V, the Levi Problem (Theorem 1.17) was first solved in 1942 by K. Oka for regions in $\mathbb{C}^2$ ([Oka], VI). The general case was proved in 1953–54, independently, by K. Oka ([Oka], IX), H. Bremermann [Bre 1], and F. Norguet [Nor 1]. Corollary 1.19 was proved—on Stein manifolds—by F. Docquier and H. Grauert [DoGr].

The first *direct* existence proof for solutions of the Cauchy–Riemann equations in pseudoconvex domains (Theorem 2.1), independent of the solution of Cousin I on Stein domains and of the solution of the Levi problem, was given in 1965 by L. Hörmander [Hör 1], who used methods from functional analysis and partial differential equations; related results are due to A. Andreotti and E. Vesentini [AnVe]. The equivalence of (ii) and (iii) in Theorem 2.4 is due to J.P. Serre [Ser 1]. The special case that a domain $D \subset \mathbb{C}^2$ on which Cousin I is solvable (i.e., with $H^1_{\bar{\partial}}(D) = 0$) is necessarily Stein was discovered in 1934 by

H. Cartan ([Car], 471–473). The other equivalences in Theorem 2.4 are, of course, a consequence of the solution of the Levi problem.

The results in §3.1–3.2 are due to Serre [Ser 1], who proved them on Stein manifolds by using sheaf cohomology and Cartan's Theorem B; the more elementary proofs for domains in $\mathbb{C}^n$ given here are taken from [Hör 2]. Theorem 3.6 was proved by Serre in 1955 [Ser 2].

As mentioned in the text, the First Cousin Problem was already solved in 1895 by P. Cousin for product domains. No progress was made on this important question for the next 40 years. As noted above, the breakthrough came in 1936 with K. Oka, who settled first the polynomially convex case ([Oka], I), and shortly thereafter the general holomorphically convex case. ([Oka], II; see Theorem 4.9 in this book; the hard part of the proof given here is, of course, contained in Theorem 1.4.) Unknown to Oka, H. Cartan had already solved the Cousin I problem on polynomially convex regions in $\mathbb{C}^2$ in 1935 ([Car], 471–473) by replacing the Cauchy Integral Formula in P. Cousin's classic proof with the Weil integral formula. A few years later Oka gave Cartan due credit ([Oka], p. 24, footnote 2). The first example—in $\mathbb{C}^3$—of a non-Stein domain on which Cousin I is solvable was discovered in 1938 by H. Cartan ([Car], 536–538), who used Laurent series; no such examples exist in $\mathbb{C}^1$ and $\mathbb{C}^2$!

The history of the multiplicative Cousin problem has been discussed in §5.1. The Oka example in §5.2 is taken from [Oka], III, where Theorem 5.2 was first proved. The proof of Theorem 5.2 given here is an explicit version of the abstract sheaf theoretic proof of J.P. Serre [Ser 1] (see also §6.4). The same goes for the discussion of the topological obstruction, called the Oka property in §5.4. This more general version (than the one considered by Oka) is needed here in order to solve the Cousin II problem for meromorphic data (Theorem 5.3), and not just the holomorphic problem Z. Theorem 5.3 does not follow directly from Oka's work, unless one uses Lemma 5.9, which depends on more detailed local algebraic information, i.e., Theorem 4.4. The proof of Theorem 5.3 is based on Serre's solution of the Cousin II problem (*op. cit.*).

The material in §6 is amply referenced within the text. The theory of coherent analytic sheaves is the natural next step for the reader who wishes to pursue the more recent developments arising from the results of K. Oka discussed in §1–§5.

CHAPTER VII

Topics in Function Theory on Strictly Pseudoconvex Domains

Every domain in the complex plane with C^2 boundary is strictly (Levi) pseudoconvex. In this chapter we generalize several classical function theoretic results from planar domains to strictly pseudoconvex domains in $\mathbb{C}^n$. In contrast to the results on arbitrary domains of holomorphy discussed in Chapter VI, the emphasis here will be on the behavior of holomorphic functions and other analytic objects *up to the boundary* of the domain. In somewhat more detail, we will present the construction and basic properties of two analogues of the Cauchy kernel for a strictly pseudoconvex domain D, and of a solution operator for $\bar{\partial}$ on D with L^p estimates for $1 \leq p \leq \infty$. Moreover, we will discuss applications of these results to uniform and L^p approximation by holomorphic functions and to ideals in the algebra $A(D)$ of holomorphic functions with continuous boundary values. The highlight will be a regularity theorem for the Bergman projection based on a rather explicit representation of the abstract Bergman kernel, and its application to the study of boundary regularity of biholomorphic maps.

These topics, of course, provide just an introduction to the numerous results on strictly pseudoconvex domains which have been obtained during the last 15 years, but they should suffice to give the reader an idea of some of the tools and techniques which are available to study function theory on strictly pseudoconvex domains. Selected references to other topics are given in the Notes at the end of the chapter.

Moreover, perhaps even more problems are awaiting an answer, even in the special case of the unit ball. Compared to the very detailed and extensive knowledge in function theory on the unit disc in the complex plane accumulated in over a century, the corresponding theory on the ball is still very young and in a somewhat rudimentary state, in spite of much progress made in recent years (see W. Rudin [Rud 3,4] for up-to-date accounts). Furthermore, because of the lack of the analogue of the Riemann Mapping Theorem in more than one variable, the natural setting for such a theory should be, at a minimum, on smoothly bounded strictly pseudoconvex domains.

The reader may well wonder about the restriction to *strictly* pseudoconvex domains in this chapter. Shouldn't a smooth pseudoconvex boundary suffice in order to obtain good control of analytic objects at the boundary? Unfortunately, many of the fundamental tools used on strictly pseudoconvex domains do not work on such more general domains (these are often called *weakly* pseudoconvex, for contrast). The failure is most often due to a lack of an analogue of the Levi polynomial, as evidenced by the example of Kohn and Nirenberg ([KoNi]; see also the comments after the proof of Theorem II.2.17 and Exercise E.II.2.6). More surprisingly, some of the results simply are no longer true on weakly pseudoconvex domains without additional assumptions. For example, N. Sibony [Sib] found a smoothly bounded pseudoconvex domain D in $\mathbb{C}^3$ and a bounded $\bar\partial$-closed (0, 1)-form f on D, such that the equation $\bar\partial u = f$ has no *bounded* solution u on D. Altogether, the topics dealt with in this chapter are still understood quite poorly in the absence of strict pseudoconvexity. (References to some partial results known for weakly pseudoconvex domains are given in the Notes.) To improve our understanding of such questions presents a major challenge and opportunity for present and future research in complex analysis.

Some general conventions. In this chapter D will usually denote a bounded domain in $\mathbb{C}^n$ which is assumed to be strictly pseudoconvex with boundary of class C^{k+2} for some $k \geq 0$; in some sections it will be necessary to assume $k \geq 1$ or $k \geq 2$ (alternatively, we assume bD of class C^{k+3} or C^{k+4} with $k \geq 0$). The symbol r denotes a fixed C^{k+2} defining function for D, which will be assumed to be strictly plurisubharmonic on a neighborhood U of bD.[1] As usual, for $\delta \in \mathbb{R}$ close to 0, we set

$$D_\delta = (D - U) \cup \{z \in U : r(z) < \delta\};$$

$|\delta|$ is always assumed to be so small that D_δ is again strictly pseudoconvex with C^{k+2} boundary.

§1. A Cauchy Kernel for Strictly Pseudoconvex Domains

In Chapter IV, §3.2, we constructed the Cauchy kernel C_D for a *convex* domain D in $\mathbb{C}^n$ by making use of the *global* geometric properties of such a domain. We shall now consider analogous kernels for strictly pseudoconvex domains. Here

[1] See Proposition II.2.14. Also, by Exercise E.II.2.8 one could assume that r is strictly plurisubharmonic on a neighborhood of $\bar D$, but we prefer to keep the main hypotheses localized near the boundary.

the geometric information is only *local*. The tool which allows us to pass to a *global* Cauchy-type kernel is the explicit integral solution operator for $\bar\partial$ on Stein compacta discussed in Chapter V, §2.3. In this section we discuss a construction due to Kerzman and Stein [KeSt], which involves a simple modification of the kernel $\Omega_0(L_D)$ which we introduced in Chapter V, §1, and which gives the desired Cauchy kernel in the most direct way.

1.1. An Analogue of the Cauchy Kernel

In Chapter V, §1.1, we constructed a special generating form $L_D = P/\Phi \in C^{k+1,\infty}_{0,1}(bD \times D)$ for every strictly pseudoconvex domain D. Recall that $\Phi(\zeta, z)$ agrees with the Levi polynomial of the defining function r for $|\zeta - z| < \varepsilon/2$ (or rather, with a second-order perturbation $F^\#$ of the Levi polynomial), and that $P = \sum_{j=1}^n P_j\, d\zeta_j$, where $\Phi = \sum_{j=1}^n P_j(\zeta_j - z_j)$ is the obvious explicit decomposition resulting from the definition of Φ. It followed from the construction that for fixed $\zeta \in bD$, L_D is holomorphic in z on $\{z \in \bar D: 0 < |z - \zeta| < \varepsilon/2\}$. To simplify notation we denote the associated Cauchy–Fantappiè form $\Omega_0(L_D)(\zeta, z)$ of order 0 by $E(\zeta, z)$.

Lemma 1.1. *If $f \in A^1(D)$, then*

$$f(z) = \int_{bD} f(\zeta)E(\zeta, z) \qquad \text{for } z \in D. \tag{1.1}$$

Proof. This follows from the case $q = 0$ of Theorem IV.3.6, with $W = L_D$. Notice that $T_1 \bar\partial f \equiv 0$ since f is holomorphic. ■

Thus the kernel E reproduces holomorphic functions from their boundary values, but E is holomorphic in z only for z close to ζ (i.e. close to the singularity). We now modify the kernel E to make it globally holomorphic in z.

The construction in V.1.1 gives $\delta > 0$ so that E is well defined with coefficients in $C^{k,\infty}$ for all (ζ, z) with $\zeta \in bD$ and $z \in D_\delta$ with $|z - \zeta| \geq \varepsilon/2$. The same is then true for $\bar\partial_z E$, and since $\bar\partial_z E \equiv 0$ for $|z - \zeta| \leq \varepsilon/2$, $\bar\partial_z E$ defines, by trivial extension, a double form on $bD \times D_\delta$ of type $(n, n-1)$ in ζ and type $(0, 1)$ in z, with coefficients in $C^{k,\infty}$.

Lemma 1.2. *There are a neighborhood $D^\#$ of $\bar D$ and a double form*

$$A(\zeta, z) \in C^{k,\infty}_{n,n-1}(bD \times D^\#) \qquad \text{such that}$$

$$\bar\partial_z A(\zeta, z) = \bar\partial_z E(\zeta, z) \qquad \text{on } bD \times D^\#.$$

Proof. We shall use the integral solution operator for $\bar\partial$ on a neighborhood of $\bar D$ given by Theorem V.2.5. Specifically, we choose open neighborhoods $D^\#$ and V of the Stein compactum $K = \bar D$, with $\bar D \subset\subset D^\# \subset\subset V \subset\subset D_\delta$, so that the operator

$$\mathbf{T}_1^{V,D^\#}: C_{0,1}^1(\bar{V}) \to C(D^\#)$$

is well defined. In particular $\bar{\partial}(\mathbf{T}_1^{V,D^\#}f) = f$ on $D^\#$ if f is a $\bar{\partial}$-closed (0, 1)-form on $\bar{V}$, and $\mathbf{T}_1^{V,D^\#}f$ has the same regularity properties as f, including dependence on parameters. Therefore, by viewing $\bar{\partial}_z E(\zeta, z)$ as a (0, 1)-form in z with coefficients being $(n, n-1)$-forms in ζ, it follows that for fixed $\zeta \in bD$, $A(\zeta, \cdot) := \mathbf{T}_1^{V,D^\#}(\bar{\partial}_z E(\zeta, \cdot))$ is a C^∞ function in $z \in D^\#$ which satisfies $\bar{\partial}_z A(\zeta, z) = \bar{\partial}_z E(\zeta, z)$. Moreover, since $\bar{\partial}_z E \in C^{k,\infty}$, so is A. ∎

Theorem 1.3. *The kernel $C(\zeta, z) = E(\zeta, z) - A(\zeta, z) \in C_{n,n-1}^{k,\infty}(bD \times \bar{D} - \{(\zeta, \zeta): \zeta \in bD\})$ has coefficients which are holomorphic in z on $\bar{D} - \{\zeta\}$. For all $f \in A(D)$ one has*

$$f(z) = \int_{bD} f(\zeta) C(\zeta, z) \qquad \textit{for } z \in D. \tag{1.2}$$

Proof. The first statement is clear, since $\bar{\partial}_z(E - A) = 0$ on $\bar{D} - \{\zeta\}$ by Lemma 1.2. In order to prove the reproducing property (1.2) we first assume that $f \in A^1(D)$. Because of (1.1) we must show that

$$\int_{bD} f(\zeta) A(\zeta, z) = 0 \qquad \text{for } z \in D. \tag{1.3}$$

By interchanging the order of integration—recall that $\mathbf{T}_1^{V,D^\#}$ is defined by integration—it follows that

$$\int_{bD} fA = \int_{bD} f\mathbf{T}_1^{V,D^\#}(\bar{\partial}_z E) = \mathbf{T}_1^{V,D^\#}\left(\int_{bD} f(\zeta)\bar{\partial}_z E(\zeta, z)\right). \tag{1.4}$$

We shall now show that the integral on the right in (1.4), to which $\mathbf{T}_1^{V,D^\#}$ is applied, is identically 0 on D_δ, so that (1.3) follows. Fix $z \in D_\delta$. Then $\bar{\partial}_z E = 0$ on $\{\zeta \in bD: |\zeta - z| \leq \varepsilon/2\}$, and on $\{\zeta \in bD: |\zeta - z| \geq \varepsilon/2\}$ $E = \Omega_0(L_D)$ is a CF-form without singularities. Therefore, Lemma IV. 3.5 implies

$$\int_{bD} f\bar{\partial}_z E = \int_{\zeta \in bD; |\zeta - z| \geq \varepsilon/2} f\bar{\partial}_z \Omega_0(L_D) = -\int_{\zeta \in bD; |\zeta - z| \geq \varepsilon/2} f\bar{\partial}_\zeta \Omega_1(L_D)$$
$$= -\int_{bD} f(\zeta)\bar{\partial}_\zeta \Omega_1(L_D),$$

where we have used $\Omega_1(L_D) \equiv 0$ on $\{\zeta \in bD: |\zeta - z| \leq \varepsilon/2\}$ in the last equality (see the proof of Proposition V.1.1).

Since Ω_1 is of type $(n, n-2)$ in ζ, Stokes' Theorem implies

$$-\int_{bD} f\bar{\partial}_\zeta \Omega_1 = -\int_{bD} fd_\zeta \Omega_1 = \int_{bD} df \wedge \Omega_1 = \int_{bD} \bar{\partial} f \wedge \Omega_1.$$

Here the last integral is zero for $f \in A^1(D)$. Thus we have proved (1.2) for $f \in A^1(D)$.

If f is only in $A(D)$, we apply the preceding arguments to the domain D_η for $\eta < 0$ sufficiently close to zero—notice that L_D is also a generating form on $bD_\eta \times D_\eta$, and that $\bar{\partial}_z E(\zeta, z)$ and $A(\zeta, z) = \mathbf{T}_1^{V, D^\#}(\bar{\partial}_z E(\zeta, z))$ are in fact in $C^{k,\infty}(U' \times D^\#)$ for a suitable neighborhood $U' \subset\subset U$ of bD. Hence, given $z \in D$, we choose $\eta < 0$ so that $z \in D_\eta$. Since $f \in A^\infty(D_\eta)$, we have $f(z) = \int_{bD_\eta} f(\zeta) C(\zeta, z)$. Now let $\eta \to 0$; the continuity of f and $C(\zeta, z)$ for $\zeta \in \bar{D}$ sufficiently close to bD implies that

$$\lim_{\eta \to 0} \int_{bD_\eta} fC(\cdot, z) = \int_{bD} fC(\cdot, z). \quad \blacksquare$$

1.2. A Regularity Property of $C = E - A$

The kernel $C = E - A$ is a useful higher dimensional analog for strictly pseudoconvex domains of the Cauchy kernel in $\mathbb{C}^1$. Notice that in the case $n = 1$, the construction in §1.1 trivially gives the Cauchy kernel: the uniqueness of generating forms in dimension 1 gives $L_D = d\zeta/(\zeta - z)$, so $E = \Omega_0(L_D) = (2\pi i)^{-1}\, d\zeta/(\zeta - z)$ (see IV.§1.2), which is already globally holomorphic in z. Hence $\bar{\partial}_z E = 0$ and therefore $A = 0$, i.e., $C(\zeta, z) = (2\pi i)^{-1}\, d\zeta/(\zeta - z)$. The "essential" part E, and, in particular, the singularity of C are completely explicit, while the term A, which is somewhat less explicit, is harmless as far as boundary regularity is concerned. Since $A(\zeta, z)$ has coefficients in $C^{0,\infty}(bD \times D^\#)$, it is obvious that $\mathbf{A}f = \int_{bD} fA$ is C^∞ on $D^\#$ for any $f \in L^1(bD)$, and that

$$(1.5) \qquad |\mathbf{A}f|_{C^l(\bar{D})} \lesssim \|f\|_{L^1(bD)} \qquad \text{for } l = 0, 1, 2, \ldots.$$

We now generalize some other classical results for the Cauchy kernel in $\mathbb{C}^1$ to the kernel C.

Theorem 1.4. *Define* $\mathbf{C}f(z) = \int_{bD} fC(\cdot, z)$. *Then*

(i) $\mathbf{C}f$ *is holomorphic on* D *for all* $f \in L^1(bD)$.
(ii) $\mathbf{C}\colon L^p(bD) \to \mathcal{O}(D)$ *is continuous for* $1 \le p \le \infty$.
(iii) $\mathbf{C}\colon \Lambda_\alpha(bD) \to \mathcal{O}\Lambda_{\alpha/2} D)$ *is bounded for* $0 < \alpha < 1$.
(iv) *If* $\chi \in \Lambda_\alpha(bD)$ *for some* $\alpha > 0$, *and if* $f \in A(D)$, *then* $\mathbf{C}(\chi f) \in A(D)$.

We first prove an estimate which will be useful in other places as well.

Lemma 1.5. *Let* Φ *be the denominator of the generating form* L_D, *and set*

$$J_\alpha(z) = \int_{bD} \frac{dS}{|\Phi(\cdot, z)|^{n+\alpha}} \qquad \text{for } z \in \bar{D}.$$

Then

(1.6) $$J_\alpha(z) \lesssim \begin{cases} 1 & \textit{if } \alpha < 0 \textit{ and } z \in \bar{D}; \\ |\log \delta_D(z)| & \textit{if } \alpha = 0 \textit{ and } z \in D; \\ \delta_D^{-\alpha}(z) & \textit{if } \alpha > 0 \textit{ and } z \in D. \end{cases}$$

PROOF. We shall use the special real coordinate system $t = (t_1, t_2, t')$ on the ball $B(z, \eta)$ introduced in Lemma V.3.4. With the notation chosen there, it is clearly enough to prove (1.6) for $\delta_D(z) \leq a$, and the region of integration replaced by $bD \cap B(z, \eta)$. Set $\delta = \delta_D(z)$. By using the estimate $|\Phi| \gtrsim |t_2| + \delta + |t'|^2$ for $\zeta \in bD \cap B(z, \eta)$ (see V(3.18)), it follows that

(1.7) $$J_\alpha(z) \lesssim \int_{0<t_2<1;|t'|\leq 1} \frac{dt_2 \dots dt_{2n}}{[t_2 + \delta + |t'|^2]^{n+\alpha}}.$$

Since the required estimate in case $n = 1$ is obvious, we shall assume that $n \geq 2$. By integrating in t_2 and introducing polar coordinates for $t' \in \mathbb{R}^{2n-2}$ in the integral in (1.7), one obtains

(1.8) $$J_\alpha(z) \lesssim \int_0^1 \frac{\rho^{2n-3}\, d\rho}{(\delta + \rho^2)^{n+\alpha-1}} \qquad (\text{if } \alpha \neq 1 - n).$$

It is now clear that $J_\alpha(z) \lesssim 1$ if $\alpha < 0$ (even if $\alpha = 1 - n$). If $\alpha = 0$, (1.8) implies

$$J_\alpha(z) \lesssim \int_0^1 \frac{\rho\, d\rho}{\delta + \rho^2} \lesssim |\log \delta|.$$

Finally, if $\alpha > 0$, we substitute $\rho = \sqrt{\delta} s$ in (1.8), obtaining

$$J_\alpha(z) \lesssim \int_0^{1/\sqrt{\delta}} \frac{\delta^{n-1}\, ds}{\delta^{n+\alpha-1}(1 + s^2)^{n+\alpha-1}}$$

$$\leq \delta^{-\alpha} \int_0^\infty \frac{ds}{(1 + s^2)^{n-1+\alpha}}.$$

Since $n - 1 + \alpha > 1$, the last integral converges, and we are done. ∎

PROOF OF THEOREM 1.4. (i) and (ii) are straightforward. In order to verify (iii), we shall prove that

(1.9) $$|d_z \int_{bD} fE(\cdot, z)| \lesssim [\delta_D(z)]^{\alpha/2-1} |f|_{\Lambda_\alpha(bD)} \qquad \text{for } z \in D.$$

Together with Lemma V.3.1, (1.9) will imply $|\int_{bD} fE|_{\Lambda_{\alpha/2}(D)} \lesssim |f|_{\Lambda_\alpha(bD)}$, and the required estimate for $\mathbf{C}f$ will then follow from (1.5). Applying Lemma 1.1 to the function $f \equiv 1$ gives $d_z \int_{bD} E(\cdot, z) \equiv 0$. If $z \in D$ is fixed we choose $z' \in bD$ with $|z - z'| = \delta_D(z)$, so that $|\zeta - z'| \leq 2|\zeta - z|$. Since $\int_{bD} f(z')\, d_z E(\cdot, z) \equiv 0$, it follows that

$$\int_{bD} f(\zeta)\, d_z E(\zeta, z) = \int_{bD} [f(\zeta) - f(z')]\, d_z E(\zeta, z),$$

and hence

$$\left|\int_{bD} f\, d_z E(\cdot, z)\right| \lesssim |f|_{\Lambda_\alpha} \int_{bD} |\zeta - z|^\alpha |d_z E|\, dS. \tag{1.10}$$

From $E = \Omega_0(P/\Phi)$ one easily obtains

$$|d_z E(\zeta, z)| \lesssim |\Phi(\zeta, z)|^{-n-1}, \tag{1.11}$$

and hence (1.10) implies

$$\left|d_z \int_{bD} fE(\cdot, z)\right| \lesssim |f|_{\Lambda_\alpha} \int_{bD} \frac{|\zeta - z|^\alpha}{|\Phi|^{n+1}}\, dS \tag{1.12}$$

$$\lesssim |f|_{\Lambda_\alpha} \int_{bD} \frac{1}{|\Phi|^{n+1-\alpha/2}}\, dS,$$

where we have used $|\Phi| \gtrsim |\zeta - z|^2$ in the last step. Equation (1.9) now follows from (1.12) and Lemma 1.5, and the proof of (iii) is complete. In order to prove (iv) it is enough to show that $\mathbf{C}(\chi f)$ extends continuously to $\bar{D}$. Without loss of generality we may assume that $\chi \in \Lambda_\alpha(\bar{D})$. From (1.2) it follows that

$$\mathbf{C}(\chi f)(z) = \chi(z) f(z) + \int_{bD} f(\zeta)[\chi(\zeta) - \chi(z)] C(\zeta, z) \tag{1.13}$$

for $z \in D$. Clearly χf is continuous on $\bar{D}$. Since $|E(\zeta, z)| \lesssim |\Phi|^{-n}$ and A is uniformly bounded on $bD \times \bar{D}$, it follows that the integrand $J(\zeta, z) = f(\zeta)[\chi(\zeta) - \chi(z)] C(\zeta, z)$ satisfies

$$|J(\zeta, z)| \lesssim \frac{|\zeta - z|^\alpha}{|\Phi(\zeta, z)|^n} \lesssim \frac{1}{|\Phi|^{n-\alpha/2}}. \tag{1.14}$$

Equation (1.14) and Lemma 1.5 show that $J(\cdot, z)$ is uniformly integrable over bD for all $z \in \bar{D}$, so that Lebesgue's Dominated Convergence Theorem can be applied to show that the integral on the right side of (1.13) is continuous in z on $\bar{D}$ as well. ∎

Remark 1.6. It is not true, even in the case $n = 1$, that $\mathbf{C}(f)$ extends continuously to $\bar{D}$ if f is only continuous on bD; (iii) and (iv) give useful sufficient conditions for the continuity of $\mathbf{C}(f)$ on $\bar{D}$ (see also the discussion of the Bochner–Martinelli transform in IV.§2). By somewhat more complicated arguments one can show that if bD is of class C^3 and $\alpha < 1$, then $\mathbf{C}$ is in fact bounded from $\Lambda_\alpha(bD)$ to $\Lambda_\alpha(D)$ and not just to $\Lambda_{\alpha/2}(D)$. The interested reader may find more details in Exercise E.1.2.

Exercises

E.1.1. Let V be an allowable vector field on $\bar{D}$ (see E.V.3.2 for the definition) and let $f \in \Lambda_\alpha(bD)$ for $\alpha < 1$. Show that

$$|(V\mathbf{C}f)(z)| \lesssim |f|_{\Lambda_\alpha} \delta_D^{-1+\alpha}(z) \qquad \text{for } z \in D.$$

E.1.2. Prove that if bD is of class C^3, then the Cauchy transform $\mathbf{C}$ is bounded from $\Lambda_\alpha(bD)$ into $\Lambda_\alpha(D)$ for $\alpha < 1$. (Hint: Near $P \in bD$ use a C^2 coordinate system $(w_1, w_2, \ldots, w_n)$ with $w_1(\zeta) = \Phi(\zeta, P)$; notice that

$$|f(w(\zeta)) - f(w(P))| \leq |f(w(\zeta)) - f(w_1(P), w'(\zeta))| + |f(w_1(P), w'(\zeta)) - f(w(P))|,$$

and estimate each term separately. The first one requires an integration by parts! See [AhSc] for more details.)

§2. Uniform Approximation on $\bar{D}$

2.1. Some Background

As usual, we assume D strictly pseudoconvex with C^2 boundary. From Theorem VI.1.16 we know that (D, D_δ) is a Runge pair for sufficiently small $\delta > 0$. Hence every $f \in \mathcal{O}(D)$ can be approximated uniformly on compact subsets of D by functions in $\mathcal{O}(D_\delta)$. We will show in this section that if $f \in A(D)$, then one can achieve uniform approximation on $\bar{D}$! (Theorem 2.1 below). This result was first proved around 1969–1970 by Henkin [Hen 1], Kerzman [Ker], and Lieb [Lie 1], independently.

In case of dimension 1, the corresponding result is much older, and it holds in greater generality. S.N. Mergelyan proved in 1952 that if $K \subset \mathbb{C}$ is compact and $\mathbb{C} - K$ has finitely many connected components, then every $f \in C(K)$ which is holomorphic on the interior of K can be approximated uniformly on K by rational functions with poles off K. ("Uniform approximation to functions of a complex variable", Uspehi Mat. Nauk **7**, 31–122 (1952). See also [Gam 1] for a modern "function algebra" proof.) The hard part is, of course, to prove uniform approximation by functions in $\mathcal{O}(K)$; the rest then follows from the Runge Approximation Theorem. We shall therefore concentrate on the analogous result in several variables.

Theorem 2.1. *Let $D \subset\subset \mathbb{C}^n$ be strictly pseudoconvex with C^2 boundary. Then every $f \in A(\bar{D})$ can be approximated uniformly on $\bar{D}$ by functions in $\mathcal{O}(\bar{D})$.*

The strong geometric hypothesis required in Theorem 2.1—in comparison to the classical one-variable result—are not only needed because of technical limitations of the method of proof, but they also reflect intriguing new phenomena which are unique to several variables, as is seen from the following example.

Example 2.2. Let $D = \{(z, w) \in \mathbb{C}^2 : 0 < |z| < |w| < 1\}$. It is easy to see that D is a Stein domain and that $f(z, w) = z^2/w$ defines a function in $A(D)$ which is completely singular at 0. Suppose there were a sequence $\{g_j : j = 1, 2, \ldots\} \subset \mathcal{O}(\bar{D})$ which converges to f uniformly on $\bar{D}$. It follows from Theorem I.1.6 that every g_j extends holomorphically to $\bar{P} = \overline{P(0, 1)}$, and since $b_0 P = \{|z| = |w| = 1\} \subset \bar{D}$, uniform convergence of $\{g_j\}$ on $b_0 P$ implies uniform convergence of $\{g_j\}$ on $\bar{P}$ to a function g (by the Maximum Principle Theorem I.1.8). g is then necessarily holomorphic on P, and $g = f$ on $D \subset P$. Thus g would give a holomorphic extension of f to 0, which is impossible. So the analogue of Mergelyan's Theorem does not hold for the compact set $\bar{D}$.

One might hope that such phenomena could not occur in case D has smooth boundary. But in 1975, quite surprisingly, K. Diederich and J. Fornaess [DiFo 2] constructed a pseudoconvex domain D in $\mathbb{C}^2$ with C^∞ boundary and a function $f \in A^\infty(D)$ which is not the uniform limit on $\bar{D}$ of functions in $\mathcal{O}(\bar{D})$. A common feature of the Diederich–Fornaess example and the one discussed above, which appears to be quite relevant for the failure of approximation, is the fact that even though D is a Stein domain, the closure $\bar{D}$ is not Stein in either case (i.e., there does not exist a neighborhood basis of Stein domains for $\bar{D}$). In spite of several other known partial results (see the Notes at the end of the chapter), the general situation is still very little understood.

2.2. Separation of Singularities

The main step in the proof of Theorem 2.1 given in this section is the following result on "separation of singularities", which is an easy consequence of the properties of the Cauchy kernel $C(\zeta, z)$ constructed in §1.1.

Lemma 2.3. *Let $\{U_\nu, 1 \leq \nu \leq l\}$ be a finite open cover of bD. Then every $f \in A(D)$ has a decomposition*

$$f = f_1 + \cdots + f_l, \tag{2.1}$$

where $f_\nu \in A(D)$ is holomorphic on $\bar{D} - U_\nu$ for $1 \leq \nu \leq l$.

Proof. Choose functions $\chi_\nu \in C_0^\infty(U_\nu)$, $1 \leq \nu \leq l$, with $\sum_{\nu=1}^{l} \chi_\nu = 1$ on bD. By Theorem 1.3,

$$f = \mathbf{C}(f) = \sum_{\nu=1}^{l} \mathbf{C}(\chi_\nu f) \qquad \text{on } D,$$

and by Theorem 1.4(iv), $f_\nu = \mathbf{C}(\chi_\nu f)$ is in $A(D)$. Also, Theorem 1.3 shows that $C(\zeta, \cdot)$ is holomorphic on $\bar{D} - \{\zeta\}$. Since $f_\nu = \mathbf{C}(\chi_\nu f)$ involves only integration over $bD \cap U_\nu$, it easily follows that $f_\nu \in \mathcal{O}(\bar{D} - U_\nu)$. ■

2.3. Proof of Theorem 2.1

By Lemma 2.3, in order to approximate $f \in A(D)$, it is clearly enough to do this for each f_ν in the decomposition (2.1), whose singularities are concentrated on $bD \cap U_\nu$. In that case, suitable "translates" of f_ν will do, provided U_ν is sufficiently small. To make this precise, we use the following elementary geometric fact.

Lemma 2.4. *Let $D \subset\subset \mathbb{R}^n$ have C^1 boundary at $P \in bD$, and let* $\mathbf{n}$ *be the unit inner normal to bD at P. Then there are a neighborhood U of P and $\tau_0 > 0$, such that*

$$(2.2) \qquad z + \tau\mathbf{n} \in D \quad \text{for all } z \in \bar{D} \cap U \quad \text{and } 0 < \tau < \tau_0.$$

Assuming the Lemma, let us complete the proof of Theorem 2.1. Fix $P \in bD$ and choose U and τ_0 as in the Lemma. Suppose $f \in A(D)$ is holomorphic on $\bar{D} - U$, i.e., there is an open neighborhood W of $\bar{D} - U$, such that $f \in \mathcal{O}(W)$. By shrinking W, we may suppose that f is uniformly continuous on W. Since $\eta = \operatorname{dist}(\bar{D} - U, bW)$ is positive, we can assume that $0 < \tau_0 < \eta$. It follows that $z + \tau\mathbf{n} \in W$ for $z \in \bar{D} - U$ and $\tau < \tau_0$. Together with (2.2) one therefore obtains

$$(2.3) \qquad z + \tau\mathbf{n} \in W \cup D \quad \text{for all } z \in \bar{D} \quad \text{and } 0 < \tau < \tau_0.$$

Hence $f_\tau(z) := f(z + \tau\mathbf{n})$ is holomorphic on $\bar{D}$ for $0 < \tau < \tau_0$, and by uniform continuity of f on $W \cup \bar{D}$, it follows that $f_\tau \to f$ uniformly on $\bar{D}$ as $\tau \to 0$. The rest is now routine: by compactness of bD, we cover bD by finitely many open neighborhoods $U_1, \ldots, U_l$, for which Lemma 2.4 holds. Given $f \in A(D)$, the argument just given shows that each f_ν in the decomposition of f given by Lemma 2.3 is a uniform limit on $\bar{D}$ of functions in $\mathcal{O}(\bar{D})$, and hence so is f. It remains to prove the Lemma.

Proof of Lemma 2.4. Without loss of generality we may assume that $P = 0$, $\mathbf{n} = (0, \ldots, -1)$, and that there is a local defining function $r^\#$ for D of the form $r^\# = x_n - \varphi(x_1, \ldots, x_{n-1})$ on a neighborhood $U^\# = U' \times (-\varepsilon, \varepsilon)$, where $\varepsilon > 0$ and U' is a neighborhood of 0 in $\mathbb{R}^{n-1}$ (see II.§2.3, formula (2.9)). Now set $\tau_0 = \varepsilon/2$ and $U = U' \times (-\tau_0, \tau_0)$, and (2.2) follows. ■

In §6 we shall give an alternate proof of Theorem 2.1 based on solving a Cousin I problem with bounds. At that time we shall also consider approximation of functions in $\mathcal{O}L^p(D)$.

Corollary 2.5. *Under the hypothesis of Theorem 2.1, if $\delta > 0$ is sufficiently small, then every $f \in A(D)$ can be approximated uniformly on $\bar{D}$ by functions in $\mathcal{O}(D_\delta)$.*

Proof. By the Theorem, $f \in A(D)$ can be approximated uniformly by functions in $\mathcal{O}(\bar{D})$. Now use the Runge property of (D, D_δ) (see VI.§1.6): $\hat{\bar{D}}_{\mathcal{O}(D_\delta)} = \bar{D}$ for δ sufficiently small. Hence every $h \in \mathcal{O}(\bar{D})$ can be approximated uniformly on $\bar{D}$ by functions holomorphic on D_δ (Theorem VI.1.10). ■

Exercises

E.2.1. Complete the missing details in the discussion of Example 2.2.

E.2.2. Let D be strictly pseudoconvex with C^2 boundary and $0 < \alpha < 1$.

(i) Show that every $f \in \mathcal{O}\Lambda_\alpha(D) := \mathcal{O}(D) \cap \Lambda_\alpha(D)$ can be approximated in $\Lambda_{\alpha/2}(D)$ by functions in $\mathcal{O}(\bar{D})$.

(ii) Show that if D has C^3 boundary, the approximation in (i) is in Λ_α norm.

§3. The Kernel of Henkin and Ramirez

In this section we follow the pioneering work of Henkin [Hen 1] and Ramirez [Ram] and construct a generating form on $bD \times D$ which, in contrast to the generating form $L_D = P/\Phi$ used in §1, is *globally* holomorphic on D. The construction involves two major steps: first one constructs a function $g(\zeta, z)$ on $bD \times \bar{D}$ which is holomorphic for z in $\bar{D}$ and vanishes only for $z = \zeta$; then one proves a decomposition $g(\zeta, z) = \sum_{j=1}^n g_j(\zeta, z)(\zeta_j - z_j)$, with g_j still holomorphic for z in $\bar{D}$. The desired generating form is then given by

$$W^{\mathrm{HR}} = \sum_{j=1}^n g_j \, d\zeta_j / g.$$

Both steps make use of the explicit integral solution operator $\mathbf{T}_1^{V,D^\#}$ for $\bar{\partial}$ in a neighborhood of $\bar{D}$ given by Theorem V.2.5. This makes it easy to establish the necessary differentiability with respect to the parameter ζ.

We then combine the generating form W^{HR} with the general integral representation formula from Chapter IV, §3 in order to obtain a Cauchy-type kernel for the strictly pseudoconvex domain D which is closer in spirit to the Cauchy kernel for convex domains discussed in IV.§3.2 than the kernel constructed in §1. By precisely identifying the principal parts of this kernel as well as of the one discussed in §1, we will see that the principal parts of the two kernels are in fact identical.

3.1. A Smooth Family of Peaking Functions

The first step in the construction of the globally holomorphic generating form involves finding a function $g(\zeta, z) \in C^1(bD \times \bar{D})$ which is holomorphic in z and satisfies $g(\zeta, z) \neq 0$ for $z \in \bar{D} - \{\zeta\}$. The proof given below is based on a parametrized version of the construction of holomorphic peaking functions in Theorem VI.1.13. We start with the function Φ on $U \times \mathbb{C}^n$, where U is a neighborhood of bD, which was defined in V.§1.1 by patching a modification $F^\#$ of the Levi polynomial of the defining function for D with $|z - \zeta|^2$. Recall that Φ is holomorphic in z for $|z - \zeta| < \varepsilon/2$ and that

$$\operatorname{Re} \Phi(\zeta, z) \gtrsim r(\zeta) - r(z) + c|\zeta - z|^2 \tag{3.1}$$

(see V(1.6) and V(1.7)). In particular, by choosing U and $\delta > 0$ sufficiently small, we may assume that $\operatorname{Re} \Phi(\zeta, z) > 0$ for all $(\zeta, z) \in U \times D_\delta$ with $|\zeta - z| \geq \varepsilon/2$ (see V(1.8)), and that $\alpha = \bar{\partial}_z(1/\Phi)$ extends trivially to a $\bar{\partial}_z$-closed $(0, 1)$-form on D_δ with coefficients in $C^{k+1,\infty}(U \times D_\delta)$. Now choose $0 < \delta^\# < \delta$ and V with $D_{\delta^\#} \subset\subset V \subset\subset D_\delta$, so that Theorem V.2.5 applies, giving the integral solution operator $\mathbf{T} = \mathbf{T}_1^{V, D_{\delta^\#}}$ for $\bar\partial$. For $\zeta \in U$ we set $v(\zeta, \cdot) = \mathbf{T}(\alpha(\zeta, \cdot))$. Then $v \in C^{k+1,\infty}(U \times D_{\delta^\#})$ and $\bar\partial_z v = \alpha$; so v is holomorphic in z for $|z - \zeta| < \varepsilon/2$. Without loss of generality we may assume that

$$m = \sup_{U \times D_{\delta^\#}} |v(\zeta, z)| < \infty. \tag{3.2}$$

It follows that $u = 1/\Phi + (m - v)$ is holomorphic in z if $\Phi(\zeta, z) \neq 0$, and that

$$\operatorname{Re} u \geq \operatorname{Re} 1/\Phi > 0 \qquad \text{on } \{(\zeta, z) \in U \times D_{\delta^\#} \colon \operatorname{Re} \Phi(\zeta, z) > 0\}. \tag{3.3}$$

Therefore

$$\begin{aligned} &g = 1/u \textit{ is holomorphic in } z \textit{ and} \\ &\operatorname{Re} g > 0 \textit{ on } \{(\zeta, z) \in U \times D_{\delta^\#} \colon \operatorname{Re} \Phi(\zeta, z) > 0\}. \end{aligned} \tag{3.4}$$

We now show that g is actually holomorphic for all $z \in D_{\delta^\#}$. Clearly $g = \Phi \cdot A$, where

$$A = [1 + \Phi(m - v)]^{-1}. \tag{3.5}$$

Since $\Phi(\zeta, \zeta) = 0$, there is γ with $0 < \gamma \leq \varepsilon/2$ so that

$$|\Phi(m - v)| \leq 1/2 \qquad \text{for } |z - \zeta| \leq \gamma.$$

It follows that A and $g = \Phi A$ are of class $C^{k+1,\infty}$ and holomorphic in z for $|z - \zeta| < \gamma$, and that $|A(\zeta, z)| \geq \frac{2}{3}$. Finally, because of (3.1), by shrinking U and $\delta^\#$ further, we can assume that $\operatorname{Re} \Phi > 0$ on $\{(\zeta, z) \in U \times D_{\delta^\#} \colon |\zeta - z| \geq \gamma\}$, so that, by (3.4), g is of class $C^{k+1,\infty}$ and holomorphic in z on this set as well.

We now summarize the basic properties of the function g so obtained (we write again δ instead of $\delta^\#$).

Proposition 3.1. *Suppose $D \subset\subset \mathbb{C}^n$ is strictly pseudoconvex with C^{k+2} boundary, $k \geq 0$. There are a neighborhood U of bD, positive constants δ, c, and γ, and a function $g \in C^{k+1,\infty}(U \times D_\delta)$ with the following properties.*

(i) *$g(\zeta, z)$ is holomorphic in z on D_δ.*
(ii) *$g(\zeta, \zeta) = 0$ for $\zeta \in U$.*
(iii) *$\operatorname{Re} g(\zeta, z) > 0$ for $(\zeta, z) \in U \times D_\delta$ with $r(\zeta) - r(z) + c|\zeta - z|^2 > 0$.*
(iv) *On $\{(\zeta, z) \in U \times D_\delta \colon |\zeta - z| \leq \gamma\}$ there is a function $A \in C^{k+1,\infty}$ with $|A(\zeta, z)| \geq \frac{2}{3}$, so that $g = F^\# \cdot A$, where $F^\#$ is the modification of the Levi polynomial introduced in* V(1.3).

Corollary 3.2. *Given D as in Proposition* 3.1, *there is a function $H \in C^{k+1}(bD \times D_\delta)$ such that for each $\zeta \in bD$ one has*

(i) *$H(\zeta, \cdot) \in \mathcal{O}(D_\delta)$,*
and
(ii) *$H(\zeta, \zeta) = 1$ and $|H(\zeta, z)| < 1$ for $z \in \bar{D} - \{\zeta\}$.*

PROOF. $H(\zeta, z) = \exp(-g(\zeta, z))$ will do. ∎

3.2. Hefer's Theorem with Parameters

The next step involves decomposing the function $g(\zeta, z)$ given in Proposition 3.1 as $\sum g_j(\zeta, z)(\zeta_j - z_j)$, with g_j holomorphic in z. For ζ fixed, this would follow immediately from Theorem V.2.2. By using the explicit integral solution operator for $\bar\partial$ given by Theorem V.2.5, we will now prove a parametrized version of that theorem. If M is a C^k manifold and $K \subset \mathbb{C}^n$ is compact, we denote by $C^{k,\infty}(M \times K)$ the set of functions f which are in $C^{k,\infty}(M \times W)$ for some open neighborhood W of K (which may depend on f).

Proposition 3.3. *Let $K \subset \mathbb{C}^n$ be a Stein compactum and let M be a C^k manifold, $k \geq 1$. Given $f \in C^{k,\infty}(M \times K)$ such that $f(x, \cdot) \in \mathcal{O}(K)$ for $x \in M$, there are functions $Q_j \in C^{k,\infty}(M \times (K \times K))$, $1 \leq j \leq n$, with $Q_j(x, \cdot) \in \mathcal{O}(K \times K)$ for $x \in M$, such that*

$$f(x, \zeta) - f(x, z) = \sum_{j=1}^{n} Q_j(x, \zeta, z)(\zeta_j - z_j)$$

for $x \in M$ and (ζ, z) in some neighborhood of $K \times K$.

The main application follows immediately.

Corollary 3.4. *Let $g \in C^{k+1,\infty}(U \times D_\delta)$ be the function given by Proposition 3.1 and let $0 < \eta < \delta$. After shrinking U, there are functions $g_j \in C^{k+1,\infty}(U \times D_\eta)$, $1 \leq j \leq n$, with $g_j(\zeta, \cdot) \in \mathcal{O}(D_\eta)$ for $\zeta \in U$, such that*

$$g(\zeta, z) = \sum_{j=1}^{n} g_j(\zeta, z)(\zeta_j - z_j) \qquad \text{on } U \times D_\eta. \tag{3.6}$$

PROOF. Since $\bar{D}_\eta$ is a Stein compactum, we can apply Proposition 3.3 to $g \in C^{k+1,\infty}(U \times \bar{D}_\eta)$, obtaining

$$g(x, \zeta) - g(x, z) = \sum_{j=1}^{n} Q_j(x, \zeta, z)(\zeta_j - z_j) \tag{3.7}$$

for $x \in U$ and $\zeta, z \in \bar{D}_\eta$, with Q_j holomorphic in $(\zeta, z) \in \bar{D}_\eta \times \bar{D}_\eta$. Now set $x = \zeta$ in (3.7), and (3.6) follows by setting $g_j(\zeta, z) = -Q_j(\zeta, \zeta, z)$. ∎

The proof of Proposition 3.3 follows closely the proof of Hefer's Theorem in V.§2.2. We shall discuss the parametrized version of the extension Lemma V.2.1, and leave the remaining details to the reader.

Lemma 3.5. *Given M and K as in Proposition 3.3, let $K_1 = K \cap \{z = (z_1, z') \in \mathbb{C}^n: z_1 = 0\}$ and suppose $f \in C^{k,\infty}(M \times K_1)$ satisfies $f(x, \cdot) \in \mathcal{O}(K_1)$ for $x \in M$. Then there is $F \in C^{k,\infty}(M \times K)$ with $F(x, \cdot) \in \mathcal{O}(K)$ for all $x \in M$, such that*

$$F(x, (0, z')) = f(x, (0, z')) \tag{3.8}$$

for $x \in M$ and all points of the form $(0, z')$ in a neighborhood of K_1.

PROOF. We proceed as in the proof of Proposition V.2.1, using the same notation and adding the parameter $x \in M$, as needed. Thus $\alpha = \alpha_1 = \bar{\partial}_z[\chi(z)f(x, \pi(z))]/z_1$ is $\bar{\partial}_z$-closed with coefficients in $C^{k,\infty}(M \times W)$, where W is a neighborhood of K. Let $\mathbf{T} = \mathbf{T}_1^{V,V_0}$ be the solution operator for $\bar{\partial}$ given by Theorem V.2.5, where $K \subset V_0 \subset\subset V \subset\subset W$, and set $g(x, \cdot) = \mathbf{T}(\alpha(x, \cdot))$. Then $g \in C^{k,\infty}(M \times V_0)$, and $F(x, z) = \chi(z)f(x, \pi(z)) - z_1 g(x, z)$ is the required function which satisfies (3.8). ∎

3.3. The Kernel of Henkin and Ramirez

Proposition 3.1 and Corollary 3.4 imply that for all δ sufficiently close to 0,

$$W^{\mathrm{HR}} = \left(\sum_{j=1}^{n} g_j \, d\zeta_j\right)\Big/ g \tag{3.9}$$

is a generating form in $C^{k+1,\infty}(bD_\delta \times D_\delta)$ which, for fixed $\zeta \in bD_\delta$, is holomorphic for all $z \in \bar{D}_\delta - \{\zeta\}$. The associated Cauchy–Fantappiè form $\Omega_0(W^{\mathrm{HR}})$ of order 0 is called the **Henkin–Ramirez Reproducing Kernel**. We state some of its basic properties.

Theorem 3.6. *Suppose $D \subset\subset \mathbb{C}^n$ is strictly pseudoconvex with boundary of class C^{k+2}, $k \geq 0$. Then $\Omega_0(W^{\mathrm{HR}})$ is of class $C^{k,\infty}$ on $bD \times D$, and, for $\zeta \in bD$ fixed, $\Omega_0(W^{\mathrm{HR}})$ is holomorphic in z on $\bar{D} - \{\zeta\}$. If one defines*

$$\mathbf{C}^{\mathrm{HR}}f(z) = \int_{bD} f(\zeta)\Omega_0(W^{\mathrm{HR}})(\zeta, z) \qquad \text{for } z \in D,$$

then

(i) $\mathbf{C}^{\mathrm{HR}}f \in \mathcal{O}(D)$ *for all $f \in L^1(bD)$.*
(ii) $\mathbf{C}^{\mathrm{HR}}f = f$ *for $f \in A(D)$.*
(iii) $\mathbf{C}^{\mathrm{HR}}$: *$L^p(bD) \to \mathcal{O}(D)$ is continuous for $1 \leq p \leq \infty$.*
(iv) $\mathbf{C}^{\mathrm{HR}}$: *$\Lambda_\alpha(bD) \to \mathcal{O}\Lambda_{\alpha/2}(D)$ is bounded for $0 < \alpha < 1$.*
(v) *If $\chi \in \Lambda_\alpha(bD)$ and $f \in A(D)$, then $\mathbf{C}^{\mathrm{HR}}(\chi f) \in A(D)$.*

PROOF. Notice that the singularities of $\Omega_0(W^{\mathrm{HR}})$ are determined by g^{-n}, and that for ζ close to z, g^{-n} is a nonzero multiple of Φ^{-n}. Hence the proof of Theorem 3.6 is analogous to the proof of the corresponding statements for the Cauchy kernel $C = E - A$ in Theorem 1.4. The details are left to the reader. ∎

3.4. The Principal Singularity

At this point the reader will probably wonder about the precise relationship between the kernel $\Omega_0(W^{\mathrm{HR}})$ and the kernel $C = E - A$ defined in §1. We shall see that the principal singularities of the two kernels are, in fact, identical. The difference $\Omega_0(W^{\mathrm{HR}}) - C$ is a kernel which is integrable over bD also for $z \in bD$, and hence is—in a sense to be made precise—negligible in comparison to $\Omega_0(W^{\mathrm{HR}})$ and C.

Let us write $Q = \sum_{j=1}^{n} g_j \, d\zeta_j$, so that

$$\Omega_0(W^{\mathrm{HR}}) = \Omega_0(Q/g) = \frac{1}{(2\pi i)^n} g^{-n} Q \wedge (\bar{\partial}_\zeta Q)^{n-1}. \tag{3.10}$$

Similarly, recall from V.§1.1 that

$$E = \Omega_0(P/\Phi) = \frac{1}{(2\pi i)^n} \Phi^{-n} P \wedge (\bar{\partial}_\zeta P)^{n-1}, \tag{3.11}$$

where $P = \sum_{j=1}^{n} P_j \, d\zeta_j$ was defined in V(1.9).

Lemma 3.7. *At all points* $(\zeta, \zeta) \in bD \times bD$ *one has*

$$P \wedge (\bar{\partial}_\zeta P)^{n-1} = Q \wedge (\bar{\partial}_\zeta Q)^{n-1} = \partial r \wedge (\bar{\partial}\partial r)^{n-1}. \tag{3.12}$$

Proof. By applying $\partial/\partial\zeta_j$ to the equation

$$g = \sum_{\nu=1}^{n} g_\nu(\zeta_\nu - z_\nu) = A \cdot F^{\#} = A \sum_{\nu=1}^{n} P_\nu(\zeta_\nu - z_\nu),$$

valid for $|\zeta - z| \le \gamma$ (see Proposition 3.1(iv)), one obtains

$$g_j + \sum_{\nu=1}^{n} \frac{\partial g_\nu}{\partial \zeta_j}(\zeta_\nu - z_\nu) = \frac{\partial A}{\partial \zeta_j} \sum P_\nu(\zeta_\nu - z_\nu) + A P_j + A \sum_{\nu=1}^{n} \frac{\partial P_\nu}{\partial \zeta_j}(\zeta_\nu - z_\nu). \tag{3.13}$$

Rewrite (3.13) in the form

$$g_j = A P_j + \sum_{\nu=1}^{n} B_\nu(\zeta_\nu - z_\nu), \tag{3.14}$$

where $B_\nu = -\partial g_\nu/\partial \zeta_j + \partial A/\partial \zeta_j \, P_\nu + A \, \partial P_\nu/\partial \zeta_j$ is of class $C^{k,\infty}$. Hence $B_\nu(\zeta_\nu - z_\nu)$ is differentiable at $\zeta = z$ (use statement II(2.8) if $k = 0$!), and $\bar{\partial}_\zeta [B_\nu(\zeta_\nu - z_\nu)](\zeta, \zeta) = 0$. Thus, by applying $\bar{\partial}_\zeta$ to (3.14) one obtains

$$\bar{\partial}_\zeta g_j = (\bar{\partial}_\zeta A) P_j + A \bar{\partial}_\zeta P_j \qquad \text{at } (\zeta, \zeta). \tag{3.15}$$

Formula (3.5) for A implies $A(\zeta, \zeta) = 1$ and $\bar{\partial}_\zeta A(\zeta, \zeta) = 0$, while the definition of P_j (see V(1.9)) implies $P_j(\zeta, \zeta) = \partial r/\partial \zeta_j$ and $(\bar{\partial}_\zeta P_j)(\zeta, \zeta) = \bar{\partial}(\partial r/\partial \zeta_j)$. Thus (3.14) and (3.15) give

$$Q = P = \partial r \qquad \text{at } (\zeta, \zeta) \tag{3.16}$$

and

$$\bar{\partial}_\zeta Q = \bar{\partial}_\zeta P = \bar{\partial}\partial r \qquad \text{at } (\zeta, \zeta), \tag{3.17}$$

and (3.12) follows. ■

We can now identify the principal singularity of the Cauchy kernel.

Proposition 3.8. *Suppose D is strictly pseudoconvex with boundary of class C^{k+2}, $k \geq 0$. Then*

(i) $\Omega_0(W^{\mathrm{HR}}) = \Omega_0(\partial r/\Phi) + \dfrac{O(|\zeta - z|)}{\Phi^n}$.

(ii) $E = \Omega_0(L_D) = \Omega_0(\partial r/\Phi) + \dfrac{O(|\zeta - z|)}{\Phi^n}$.

We shall see below that

$$\iota^*[\Omega_0(\partial r/\Phi)] = \frac{h(\zeta)\, dS_\zeta}{\Phi^n}, \qquad \text{with } h(\zeta) > 0, \tag{3.18}$$

where $\iota\colon bD \hookrightarrow \mathbb{C}^n$ is the inclusion. Thus Proposition 3.8 indeed shows that $\Omega_0(\partial r/\Phi)$ is the principal singularity for either Cauchy kernel.

PROOF. (ii) is an immediate consequence of Lemma 3.7. In fact, since $\Omega_0(P)$ is of class $C^{k,\infty}$ on $bD \times D_\delta$, all partial derivatives of $\Omega_0(P)$ with respect to z are continuous on $bD \times D_\delta$ (even if $k = 0$!), and hence bounded on $bD \times D_{\delta'}$, if $\delta' < \delta$. So an application of the Mean Value Theorem with respect to z gives $\Omega_0(P)(\zeta, z) = \Omega_0(P)(\zeta, \zeta) + O(|\zeta - z|)$. A similar argument, combined with the fact that $g = A\Phi$ for $|z - \zeta| \leq \gamma$, with $A(\zeta, \zeta) = 1$, gives (i). ∎

Remark. Notice the formal analogy of the principal part $\iota^*\Omega_0(\partial r/\Phi)$ of the Cauchy kernel for strictly pseudoconvex domains with the Cauchy kernel $C_D = \iota^*\Omega_0(\partial r/\langle \partial r, \zeta - z\rangle)$ for a convex domain D (see IV.§3.2), where

$$\langle \partial r, \zeta - z\rangle = \sum_{j=1}^{n} \frac{\partial r}{\partial \zeta_j}(\zeta_j - z_j)$$

is the *linear* part of the Levi polynomial of r, which is, locally, the linear part of Φ as well. We see that at the local level, passing from convex domains to strictly pseudoconvex ones is essentially accomplished by adding the second-order terms of the Levi polynomial F. The fact that we have used the modification $F^{\#}$ instead of F is a minor technical point, necessary only in the case of C^2 boundaries. If bD is of class C^{k+3}, $k \geq 0$, one can use the Levi polynomial F itself. On the other hand, the fact that $h(\zeta) > 0$ in (3.18) turns out to be equivalent to strict pseudoconvexity, and does not hold for arbitrary convex domains. (See Exercise E.3.4 for a specific example.)

3.5. A Geometric Formula for $\Omega_0(\partial r)$

We shall now prove (3.18) by finding a precise representation of the numerator of the Cauchy kernel in terms of geometric quantities. The Levi form of the defining function r for D induces a Hermitian form

$$\mathscr{L}_P(r)\colon T_P^{\mathbb{C}}(bD) \times T_P^{\mathbb{C}}(bD) \to \mathbb{C}$$

on the complex tangent space $T_P^{\mathbb{C}}(bD)$ at $P \in bD$, defined by

$$\mathscr{L}_P(r)(t, t') = \sum_{j,k=1}^{n} \frac{\partial^2 r}{\partial \zeta_j \partial \bar{\zeta}_k}(P) t_j \bar{t}'_k \tag{3.19}$$

for $t, t' \in T_P^{\mathbb{C}}(bD)$. We shall denote the determinant of $\mathscr{L}_P(r)$ by $\mathscr{D}_P(r)$. Notice that D is strictly Levi pseudoconvex at P if and only if $\mathscr{L}_P(r)$ is positive definite, which implies $\mathscr{D}_P(r) > 0$. After a unitary change of coordinates in $\mathbb{C}^n$ we may assume that $\partial r/\partial \zeta_j(P) = 0$ for $1 \le j \le n-1$, so that

$$\partial r(P) = (\partial r/\partial \zeta_n)(P)\, d\zeta_n \tag{3.20}$$

and $T_P^{\mathbb{C}}(bD) = \{t = (t_1, \ldots, t_n) \in \mathbb{C}^n : t_n = 0\}$, and furthermore, that $\mathscr{L}_P(r)$ is in diagonal form, i.e.,

$$(\partial^2 r/\partial \zeta_j \partial \bar{\zeta}_k)(P) = \delta_{jk} \lambda_j \qquad \text{for } 1 \le j, k \le n-1,$$

where $\lambda_1, \ldots, \lambda_{n-1}$ are the (real) eigenvalues of $\mathscr{L}_P(r)$. It follows that $\mathscr{D}_P(r) = \lambda_1 \cdot \ldots \cdot \lambda_{n-1}$.

Lemma 3.9. *If $\iota: bD \to \mathbb{C}^n$ is the inclusion, then*

$$\iota^* \Omega_0(\partial r) = \frac{(n-1)!}{4\pi^n} \mathscr{D}(r) \| dr \| \, dS \tag{3.21}$$

on bD, where dS is the surface element of bD.

Proof. It is enough to verify (3.21) at an arbitrary point $P \in bD$. Fix P and choose the coordinates of $\mathbb{C}^n$ as above, so that $\mathscr{L}_P(r)$ is diagonal. All forms below are evaluated at P. Then

$$\bar{\partial}\partial r = \sum_{j=1}^{n-1} \lambda_j \, d\bar{\zeta}_j \wedge d\zeta_j + \omega_1 \wedge d\zeta_n + \omega_2 \wedge d\bar{\zeta}_n \tag{3.22}$$

for certain 1-forms ω_1 and ω_2, and since $\iota^*(\partial r) = -\iota^*(\bar{\partial} r)$, (3.20) and (3.22) imply

$$\iota^*(\partial r \wedge \bar{\partial}\partial r) = \iota^*\left(\partial r \wedge \sum_{j=1}^{n-1} \lambda_j \, d\bar{\zeta}_j \wedge d\zeta_j\right). \tag{3.23}$$

It follows that

$$\begin{aligned} \iota^* \Omega_0(\partial r) &= \frac{(n-1)!}{(2\pi i)^n} \mathscr{D}_P(r) \iota^* \left[\frac{\partial r}{\partial \zeta_n} d\zeta_n \wedge \left(\bigwedge_{j<n} d\bar{\zeta}_j \wedge d\zeta_j \right) \right] \\ &= \frac{(n-1)!}{2\pi^n} \mathscr{D}_P(r) \iota^* \left[* \frac{\partial r}{\partial \zeta_n} d\zeta_n \right], \end{aligned} \tag{3.24}$$

where we have used Lemma III.3.3 in the computation of the $*$-operator. By Corollary III.3.5, $\iota^*(*\partial r) = 1/2 \| dr \| \, dS$. After inserting this into (3.24) (use (3.20) again!), the desired conclusion follows. ∎

Corollary 3.10. *The Henkin–Ramirez kernel* $\Omega_0(W^{\mathrm{HR}})$ *on a strictly pseudoconvex domain* $D \subset \mathbb{C}^n$ *with* C^{k+2} *boundary satisfies*

$$\iota^*\Omega_0(W^{\mathrm{HR}}) = \frac{h(\zeta)}{g^n}\,dS + \frac{O(|\zeta - z|)}{g^n}, \tag{3.25}$$

where $h \in C^k(bD)$ *and* $h(\zeta) > 0$ *for all* $\zeta \in bD$.

EXERCISES

E.3.1. Give a complete proof of Proposition 3.3 by using Lemma 3.5.

E.3.2. Prove the properties of the Cauchy kernel $\mathbf{C}^{\mathrm{HR}}$ stated in Theorem 3.6.

E.3.3. By (3.10), the Henkin–Ramirez kernel is given by $g^{-n}\Omega_0(Q)$. Use the fact that for every $\zeta \in bD$ there is a peaking function $h_\zeta \in \mathcal{O}(\bar{D})$ for ζ (see Corollary 3.2) to show that $\iota^*\Omega_0(Q)(\zeta, \zeta) \neq 0$, where $\iota: bD \to \mathbb{C}^n$ is the inclusion. (Hint: If not, then $\Omega_0(W^{\mathrm{HR}})(\cdot, p)$ would be in $L^1(bD)$ for some $p \in bD$! Why?)

E.3.4. For $m \in \mathbb{N}^+$, let $D^m = \{|z_1|^2 + |z_2|^{2m} < 1\}$. Show by direct computation that for $m > 1$ the Cauchy kernel C_{D^m} does not have the property stated in E.3.3 (with $g = \langle \partial r^{(m)}, \zeta - z\rangle$ and $r^{(m)} = |z_1|^2 + |z_2|^{2m} - 1$).

§4. Gleason's Problem and Decomposition in $A(D)$

4.1. Some Background

In case of one complex variable it is obvious that every $f \in \mathcal{O}(D)$ with $f(a) = 0$ at a point $a \in D \subset \mathbb{C}^1$ has a factorization $f = (z - a)g$, where $g = (z - a)^{-1}f$ is holomorphic on D. Thus $z - a$ generates the ideal $I_a(D) = \{f \in \mathcal{O}(D): f(a) = 0\}$ over $\mathcal{O}(D)$. Furthermore, it is clear that if $f \in A(D)$, then $g \in A(D)$ as well. We shall now consider analogous results in several variables. It should not come as a surprise that matters become quite a bit more complicated. For example, if $D \subset \mathbb{C}^n$ and $a \in D$, it follows from the power series expansion of $f \in I_a(D)$, that $f = \sum_{j=1}^n (z_j - a_j)g_j$ for some functions $g_1, \ldots, g_n$ in the ring $\mathcal{O}_a$ of functions holomorphic at a, but it is not at all clear whether the functions $g_1, \ldots, g_n$ can be chosen globally holomorphic on D, except in special cases like a polydisc.

We are already familiar with one major positive result: Hefer's Theorem V.2.2 implies that if $K \subset \mathbb{C}^n$ is a Stein compactum and $a \in K$, then every $f \in \mathcal{O}(K)$ with $f(a) = 0$ has a decomposition $f = \sum_{j=1}^n (z_j - a_j)g_j$, with $g_j \in \mathcal{O}(K)$. The same method of proof, combined with the fact that $H^1_{\bar\partial}(D) = 0$ for a Stein domain D, yields the following result.

Theorem 4.1. *If* $D \subset \mathbb{C}^n$ *is Stein and* $f \in \mathcal{O}(D)$, *there are* $Q_1, \ldots, Q_n \in \mathcal{O}(D \times D)$ *such that*

$$f(z) - f(w) = \sum_{j=1}^{n} (z_j - w_j) Q_j(z, w)$$

for all $z, w \in D$.

Theorem 4.1 implies that the ideal $I_a(D)$ of functions in $\mathcal{O}(D)$ vanishing at a is generated over $\mathcal{O}(D)$ by the coordinate functions $z_1 - a_1, \ldots, z_n - a_n$. The theory of coherent analytic sheaves, and, in particular, Cartan's Theorems A and B (see VI.§6) provide a powerful tool to deal—in much greater generality—with many similar questions of a global nature. However, this general theory is of little use when one is interested in boundary behavior, as is the case if one wants to study generators for ideals in the algebra $A(D)$. Interest in this latter sort of question arose in the 1950s and early 1960s, as the theory of uniform algebras was developed. Specifically, when A. Gleason ("Finitely generated ideals in Banach Algebras", J. Math. Mech. **13** (1964), 125–132) proved the existence of complex analytic structure in a neighborhood of a finitely generated maximal ideal in the spectrum of a uniform algebra, he was looking for examples of such finitely generated ideals among the standard uniform algebras of holomorphic functions, and he immediately ran into the following concrete question: if $B = \{(z, w) \in \mathbb{C}^2 : |z|^2 + |w|^2 < 1\}$ is the unit ball in $\mathbb{C}^2$ and $f \in A(B)$ satisfies $f(0) = 0$, are there g_1 and $g_2 \in A(B)$ so that $f = zg_1 + wg_2$? The reader should try his "bare hands" at answering this apparently simple question in order to appreciate its nontrivial nature.

During the 1960s, this question, as well as its analogue for more general domains, became known as "Gleason's Problem". Eventually, Z.L. Leibenson (unpublished; see G.M. Henkin [Hen 3], and also Exercise E.4.3) gave an affirmative answer based upon elementary arguments. However, his proof was limited to Euclidean convex domains, and thus was of no use for more general domains. Finally, Gleason's problem was solved in reasonable generality around 1970 by the methods of integral representations and precise estimates for solutions of $\bar\partial$ (Henkin [Hen 3], Kerzman and Nagel [KeNa], Lieb [Lie 2], and Øvrelid [Øvr 2]]). In the next section we will solve Gleason's problem—in more general form—by proving an $A(D)$-version of Hefer's decomposition theorem on strictly pseudoconvex domains.

4.2. Decomposition of Functions in $A(D)$

We state the principal result of this section.

Theorem 4.2. *Let* $D \subset\subset \mathbb{C}^n$ *be strictly pseudoconvex with boundary of class* ≥ 3. *There are linear operators* $L_j\colon A(D) \to \mathcal{O}(D \times D)$, $1 \leq j \leq n$, *such that for* $f \in A(D)$ *the following hold*:

(4.1) $$f(z) - f(w) = \sum_{j=1}^{n} (z_j - w_j)(L_j f)(z, w) \qquad \text{for } z, w \in D.$$

(4.2) *For fixed* $a \in D$, *the functions* $(L_j f)(a, \cdot)$ *and* $(L_j f)(\cdot, a)$ *are in* $A(D)$.

The Theorem immediately solves Gleason's problem for a strictly pseudoconvex domain D.

Corollary 4.3. *If $f \in A(D)$ and $f(a) = 0$ at the point $a \in D$, then there are functions $g_1, \ldots, g_n \in A(D)$ such that*

$$f(z) = \sum_{j=1}^{n} (z_j - a_j) g_j(z).$$

PROOF. Take $g_j = (L_j f)(\cdot, a)$. ∎

PROOF OF THEOREM 4.2. We shall use the reproducing kernel $\Omega_0(W^{\text{HR}}) = \Omega_0(Q/g) = g^{-n}\Omega_0(Q)$ of Henkin and Ramirez, and write $\iota^*\Omega_0(Q) = N(\zeta, z)\, dS_\zeta$, with $N \in C^{1,\infty}(bD \times D_\delta)$ holomorphic in z. For $f \in A(D)$ and $z, w \in D$ one obtains

$$\begin{aligned} f(z) - f(w) &= \int_{bD} f(\zeta) \left[\frac{N(\zeta, z)}{g^n(\zeta, z)} - \frac{N(\zeta, w)}{g^n(\zeta, w)} \right] dS_\zeta \\ &= \int_{bD} f(\zeta) \frac{g^n(\zeta, w) N(\zeta, z) - g^n(\zeta, z) N(\zeta, w)}{g^n(\zeta, z) g^n(\zeta, w)} dS_\zeta. \end{aligned} \tag{4.3}$$

The function $H(\zeta, z, w) = g^n(\zeta, w) N(\zeta, z)$ is in $C^{1,\infty}(bD \times (\bar{D} \times \bar{D}))$ and holomorphic in $u = (z, w)$. We apply Proposition 3.3. to H and the Stein compactum $K = \bar{D} \times \bar{D} \subset \mathbb{C}^{2n}$, and obtain functions $Q_1, \ldots, Q_{2n} \in C^{1,\infty}(bD \times (K \times K))$, holomorphic in (u, v) on $K \times K$, so that

$$H(\zeta, u) - H(\zeta, v) = \sum_{i=1}^{2n} (u_j - v_j) Q_j(\zeta, u, v). \tag{4.4}$$

By taking $u = (z, w)$ and $v = (w, z)$ in (4.4), it follows that

$$H(\zeta, z, w) - H(\zeta, w, z) = \sum_{j=1}^{n} (z_j - w_j) M_j(\zeta, z, w), \tag{4.5}$$

where

$$M_j(\zeta, z, w) = Q_j(\zeta, z, w, w, z) - Q_{n+j}(\zeta, z, w, w, z) \qquad \text{for } j = 1, \ldots, n.$$

We then define

$$L_j f(z, w) = \int_{bD} f(\zeta) \frac{M_j(\zeta, z, w)}{g^n(\zeta, z) g^n(\zeta, w)} dS_\zeta.$$

It is clear that $L_j f \in \mathcal{O}(D \times D)$, and that (4.3) and (4.5) imply (4.1). We now prove (4.2). By Corollary 3.10, $N(\zeta, z) = h(\zeta) + O(|\zeta - z|)$, where $h(\zeta) \neq 0$ for all $\zeta \in bD$. Hence, for $a \in D$ we may write

$$L_j f(z, a) = \int_{bD} f(\zeta) \chi(\zeta, z, a) \frac{h(\zeta)}{g^n(\zeta, z)} dS_\zeta \tag{4.6}$$

where

$$\chi(\zeta, z, a) = \frac{M_j(\zeta, z, a)}{h(\zeta)g^n(\zeta, a)}.$$

By (4.6) and (3.25),

$$L_j f(z, a) = \int_{bD} f(\zeta)\chi(\zeta, z, a)\Omega_0(W^{\mathrm{HR}}) + \int_{bD} f\chi \frac{O(|\zeta - z|)}{g^n}\, dS, \tag{4.7}$$

and by setting $\chi(\zeta, z, a) = \chi(\zeta, \zeta, a) + O(|\zeta - z|)$, we obtain

$$\begin{aligned} L_j f(z, a) = &\int_{bD} f(\zeta)\chi(\zeta, \zeta, a)\Omega_0(W^{\mathrm{HR}}) \\ &+ \int_{bD} f\, O(|\zeta - z|)\Omega_0(W^{\mathrm{HR}}) + \int_{bD} f\chi\frac{O(|\zeta - z|)}{g^n}\, dS. \end{aligned} \tag{4.8}$$

Since $\chi(\zeta, \zeta, a)$ is of class C^1 on bD, (this is where bD of class C^3 is used), Theorem 3.6(v) implies that the first integral in (4.8) is in $A(D)$. The kernels in the remaining two integrals in (4.8) are uniformly integrable over $\zeta \in bD$ for all $z \in \bar{D}$ (see Lemma 1.5), so that these integrals extend continuously to $\bar{D}$. Hence (4.8) implies that $L_j f(., a) \in A(D)$. By interchanging the roles of z and w in the above proof one obtains that $L_j f(a, \cdot) \in A(D)$ as well. ■

Remark. We leave it to the reader to show that $L_j f$ even extends continuously to $\bar{D} \times \bar{D} - \{(\zeta, \zeta): \zeta \in bD\}$.

Exercises

E.4.1. Let $P = P(0, 1)$ be the unit polydisc in $\mathbb{C}^n$. Show by elementary arguments that if $a \in P$ and $f \in A(P)$ is 0 at a, then

$$f(z) = \sum_{j=1}^{n} (z_j - a_j)g_j(z) \qquad \text{where } g_1, \ldots, g_n \in A(P).$$

E.4.2. Let $D \subset\subset \mathbb{C}^n$ be convex.

(i) Show that if $f \in \mathcal{O}(D)$ and $a \in D$, then

$$f(z) = f(a) + \sum_{j=1}^{n} (z_j - a_j) \int_0^1 \frac{\partial f}{\partial z_j}(a + t(z - a))\, dt.$$

(ii) Use (i) to obtain an elementary proof that

$$f(z) = f(a) + \sum_{j=1}^{n} (z_j - a_j)g_j(z) \qquad \text{where } g_1, \ldots, g_n \in \mathcal{O}(D).$$

E.4.3. Use the explicit method in E.4.2 to show that in case $D = B$ is the unit ball and $f \in A(B)$, then the functions $g_1, \ldots, g_n$ are in $A(B)$ as well, i.e., Gleason's problem is solved for B.

E.4.4. Prove Theorem 4.1.

E.4.5. Prove that the functions $L_j f$, $1 \le j \le n$, $f \in A(D)$, given in the proof of Theorem 4.2 extend continuously to $\bar{D} \times \bar{D} - \{(\zeta, \zeta): \zeta \in bD\}$.

E.4.6. Let D be strictly pseudoconvex with boundary of class C^3. Show that if $a \in D$ and $f \in \mathcal{O}\Lambda_\alpha(D)$ for some $0 < \alpha < 1$, then

$$f(z) - f(a) = \sum_{j=1}^{n} (z_j - a_j) g_j(z) \qquad \text{for } z \in D,$$

where $g_1, \ldots, g_n \in \mathcal{O}\Lambda_{\alpha/2}(D)$. (By using E.1.2, extended to $\mathbf{C}^{\mathrm{HR}}$, one can show that $g_1, \ldots, g_n \in \mathcal{O}\Lambda_\alpha(D)$.)

E.4.7. Let D be strictly pseudoconvex with C^3 boundary. Fix $a \in D$.

(i) Show that there is $M < \infty$ such that for all $f \in A(D)$ there are $g_1, \ldots, g_n \in A(D)$ with $|g_j|_D \le M |f|_D$ for $j = 1, \ldots, n$ and

$$f(z) - f(a) = \sum (z_j - a_j) g_j(z) \qquad \text{for } z \in D.$$

(Hint: Use the Open Mapping Theorem.)

(ii) Iterate (i) to prove: there is $\varepsilon > 0$ such that every $f \in A(D)$ has a power series expansion $f(z) = \sum_{\alpha \in \mathbb{N}^n} c_\alpha (z - a)^\alpha$ which converges on $P(a, \varepsilon)$. Do *not* use Theorem I.1.18!

Remark. A. Gleason used this sort of argument to introduce an analytic structure in a suitable neighborhood of a finitely generated maximal ideal in a Banach algebra.

§5. L^p Estimates for Solutions of $\bar{\partial}$

The integral solution operator $\mathbf{S}_q$ for $\bar{\partial} u = f$ on a strictly pseudoconvex domain D given by Theorem V.2.7 satisfies Lipschitz estimates for $u = \mathbf{S}_q f$ up to the boundary of D, but since $\mathbf{S}_q$ involves integration over bD as well as over D, it can only be applied to forms defined on $\bar{D}$. This limitation could be lifted by introducing a suitable exhaustion of D from the interior, but instead of pursuing this approach, we shall now discuss a modification of $\mathbf{S}_q$ in which the boundary integrals are replaced by integration over D. The additional advantage of such a modification is that it makes it easy to estimate solutions of $\bar{\partial} u = f$ in terms of the $L^p(D)$ norm of f for $1 \le p \le \infty$. Such estimates are useful in applications, as we shall see in §6.

5.1. Extension to the Interior of D

The essential step in the modification to be carried out involves an application of Stokes' Theorem in order to eliminate all boundary integrals in the fundamental representation

$$(5.1) \qquad \begin{aligned} f &= \mathbf{E}_q f + \bar{\partial} \mathbf{T}_q f \qquad (q \ge 1), \\ f &= \mathbf{E}_0 f \qquad (q = 0) \end{aligned}$$

for $\bar\partial$-closed forms f in $C^1_{0,q}(\bar D)$ obtained in V(1.11). Formula (5.1) is a special case of Theorem IV.3.6, which, in the case at hand states

$$(5.2)\quad f = \int_{bD} f\Omega_q(L_D) + \bar\partial\left[\int_{bD\times I} f\wedge\Omega_{q-1}(\hat L_D) - \int_D f\wedge\Omega_{q-1}(B)\right],$$

where $L_D = P/\Phi$, $B = \partial\beta/\beta$, and $\hat L_D = \lambda L_D + (1-\lambda)B$. Recall that

$$\mathbf{E}_q f = \int_{bD} f\wedge\Omega_q(L_D),$$

and that $\mathbf{T}_q f$ is the expression in [ ] in (5.2), where $\mathbf{T}_0 f \equiv 0$ since $\Omega_{-1}\equiv 0$ by definition. In Lemma V.3.2 we introduced a double form $A_q(L_D, B)$ on $bD\times D$ so that

$$(5.3)\qquad \int_{bD\times I} f\wedge\Omega_q(\hat L_D) = \int_{bD} f\wedge A_q(L_D, B).$$

Before applying Stokes' Theorem we must extend the generating forms L_D and B from $bD\times D$ to $\bar D\times D$ without singularities. Let us first consider $L_D = P/\Phi$. The extension we shall choose is motivated by the computation of the Bergman kernel for the unit ball $B = B(0, 1)$ in Chapter IV, §4.4, and it will play an important role later in §7, when we consider the Bergman kernel for an arbitrary strictly pseudoconvex domain. Recall that the Levi polynomial F of $r(z) = |z|^2 - 1$ is given by

$$F(\zeta, z) = \sum_{j=1}^n \bar\zeta_j(\zeta_j - z_j) = |\zeta|^2 - (z, \zeta).$$

For $\zeta\in bB$ this agrees with $1-(z,\zeta) = F(\zeta, z) - r(\zeta)$, a function without zeroes on $B\times B$. Returning to the general case with strictly plurisubharmonic defining function r, it thus appears that $F(\zeta, z) - r(\zeta)$ is a natural choice for extending the Levi polynomial from the boundary bD to D without introducing new zeroes—at least locally. Since $\Phi = F$ locally, we therefore consider

$$\hat\Phi(\zeta, z) = \Phi(\zeta, z) - r(\zeta).$$

We fix $\varepsilon_0 > 0$ so that $\{\zeta\in U: |r(\zeta)| < 2\varepsilon_0\}\subset\subset U$, and choose $\varphi\in C^\infty(\bar D)$, with $\varphi(\zeta)\equiv 1$ for $\zeta\in U$ with $r(\zeta)\ge -\varepsilon_0$ and $\varphi(\zeta)\equiv 0$ for $\zeta\in D$ with $r(\zeta)\le -2\varepsilon_0$. For $0\le q\le n$ we then define the double form

$$(5.4)\qquad \hat E_q = \varphi\frac{\Omega_q(P)}{\hat\Phi^n}.$$

We will denote the "*boundary diagonal*" $\{(\zeta,\zeta): \zeta\in bD\}$ by Δ_{bD}. The following result holds.

Lemma 5.1. *Suppose r is of class C^{k+2}. The neighborhood U of bD, and the positive constants c, ε, and δ can be chosen so that*

$$(5.5)\quad \text{for } \zeta\in U,\ 2\operatorname{Re}\hat\Phi(\zeta, z) \ge \begin{cases} -r(\zeta) - r(z) + c|\zeta - z|^2 & \text{if } |z-\zeta|\le\varepsilon/2;\\ -r(\zeta) + \delta & \text{if } |z-\zeta|\ge\varepsilon/2 \text{ and } r(z)\le\delta.\end{cases}$$

(5.6) $$\hat{E}_q \in C^{k,\infty}(\bar{D} \times \bar{D}_\delta) \qquad \textit{for } 1 \leq q \leq n.$$

(5.7) $\hat{E}_0 = C^{k,\infty}(\bar{D} \times \bar{D} - \Delta_{bD})$, *and* $\hat{E}_0$ *is holomorphic in* z *for* $|\zeta - z| < \varepsilon/2$.

(5.8) $$\textit{For } \zeta \in bD \qquad (\textit{i.e.}, r(\zeta) = 0),\ \hat{E}_q = \frac{\Omega_q(P)}{\Phi^n} = \Omega_q(L_D).$$

PROOF. Equation (5.5) is an immediate consequence of the estimates V(1.6) and V(1.7) for Φ; (5.7) and (5.8) then follow from (5.5) and (5.4). The fact that $\hat{E}_q$ is nonsingular and extends to $\bar{D}_\delta$ if $q \geq 1$ follows as the corresponding fact for $\Omega_q(L_D)$ in Proposition V.1.1. Just use (5.4) and (5.5), and recall that P is holomorphic in z for $|z - \zeta| < \varepsilon/2$, so that $\Omega_q(P) \equiv 0$ for $|z - \zeta| \leq \varepsilon/2$ if $q \geq 1$. ∎

Next we set $\hat{\beta} = \beta + r(\zeta)r(z) = |\zeta - z|^2 + r(\zeta)r(z)$ and define $\hat{B} = \partial\beta/\hat{\beta}$. Clearly $\hat{B} \in C^\infty(\bar{D} \times \bar{D} - \Delta_{bD})$, and $\hat{B} = B$ for $\zeta \in bD$. The particular extension $\hat{B}$ chosen here is motivated by symmetry considerations which are important in other contexts, but for the present purposes, any other nonsingular extension of B would serve just as well.

By using the extensions $\hat{\Phi}$ and $\hat{\beta}$ in the denominator of $A_q(L_D, B)$ we now obtain an extension $\hat{A}_q(L_D, B) \in C^{k,\infty}(\bar{D} \times \bar{D} - \Delta_{bD})$ of $A_q(L_D, B)$. More precisely, we set

(5.9) $$\hat{A}_q^{j,k}(L_D, B) = \varphi(\zeta)\frac{A_q^{j,k}(P, \partial\beta)}{\hat{\Phi}^{j+k+1}\hat{\beta}^{n-(j+k+1)}}$$

and

(5.10) $$\hat{A}_q(L_D, B) = \sum_{j=0}^{n-q-2} \sum_{k=0}^{q} a_q^{j,k} \hat{A}_q^{j,k}(L_D, B),$$

where $a_q^{j,k}$ are the numerical constants in Lemma V.3.2—the reader should review that result together with V(3.13) in order to recall how we were led to these particular forms.

It follows directly from the definition that

(5.11) $$\hat{A}_q(L_D, B) = A_q(L_D, B) \qquad \text{on } bD \times D.$$

5.2. A Modified Parametrix for $\bar{\partial}$

We now assume that bD is of class C^{k+2} with $k \geq 1$, so that the kernels $\hat{E}_q$ and $\hat{A}_q$ are at least of class C^1 and Stokes' Theorem may be applied. We define the integral operators ${}_{\bar{\partial}}\hat{\mathbf{T}}_q$: $L^1_{0,q}(D) \to L^1_{0,q-1}(D)$, $1 \leq q \leq n$, by

(5.12) $${}_{\bar{\partial}}\hat{\mathbf{T}}_q f = (-1)^q \int_D f \wedge \bar{\partial}_\zeta \hat{A}_{q-1}(L_D, B) - \int_D f \wedge \Omega_{q-1}(B),$$

and

$$_{\bar\partial}\hat{\mathbf{E}}_q\colon L^1_{0,q}(D) \to C^\infty_{0,q}(D), \qquad 0 \le q \le n,$$

by

$$(5.13) \qquad _{\bar\partial}\hat{\mathbf{E}}_q f = (-1)^q \int_D f \wedge \bar\partial_\zeta \hat{E}_q.$$

It is clear that $_{\bar\partial}\hat{\mathbf{T}}_q$ and $_{\bar\partial}\hat{\mathbf{E}}_q$ are defined for $f \in L^1_{0,q}(D)$, and that $\hat{\mathbf{E}}_q f \in C^\infty_{0,q}(D)$ (see Lemma 5.1), but the fact that $_{\bar\partial}\hat{\mathbf{T}}_q f$ is in L^1 is more delicate and will be a special case of Theorem 5.4 below. We first state some immediate consequences of the definitions.

Lemma 5.2. (i) *For $f \in C^1_{0,q}(\bar{D})$ with $\bar\partial f = 0$ one has*

$$_{\bar\partial}\hat{\mathbf{E}}_q f = \mathbf{E}_q f \qquad \textit{and } {}_{\bar\partial}\hat{\mathbf{T}}_q f = \mathbf{T}_q f.$$

(ii) *If $q \ge 1$, then $_{\bar\partial}\hat{\mathbf{E}}_q f \in C^\infty_{0,q}(\bar{D}_\delta)$ and*

$$(5.14) \qquad |_{\bar\partial}\hat{\mathbf{E}}_q f|_{C^l(\bar{D}_\delta)} \lesssim \|f\|_{L^1(D)} \qquad \textit{for } l = 0, 1, 2, \ldots.$$

Proof. (i) follows from (5.3) and (5.11), and an application of Stokes' Theorem—note that $\bar\partial f = 0$! (ii) follows from (5.13) and (5.6) by standard estimations. ∎

Because of Lemma 5.2(i) we may rewrite (5.1) as follows:

$$(5.15) \qquad f = {}_{\bar\partial}\hat{\mathbf{E}}_0 f \qquad \textit{for } f \in A^1(D);$$

$$(5.16) \qquad f = {}_{\bar\partial}\hat{\mathbf{E}}_q f + \bar\partial({}_{\bar\partial}\hat{\mathbf{T}}_q f) \qquad \textit{if } q \ge 1,\ f \in C^1_{0,q}(\bar{D}), \textit{ and } \bar\partial f = 0.$$

Proposition 5.3. (i) *The representation* (5.15) *holds for all $f \in \mathcal{O}L^1(D)$.*
(ii) *The representation* (5.16) *holds for all $f \in C^1_{0,q}(D) \cap L^1$ with $\bar\partial f = 0$ on D.*
(iii) *If $q \ge 1$, then $_{\bar\partial}\hat{E}_q f$ is $\bar\partial$-closed on D_δ for all f as in* (ii).

Remark. The main new feature is that no hypotheses about boundary values of f are required; the integrability of f over D is, of course, necessary for the definition of the integral operators in (5.12) and (5.13). On the other hand, the interior C^1 regularity of f could be dropped by introducing derivatives in the distribution sense, but we shall not pursue such technical generalizations.

Proof of Proposition 5.3. Let us first consider (ii) and fix $f \in C^1_{0,q}(D)$, $q \ge 1$, with $\bar\partial f = 0$. For $\eta < 0$ sufficiently small, the double forms $\Omega_q(L_D)$ and $A_q(L_D, B)$ are defined on $bD_\eta \times D_\eta$, and since $f \in C^1_{0,q}(\bar{D}_\eta)$, (5.1) holds on D_η, i.e.,

$$(5.17)\ f = \int_{bD_\eta} f \wedge E_q + \bar\partial_z \left[\int_{bD_\eta} f \wedge A_{q-1}(L_D, B) - \int_{D_\eta} f \wedge \Omega_{q-1}(B) \right].$$

In (5.17) we now replace E_q and A_{q-1} by their extensions $\hat{E}^{(\eta)}_q$ and $\hat{A}^{(\eta)}_{q-1}$ from $bD_\eta \times D_\eta$ to $\bar{D}_\eta \times D_\eta$, which are defined just as $\hat{E}_q$ and $\hat{A}_{q-1}$, except that the defining function r for D is replaced by the defining function $r - \eta$ for D_η. Next, we apply Stokes' Theorem (use $\bar\partial f = 0$!), and obtain

$$f = (-1)^q \int_{D_\eta} f \wedge \bar{\partial}_\zeta \hat{E}_q^{(\eta)} \tag{5.18}$$
$$+ \bar{\partial}_z \left[(-1)^q \int_{D_\eta} f \wedge \bar{\partial}_\zeta \hat{A}_{q-1}^{(\eta)} - \int_{D_\eta} f \wedge \Omega_{q-1}(B) \right].$$

Since f is assumed to be in $L^1_{0,q}(D)$, the representation (5.16) now follows by standard analysis techniques by fixing $z \in D$ and letting $\eta \to 0$ in (5.18). This same argument proves also (i), the only difference being that A_{-1} and Ω_{-1} are zero. Finally, in order to prove (iii), notice that

$$G_\eta f := (-1)^q \int_{D_\eta} f \wedge \bar{\partial}_\zeta \hat{E}_q^{(\eta)} = \int_{bD_\eta} f \wedge E_q = \int_{bD_\eta} f \wedge \Omega_q(L_D)$$

is $\bar{\partial}$-closed on $\bar{D}_\delta$ by Lemma V.1.2 applied to the domain D_η instead of D. But ${}_{\bar{\partial}}\hat{\mathbf{E}}_q f = \lim_{\eta \to 0} G_\eta f$, the convergence being uniform for all derivatives (this uses $f \in L^1_{0,q}(D)$), and hence ${}_{\bar{\partial}}\hat{\mathbf{E}}_q f$ is $\bar{\partial}$-closed on $\bar{D}_\delta$ as well. ∎

5.3. The L^p Boundedness of ${}_{\bar{\partial}}\hat{\mathbf{T}}_q$

Notice that by Lemma 5.2(i) ${}_{\bar{\partial}}\hat{\mathbf{T}}_q$ agrees with $\mathbf{T}_q$ on $\bar{\partial}$-closed forms which are smooth up to the boundary, so not much seems to have been gained. We now prove the result which justifies our work in §5.1 and §5.2.

Theorem 5.4. *Suppose D is strictly pseudoconvex in $\mathbb{C}^n$ with boundary of class at least C^3. Then there is a constant $C < \infty$ such that the operators ${}_{\bar{\partial}}\hat{\mathbf{T}}_q$ defined by* (5.12) *for $1 \le q \le n$ satisfy the following*:

(i) $\|{}_{\bar{\partial}}\hat{\mathbf{T}}_q f\|_{L^p(D)} \le C \|f\|_{L^p(D)}$ *for all* $1 \le p \le \infty$.
(ii) $|{}_{\bar{\partial}}\hat{\mathbf{T}}_q f|_{\Lambda_{1/2}(D)} \lesssim C \|f\|_{L^\infty(D)}$.
(iii) *For* $l = 0, 1, 2, \ldots$, ${}_{\bar{\partial}}\hat{\mathbf{T}}_q f \in C^l_{0,q-1}(D)$ *if* $f \in L^1_{0,q}(D) \cap C^l_{0,q}(D)$.

The main new result is the L^p estimate (i), since (ii) and (iii) are already known for $\mathbf{T}_q$ (see the proof of Theorem V.2.7), although a separate proof will be needed for the operator ${}_{\bar{\partial}}\hat{\mathbf{T}}_q$. The proof of (i) will be a consequence of standard results in analysis once we prove the following estimate.

Lemma 5.5. *Let $M(\zeta, z)$ denote any of the coefficients of the double forms $\bar{\partial}_\zeta \hat{A}_{q-1}(L_D, B)$ and $\Omega_{q-1}(B)$ which make up the kernels of ${}_{\bar{\partial}}\hat{\mathbf{T}}_q$. For each s with $1 \le s < (2n+2)/(2n+1)$ there is a constant $C_s < \infty$ such that*

$$\int_D |M(\zeta, z)|^s \, dV(\zeta) \le C_s \qquad \textit{for all } z \in D, \tag{5.19}$$

and

$$\int_D |M(\zeta, z)|^s \, dV(z) \le C_s \qquad \textit{for all } \zeta \in D. \tag{5.20}$$

PROOF. For simplicity we shall only consider the case $s = 1$, which suffices for the L^p estimates (i). The estimate for general $s < (2n + 2)/(2n + 1)$ allows to prove stronger results for ${}_{\bar\partial}\hat{\mathbf{T}}_q$ (see Remark 5.7 below), and is left as an Exercise for the reader. If M is a coefficient of $\Omega_{q-1}(B)$, then $|M(\zeta, z)| \lesssim |\zeta - z|^{-2n+1}$, and the desired estimates are straightforward. So assume that M is a coefficient of $\bar\partial_\zeta \hat{A}_{q-1}$. Clearly it is enough to consider integration over $U \cap D$. From (5.9) and (5.10) we see that $\hat{A}_{q-1}$ is a linear combination of terms

$$\frac{N_1(\zeta, z)}{\hat\Phi^l \hat\beta^{n-l}} \qquad \text{with } 1 \le l \le n - 1, \tag{5.21}$$

where $N_1 \in C^{1,\infty}(\bar D \times \bar D)$ and $N_1 = O(|\zeta - z|)$. Hence M is a linear combination of terms

$$\frac{\bar\partial_\zeta N_1}{\hat\Phi^l \hat\beta^{n-l}} - \frac{l(\bar\partial_\zeta \Phi) N_1}{\hat\Phi^{l+1} \hat\beta^{n-l}} - (n-l)\frac{[\bar\partial_\zeta \beta + (\bar\partial_\zeta r) r(z)] N_1}{\hat\Phi^l \hat\beta^{n-l+1}}, \tag{5.22}$$

with $1 \le l \le n - 1$. Suppose $\zeta \in U \cap \bar D$ and $z \in \bar D$. By using the estimates $|\hat\Phi| \gtrsim |r(z)| + |\zeta - z|^2$, which follows from (5.5), and $|\hat\beta| \ge |\zeta - z|^2$, (5.22) implies

$$\begin{aligned} |M(\zeta, z)| &\lesssim \frac{1}{|\hat\Phi||\zeta - z|^{2n-2}} + \frac{1}{|\hat\Phi|^2 |\zeta - z|^{2n-3}} \\ &\quad + \frac{1}{|\zeta - z|^{2n-1}} \qquad \text{for } \zeta, z \in D. \end{aligned} \tag{5.23}$$

Clearly the third term on the right in (5.23) is uniformly integrable over D in z or in ζ. We now prove the necessary estimates for the second term. The corresponding estimates for the first term are handled by analogous methods and will be left to the reader. Let us first prove

$$\int_{D \cap U} \frac{dV(\zeta)}{|\hat\Phi(\zeta, z)|^2 |\zeta - z|^{2n-3}} \lesssim 1 \qquad \text{for all } z \in D. \tag{5.24}$$

As in similar situations, we fix $z \in D$ close to bD, and we use the coordinate system $t = (t_1, t_2, t')$, $t' \in \mathbb{R}^{2n-2}$, given by Lemma V.3.4 on the neighborhood $B(z, \eta)$, where $\eta > 0$ is independent of z. Recall that $t_1 = r(\zeta)$ and $t_2 = \operatorname{Im} \Phi(\cdot, z) = \operatorname{Im} \hat\Phi(\cdot, z)$. Since the integrand in (5.24) is bounded uniformly for $|\zeta - z| \ge \eta$, independently of $z \in \bar D$, in order to prove (5.24) it suffices to prove the estimate with the region of integration $D \cap U$ replaced by $D \cap B(z, \eta)$. From (5.5) it follows that for $\zeta \in B(z, \eta) \cap D$

$$\operatorname{Re} \hat\Phi(\zeta, z) \gtrsim -t_1 + \delta_D(z) + |t|^2, \tag{5.25}$$

and hence

$$|\hat\Phi(\zeta, z)| \gtrsim |t_1| + |t_2| + \delta_D(z) + |t|^2, \tag{5.26}$$

where $t = t(\zeta)$. It is now clear that (5.24) will follow from

$$I = \int_{0 < t_1 < 1;\, 0 < t_2 < 1;\, |t'| \le 1} \frac{dt_1\, dt_2 \ldots dt_{2n}}{[t_1 + t_2 + |t'|^2]^2 |t'|^{2n-3}} < \infty. \tag{5.27}$$

By integrating first in t_1 and then in t_2 one obtains

$$I \lesssim \int_{|t'|\leq 1} \frac{|\log|t'|^2|\, dt_3 \ldots dt_{2n}}{|t'|^{2n-3}} \tag{5.28}$$

and the latter integral is easily seen to be finite; so (5.27), and hence (5.24), is proved.

The proof of the estimate

$$\int_D \frac{dV(z)}{|\hat{\Phi}(\zeta, z)|^2 |\zeta - z|^{2n-3}} \lesssim 1 \qquad \text{for all } \zeta \in D \cap U, \tag{5.29}$$

which is needed for the proof of (5.20), is basically the same: A proof identical to the one of Lemma V.3.4, except for obvious modifications, shows that for $\zeta \in D$ fixed near bD there is a coordinate system $u = (u_1, u_2, u')$ for $z \in B(\zeta, \tilde{\eta})$, where $\tilde{\eta} > 0$ is independent of ζ, such that $u_1(z) = r(z)$, $u_2(z) = \operatorname{Im} \Phi(\zeta, z)$, and $u(\zeta) = (r(\zeta), 0, \ldots, 0)$, which has properties analogous to those of the coordinate system t on $B(z, \eta)$. From (5.5) we then obtain

$$|\hat{\Phi}(\zeta, z)| \gtrsim |u_1| + |u_2| + \delta_D(\zeta) + |u|^2 \tag{5.30}$$

for $u = u(z)$ and $z \in B(\zeta, \tilde{\eta}) \cap \bar{D}$, so that one can proceed as in the proof of (5.24). ■

Proof of Theorem 5.4. Part (i) is an immediate consequence of Lemma 5.5, with $s = 1$, and the generalized Young's inequality (see Appendix B). Part (iii) is obvious from the definition of ${}_{\bar{\partial}}\mathbf{T}_q$ (recall also Theorem IV.1.14 for the required regularity result for $\Omega_q(B)$). Finally, the estimate (ii) follows from Theorem IV.1.14 and from Lemma V.3.1 combined with

$$\left| d_z \int_D f \wedge \bar{\partial}_\zeta \hat{A}_{q-1}(\cdot, z) \right| \lesssim \delta_D(z)^{-1/2} \|f\|_{L^\infty(D)}. \tag{5.31}$$

The proof of (5.31) is obtained by combining the methods used in the proof of Proposition V.3.3, the representation (5.22) for the typical term in a coefficient of $\bar{\partial}_\zeta \hat{A}_{q-1}$, and the estimate (5.26) for $|\hat{\Phi}|$. The details are left to the reader. ■

5.4. A Solution Operator for $\bar{\partial}$ with L^p Estimates

By combining the parametrix ${}_{\bar{\partial}}\mathbf{T}_q$ for $\bar{\partial}$ with the construction in Chapter V, §2.4, we easily obtain a solution operator for $\bar{\partial}$ with the same regularity properties as ${}_{\bar{\partial}}\hat{\mathbf{T}}_q$, as follows.

Theorem 5.6. *Let $D \subset\subset \mathbb{C}^n$ be strictly pseudoconvex with boundary of class C^3. For $1 \leq q \leq n$ there are linear (integral) operators*

$$\hat{\mathbf{S}}_q \colon L^1_{0,q}(D) \to L^1_{0,q-1}(D)$$

and a constant C with the following properties:

(i) $\|\hat{\mathbf{S}}_q f\|_{L^p(D)} \leq C\|f\|_{L^p(D)}$ *for* $1 \leq p \leq \infty$.
(ii) $|\hat{\mathbf{S}}_q f|_{\Lambda_{1/2}(D)} \leq C\|f\|_{L^\infty(D)}$.
(iii) *For* $l = 0, 1, 2, \ldots$, *if* $f \in L^1_{0,q} \cap C^l$, *then* $\hat{\mathbf{S}}_q f \in C^l_{0,q-1}(D)$.
(iv) *If* $f \in C^1_{0,q}(D) \cap L^1_{0,q}$ *and* $\bar{\partial} f = 0$, *then* $\bar{\partial}(\hat{\mathbf{S}}_q f) = f$ *on D.*

Remark. Except for (i), this result is identical with Theorem V.2.7. In fact, the definition of $\hat{\mathbf{S}}_q$ below and Lemma 5.2(i) show that the operators $\mathbf{S}_q$ and $\hat{\mathbf{S}}_q$ agree on $\bar{\partial}$-closed forms which are C^1 up to the boundary.

PROOF. By Lemma 5.2(ii), ${}_{\bar{\partial}}\hat{\mathbf{E}}_q\colon L^1_{0,q}(D) \to C^\infty_{0,q}(\bar{D}_\delta)$ is continuous for $1 \leq q \leq n$. As in the proof of Theorem V.2.7, we introduce the operator $\mathbf{T}_q^{V,V_0}$ given by Theorem V.2.5, where $D \subset\subset V_0 \subset\subset V \subset\subset D_\delta$, and define

$$\hat{\mathbf{S}}_q = {}_{\bar{\partial}}\hat{\mathbf{T}}_q + \mathbf{T}_q^{V,V_0} \circ {}_{\bar{\partial}}\hat{\mathbf{E}}_q. \tag{5.32}$$

Suppose $f \in L^1_{0,q}(D) \cap C^1_{0,q}(D)$ is $\bar{\partial}$-closed. By Proposition 5.3(ii),

$$f = {}_{\bar{\partial}}\hat{\mathbf{E}}_q f + \bar{\partial}({}_{\bar{\partial}}\hat{\mathbf{T}}_q f), \tag{5.33}$$

and since by Proposition 5.3(iii), ${}_{\bar{\partial}}\hat{\mathbf{E}}_q f$ is $\bar{\partial}$-closed on D_δ, it follows from Theorem V.2.5 that

$$\bar{\partial}(\mathbf{T}_q^{V,V_0} \circ {}_{\bar{\partial}}\hat{\mathbf{E}}_q f) = {}_{\bar{\partial}}\hat{\mathbf{E}}_q f \qquad \text{on } D. \tag{5.34}$$

Clearly (5.32), (5.33), and (5.34) imply (iv) in Theorem 5.6. By (5.14) and the definition of $\mathbf{T}_q^{V,V_0}$, it is clear that (i), (ii), and (iii) hold for the operator $\mathbf{T}_q^{V,V_0} \circ {}_{\bar{\partial}}\mathbf{E}_q$, and since by Theorem 5.4 they hold for ${}_{\bar{\partial}}\hat{\mathbf{T}}_q$, by (5.32) they hold for $\hat{\mathbf{S}}_q$ as well. ∎

Remark 5.7. By using Lemma 5.5 for arbitrary $s < (2n+2)/(2n+1)$, Appendix B implies that ${}_{\bar{\partial}}\hat{\mathbf{T}}_q$ is a bounded operator from L^p to L^r for any $1 \leq p, r \leq \infty$ which satisfy $1/r > 1/p - 1/(2n+2)$. One easily checks, by (5.14), that

$$|\mathbf{T}_q^{V,V_0} \circ {}_{\bar{\partial}}\hat{\mathbf{E}}_q f|_{\Lambda_\alpha(D)} \lesssim \|f\|_{L^1(D)} \tag{5.35}$$

for any $\alpha < 1$. Hence it follows that

$$\|\hat{\mathbf{S}}_q f\|_{L^r(D)} \lesssim \|f\|_{L^p(D)} \qquad \text{if } \frac{1}{r} > \frac{1}{p} - \frac{1}{2n+2}. \tag{5.36}$$

In particular,

$$\|\hat{\mathbf{S}}_q f\|_{L^\infty(D)} \lesssim \|f\|_{L^p(D)} \qquad \text{if } p > 2n+2. \tag{5.37}$$

More precisely, one can show that

$$|\hat{\mathbf{S}}_q f|_{\Lambda_{1/2-(n+1)/p}} \lesssim \|f\|_{L^p(D)} \qquad \text{for } p > 2n+2. \tag{5.38}$$

The reader interested in such refinements may consult [Kra 1].

Remark 5.8. All the estimates for the operator $\hat{\mathbf{S}}_q$ given in Theorem 5.6 are stable under sufficiently small C^3 perturbations of the defining function r. This can be verified by arguments analogous to those used in the proof of the corresponding stability result for the operator $\mathbf{S}_q$ in Theorem V.3.6. We shall not go into these details here, but consider only the special case of the domain D_η with defining function $r^{(\eta)} = r - \eta$ for $|\eta| \leq \eta_0$. If η_0 is sufficiently small, the generating form L_D is well defined on $bD_\eta \times D_\eta$ for $|\eta| \leq \eta_0$. Similarly, the kernels $\hat{E}_q^{(\eta)}$ and $\hat{A}_q^{(\eta)}(L_D, B)$ are well defined (we had already used these for $\eta < 0$ in the proof of Proposition 5.3), and we can define

$$\hat{\mathbf{S}}_q^{(\eta)} = {}_{\bar{\partial}}\hat{\mathbf{T}}_q^{(\eta)} + \mathbf{T}_q^{V,V_0} \circ {}_{\bar{\partial}}\hat{\mathbf{E}}_q^{(\eta)} \colon L^1_{0,q}(D_\eta) \to L^1_{0,q-1}(D_\eta),$$

where the integrals in ${}_{\bar{\partial}}\hat{\mathbf{T}}_q^{(\eta)}$ and ${}_{\bar{\partial}}\hat{\mathbf{E}}_q^{(\eta)}$ are now taken over D_η. Since the derivatives of $r^{(\eta)}$ are independent of η, and since the volume of D_η is bounded by a constant independent of η for $|\eta| \leq \eta_0$, it follows that all the estimates involved in the proof of Theorem 5.6 are independent of η for $|\eta| \leq \eta_0$. We thus have:

Corollary 5.9. *There is $\eta_0 > 0$ such that the operator $\hat{\mathbf{S}}_q^{(\eta)}$ given by Theorem 5.6 for the domain D_η for $|\eta| \leq \eta_0$ satisfies* (i)–(iv) *with a constant C independent of η.*

As mentioned earlier, on an arbitrary smoothly bounded pseudoconvex domain D, in general there does not exist a solution operator $\mathbf{S}_q$ for $\bar{\partial}$, such that

$$\|\mathbf{S}_q f\|_{L^\infty(D)} \lesssim \|f\|_{L^\infty(D)}$$

(see N. Sibony [Sib]). On the other hand, L. Hörmander [Hör 1] has proved the remarkable result that for every bounded pseudoconvex domain D in $\mathbb{C}^n$ there is a solution operator $\mathbf{S}_q^H\colon L^2_{0,q}(D) \to L^2_{0,q-1}(D)$ with

$$\bar{\partial}(\mathbf{S}_q^H f) = f \qquad \text{if } \bar{\partial} f = 0$$

and

$$\|\mathbf{S}_q^H f\|_{L^2(D)} \leq C_D \|f\|_{L^2(D)},$$

where the constant C_D depends only on the diameter of D. Hörmander's techniques are based on functional analysis and *a priori* estimates in Hilbert spaces, and they are nonconstructive. Nothing seems to be known regarding L^p estimates for solutions of $\bar{\partial}$ on arbitrary pseudoconvex domains for $p < \infty$ except in case $p = 2$.

Exercises

E.5.1. Let $D \subset\subset \mathbb{C}^n$. Show that the BMK kernel $K_q = \Omega_q(B)$ satisfies

$$\int_D |K_q(\zeta, z)|^s \, dV(\zeta) \leq C_s < \infty \qquad \text{for all } z \in D$$

and all s with $1 \leq s < 2n/(2n-1)$.

E.5.2. In the setting of §5.3, show that

$$\sup_{z\in D}\int_D \frac{dV(\zeta)}{|\hat{\Phi}||\zeta - z|^{2n-2}} < \infty$$

E.5.3. Prove Lemma 5.5 for $1 < s < (2n+2)/(2n+1)$.

E.5.4. Give a complete proof of the estimate (5.31).

§6. Approximation of Holomorphic Functions in L^p Norm

We shall now consider the problem of approximating functions in $\mathcal{O}L^p(D)$ by functions in $\mathcal{O}(\bar{D})$ in L^p norm. In particular, we will obtain a new proof for the uniform approximation result for functions in $A(D)$ discussed in §2. The basic idea of the proof is very simple: locally, the approximation is achieved by translation; the global obstruction to fitting together the locally defined approximating functions is then removed by solving an additive Cousin problem with bounds via estimates for solutions of $\bar{\partial}$.

6.1. Local Approximation

We choose an open covering $\{U_j, j = 1, \ldots, N\}$ of bD by neighborhoods U_j of points $P_j \in bD$, so that Lemma 1.9 holds. Thus, if $\mathbf{n}_j$ is the inner unit normal to bD at P_j, and if for $0 < \tau < \tau_0$ we set $U_0^\tau = D$ and

(6.1) $$U_j^\tau = \{z = w - \tau\mathbf{n}_j : w \in U_j \cap D\} \cap U_j$$

for $j = 1, \ldots, N$, then $\{U_j^\tau : j = 0, \ldots, N\}$ is an open cover of $\bar{D}$ provided τ_0 is sufficiently small, and the following result holds.

Lemma 6.1. *Suppose $1 \le p \le \infty$ and let $f \in \mathcal{O}L^p(D)$. For $0 < \tau < \tau_0$ define $f_0^\tau = f$ and*

(6.2) $$f_j^\tau(z) = f(z + \tau\mathbf{n}_j) \qquad \text{for } j = 1, \ldots, N.$$

Then

(6.3) $$f_j^\tau \in \mathcal{O}L^p(U_j^\tau) \qquad \text{for } j = 0, 1, \ldots, N.$$

(6.4) $$\lim_{\tau\to 0} f_j^\tau = f \qquad \text{pointwise on } D \cap U_j.$$

(6.5) $$\lim_{\tau\to 0} \|f_j^\tau - f\|_{L^p(U_j \cap D)} = 0 \qquad \text{if either } p < \infty, \text{ or if } p = \infty \text{ and } f \in A(D).$$

PROOF. Equations (6.3) and (6.4) are obvious from the definitions of U_j^τ and f_j^τ. Equation (6.5) for $p < \infty$ follows from the fact that translation defines a continuous operator on L^p spaces if $p < \infty$; the case $p = \infty$ follows from the uniform continuity of f. ∎

Unfortunately, the *local* approximating functions f_j^τ do not match up to form a global holomorphic function on $\bar{D}$, since

$$(6.6) \qquad g_{ij}^\tau := f_j^\tau - f_i^\tau \qquad \text{on } U_j^\tau \cap U_i^\tau$$

will in general be different from zero if $i \neq j$. However, Lemma 6.1 implies that g_{ij}^τ is "almost" zero, as follows.

Corollary 6.2. *Define*

$$(6.7) \qquad M_p^\tau(f) = \max\{\|g_{ij}^\tau\|_{L^p(U_i^\tau \cap U_j^\tau)}: \quad 0 \leq i, j \leq N\}.$$

Then one has

$$(6.8) \qquad \lim_{\tau \to 0} M_p^\tau(f) = 0 \qquad \textit{if } p < \infty \textit{ or if } p = \infty \textit{ and } f \in A(D),$$

and

$$(6.9) \qquad M_\infty^\tau(f) \lesssim \|f\|_{L^\infty(D)} \qquad \textit{if } f \in L^\infty(D).$$

Proof. Equation (6.9) is obvious. For (6.8), fix i and j and let $W = U_i \cap U_j \cap D$. Then

$$\|g_{ij}^\tau\|_{L^p(W)} \leq \|f_j^\tau - f\|_{L^p(W)} + \|f_i^\tau - f\|_{L^p(W)},$$

and (6.5) implies that

$$(6.10) \qquad \|g_{ij}^\tau\|_{L^p(W)} \to 0 \qquad \text{as } \tau \to 0.$$

Since $\operatorname{vol}[(U_i^\tau \cap U_j^\tau) - W] \to 0$ as $\tau \to 0$, (6.10) implies (6.8) for $p < \infty$. The case $p = \infty$ follows again by the uniform continuity of $f \in A(D)$. ∎

6.2. An Additive Cousin Problem with Bounds

The obstruction in §6.1 to finding a *global* approximating function is given by $\{g_{ij}^\tau\}$, defined in (6.6); clearly this data satisfies the necessary relations

$$(6.11) \qquad g_{ij}^\tau = -g_{ji}^\tau; \qquad g_{ij}^\tau + g_{jk}^\tau + g_{ki}^\tau = 0$$

for the solution of an additive Cousin problem (see Chapter VI, §4). Of course the obvious solution $\{f_j^\tau\}$ is of no use here; instead, what is needed is a solution $\{g_j^\tau\}$ which is bounded by $\{g_{ij}^\tau\}$, i.e., by $M_p^\tau(f)$. Since the relevant result is of independent interest, we state it separately.

Theorem 6.3. *Let D be strictly pseudoconvex with C^3 boundary and let $\{U_j: j = 0, \ldots, N\}$ be an open covering of $\bar{D}$. Set $V_j = U_j \cap D$. Then there is a constant $C < \infty$ with the following property. If the functions $g_{ij} \in \mathcal{O}(V_i \cap V_j)$ satisfy* (6.11) *for all* $i, j = 0, \ldots, N$, *and if*

$$M_p = M_p(\{g_{ij}\}) = \max\{\|g_{ij}\|_{L^p(V_i \cap V_j)} : 0 \le i, j \le N\},$$

then there are functions $g_j \in \mathcal{O}(V_j)$, $j = 0, \ldots, N$, *such that*

$$g_j - g_i = g_{ij} \qquad \text{on } V_i \cap V_j \tag{6.12}$$

and

$$\|g_j\|_{L^p(V_j)} \le CM_p(\{g_{ij}\}) \qquad \text{for } 1 \le p \le \infty. \tag{6.13}$$

PROOF. The proof is analogous to the proof of Theorem VI.4.8, so we will be brief. Choose C^∞ functions $\chi_j \in C_0^\infty(U_j)$, $j = 0, \ldots, N$, such that $\sum \chi_j \equiv 1$ on $\bar{D}$, and set $v_j = \sum_{\nu=0}^N \chi_\nu g_{\nu j}$. Then $v_j \in C^\infty(V_j)$ and

$$\|v_j\|_{L^p(V_j)} \lesssim M_p; \tag{6.14}$$

moreover, by (6.11),

$$v_j - v_i = \sum_{\nu=0}^N \chi_\nu(g_{\nu j} - g_{\nu i}) = \sum_{\nu=0}^N \chi_\nu(g_{ij}) = g_{ij}. \tag{6.15}$$

Hence $\bar\partial v_j - \bar\partial v_i = \bar\partial g_{ij} = 0$ on $V_i \cap V_j$, so that the (0, 1)-form α defined locally on V_j by $\bar\partial v_j$ is globally well defined and C^∞ on D. Clearly $\bar\partial\alpha = 0$, and since $\bar\partial v_j = \sum(\bar\partial\chi_\nu)g_{\nu j}$, one has

$$\|\alpha\|_{L^p_{0,1}(D)} \le \sum_{j=0}^N \|\bar\partial v_j\|_{L^p(V_j)} \lesssim M_p. \tag{6.16}$$

By Theorem 5.6 and (6.16), $u = \hat{\mathbf{S}}_1\alpha \in C^\infty(D)$ satisfies $\bar\partial u = \alpha$ and

$$\|u\|_{L^p} \le C\|\alpha\|_{L^p} \le C'M_p. \tag{6.17}$$

It follows that $g_j = v_j - u \in C^\infty(V_j)$ satisfies $\bar\partial g_j = 0$, i.e., g_j is holomorphic, and by (6.15), $g_j - g_i = (v_j - u) - (v_i - u) = v_j - v_i = g_{ij}$ on $V_i \cap V_j$. Finally, (6.14) and (6.17) imply (6.13) with a new constant C independent of $\{g_{ij}\}$ and p. ∎

6.3. Global Approximation

We now apply Theorem 6.3 to the data $\{g_{ij}^\tau\}$, $0 < \tau < \tau_0$, defined by (6.6) on a suitable neighborhood $D_{\eta(\tau)}$ of $\bar{D}$, as follows. Given the covering $\{U_j : j = 1, \ldots, N\}$ of bD introduced at the beginning of section 6.1, we choose $\chi_j \in C_0^\infty(U_j)$ and $\chi_0 \in C_0^\infty(D)$, so that $\sum_{j=0}^N \chi_j \equiv 1$ on a neighborhood Ω of $\bar{D}$. For each $\tau < \tau_0$ choose $\eta(\tau) > 0$ so that

$$D_{\eta(\tau)} \subset \Omega \cap \left(\bigcup_{j=0}^N U_j^\tau\right).$$

Clearly $\eta(\tau) \to 0$ as $\tau \to 0$. By choosing τ_0 sufficiently small, we may assume that

$$\operatorname{supp} \chi_j \cap \bar{D}_{\eta(\tau)} \subset U_j^\tau \qquad \text{for all } 0 < \tau < \tau_0 \text{ and } j = 0, \ldots, N,$$

and that the estimates for the operator $\hat{\mathbf{S}}_1^{(\eta(\tau))}$ given by Theorem 5.6 on the domain $D_{\eta(\tau)}$ are independent of τ for $0 < \tau < \tau_0$ (see Corollary 5.9). Now set

$$V_j^\tau = U_j^\tau \cap D_{\eta(\tau)},$$

and apply Theorem 6.3 to the functions $g_{ij}^\tau \in \mathcal{O}(V_i^\tau \cap V_j^\tau)$. We thus obtain functions $g_j^\tau \in \mathcal{O}(V_j^\tau)$, $0 \le j \le N$, such that

$$g_j^\tau - g_i^\tau = g_{ij}^\tau \qquad \text{on } V_i^\tau \cap V_j^\tau \tag{6.18}$$

and

$$\| g_j^\tau \|_{L^p(V_j^\tau)} \le C M_p^\tau(f) \qquad \text{for } 1 \le p \le \infty. \tag{6.19}$$

The remarks just made about $\hat{\mathbf{S}}_1^{(\eta(\tau))}$, and the fact that the functions χ_j introduced above—which are used in the proof of Theorem 6.3—are independent of τ, imply that the constant C in (6.19) is independent of τ.

We are now ready to complete the construction of the *global* approximating functions. From (6.6) and (6.18) it follows that

$$f_j^\tau - g_j^\tau = f_i^\tau - g_i^\tau \qquad \text{on } V_i^\tau \cap V_j^\tau.$$

Hence there is $f^\tau \in \mathcal{O}(D_{\eta(\tau)})$, such that

$$f^\tau = f_j^\tau - g_j^\tau \qquad \text{on } V_j^\tau. \tag{6.20}$$

Finally, we estimate $f - f^\tau$ as $\tau \to 0$. Since $f_0^\tau = f$, (6.20) and (6.19) imply

$$\| f - f^\tau \|_{L^p(D)} \le \sum_{j=1}^N \| f - f_j^\tau \|_{L^p(U_j \cap D)} + (N+1) C M_p^\tau(f). \tag{6.21}$$

If $p < \infty$, or if $p = \infty$ and $f \in A(D)$, then (6.5), (6.8), and (6.21) imply that

$$\| f - f^\tau \|_{L^p(D)} \to 0 \qquad \text{as } \tau \to 0. \tag{6.22}$$

In case f is only in $H^\infty(D)$, the above arguments do *not* imply $\| f - f^\tau \|_{L^\infty(D)} \to 0$. (The reader should make sure to see where the argument breaks down!) However, we can draw the following conclusions: from (6.20), (6.19), and (6.9) one obtains

$$\| f^\tau \|_{L^\infty(D_{\eta(\tau)})} \le C \| f \|_{L^\infty(D)} \tag{6.23}$$

for all $0 < \tau < \tau_0$, where C is independent of τ; and since $H^\infty(D) \subset \mathcal{O}L^p(D)$ for any $p < \infty$, we still have (6.22) for any $p < \infty$.

We now summarize the above results.

Theorem 6.4. *Let D be strictly pseudoconvex with C^3 boundary. For $1 \le p < \infty$, every function in $\mathcal{O}L^p(D)$ can be approximated in $L^p(D)$ norm by functions in $\mathcal{O}(\bar{D})$. Moreover, there is a constant $C < \infty$, such that every $f \in H^\infty(D)$ can be approximated in $\mathcal{O}(D)$ by a sequence $\{f_\nu : \nu = 1, 2, \ldots\} \subset \mathcal{O}(\bar{D})$ with the following properties:*

(i) $\| f_\nu \|_{H^\infty(D)} \le C \| f \|_{H^\infty(D)}$ *for all* $\nu \ge 1$.

(ii) $\lim_{\nu\to\infty} \|f_\nu - f\|_{L^p(D)} = 0$ *for every* $1 \le p < \infty$.
(iii) *If* f *extends continuously to* $\bar{D}$ (*i.e.*, $f \in A(D)$), *then* $f_\nu \to f$ *uniformly on* $\bar{D}$.

Remark 6.5. The question of "pointwise bounded approximation" of functions in $H^\infty(D)$ by functions in $A(D)$ or $\mathcal{O}(\bar{D})$ has been studied extensively for domains $D \subset \mathbb{C}^1$ by a combination of techniques from complex analysis and functional analysis (see, for example, A.M. Davie, T.W. Gamelin, and J. Garnett: "Distance estimates and pointwise bounded density", Trans. A.M.S. **175** (1973), 37–68.). Motivated by such work in one variable, B. Cole and R.M. Range proved that the constant C in Theorem 6.4(i) can be chosen to be 1. [CoRa].

Exercises

E.6.1. Let $D \subset\subset \mathbb{C}^n$ be bounded and suppose $f \in H^\infty(D)$ and $f_\nu \in H^\infty(D)$, $\nu = 1, 2, \ldots,$ satisfy

(i) $f_\nu \to f$ pointwise on D as $\nu \to \infty$;
(ii) $\|f_\nu\|_{L^\infty(D)} \le C\|f\|_{L^\infty(D)}$, $\nu = 1, 2, \ldots,$

where $C < \infty$. Show that $f_\nu \to f$ in $L^p(D)$ for all p with $1 \le p < \infty$, but not necessarily for $p = \infty$.

In the remaining exercises assume that D is strictly pseudoconvex with C^3 boundary.

E.6.2. Given $f \in A(D)$, show that there is a sequence $\{f_\nu\} \subset A(D)$, such that $|f - f_\nu|_D \to 0$ as $\nu \to \infty$ and $|f_\nu|_D \le |f|_D$ for all ν.

Remark. See Remark 6.5 for the corresponding statement for $f \in H^\infty(D)$.

E.6.3. Show that there is $\eta > 0$, such that for every $f \in \mathcal{O}L^p(D)$ there is a sequence $\{f_\nu\} \in \mathcal{O}L^p(D_\eta)$ such that

$$\lim_{\nu\to\infty} \|f - f_\nu\|_{L^p(D)} = 0.$$

Is it possible to also require that $\|f_\nu\|_{L^p(D_\eta)} \le C\|f\|_{L^p(D)}$ for some constant $C < \infty$?

E.6.4. Suppose $f \in H^\infty(D)$ extends continuously to a point $P \in bD$. Show that there is a sequence $\{f_\nu\} \in H^\infty(D)$ of functions which extend holomorphically across P such that $f_\nu \to f$ uniformly on D as $\nu \to \infty$.

Remark. The result remains true with P replaced by an arbitrary subset E of bD; see [Ran 1] for details.

§7. Regularity Properties of the Bergman Projection

In this section we will use ideas of N. Kerzman and E.M. Stein [KeSt] and E. Ligocka [Lig 2] in order to establish a connection between the Bergman kernel K_D of a strictly pseudoconvex domain D and explicit integral representations. This will allow us to prove some estimates and regularity properties for the Bergman projection which are of major importance in the study of biholomorphic mappings, as we shall see in the next paragraph. Our approach is motivated by the computation of the Bergman kernel K_B for the unit ball B in Chapter IV, §4.4. By using the techniques developed in §1 and §5 we first find a rather explicit kernel $G_D(\zeta, z)$ on $D \times D$ holomorphic in $z \in D$, so that

$$f(z) = \int_D f(\zeta) G(\zeta, z)\, dV(\zeta) = (f, \overline{G(\cdot, z)})_D \qquad \textit{for } f \in \mathscr{H}^2 = \mathcal{O}L^2(D) \textit{ and } z \in D.$$

On the ball B we were able to find $G_B(\zeta, z)$ which was Hermitian symmetric, i.e., $G_B^*(\zeta, z) := \overline{G_B(z, \zeta)} = G_B(\zeta, z)$ (cf. Chapter IV, formulas (4.19) and (4.20)); hence $\overline{G_B(\zeta, z)}$ is holomorphic in ζ, and this allowed us to conclude that $\overline{G_B(\zeta, z)}$ is the Bergman kernel. On arbitray domains this symmetry does no longer hold, but if we are careful, it remains true "approximately". This will allow us to use $\overline{G_D}$ or G_D^* as an approximation for the Bergman kernel K_D and to obtain estimates for K_D and the Bergman projection by proving such estimates for the explicit kernel $\overline{G_D}$.

7.1. A Reproducing Kernel for Functions in $\mathscr{H}^2$

Recall the construction of the "Cauchy kernel" $C(\zeta, z)$ in §1. C is globally holomorphic in z, and $C(\zeta, z) = E(\zeta, z) - A(\zeta, z)$, where $E = E_0 = \Omega_0(L_D)$, and $A(\zeta, \cdot) = \mathbf{T}_1^{V, D^\#}(\bar\partial_z E(\zeta, \cdot))$ has no singularities on $bD \times \bar D$. We now use the extension $\hat E = \hat E_0$ of E from $bD \times D$ to $\bar D \times D$ given in (5.4) in order to find an extension $\hat C$ for C. For $\zeta \in \bar D$ we set $\hat A(\zeta, \cdot) = \mathbf{T}_1^{V, D^\#}(\bar\partial_z \hat E_0(\zeta, \cdot))$ and $\hat C(\zeta, \cdot) = \hat E_0(\zeta, \cdot) - \hat A(\zeta, \cdot)$.

Theorem 7.1. *Suppose the defining function r for D is of class C^{k+3}. Then*

(i) *$\hat A$ and $\bar\partial_\zeta \hat A \in C^{k+1,\infty}(\bar D \times D^\#)$;*
(ii) *$\bar\partial_\zeta \hat C \in C^{k+1,\infty}(\bar D \times \bar D - \Delta_{bD})$, and for ζ fixed in $\bar D$, $\bar\partial_\zeta \hat C(\zeta, z)$ is holomorphic in z on $\bar D$ if $\zeta \in D$, and on $\bar D - \{\zeta\}$ if $\zeta \in bD$.*
(iii) *For all $f \in \mathcal{O}L^1(D)$ one has*

$$f(z) = \int_D f(\zeta) \bar\partial_\zeta \hat C(\zeta, z) \qquad \textit{for } z \in D. \tag{7.1}$$

Proof. Since $E = \Omega_0(L_D) = \Omega_0(P)/\Phi^n$, it follows that

$$\bar\partial_\zeta \hat E = \varphi \left[\frac{\bar\partial_\zeta \Omega_0(P)}{[\Phi - r(\zeta)]^n} - n \frac{\bar\partial_\zeta \hat\Phi \wedge \Omega_0(P)}{\hat\Phi^{n+1}} \right] + \bar\partial\varphi \wedge \frac{\Omega_0(P)}{\hat\Phi^n},$$

which shows that $\bar\partial_\zeta \hat E$ is of class $C^{k+1,\infty}$, as $\bar\partial_\zeta \Omega_0(P) = (2\pi i)^{-n}(\bar\partial_\zeta P)^n$. Hence $\bar\partial_z \bar\partial_\zeta \hat E = \bar\partial_\zeta \bar\partial_z \hat E$ and $\bar\partial_z \hat E$ are of class $C^{k+1,\infty}$ on $\bar D \times D^\#$, and the same then follows for $\hat A = \mathbf{T}_1^{V, D^\#}(\bar\partial_z \hat E)$ and $\bar\partial_\zeta \hat A = \mathbf{T}_1^{V, D^\#}(\bar\partial_\zeta \bar\partial_z \hat E)$. This proves (i). (ii) is then clear from the construction of $\hat{C}$. Finally, since $\hat{C}(\zeta, \cdot) = C(\zeta, \cdot)$ for $\zeta \in bD$, (7.1) is an immediate consequence of Theorem 1.3 and Stokes' Theorem if $f \in A^1(D)$. The general case $f \in \mathcal{O}L^1(D)$ follows by applying (7.1) to the domains $D_\eta \subset\subset D$, $\eta < 0$, and taking the limit $\eta \to 0$. The details of this argument are essentially the same as those used in the proof of Proposition 5.3, and hence will not be repeated here. ■

We now define

$$G_D(\zeta, z) = *_\zeta \bar{\partial}_\zeta \hat{C}(\zeta, z) \qquad \text{for } \zeta, z \in D; \tag{7.2}$$

then (7.1) can be written as

$$f(z) = (f, \overline{G_D(\cdot, z)})_D = \int_D f(\zeta) G_D(\zeta, z)\, dV(\zeta) \tag{7.3}$$

for $f \in \mathcal{O}L^1(D)$, and hence, in particular, for $f \in \mathscr{H}^2(D)$. By Theorem 7.1(ii) and (7.2), $G_D(a, \cdot)$ is in $\mathcal{O}(\bar{D}) \subset \mathscr{H}^2(D)$ for fixed $a \in D$. The formal adjoint $G_D^*(\zeta, z) = \overline{G_D(z, \zeta)}$ is then conjugate holomorphic in ζ, i.e., $\overline{G^*(\cdot, z)} \in \mathscr{H}^2(D)$ for fixed $z \in D$. If we had $G_D^* \equiv G_D$ on $D \times D$, then $\overline{G_D(\zeta, z)}$ would be holomorphic in ζ, and therefore (7.3) would imply that $\overline{G_D}$ is the Bergman kernel K_D. We are thus led to investigate the kernel $G_D^* - G_D$.

7.2. Cancellation of Singularities

The regularity properties of the mapping $f \to \mathbf{G}_D f$ defined by

$$(\mathbf{G}_D f)(z) = \int_D f(\zeta) G_D(\zeta, z)\, dV(\zeta) \qquad \text{for } z \in D$$

are determined by the singularities of G_D at $\zeta = z \in bD$, which depend on the zeroes of $\hat{\Phi} = \Phi - r(\zeta)$, as follows.

Lemma 7.2. *Suppose bD is of class C^{k+3}. There are $\varepsilon_0 > 0$ and functions N_0, N_1, $N_2 \in C^{k+1,\infty}(\bar{D} \times \bar{D})$ so that*

$$G_D(\zeta, z) = \frac{N_0}{[\hat{\Phi}(\zeta, z)]^{n+1}} + \frac{N_1}{[\hat{\Phi}(\zeta, z)]^{n}} + N_2 \tag{7.4}$$

and

$$N_0(\zeta, \zeta) \in \mathbb{R} \tag{7.5}$$

for ζ with $-\varepsilon_0 < r(\zeta) \le 0$ and $z \in \bar{D}$.

Proof. We choose $\varepsilon_0 > 0$ as in the construction of $\hat{E}_q$ in §5.1, so that $\varphi(\zeta) = 1$ for $-\varepsilon_0 < r(\zeta) \le 0$. Since $\hat{C} = \hat{E}_0 - \hat{A}$, it follows from (5.4) that

$$\begin{aligned}\bar{\partial}_\zeta \hat{C} = \bar{\partial}_\zeta \hat{E}_0 - \bar{\partial}_\zeta \hat{A} &= \bar{\partial}_\zeta\left[\frac{\Omega_0(P)}{\hat{\Phi}^n}\right] - \bar{\partial}_\zeta \hat{A} \\ &= -n\frac{\bar{\partial}_\zeta \hat{\Phi} \wedge \Omega_0(P)}{\hat{\Phi}^{n+1}} + \frac{N_1\, dV}{\hat{\Phi}^n} + N_2\, dV,\end{aligned}$$

so that (7.4) follows by setting $N_0 = -n*[\bar{\partial}_\zeta \hat{\Phi} \wedge \Omega_0(P)]$. To see (7.5), recall that $\hat{\Phi} = \Phi - r(\zeta)$, and that the definition of $P = \sum_{j=1}^n P_j\, d\zeta_j$ in V(1.9) implies

$$P(\zeta, \zeta) = \partial r(\zeta) \quad \text{and } \bar{\partial}_\zeta P(\zeta, \zeta) = \bar{\partial}\partial r(\zeta).$$

Hence

(7.6) $$\bar{\partial}_\zeta \hat{\Phi} \wedge \Omega_0(P) = -(2\pi i)^{-n} \bar{\partial}_\zeta r \wedge \partial_\zeta r \wedge (\bar{\partial}\partial r)^{n-1} \qquad \text{at } (\zeta, \zeta).$$

The right side in (7.6) is clearly a real valued (n, n) form, and since the $*$-operator is real, (7.5) follows. ∎

Remark. A more careful examination of (7.6) reveals that, as a consequence of the strict pseudoconvexity of D, $N_0(\zeta, \zeta) \neq 0$ for ζ sufficiently close to bD (see Exercise E.7.2).

Lemma 7.3. *The function* $\hat{\Phi}$ *satisfies*

$$\int_{\{\zeta \in D : r(\zeta) \geq -\varepsilon_0\}} |\hat{\Phi}(\zeta, z)|^{-(n+1+\alpha)}\, dV(\zeta) \lesssim \begin{cases} 1 & \text{if } \alpha < 0 \\ |\log \delta_D(z)| & \text{if } \alpha = 0 \\ [\delta_D(z)]^{-\alpha} & \text{if } \alpha > 0 \end{cases}$$

for all $z \in D$.

Proof. The proof of Lemma 7.3 is very similar to the one of Lemma 1.5, and the details are left to the reader. The modification required to estimate a volume integral involving $\hat{\Phi}$, rather than a boundary integral, has already been used in the proof of Lemma 5.5—see the estimate (5.26) for $|\hat{\Phi}|$ in terms of special local coordinates. ∎

It now follows from Lemma 7.2 that the principal term of $G_D(\zeta, z)$ is given by $N_0/\hat{\Phi}^{n+1}$, the other two terms in the representation (7.4) for G_D being uniformly integrable according to Lemma 7.3. In order to estimate $G_D - G_D^*$ we therefore compute

(7.7) $$\begin{aligned}\frac{N_0}{\hat{\Phi}^{n+1}} - \frac{N_0^*}{\hat{\Phi}^{*n+1}} &= \frac{N_0(\zeta, z)}{\hat{\Phi}(\zeta, z)^{n+1}} - \frac{\overline{N_0(z, \zeta)}}{\overline{\hat{\Phi}(z, \zeta)}^{n+1}} \\ &= N_0(\zeta, \zeta)\left[\frac{1}{\hat{\Phi}^{n+1}} - \frac{1}{\hat{\Phi}^{*n+1}}\right] + \frac{O(|\zeta - z|)}{\hat{\Phi}^{n+1}} + \frac{O(|\zeta - z|)}{\hat{\Phi}^{*n+1}}\end{aligned}$$

where we have used that $N_0(\zeta, z) = N_0(\zeta, \zeta) + O(|\zeta - z|)$ and $\overline{N_0(z, \zeta)} = N_0(\zeta, \zeta) + O(|\zeta - z|)$, and the fact that $N_0(\zeta, \zeta)$ is real valued (Lemma 7.2).

Next, observe that

$$\frac{1}{\hat{\Phi}^t} - \frac{1}{\hat{\Phi}^{*t}} = \frac{(\hat{\Phi}^* - \hat{\Phi})\sum_{\nu=0}^{t-1}\hat{\Phi}^{\nu}\hat{\Phi}^{*t-1-\nu}}{\hat{\Phi}^t\hat{\Phi}^{*t}}$$

(7.8)

$$= (\hat{\Phi}^* - \hat{\Phi})\sum_{\nu=0}^{t-1}\hat{\Phi}^{-(t-\nu)}\hat{\Phi}^{*-(\nu+1)}$$

for any $t = 1, 2, \ldots$. Recall that in case D is the unit ball, $\hat{\Phi} = 1 - \langle z, \zeta\rangle$ obviously satisfies $\hat{\Phi}^* = \hat{\Phi}$. We now show that for an arbitrary strictly pseudoconvex domain one can achieve

(7.9) $$\hat{\Phi}^* - \hat{\Phi} = O(|\zeta - z|^3).$$

As we shall see, this approximate symmetry is a crucial ingredient in our study of the Bergman kernel.

In order to prove (7.9), Φ must be chosen in a special way. Recall from V(1.5) that $\Phi = F^{\#}$ for $|\zeta - z| < \varepsilon/2$, where $F^{\#}$ is a suitable small perturbation of the Levi polynomial $F = F^{(r)}$ of the defining function r, which we had introduced in order to carry out the constructions in Chapter V with minimal differentiability assumptions on bD. However, in order to prove (7.9) we must choose Φ *equal* to the Levi polynomial F of r for $|\zeta - z| \leq \varepsilon/2$.

Lemma 7.4. *If r is of class C^3 and $F(\zeta, z)$ is the Levi polynomial of r, then*

$$[F(\zeta, z) - r(\zeta)] - [\overline{F(z, \zeta)} - r(z)] = O(|\zeta - z|^3).$$

Corollary 7.5. *If $\Phi(\zeta, z) = F$ locally, then*

(i) $\hat{\Phi} - \hat{\Phi}^* = O(|\zeta - z|^3)$, *and*
(ii) $|\hat{\Phi}^*| \gtrsim |\hat{\Phi}|$ *for $\zeta, z \in \bar{D}$ close to bD.*

Proof. (i) is obvious; (ii) is then an immediate consequence, since $|\hat{\Phi}^*| \geq |\hat{\Phi}| - |\hat{\Phi}^* - \hat{\Phi}|$, by the triangle inequality, and $|\hat{\Phi}(\zeta, z)| \gtrsim |\zeta - z|^2$ for $\zeta \in U \cap \bar{D}$ and $z \in \bar{D}$ by Lemma 5.1. ∎

Proof of Lemma 7.4. To simplify notation we shall write $\partial r/\partial\zeta_j = r_j$, $\partial^2 r/\partial\zeta_j\partial\bar{\zeta}_k = r_{j\bar{k}}$, etc. In the definition of the Levi polynomial

$$F(\zeta, z) = \sum_{j=1}^{n} r_j(\zeta)(\zeta_j - z_j) - 1/2 \sum_{j,k=1}^{n} r_{jk}(\zeta)(\zeta_j - z_j(\zeta_k - z_k)$$

we substitute $r_j(\zeta)$ by its first order Taylor expansion at z,

$$r_j(\zeta) = r_j(z) + \sum_{k=1}^{n} r_{jk}(z)(\zeta_k - z_k) + \sum_{k=1}^{n} r_{j\bar{k}}(z)(\bar{\zeta}_k - \bar{z}_k) + O(|\zeta - z|^2),$$

and

$$r_{jk}(\zeta) = r_{jk}(z) + O(|\zeta - z|).$$

It follows that

$$F(\zeta, z) = \dot{-}F(z, \zeta) + L_z(r; \zeta - z) + O(|\zeta - z|^3),$$

and since $r_{j\bar{k}}(z) = r_{j\bar{k}}(\zeta) + O(|\zeta - z|)$, this implies

$$F(\zeta, z) = -F(z, \zeta) + L_\zeta(r; \zeta - z) + O(|\zeta - z|^3). \tag{7.10}$$

Now recall the Taylor expansion

$$r(z) = r(\zeta) - F(\zeta, z) - \overline{F(\zeta, z)} + L_\zeta(r; \zeta - z) + O(|\zeta - z|^3) \tag{7.11}$$

of r (see II(2.31)). By replacing $\overline{F(\zeta, z)}$ in (7.11) with (7.10)—note that $L_\zeta = \bar{L}_\zeta$—and rearranging, the desired conclusions follows. ∎

From now on we shall assume that $\Phi = F$ locally, so that Corollary 7.5 holds. Since the proof of Theorem 7.1 requires Φ to be of class $C^{k+2,\infty}$, we will also assume that bD is of class at least C^{k+4}. It then follows that G_D is of class $C^{k+1,\infty}$ (see Lemma 7.2).

Theorem 7.6. *The kernel*

$$B(\zeta, z) = G_D(\zeta, z) - G_D^*(\zeta, z) \tag{7.12}$$

satisfies

$$\int_D |B(\zeta, z)|^s \, dV(\zeta) \lesssim 1 \qquad \text{for all } z \in D \tag{7.13}$$

and

$$\int_D |B(\zeta, z)|^s \, dV(z) \lesssim 1 \qquad \text{for all } \zeta \in D, \tag{7.14}$$

for every $s < (2n + 2)/(2n + 1)$.

Proof. From (7.4), (7.7), and (7.8) with $t = n + 1$ and Corollary 7.5 it follows that

$$|G_D(\zeta, z) - G_D^*(\zeta, z)| \lesssim |\hat{\Phi}(\zeta, z)|^{-(n+1)+1/2} \qquad \text{for } \zeta, z \in D \text{ close to } bD,$$

where we have used $|\hat{\Phi}(\zeta, z)|^{1/2} \gtrsim |\zeta - z|$. Thus $|B|^s \lesssim |\hat{\Phi}|^{-s(n+1/2)}$ for $\zeta \in D$ close to bD, and (7.13) follows from Lemma 7.3, since $s(n + 1/2) < n + 1$ for $s < (2n + 2)/(2n + 1)$. Finally, since $B(\zeta, z) = -\overline{B(z, \zeta)}$, (7.14) follows from (7.13). ∎

Corollary 7.7. *Define the operator* $\mathbf{B}: f \mapsto \mathbf{B}f$ *by*

$$\mathbf{B}f(z) = \int_D f(\zeta) B(\zeta, z) \, dV(\zeta) = (f, \overline{B(\cdot, z)})_D.$$

Then

(i) $\mathbf{B}$ *is a bounded operator from* $L^p(D)$ *to* $L^q(D)$ *for any* $1 \le p, q \le \infty$ *with* $1/q > 1/p - 1/(2n+2)$.
(ii) $\mathbf{B}: L^2(D) \to L^2(D)$ *is compact.*
(iii) *The adjoint* $\mathbf{B}^*$ *of* $\mathbf{B}$ *has kernel* $B^*(\zeta, z) = \overline{B(z, \zeta)}$.

Proof. (i) follows from the Theorem and Appendix B, and (ii) follows from the Theorem with $s = 1$ and Appendix C. For (iii), define $\mathbf{B}^*f(z) = (f, \overline{B^*(\cdot, z)})_D$; if $f, g \in C_0(D)$, it readily follows that $(f, \mathbf{B}g)_D = (\mathbf{B}^*f, g)$, implying that $\mathbf{B}^*$ is indeed the adjoint of $\mathbf{B}$. ■

7.3. A Representation for the Bergman Projection

We now combine the integral representation formula obtained in §7.1 with the results in §7.2 in order to obtain a representation of the Bergman projection $\mathbf{P}_D: L^2(D) \to \mathscr{H}^2(D)$ in terms of explicit kernels.

Let f be arbitrary in $L^2(D)$. Then $\mathbf{P}_D f \in \mathscr{H}^2(D)$, and by (7.3) we have

$$\mathbf{P}_D f(a) = (\mathbf{P}_D f, \overline{G_D(\cdot, a)})_D \qquad \text{for } a \in D. \tag{7.15}$$

Since by (7.12), $G_D = G_D^* + B$, (7.15) implies that

$$\begin{aligned} \mathbf{P}_D f(a) &= (\mathbf{P}_D f, \overline{G_D^*(\cdot, a)})_D + (\mathbf{P}_D f, \overline{B(\cdot, a)})_D \\ &= (f, \mathbf{P}_D G_D(a, \cdot)) + (\mathbf{P}_D f, \overline{B(\cdot, a)}), \end{aligned} \tag{7.16}$$

since $\overline{G_D^*(\cdot, a)} = G_D(a, \cdot)$ and $\mathbf{P}_D$ is Hermitian. Now recall that by (7.2) and Theorem 7.1, $G_D(a, \cdot)$ is in $\mathscr{H}^2(D)$ for fixed $a \in D$, and hence $\mathbf{P}_D G_D(a, \cdot) = G_D(a, \cdot)$. Thus (7.16) implies

$$\mathbf{P}_D f(a) = (f, \overline{G_D^*(\cdot, a)}) + (\mathbf{P}_D f, \overline{B(\cdot, a)}) \tag{7.17}$$

for all $a \in D$.

We summarize the main conclusions.

Theorem 7.8. *Define the operator* $f \mapsto \mathbf{G}_D^* f$ *by*

$$\mathbf{G}_D^* f(z) = \int_D f(\zeta) G_D^*(\zeta, z)\, dV(\zeta) = (f, \overline{G_D^*(\cdot, z)}).$$

Then

$$\mathbf{P}_D = \mathbf{G}_D^* + \mathbf{B} \circ \mathbf{P}_D \qquad \textit{as operators on } L^2(D). \tag{7.18}$$

Moreover, $\mathbf{G}_D$ *and* $\mathbf{G}_D^*$ *are bounded in* $L^2(D)$, $\mathbf{I} + \mathbf{B}$ *and* $\mathbf{I} - \mathbf{B}$ *are invertible bounded operators in* $L^2(D)$, *and*

$$\mathbf{P}_D = (\mathbf{I} - \mathbf{B})^{-1} \circ \mathbf{G}_D^*. \tag{7.19}$$

Proof. Equation (7.18) is just a reformulation of (7.17). Solving for $\mathbf{G}_D^*$ gives

$$\mathbf{G}_D^* = \mathbf{P}_D - \mathbf{B} \circ \mathbf{P}_D = (\mathbf{I} - \mathbf{B}) \circ \mathbf{P}_D. \tag{7.20}$$

Since $\mathbf{B}$ and $\mathbf{P}_D$ are bounded, (7.20) shows that $\mathbf{G}_D^*$ and $\mathbf{G}_D = \mathbf{G}_D^* + \mathbf{B}$ are bounded operators on $L^2(D)$. By Corollary 7.7(ii), $\mathbf{B}$ is in fact compact, and since $\mathbf{B}^* = -\mathbf{B}$ (cf. Corollary 7.7(iii)), the eigenvalues of $\mathbf{B}$ are purely imaginary, implying that the kernels of $\mathbf{I} - \mathbf{B}$ and $\mathbf{I} + \mathbf{B}$ are $\{0\}$. The Fredholm Theory of compact operators (see [Rud 2]) then implies that $\mathbf{I} - \mathbf{B}$ and $\mathbf{I} + \mathbf{B}$ are invertible in the algebra of bounded operators in $L^2(D)$. Equation (7.19) then follows from (7.20). ■

Corollary 7.9. *The Bergman kernel K_D for D satisfies*

$$K_D(\zeta, z) = \overline{G_D^*(\zeta, z)} + \mathscr{R}(\zeta, z) \qquad \text{on } D \times D,$$

where the kernel $\mathscr{R}(\zeta, z)$ defines a compact operator

$$\mathbf{R}\colon L^2(D) \to L^2(D) \quad \text{via } \mathbf{R}f(z) = (f, \mathscr{R}(\cdot, z))_D.$$

This result makes precise the statement that $\overline{G_D^*}$ is the principal term of the Bergman kernel K_D. Since $\bar{G}_D = \bar{G}_D^* + \bar{B}$ and $\mathbf{B}$ is compact, the principal term of K_D—modulo compact operators—is also given by $\bar{G}_D$.

7.4. The Hölder Continuity of the Bergman Projection

We shall now use the representation (7.19) for the Bergman projection $\mathbf{P}_D$ in terms of the explicit operators $\mathbf{B}$ and $\mathbf{G}_D^*$ in order to prove the following important regularity result. For $k = 0, 1, 2, \ldots$, and $0 < \alpha < 1$ we denote by $C^{k+\alpha}(\bar{D})$ the space of those functions $f \in C^k(\bar{D})$ all whose partial derivatives of order k are in $\Lambda_\alpha(D)$. For $f \in C^{k+\alpha}(D)$, the $(k + \alpha)$-norm on D is defined by

$$(7.21) \qquad |f|_{k+\alpha, D} = |f|_{k, D} + \max\{|D^\gamma f|_{\alpha, D}\colon |\gamma| = k\}.$$

It is straightforward to check that $C^{k+\alpha}(\bar{D})$ with the norm (7.21) is a Banach space.

Theorem 7.10. *Suppose D is strictly pseudoconvex with boundary of class C^{2k+4} for some integer $k \geq 0$. Then the Bergman projection $\mathbf{P}_D$ maps $C^{k+\alpha}(\bar{D})$ boundedly into $C^{k+\alpha/2}(\bar{D})$ for any $0 < \alpha < 1$.*

Remark 7.11. One can show that for $l \in \mathbb{N}$ with $0 \leq l < k$ and $0 < \alpha < 1$ the Bergman projection is bounded from $C^{l+\alpha}$ into $C^{l+\alpha}$ (and not only into $C^{l+\alpha/2}$). The spaces $C^{l+\alpha}(\bar{D})$ are also denoted by $\Lambda_{l+\alpha}(D)$, as they are a natural generalization of the Hölder spaces (or Lipschitz spaces) $\Lambda_\alpha(D)$, $\alpha < 1$, to higher order. In many areas of analysis one considers also Hölder spaces Λ_l of integer order $l \in \mathbb{N}$. Without going into the details, let us mention that $\Lambda_1(D)$ is defined in terms of second differences, and that $\Lambda_1(D)$ does *not* agree with $C^1(\bar{D})$ or with $\text{Lip}_1(D) = \{f \in C(D)\colon |f(z) - f(w)| \leq C_f|z - w|\}$, but is strictly larger than either space! (See E.M. Stein [SteE 3] for details.) It turns out that the Bergman projection $\mathbf{P}_D$ is bounded from $\Lambda_l(D)$ to $\Lambda_l(D)$ for any $l \in \mathbb{N}$, $0 < l \leq k$, but $\mathbf{P}_D$ is *not* bounded from $\mathrm{Lip}_1(D)$ to $\mathrm{Lip}_1(D)$, or from $C^l(\bar{D})$ to $C^l(\bar{D})$ for any integer $l \geq 0$. This sort of phenomenon is a familiar one in the theory of singular integral operators (see E.M. Stein, op. cit.). The proof of the more general continuity result for the Bergman projection stated above can be obtained by combining the methods we shall use in the proof of Theorem 7.10 with the techniques of Ahern and Schneider [AhSc]. However, in this book we shall limit the discussion to the proof of Theorem 7.10, not only because this result is quite sufficient for the applications in the next paragraph, but mainly because its proof is technically somewhat simpler than the one for the more general result.

The proof of Theorem 7.10 will be an easy consequence of the following regularity results for the explicit operators $\mathbf{G}_D^*$ and $\mathbf{B}$.

Main Lemma. *If bD is of class C^{2k+4}, then*

(7.22) $\mathbf{G}_D$ *and* $\mathbf{G}_D^*$ *are bounded from* $C^{k+\alpha}(\bar{D})$ *to* $C^{k+\alpha/2}(\bar{D})$ *for any* $0 < \alpha < 1$,

and

(7.23) $\mathbf{B}$ *is bounded from* $C^k(\bar{D})$ *to* $C^{k+1/2}(\bar{D})$.

Proof of Theorem 7.10. (Assuming the Main Lemma). Suppose $0 < \alpha < \frac{1}{2}$. Then (7.23) implies that $\mathbf{I} - \mathbf{B}$: $C^{k+\alpha}(\bar{D}) \to C^{k+\alpha}(\bar{D})$ is bounded. Moreover, it is a consequence of the Ascoli–Arzela Theorem that for $\alpha > 0$ the embedding $C^{k+\alpha} \to C^k$ is a compact map (see Exercise E.7.6). Hence, by (7.23),

(7.24) $\mathbf{B}$: $C^{k+\alpha}(\bar{D}) \to C^{k+\alpha}(\bar{D})$ is compact.

Since $\mathbf{I} - \mathbf{B}$ has kernel $= \{0\}$ on $L^2(D)$, so does its restriction to $C^{k+\alpha}$. Therefore, by the Fredholm Theory (see [Rud 2], Chapter 4), (7.24) implies that $\mathbf{I} - \mathbf{B}$ is invertible in the algebra of bounded operators on $C^{k+\alpha}(\bar{D})$, i.e., $(\mathbf{I} - \mathbf{B})^{-1}$ is bounded from $C^{k+\alpha}$ to $C^{k+\alpha}$. The Theorem now follows from this statement combined with (7.22) and the representation $\mathbf{P}_D = (\mathbf{I} - \mathbf{B})^{-1} \circ \mathbf{G}_D^*$ given in Theorem 7.8. ■

The remainder of this section will be devoted to the proof of the Main Lemma. Even though the proof is elementary insofar as it is based on integration by parts and concrete estimations of integrals, the details are quite long and technical. The reader is advised to skip this proof on first reading, and to proceed directly with the applications of Theorem 7.10 in §8.

7.5. Admissible Kernels

In order to prove the Main Lemma we first introduce a class of kernels on $D \times D$ which includes the kernels G_D and B_D, as well as the kernels of the commutators of operators like $\mathbf{G}_D$ or $\mathbf{B}$ with differentiation operators. We shall assume throughout §7.5 and §7.6 that D is strictly pseudoconvex in $\mathbb{C}^n$ with boundary bD of class C^{k+4}, so that $\hat{\Phi}$ is of class $C^{k+2,\infty}$. The hypothesis bD of class C^{2k+4}, which appears in the Main Lemma, will only be used in §7.7.

Definition. A kernel $\mathscr{A}(\zeta, z)$ on $D \times D$ is said to be **simple admissible of class** C^l, $0 \leq l \leq k+2$, if $\mathscr{A}$ is of class C^l on $\bar{D} \times \bar{D} - \Delta_{bD}$, and if for each $P \in bD$ there is a neighborhood U of P, such that on $U \times U$ $\mathscr{A}$ has a representation

$$(7.26) \qquad \mathscr{A}(\zeta, z) = \frac{\mathscr{E}_j(\zeta, z)}{[\hat{\Phi}(\zeta, z)]^t} \qquad \text{or } \mathscr{A} = \frac{\mathscr{E}_j}{\hat{\Phi}^t},$$

where j and $t \in \mathbb{N}$, and $\mathscr{E}_j$ is of class C^l satisfying $|\mathscr{E}_j(\zeta, z)| \lesssim |\zeta - z|^j$ on $U \times U$. We say that the representation (7.26) is of (weighted) order λ if

$$(7.27) \qquad \lambda = 2n + j + \min(2, t) - 2t.$$

$\mathscr{A}$ is of order $\geq \lambda$ if the representations (7.26) can be chosen of order $\geq \lambda$.

The definition of the order λ in (7.27) is motivated by the fact that in suitable local coordinates the function $\hat{\Phi}$ vanishes to first order at $\zeta = z \in bD$ in two directions, while it vanishes to order 2 in all other directions (see the estimate (5.26)!). Thus when counting the order of zeroes in (7.26), up to two factors $\hat{\Phi}$ in the denominator can be counted as if they were of order 1, while the remaining factors are of order 2.

Next we enlarge our class of kernels by considering "asymptotic expansions" of simple admissible kernels, as follows.

Definition. A kernel $\mathscr{A}(\zeta, z)$ on $D \times D$ is said to be **admissible of class** C^l, $0 \leq l \leq k+2$, if $\mathscr{A}$ is of class C^l on $D \times D$ and if there are simple admissible kernels $\mathscr{A}^{(0)}, \ldots, \mathscr{A}^{(N-1)}$ of class C^l (N depends on $\mathscr{A}$ and l) such that

$$(7.28) \qquad \mathscr{A} = \sum_{s=0}^{N-1} \mathscr{A}^{(s)} + \mathscr{R}^{(N)},$$

where the kernel $\mathscr{R}^{(N)} = \mathscr{A} - \sum_{s=0}^{N-1} \mathscr{A}^{(s)}$ satisfies

$$(7.29) \qquad |\mathscr{R}^{(N)} f|_{l,D} = \left| \int_D f(\zeta) \mathscr{R}^{(N)}(\zeta, \cdot)\, dV(\zeta) \right|_{l,D} \lesssim \|f\|_{L^2}$$

for all $f \in L^2(D)$. If $k = \infty$, $\mathscr{A}$ is of class C^∞ if it is of class C^l for each $l < \infty$. The admissible kernel $\mathscr{A}$ is said to be of order $\geq \lambda$ if a representation (7.28) exists, in which all $\mathscr{A}^{(s)}$ are of order $\geq \lambda$.

Notice that Lemma 7.3 implies the following result.

Lemma 7.12. *If $\mathscr{A}_\lambda(\zeta, z)$ is a simple admissible kernel of order $\geq \lambda$, then*

$$\int_D |\mathscr{A}_\lambda(\zeta, z)|\, dV(\zeta) \lesssim \begin{cases} 1 & \text{if } \lambda > 0 \\ |\log \delta_D(z)| & \text{if } \lambda = 0 \\ [\delta_D(z)]^{\lambda/2} & \text{if } \lambda < 0. \end{cases}$$

In the following we shall denote a generic admissible kernel of order $\geq \lambda$ by $\mathscr{A}_\lambda$. The precise form of $\mathscr{A}_\lambda$ will generally differ from place to place. Notice that $\mathscr{A}_\lambda$ is also of order $\geq \lambda - j$ for any $j \in \mathbb{N}$, and that it is possible for a kernel $\mathscr{A}_\lambda$ to actually be of order $\geq \lambda + j$ for some $j > 0$. If we want to indicate in the notation that $\mathscr{A}_\lambda$ is of class C^l, we shall write $A_{\lambda,l}$.

It follows from Lemma 7.2 that the kernel $G_D(\zeta, z)$ is admissible of class C^{k+1} (recall that $\Phi = F$ locally and that we assume bD of class C^{k+4}!), and of order ≥ 0. The fact that $N_0(\zeta, \zeta) \neq 0$ on bD (see Exercise E.7.2) implies that G_D is *not* of order $\geq \lambda$ for any $\lambda > 0$. In order to see that also the kernels G_D^* and $B = G_D - G_D^*$ are admissible we need the following Lemma.

Lemma 7.13. *Suppose $\mathscr{E}_j$, $j \geq 0$, is of class C^l on $\bar{D} \times \bar{D}$, $l \leq k+2$, and $|\mathscr{E}_j| \lesssim |\zeta - z|^j$. Suppose $t_1, t_2 \in \mathbb{N}$. Then*

$$\mathscr{A} = \frac{\mathscr{E}_j}{\hat{\Phi}^{t_1}\hat{\Phi}^{*t_2}} \tag{7.30}$$

is admissible of class C^l and of order $\geqq \lambda$, where

$$\lambda = 2n + j + \min(2, t_1 + t_2) - 2(t_1 + t_2).$$

Proof. By (7.8), for any integer $t \geq 0$ we have

$$\frac{1}{\hat{\Phi}^{*t}} = \frac{1}{\hat{\Phi}^t} + (\hat{\Phi} - \hat{\Phi}^*) \sum_{\nu=0}^{t-1} \hat{\Phi}^{-(t-\nu)}\hat{\Phi}^{*-(\nu+1)}. \tag{7.31}$$

By introducing (7.31) with $t = t_2$ in (7.30), and iterating this step with appropriate choices of t, one obtains

$$\begin{aligned}\mathscr{A} &= \sum_{s=0}^{N-1} \frac{\mathscr{E}_j^{(s)}(\hat{\Phi} - \hat{\Phi}^*)^s}{\hat{\Phi}^{(t_1+t_2+s)}} \\ &\quad + (\hat{\Phi} - \hat{\Phi}^*)^N \sum_{\nu=0}^{t_2-1} \tilde{\mathscr{E}}_j^{(\nu)} \hat{\Phi}^{-(t_1+t_2+N-1-\nu)} \cdot \hat{\Phi}^{*-(\nu+1)}\end{aligned} \tag{7.32}$$

for any $N \geq 1$, with new terms $\mathscr{E}_j^{(s)}$ and $\tilde{\mathscr{E}}_j^{(\nu)}$ of class C^l. Denote the second summand in (7.32) by $\mathscr{R}^{(N)}$. By Corollary 7.5, each term in the first sum in (7.32) is simple admissible of class C^l and of order $\geq \lambda$, while $\mathscr{R}^{(N)}$ is of class C^l on $D \times D$ and all its partial derivatives of order $\leq l$ with respect to z are sums of expressions which are dominated by terms of the form

$$\frac{|\zeta - z|^{j+3N-\nu}}{|\hat{\Phi}|^{t_1+t_2+N+l-\nu}}, \qquad 0 \leq \nu \leq l. \tag{7.33}$$

If N is chosen sufficiently large ($N \geq 2(t_1 + t_2 + l) - j$ will do), the kernels in (7.33) are uniformly bounded on $\bar{D} \times \bar{D}$. Thus the required estimate (7.29) for $\mathscr{R}^{(N)}$ follows in a standard way by differentiation under the integral sign. ■

Corollary 7.14. *If $\mathscr{A}$ is admissible of class C^l and of order $\geq \lambda$, then its conjugate $\bar{\mathscr{A}}$ and its formal adjoint $\mathscr{A}^*(\zeta, z) = \overline{\mathscr{A}(z, \zeta)}$ are admissible of class C^l and of order $\geq \lambda$.*

Corollary 7.15. *The kernel* $B = G_D - G_D^*$ *is admissible of class* C^{k+1} *and of order* ≥ 1.

PROOF. It is clear from the above that B is admissible of class C^{k+1}. From (7.4) and (7.7) it follows that

(7.34) $$B = N_0(\zeta, \zeta)\left[\frac{1}{\hat{\Phi}^{n+1}} - \frac{1}{\hat{\Phi}^{*n+1}}\right] + \mathscr{A}_1.$$

Now (7.8) with $t = n + 1$, Corollary 7.5(i), and Lemma 7.13 imply that the first term on the right side of (7.34) is admissible of order ≥ 1 as well. ∎

The proof of the following result is a straightforward consequence of standard differentiation rules, the definition of admissible kernels, and the fact that $\hat{\Phi} \in C^{k+2,\infty}(\bar{D} \times \bar{D})$.

Lemma 7.16. *Suppose* $\mathscr{A}_{\lambda,l}$ *is an admissible kernel on* $D \times D$ *of class* C^l *with* $1 \leq l \leq k + 2$, *and of order* $\geq \lambda$. *Let* $V^{(z)}$ *be a vector field of class* C^l *on* $\bar{D}$ *acting in the z-variable. Then*

$$V^{(z)}\mathscr{A}_{\lambda,l} = \mathscr{A}_{\lambda-1,l-1} + \mathscr{A}_{\lambda-2,l}.$$

7.6. Integration by Parts

Lemma 7.16 shows how the kernel changes when one differentiates an integral

$$\int_D f(\zeta)\mathscr{A}_\lambda(\zeta, z)\, dV(\zeta) = (f, \overline{\mathscr{A}_\lambda(\cdot, z)})_D$$

with respect to the parameter z. We now introduce a suitable integration by parts which allows to reverse the "negative" effect (i.e., the lowering of order by 1, respectively 2) by moving the differentiation onto the function f. The heart of the matter is the existence of a special vector field Y on $\bar{D}$, acting ζ, with the properties

(7.35) $$\mathscr{A}_{\lambda-1} = Y\mathscr{A}_\lambda^{(1)} + \mathscr{A}_\lambda^{(2)}$$

and

(7.36) $$(f, Yg)_D = (Y^*f, g)_D$$

for some first order partial differential operator Y^*.[1]

In order to avoid boundary integrals in the integration by parts (7.36), the vector field Y must be "tangential" in the following sense: a vector field L on $\bar{D}$ is said to be **tangential (to** bD**)**, if $L_\zeta \in T_\zeta bD$ for $\zeta \in bD$. This is equivalent to $Lr = 0$ on bD for any defining function r for D.

[1] A first order partial differential operator L of class C^l is a sum $L = V + A$, where V is a vector field and A a function of class C^l; Lf is defined by $(Lf)(\zeta) = V_\zeta f + A(\zeta)f(\zeta)$ for $f \in C^1$.

Lemma 7.17. *Suppose $D \subset \mathbb{R}^n$ with bD of class C^{l+1}, $l \geq 1$, and let L be a tangential vector field on $\bar{D}$ of class C^l. Then there is a first order (tangential) partial differential operator L^* on $\bar{D}$ of class C^{l-1} such that*

$$(f, Lg)_D = (L^*f, g)_D \tag{7.37}$$

for all $f, g \in C^1(\bar{D})$.

It is clear that (7.37) defines L^* uniquely; we call L^* the *adjoint operator to L.*

Proof. We first consider the case that L is supported in a suitable open set U. If $U \subset\subset D$, then the existence of L^* satisfying (7.37) follows by standard integration by parts. No boundary integral appears since L, and hence also L^* are 0 on bD. Next, assume that $U \cap bD \neq \emptyset$ and that U is so small that on U there is a positively oriented C^{l+1} coordinate system (x_1, x'), where $x_1 = r$ is a C^{l+1} defining function for D and $x' = (x_2, \ldots, x_n)$. Then

$$L = \sum_{j=1}^{n} a_j \frac{\partial}{\partial x_j} \qquad \text{with } a_j \in C_0^l(U),$$

and L is tangential if and only if $a_1(0, x') = 0$. The volume element dV satisfies $dV = \gamma(x)\, dx_1 \wedge dx_2 \ldots \wedge dx_n$, with $\gamma(x) > 0$ for $x \in U$. Let $f, g \in C^1(U)$. By integration by parts one obtains

$$\begin{aligned} (f, Lg)_{D \cap U} &= \sum_{j=1}^{n} \int_{x \in U, x_1 \leq 0} f \bar{a}_j \frac{\partial \bar{g}}{\partial x_j} \gamma(x)\, dx_1 \ldots dx_n \\ &= -\sum_{j=1}^{n} \int_{x \in U, x_1 \leq 0} \frac{\partial}{\partial x_j} (f \bar{a}_j \gamma) \bar{g}\, dx_1 \ldots dx_n \\ &\quad + \int_{x \in U, x_1 = 0} f \bar{a}_1 \gamma \bar{g}\, dx_2 \ldots dx_n. \end{aligned} \tag{7.38}$$

Here the boundary integral vanishes since L is tangential (i.e., $a_1(0, x') = 0$). Now observe that

$$\begin{aligned} &\int_{x \in U, x_1 \leq 0} \frac{\partial}{\partial x_j} (f \bar{a}_j \gamma) \bar{g}\, dx_1 \ldots dx_n \\ &= \int_{x \in U, x_1 \leq 0} \left[\bar{a}_j \frac{\partial f}{\partial x_j} + \gamma^{-1} \frac{\partial}{\partial x_j} (\bar{a}_j \gamma) f \right] \bar{g}\, dV(x). \end{aligned} \tag{7.39}$$

Hence, if we set

$$L^* = -\sum_{j=1}^{n} \bar{a}_j \frac{\partial}{\partial x_j} - \gamma^{-1} \sum_{j=1}^{n} \frac{\partial}{\partial x_j} (\bar{a}_j \gamma),$$

(7.38) and (7.39) imply (7.37). Finally, the general case follows from the special cases just considered by a straightforward partition of unity argument. The details are left to the reader. ∎

We now return to the strictly pseudoconvex domain D in $\mathbb{C}^n$ with C^{k+4} defining function r and define the vector field Y.

Lemma 7.18. *The vector field*

$$Y = Y^{(r)} = \sum_{j=1}^{n} \frac{\partial r}{\partial \bar{\zeta}_j} \frac{\partial}{\partial \zeta_j} - \sum_{j=1}^{n} \frac{\partial r}{\partial \zeta_j} \frac{\partial}{\partial \bar{\zeta}_j}$$

of class C^{k+3} on $\bar{D}$ is tangential and satisfies $(Y\hat{\Phi})(\zeta, \zeta) \neq 0$ for $\zeta \in bD$.

PROOF. It is obvious that $Yr = 0$, so Y is tangential. Hence $Y\hat{\Phi} = Y\Phi - Yr = YF$ at points (ζ, ζ), where F is the Levi polynomial of r. It is now easy to check that

$$(YF)(\zeta, \zeta) = \sum_{j=1}^{n} \left| \frac{\partial r}{\partial \zeta_j}(\zeta) \right|^2 \neq 0 \qquad \text{for } \zeta \in bD. \quad \blacksquare$$

Thus Lemma 7.17 applies to Y, and property (7.36) holds. Next, we establish property (7.35) for the vector field Y. For our purposes it is important to also keep track of the differentiability class of the kernels involved.

Lemma 7.19. *Let $\mathscr{A}_{\lambda,l}$ be admissible of class C^l, with $1 \leq l \leq k+2$, and of order $\geq \lambda$. Then*

$$\mathscr{A}_{\lambda,l} = Y\mathscr{A}_{\lambda+2,l} + \mathscr{A}_{\lambda+1,l-1} \qquad \text{if } \lambda \leq -1, \tag{7.40}$$

and

$$\mathscr{A}_{0,l} = Y\mathscr{A}_{1,l} + \mathscr{A}_{1,l-1} \qquad \text{if } \lambda = 0. \tag{7.41}$$

PROOF. Consider the representation $\mathscr{A}_{\lambda,l} = \sum_{s=0}^{N-1} \mathscr{A}^{(s)} + \mathscr{R}^{(N)}$ according to (7.28) and (7.29), with $\mathscr{A}^{(s)}$ simple admissible of order $\geq \lambda$. The remainder $\mathscr{R}^{(N)}$, and any term $\mathscr{A}^{(s)}$, for which the local representations can be chosen of order $> \lambda$ (i.e., $\geq \lambda + 1$), can be included in the kernel $\mathscr{A}_{\lambda+1,l-1}$ in (7.40) or (7.41) for any $\lambda \geq 0$. So the proof is reduced to the case where $\mathscr{A}_{\lambda,l}$ is simple admissible, with a representation

$$\mathscr{A}_{\lambda,l} = \frac{\mathscr{E}_j}{\hat{\Phi}^t},$$

of order λ, i.e.,

$$2n + j + \min(2, t) - 2t = \lambda. \tag{7.42}$$

Since $t \in \mathbb{N}$, (7.42) implies $t \geq 2$ if $j = \lambda = 0$, and $t \geq 3$ if $j > 0$ or $\lambda \leq -1$.

Let W be a neighborhood of the boundary diagonal $\Delta_{bD} = \{(\zeta, \zeta) \in \mathbb{C}^{2n}: \zeta \in bD\}$ such that $(Y\hat{\Phi})(\zeta, z) \neq 0$ for $(\zeta, z) \in W$, and choose $\varphi \in C_0^\infty(W)$ so that $0 \leq \varphi \leq 1$ and $\varphi \equiv 1$ on a neighborhood $W' \subset\subset W$ of $\Delta_{b\Delta}$. Then

$$\mathscr{A}_\lambda = \varphi\mathscr{A}_\lambda + (1 - \varphi)\mathscr{A}_\lambda. \tag{7.43}$$

On the open set $W - \Delta_{bD}$ we can write

$$\varphi\mathscr{A}_\lambda = \varphi\frac{\mathscr{E}_j}{\hat{\Phi}^t} = \varphi\left[-(t-1)^{-1}Y\left(\frac{\mathscr{E}_j}{\hat{\Phi}^{t-1}}\right)(Y\hat{\Phi})^{-1}\right] + \varphi(t-1)^{-1}\frac{Y\mathscr{E}_j}{\hat{\Phi}^{t-1}}(Y\hat{\Phi})^{-1}$$

$$= Y\left(\frac{\tilde{\mathscr{E}}_j}{\hat{\Phi}^{t-1}}\right) + \frac{\mathscr{E}_{j-1}}{\hat{\Phi}^{t-1}},$$

where $\tilde{\mathscr{E}}_j$ is of class C^l, $\mathscr{E}_{j-1}$ is of class C^{l-1}, and satisfies $|\mathscr{E}_{j-1}| \lesssim |\zeta - z|^{j-1}$ if $j \geq 1$, and both $\tilde{\mathscr{E}}_j$ and $\mathscr{E}_{j-1}$ have compact support in W. By using (7.42) it follows that $\tilde{\mathscr{E}}_j/\hat{\Phi}^{t-1} = \mathscr{A}_{\lambda+2,l}$ if $\lambda \leq -1$ ($t \geq 3$ in this case!), and $\tilde{\mathscr{E}}_j/\hat{\Phi}^{t-1} = \mathscr{A}_{1,l}$ if $\lambda = 0$, and furthermore $\mathscr{E}_{j-1}/\hat{\Phi}^{t-1} = \mathscr{A}_{\lambda+1,l-1}$ (use $t \geq 3$ if $j \geq 1$!). Since outside W we have $\varphi\mathscr{A}_\lambda \equiv 0$, we have thus shown that

$$\varphi\mathscr{A}_\lambda = Y\mathscr{A}_{\lambda+2,l} + \mathscr{A}_{\lambda+1,l-1} \qquad \text{if } \lambda \leq -1,$$

and

$$\varphi\mathscr{A}_0 = Y\mathscr{A}_{1,l} + \mathscr{A}_{1,l-1} \qquad \text{if } \lambda = 0.$$

The other term $(1-\varphi)\mathscr{A}_\lambda$ in (7.43) is of class C^l on $\bar{D} \times \bar{D}$ since $1 - \varphi \equiv 0$ near the singularities of $\mathscr{A}_\lambda$. Hence it is simple admissible of order $\geq 2n$, and thus it can be included in the admissible kernel $\mathscr{A}_{\lambda+1,l-1}$ in (7.40) or (7.41) ∎

We can now prove the main result of this section.

Proposition 7.20. *Suppose bD is of class C^{k+4}, with $k \geq 1$, and let $\mathscr{A}_{\lambda,l}$ be admissible of class C^l, $2 \leq l \leq k+1$, and of order $\geq \lambda$ with $\lambda = 0$ or 1. If $V^{(z)}$ is a vector field in z of class C^l on $\bar{D}$ and $f \in C^1(\bar{D})$, then*

$$V^{(z)}\int_D f\mathscr{A}_{\lambda,l}\,dV = \int_D (Y^*f)\mathscr{A}_{\lambda,l-1}\,dV + \int_D f\mathscr{A}_{\lambda,l-2}\,dV. \tag{7.44}$$

Proof. By differentiating under the integral sign and applying Lemma 7.16 to $\overline{\mathscr{A}_{\lambda,l}}$ one obtains

$$V^{(z)}\int_D f\mathscr{A}_{\lambda,l}\,dV = (f, \mathscr{A}_{\lambda-2,l})_D + (f, \mathscr{A}_{\lambda-1,l-1})_D. \tag{7.45}$$

By applying Lemma 7.19 to the first term on the right side of (7.45) (notice that $\lambda - 2 \leq -1$ if $\lambda = 0$ or 1!) and integration by parts, it follows that

$$\begin{aligned}(f, \mathscr{A}_{\lambda-2,l}) &= (f, Y\tilde{\mathscr{A}}_{\lambda,l}) + (f, \mathscr{A}_{\lambda-1,l-1})\\ &= (Y^*f, \tilde{\mathscr{A}}_{\lambda,l}) + (f, \mathscr{A}_{\lambda-1,l-1}).\end{aligned} \tag{7.46}$$

To the last terms in (7.45) and (7.46) we apply Lemma 7.19 once more (this time allowing for $\lambda - 1 = 0$), and we obtain

$$\begin{aligned}(f, \mathscr{A}_{\lambda-1,l-1}) &= (f, Y\mathscr{A}_{\lambda,l-1}) + (f, \mathscr{A}_{\lambda,l-2})\\ &= (Y^*f, \mathscr{A}_{\lambda,l-1}) + (f, \mathscr{A}_{\lambda,l-2}).\end{aligned} \tag{7.47}$$

Equation (7.44) now follows by combining (7.45)–(7.47). ∎

Remark 7.21. Notice that the differentiability class of the kernel drops by 2 in Proposition 7.20, even though $V^{(z)}$ involves only differentiation of order 1. If $\mathscr{A}$ has the special form $\mathscr{A} = \mathscr{E}_0/\hat{\Phi}^{n+1}$ with $\mathscr{E}_0 \in C^{l,\infty}$, one can show that Proposition (7.20) holds with a loss of only one order of differentiability (see Exercise E.7.9). Notice that the kernel G_D is of this form. However, the study of the operator $\mathbf{B} = \mathbf{G}_D - \mathbf{G}_D^*$ of order ≥ 1 requires the general version of Proposition 7.20. Thus, in order to estimate k derivatives of $\mathbf{B}f$ the kernel B must be at least of class C^{2k}. This explains the particular differentiability hypothesis in the Main Lemma and in Theorem 7.10.

7.7. Proof of the Main Lemma

We have now collected all the ingredients necessary for the proof of the Main Lemma in §7.4. As suggested by Remark 7.21, we assume that bD is of class C^{2k+4}. The kernels G_D, G_D^*, and $B = G_D - G_D^*$ are then admissible of class C^{2k+1} and of order $\geq \lambda$ for $\lambda = 0$, respectively $\lambda = 1$. By applying Proposition 7.20 with $V^{(z)} = \partial/\partial z_j$ or $\partial/\partial \bar{z}_j$, $1 \leq j \leq n$, to an admissible kernel $\mathscr{A}_{\lambda,2k+1}$, and using induction over $m \in \mathbb{N}$, it follows that

$$D_z^{\alpha\bar{\beta}}(f, \mathscr{A}_{\lambda,2k+1})_D = \sum_{\nu=0}^{m} (Y^{*\nu} f, \mathscr{A}_{\lambda,2(k-m)+1}^{(\nu,m)})_D$$

for all multi-indices $\alpha, \beta \in \mathbb{N}^n$ with $|\alpha| + |\beta| = m$ and for all m with $0 \leq m \leq k$; in particular, for $m = k$, we obtain

$$D_z^{\alpha\bar{\beta}}(f, \mathscr{A}_{\lambda,2k+1}) = \sum_{\nu=0}^{k} (Y^{*\nu} f, \mathscr{A}_{\lambda,1}^{(\nu)}) \qquad \text{for } |\alpha| + |\beta| = k. \tag{7.48}$$

The Main Lemma is an immediate consequence of the remarks above, the representation (7.48), and the following Lemma.

Lemma 7.22. *Define the operator* $f \mapsto \mathscr{A}_\lambda f$ *by*

$$\mathscr{A}_\lambda f(z) = \int_D f(\zeta) \mathscr{A}_{\lambda,1}(\zeta, z)\, dV(\zeta).$$

Then

(i) $\mathscr{A}_0$ *is bounded from* $\Lambda_\alpha(D)$ *to* $\Lambda_{\alpha/2}(D)$ *for any* $0 < \alpha < 1$.
(ii) $\mathscr{A}_1$ *is bounded from* $L^\infty(D)$ *to* $\Lambda_{1/2}(D)$.

PROOF. Let us first prove (ii), which is the simpler case. By Lemma V.3.1, it is enough to show that

$$|d_z \mathscr{A}_1 f(z)| \lesssim |f|_D \delta_D^{-1/2}(z) \qquad \text{for } z \in D. \tag{7.49}$$

Equation (7.49) follows by differentiation under the integral sign, Lemma 7.16, and a straightforward estimation based on Lemma 7.12—notice that the relevant kernels are of order ≥ -1 in this case.

Similarly, in order to prove (i) it suffices to show that

$$|d_z(\mathscr{A}_0 f)(z)| \lesssim |f|_{\alpha,D}\delta_D^{-1+\alpha/2}(z) \qquad \textit{for } z \in D. \tag{7.50}$$

By Lemma 7.16, with $V^{(z)} = \partial/\partial z_j$ or $\partial/\partial \bar{z}_j$, one has

$$V^{(z)}(\mathscr{A}_0 f) = (f, \mathscr{A}_{-1,0})_D + (f, \mathscr{A}_{-2,1})_D. \tag{7.51}$$

By Lemma 7.12, the first term on the right in (7.51) is estimated by $|f|_D \delta_D^{-1/2}(z)$, which is dominated by the right side of (7.50) for any $\alpha \le 1$. In order to estimate the second term in (7.51) we may assume without loss of generality that $\mathscr{A}_{-2,1}$ is simple admissible. We then write

$$\begin{aligned}\int_D f(\zeta)\bar{\mathscr{A}}_{-2,1}(\zeta, z)\, dV(\zeta) &= \int_D [f(\zeta) - f(z)]\overline{\mathscr{A}_{-2,1}}\, dV(\zeta) \\ &\quad + f(z)(1, \mathscr{A}_{-2,1}).\end{aligned} \tag{7.52}$$

The formula (7.27) for the order implies in case $\lambda = -2$ that $t = n + 2 + j/2$, so that

$$\begin{aligned}|[f(\zeta) - f(z)]\mathscr{A}_{-2,1}(\zeta, z)| &\lesssim |f|_\alpha \frac{|\zeta - z|^{\alpha+j}}{|\hat{\Phi}|^{n+2+j/2}} \\ &\lesssim |f|_\alpha |\hat{\Phi}|^{-n-2+\alpha/2}\end{aligned}$$

for $\zeta, z \in D$ near bD (recall $|\hat{\Phi}| \gtrsim |\zeta - z|^2$!). Hence, by Lemma 7.3, the first term on the right in (7.52) is estimated by the right side of (7.50). Finally, for the remaining term in (7.52) we use Lemma 7.19 and integration by parts to obtain

$$\begin{aligned}(1, \mathscr{A}_{-2,1})_D &= (1, Y\mathscr{A}_{0,1}) + (1, \mathscr{A}_{-1,0}) \\ &= (Y^*1, \mathscr{A}_{0,1}) + (1, \mathscr{A}_{-1,0}) \\ &= (1, \tilde{\mathscr{A}}_{-1,0}).\end{aligned}$$

Hence, again by Lemma 7.12, it follows that

$$|f(z)(1, \mathscr{A}_{-2,1}(\cdot, z))_D| \lesssim \|f\|_{L^\infty} \delta_D^{-1/2}(z),$$

so that this term is also estimated by the right side of (7.50). We have thus verified (7.50), and the proof of the Lemma is complete. ∎

Exercises

E.7.1. Carry out the details of the reproducing property (7.1) of $\bar{\partial}_\zeta \hat{C}$ in case $f \in \mathcal{O}L^1(D)$.

E.7.2. Prove that the function N_0 in Lemma 7.2 satisfies

$$N_0(\zeta, \zeta) = \frac{n!}{4\pi^n}\mathscr{D}_\zeta(r)\|dr\|^2\, dV(\zeta),$$

where $\mathscr{D}_\zeta(r)$ is the determinant of the Leviform on $T_\zeta^{\mathbb{C}}(bD)$ (cf. §3.5 for the precise definition).

E.7.3. Carry out the details of the proof of Lemma 7.3.

E.7.4. (i) Show that for any ε with $0 < \varepsilon < 1$ one has

$$\sup_{z \in D} \int_D |\hat{\Phi}(\zeta, z)|^{-(n+1)} |r(z)|^{\varepsilon} |r(\zeta)|^{-\varepsilon} dV(\zeta) < \infty.$$

(r is the strictly plurisubharmonic defining function for D.)

(ii) Use (i) to show that any continuous kernel $G(\zeta, z)$ on $D \times D$ which satisfies $|G(\zeta, z)| \leq |\hat{\Phi}(\zeta, z)|^{-(n+1)}$ on $D \times D$ defines, by integration, a *bounded* linear map $\mathbf{G}: L^p(D) \to L^p(D)$ for any p with $1 < p < \infty$. (See also [PhSt].)

E.7.5. (i) Let G_D be the kernel on $D \times D$ defined in the text, and let $G_D^*(\zeta, z) = \overline{G_D(z, \zeta)}$. Show that the corresponding operator $\mathbf{G}_D^*$ with kernel G_D^* is the Hilbert space adjoint of $\mathbf{G}_D$ on $L^2(D)$.

(ii) With the notations of the text, show that one also has the representation $\mathbf{P}_D = \mathbf{G}_D \circ (\mathbf{I} + \mathbf{B})^{-1}$.

E.7.6. Suppose $0 < \alpha < \beta < 1$. Show that the embedding $C^{k+\beta}(\bar{D}) \to C^{k+\alpha}(\bar{D})$ is a compact operator. (Hint: Consider $k = 0$ first.)

E.7.7. Show that if $D \subset\subset \mathbb{R}^n$ has C^l boundary and $\mathscr{E}_j \in C^l(\bar{D} \times \bar{D})$ satisfies $|\mathscr{E}_j(x, y)| = 0(|x - y|^j)$ for some integer j with $1 \leq j \leq l$, then $|d_x \mathscr{E}_j(x, y)|$ and $|d_y \mathscr{E}_j(x, y)|$ are $0(|x - y|^{j-1})$.

E.7.8. Let $D \subset\subset \mathbb{C}^n$ be strictly pseudoconvex with C^2 boundary. Allowable vector fields on $\bar{D}$ where defined in E.V.3.2.

(i) Show that the vector field Y in Lemma 7.18 is *not* allowable.

(ii) Show that for every $P \in bD$ there are a neighborhood U of P, a function g on U, and allowable vector fields W_1, W_2, W_3 on U, such that $Y = g[W_1, W_2] + W_3$ on U. ($[W_1, W_2]$ is the commutator $W_1 W_2 - W_2 W_1$.)

(iii) Improve Lemma 7.16 as follows: if the vector field V is allowable, then

$$V^{(z)} \mathscr{A}_{\lambda, l} = \mathscr{A}_{\lambda-1, l-1}.$$

E.7.9. (i) Let $\mathscr{A}_{0,l} = \mathscr{E}_0 / \hat{\Phi}^{n+1}$ be a simple admissible kernel with $\mathscr{E}_0$ of class $C^{l,\infty}$. Let V be any C^∞ vector field. Show that

$$V^{(z)} \mathscr{A}_{0,l} = Y\tilde{\mathscr{A}}_{0,l} + \tilde{\tilde{\mathscr{A}}}_{0,l-1},$$

where the kernels $\tilde{\mathscr{A}}$ and $\tilde{\tilde{\mathscr{A}}}$ are of the same type as $\mathscr{A}_{0,l}$, with $\tilde{\mathscr{E}}_0$ and $\tilde{\tilde{\mathscr{E}}}_0$ of class $C^{l,\infty}$ and $C^{l-1,\infty}$, respectively,

(ii) Use (i) to show that if bD is of class C^{k+4}, then the operator $\mathbf{G}_D$ defined in the text is bounded from $C^{k+\alpha}(\bar{D})$ to $C^{k+\alpha/2}(\bar{D})$ for any $0 < \alpha < 1$.

§8. Boundary Regularity of Biholomorphic Maps

In this section we show how the property that the Bergman projection operator preserves differentiability of functions up to the boundary can be used in the study of boundary regularity of biholomorphic maps. By combining these methods with Theorem 7.10 we obtain as the main application a complete proof of the following fundamental theorem: *Any biholomorphic map between two bounded strictly pseudoconvex domains extends smoothly to the boundary*. This result is classical in one complex variable,[1] but in several variables it had been a major outstanding conjecture for many years.

The first—and very complicated—proof of this theorem was given by Charles Fefferman in 1974 [Fef]. The proof presented here is quite different and more elementary than the original one. More significantly, the methods are applicable in much more general situations whenever the necessary regularity property for the Bergman projection can be established. The main ideas of this proof were developed in the late 1970s, beginning with a paper of S. Webster [Web] and culminating in the work of S. Bell and E. Ligocka [BeLi]. Since then, these methods have proven very fruitful in numerous investigations (see the Notes for some additional references).

Fefferman's Mapping Theorem is of major importance in the classification of strictly pseudoconvex domains—comparable to the role of the Riemann Mapping Theorem in one variable. In fact, by Fefferman's result, the question of whether two smoothly bounded strictly pseudoconvex domains D_1 and D_2 in $\mathbb{C}^n$ are biholomorphically equivalent is reduced to the question of whether there is a diffeomorphism $\hat{F}: bD_1 \to bD_2$ which satisfies the tangential Cauchy–Riemann equations. It is known that in case $n \geq 2$ there is an infinite sequence of differential obstructions to the existence of such a map $\hat{F}$, even at the local level—notice that by the Riemann Mapping Theorem locally there are no obstructions if $n = 1$; moreover, these obstructions are, at least in principle, computable. This was discovered already in 1932 by E. Cartan in case $n = 2$ ("Sur la géométrie pseudoconforme des hypersurfaces de deux variables complexes", Oeuvres II, 2, 1231–1304, and III, 2, 1217–1238). The general case was studied more recently by N. Tanaka [Tan] in 1967 and by S.S. Chern and J. Moser [ChMo] in 1974. Even though the discussion of these results is beyond the scope of this book, these brief remarks should help to put Fefferman's Mapping Theorem in the proper perspective.

8.1. A Regularity Condition and Differentiability of the Bergman Kernel

We begin by singling out the regularity property which has played a fundamental role in recent investigations on boundary regularity of holomorphic maps.

Definition. A bounded domain D in $\mathbb{C}^n$ is said to satisfy **condition** $(\mathbf{R}_k)$ for some $k \in \mathbb{N}$ if there is an integer $m_k \in \mathbb{N}$ such that the Bergman projection $\mathbf{P}_D$: $L^2(D) \to \mathscr{H}^2(D)(=\mathcal{O}L^2(D))$ is a bounded map from $C^{m_k}(\bar{D})$ into $C^k(\bar{D})$, i.e., if the following hold:

[1] The first proof seems to be due to P. Painlevé "Sur la théorie de la représentation conforme", C.R. Acad. Science Paris **112**, 653–657 (1891). Other proofs were given by O.D. Kellogg, Trans. Amer. Math. Soc. **13**, 109–132 (1912), and S. Warschawski, Math. Zeitschrift **35**, 321–456 (1932).

(i) $\mathbf{P}_D f \in C^k(\bar{D})$ if $f \in C^{m_k}(\bar{D})$, and
(ii) there is a constant $c_k < \infty$ such that

$$|\mathbf{P}_D f|_{k,D} \leq c_k |f|_{m_k,D}$$

for all $f \in C^{m_k}(\bar{D})$.

D is said to satisfy **condition (R)** if it satisfies (R_k) for every $k \in \mathbb{N}$.

In general, the precise relationship between k and m_k will not matter, but we shall write (R_{k,m_k}) instead of (R_k) whenever we want to emphasize m_k as well. The essential conclusion of Theorem 7.10 can now be stated as follows.

Theorem 8.1. *A bounded strictly pseudoconvex domain D in $\mathbb{C}^n$ with C^{2k+4} boundary satisfies condition $(R_{k,k+1})$.*

Condition (R) holds for many other domains; for example, we will show in §8.6 that it holds for smoothly bounded complete Reinhardt domains. It also holds for (weakly) pseudoconvex domains with real analytic boundary. This is a simple consequence of deep regularity results for the $\bar{\partial}$-equation on such domains proved by J.J. Kohn in 1977 [Koh 3]. On the other hand, quite recently D. Barrett [Bar] found a smoothly bounded domain in $\mathbb{C}^2$ which does not have property (R_k) for any k. Barrett's example is not pseudoconvex; it is still unknown whether condition (R)—or just (R_k) for some k—holds for every smoothly bounded pseudoconvex domain.

Theorem 8.2. *Suppose $D \subset\subset \mathbb{C}^n$ satisfies condition (R_k). Then $K_D(\cdot, a) \in C^k(\bar{D})$ for every $a \in D$.*

The proof is an immediate consequence of the following useful representation for the Bergman kernel.

Lemma 8.3. *Let $a \in D$ and suppose $\varphi_a \in C_0^\infty(D)$ is radially symmetric about a (i.e., φ_a depends only on $|z - a|$), and $\int \varphi_a \, dV = 1$. Then*

$$K_D(\cdot, a) = \mathbf{P}_D \overline{\varphi_a}.$$

Proof of Lemma 8.3. We may assume that φ_a is supported in a ball $B(a, \varepsilon) \subset\subset D$. By the mean value property (see Exercise E.8.2) $f \in \mathcal{O}(D)$ satisfies

$$f(a) \int_{bB(a,\rho)} dS = \int_{bB(a,\rho)} f \, dS \tag{8.1}$$

for $0 < \rho \leq \varepsilon$. Since φ_a is constant on $bB(a, \rho)$, (8.1) implies

$$f(a) \int_{bB(a,\rho)} \varphi_a \, dS = \int_{bB(a,\rho)} f \varphi_a \, dS. \tag{8.2}$$

Integrating (8.2) with respect to ρ from 0 to ε gives

$$f(a)\int_{B(a,\varepsilon)}\varphi_a\,dV=\int_{B(a,\varepsilon)}f\varphi_a\,dV,$$

which, because of $\int\varphi_a\,dV=1$, implies

$$f(a)=(f,\bar{\varphi}_a)_D\qquad\text{for } f\in\mathcal{O}(D). \tag{8.3}$$

If $f\in\mathcal{H}^2(D)$, then $f=\mathbf{P}_Df$, and since $\mathbf{P}_D$ is Hermitian, it follows from (8.3) that

$$f(a)=(\mathbf{P}_Df,\bar{\varphi}_a)=(f,\mathbf{P}_D\bar{\varphi}_a)\qquad\text{for } f\in\mathcal{H}^2. \tag{8.4}$$

Since $\mathbf{P}_D\bar{\varphi}_a\in\mathcal{H}^2$, (8.4) implies the desired result. ■

8.2. Homogeneous Bergman Kernel Coordinates

In this section we give an overview of the method we shall use in order to study the behavior of a biholomorphic map $F\colon D_1\to D_2$ at the boundary.

Given points $a_0,a_1,\ldots,a_n\in D_1$, we introduce (a priori meromorphic) functions

$$u_j(z)=\frac{K_{D_1}(z,a_j)}{K_{D_1}(z,a_0)}\qquad\text{on } D_1 \tag{8.5}$$

and

$$v_j(w)=\frac{K_{D_2}(w,F(a_j))}{K_{D_2}(w,F(a_0))}\qquad\text{on } D_2 \tag{8.6}$$

for $j=1,\ldots,n$. The transformation formula for the Bergman kernel (Theorem IV.4.9) implies that

$$u_j(z)=v_j(F(z))\overline{\left(\frac{\det F'(a_j)}{\det F'(a_0)}\right)}. \tag{8.7}$$

If the points $a_0,a_1,a_2,\ldots,a_n\in D_1$ can be chosen so that the corresponding functions $u_1,\ldots,u_n$ and $v_1,\ldots,v_n$ form (local) holomorphic coordinate systems, then (8.7) shows that the representation of F in these coordinates is a complex *linear* transformation! This suggests that in order to study the behavior of F near a boundary point $P\in bD_1$ one should try to find such coordinates $u_1,\ldots,u_n$ in a neighborhood $U\cap\bar{D}_1$ of P and corresponding coordinates $v_1,\ldots,v_n$ near "$Q=F(P)$".

This is, of course, quite vague as long as F is not defined on bD_1. But assuming condition (R_1), we at least know that $K_{D_1}(\cdot,a)$ extends as a C^1 function to the boundary bD_1 for fixed $a\in D_1$ (Theorem 8.2), so one has something to start with. The immediate obvious question then, is whether, given $P\in bD_1$, one can find $a_0\in D_1$ so that $K_{D_1}(P,a_0)\neq 0$. Assuming this to be the case, the next, and apparently much more difficult question is whether one can find additional points $a_1,\ldots,a_n\in D_1$, so that

(8.8) $$\det\left[\frac{\partial}{\partial z_l}\left(\frac{K_{D_1}(\cdot, a_j)}{K_{D_1}(\cdot, a_0)}\right)\right]_{\substack{l=1,\ldots,n\\ j=1,\ldots,n}}(p) \neq 0$$

Quite surprisingly, affirmative answers to both these questions are rather elementary, though nontrivial consequences of condition (R_1). And once the existence of the special coordinate systems (8.5) and (8.6) at the boundary has been established, it will be a fairly simple matter to turn the vague ideas sketched above into a rigorous argument.

In the next few sections we shall carry out the details of this program.

8.3. Bell's "Density Lemma"

We first discuss the result which will be the key ingredient in the proof of the nonvanishing of the determinant (8.8).

Theorem 8.4. *Suppose $D \subset\subset \mathbb{C}^n$ has C^∞ boundary and satisfies condition (R_1). Then $A^\infty(D)$ is contained in the closure of the linear span of $\{K_D(\cdot, a): a \in D\}$ in the C^1 norm over $\bar{D}$. The same conclusion holds for a strictly pseudoconvex domain with boundary of class at least C^6.*

PROOF. We shall discuss the proof in case D is strictly pseudoconvex with boundary of class C^6, so that D satisfies condition $(R_{1,2})$ by Theorem 8.1. The modifications necessary for the other case will be left to the reader (see also Lemma 8.13). The proof will involve two technical lemmas, but the basic scheme is quite simple, as follows. Given $f \in A^\infty(D)$, we have $f = \mathbf{P}_D f$; by Lemma 8.5, below, there is $q \in C^2(\bar{D})$ such that (i) $f = \mathbf{P}_D(f - q)$, and (ii) $f - q$ and all its partial derivatives of order ≤ 2 vanish on bD. It then follows by standard real analysis techniques (Lemma 8.6 below) that $f - q = \lim_{j\to\infty} g_j$ in $C^2(\bar{D})$, where g_j is in the linear span of functions $\varphi_a \in C_0^\infty(D)$, $a \in D$, which satisfy the hypothesis of Lemma 8.3. Thus $\mathbf{P}_D g_j$ is in the linear span of $\{\mathbf{P}_D\varphi_a: a \in D\} = \{K_D(\cdot, a): a \in D\}$, and by Theorem 8.1 with $k = 1$ it follows that $f = \mathbf{P}_D(f - q) = \lim_{j\to\infty} \mathbf{P}_D g_j$ in $C^1(\bar{D})$. ■

We now prove the two lemmas which were used in the proof of Theorem 8.4.

Lemma 8.5. *Suppose $D \subset\subset \mathbb{C}^n$ has boundary of class at least C^6 and satisfies (R_0). Given $f \in C^6(\bar{D})$, there is $q \in C^2(\bar{D})$ with $\mathbf{P}_D q = 0$ and such that*

(8.9) $$D^{\alpha\bar{\beta}}(f - q)|_{bD} = 0 \qquad \text{for all } \alpha, \beta \in \mathbb{N}^n \text{ with } |\alpha| + |\beta| \leq 2.$$

PROOF. By a partition of unity argument it is enough to prove the Lemma for f with support in $U \cap \bar{D}$, where U is a neighborhood of an arbitrary point $P \in bD$. Let r be a C^6 defining function for D. We choose $U = U(P)$ so small that for some j between 1 and n one has $\partial r/\partial z_j \neq 0$ on U. In order to achieve $\mathbf{P}_D q = 0$, it is enough to construct q of the form

$$(8.10) \qquad q = \frac{\partial}{\partial z_j} g, \qquad \text{with } g \in C^3(\bar{D}) \text{ and } g|_{bD} = 0.$$

In fact, if (8.10) holds, one has

$$\mathbf{P}_D q(z) = (q, K_D(\cdot, z))_D = \int_D \frac{\partial g}{\partial \zeta_j} \overline{K_D(\zeta, z)}\, dV$$

$$= -\int_D g \overline{\frac{\partial}{\partial \bar{\zeta}_j} K_D(\zeta, z)}\, dV$$

by integration by parts—the boundary integral vanishes since $g|_{bD} = 0$ and $K(\cdot, z) \in C(\bar{D})$ by condition (R_0) and Theorem 8.2—and the last integral is 0 since $K_D(\zeta, z)$ is holomorphic in ζ.

We may assume that U is so small that on U there is a C^6 coordinate system (x_1, x') with $x_1 = r$ and that $U = \{(x_1, x'): -\varepsilon < x_1 < \varepsilon, x' \in U'\}$ for some neighborhood U' of 0 in $\mathbb{R}^{2n-1}$. In these coordinates the vector field $\partial/\partial z_j$ transforms into a complex valued vector field $V = \sum_{\nu=1}^{2n} a_\nu(x)(\partial/\partial x_\nu)$ of class C^5 which satisfies $Vx_1 \neq 0$ on U. In order to satisfy (8.9) and (8.10) we must construct g on U so that

$$(8.11) \qquad g(0, x') = 0 \text{ and } D_x^\alpha(f - Vg)(0, x') = 0 \qquad \text{for all } x' \in U' \text{ and } \alpha \in \mathbb{N}^{2n} \text{ with } |\alpha| \leq 2.$$

We set

$$(8.12) \qquad g(x_1, x') = \sum_{\nu=0}^{2} x_1^{\nu+1} u_\nu(x')$$

with functions u_ν on U' which we now specify. Clearly g satisfies $g(0, x') = 0$. The case $\alpha = 0$ in (8.11) gives

$$0 = (f - Vg)(0, x') = f(0, x') - (Vx_1)(0, x')u_0(x'),$$

i.e., $u_0(x') = (f/Vx_1)(0, x')$.

Similarly, by recursively applying (8.11) with $D^\alpha = \partial/\partial x_1$ and $\partial^2/\partial x_1^2$ one obtains

$$u_1(x') = \frac{\dfrac{\partial}{\partial x_1}[f - (Vx_1)u_0]}{2Vx_1}(0, x')$$

and

$$u_2(x') = \frac{\dfrac{\partial^2}{\partial x^2}[f - V(x_1 u_0 + x_1^2 u_1)]}{3!\, Vx_1}(0, x').$$

The above equations define u_0, u_1, u_2 as functions on U' of class C^5, C^4, and C^3, respectively. Hence g given by (8.12) is of class C^3 on U, and it easily follows from the above that (8.11) holds. Finally we choose $\varphi \in C_0^\infty(U)$ with $\varphi \equiv 1$ on support $f \subset\subset U$, and replace g by φg. After returning to the original coordinates, the function $q = \partial/\partial z_j g$ is then in $C^2(\bar{D})$ and it satisfies (8.9), as required. ∎

The other lemma used in the proof of Theorem 8.4 is independent of the complex structure; we therefore state it for regions on $\mathbb{R}^n$.

Lemma 8.6. *Suppose $D \subset\subset \mathbb{R}^n$ has boundary of class C^k, $k \geq 1$, and denote by $\mathscr{R}(D)$ the linear span of all functions $\varphi \in C_0^\infty(D)$ which are radially symmetric around points $a \in D$ and satisfy $\int \varphi\, dV = 1$. Then every $f \in C^k(\bar{D})$ which satisfies*

$$(8.13) \qquad (D^\alpha f)(x) = 0 \qquad \text{for } x \in bD \text{ and all } \alpha \in \mathbb{N}^n \text{ with } |\alpha| \leq k,$$

is the limit in $C^k(\bar{D})$ norm of a sequence of functions on $\mathscr{R}(D)$.

PROOF. We first show that every function $f \in C^k(\bar{D})$ which satisfies (8.13) can be approximated in $C^k(\bar{D})$ norm by functions with compact support in D. By a partition of unity argument, we may assume that f has support in U, where U is a neighborhood of $P \in bD$ for which Lemma 2.4 holds. If we define $\tilde{f}$ by $\tilde{f}(x) = f(x)$ for $x \in \bar{D} \cap U$ and $\tilde{f}(x) = 0$ for $x \in U - \bar{D}$, then (8.13) implies that $\tilde{f} \in C_0^k(U)$. Moreover, if $\mathbf{n}$ is the inner unit normal at P, then $f_\tau(x) := f(x - \tau\mathbf{n})$ has compact support in $D \cap U$ for sufficiently small $\tau > 0$, and $|f - f_\tau|_{k, D\cap U} \to 0$ as $\tau \to 0$.

Thus, by replacing f with f_τ, we may assume that $f \in C_0^k(D)$. Let $\varphi \in C_0^\infty(B(0, 1))$ be a radially symmetric function around 0 with $\int \varphi\, dV = 1$, and set $\varphi_j(x) = j^n \varphi(jx)$ for $j = 1, 2, \ldots$. Define

$$(8.14) \qquad f_j(x) = \int f(y)\varphi_j(x - y)\, dV(y) = \int f(x - y)\varphi_j(y)\, dV(y).$$

Then $f_j \in C_0^\infty(D)$ for j sufficiently large, and it follows by standard analysis techniques (see also the proof of Theorem II.4.12) that

$$(8.15) \quad D^\alpha f_j(x) = \int f(y) D_x^\alpha \varphi_j(x - y)\, dV(y) = \int (D^\alpha f)(y)\varphi_j(x - y)\, dV(y)$$

for all $\alpha \in \mathbb{N}^n$ with $|\alpha| \leq k$, and that $|f - f_j|_{k,D} \to 0$ as $j \to \infty$.[1]

Finally, for fixed j, we approximate the integral (8.14) defining $f_j(x)$ by Riemann sums. These are of the form

$$\sum_{\nu=1}^{N} c_\nu f(\eta_\nu)\varphi_j(x - \eta_\nu)$$

with certain constants c_ν, and hence are elements of $\mathscr{R}(D)$ if j is sufficiently large. Since the modulus of continuity of the functions $y \to f(y) D_x^\alpha \varphi_j(x - y)$ can be bounded independently of x and $\alpha \in \mathbb{N}^n$ for $|\alpha| \leq k$, it follows from the representation (8.15) that the approximation of the integral (8.14) by Riemann sums will converge to f_j in $C^k(\bar{D})$ norm. ∎

[1] The procedure used here is commonly known as *convolution of f with an "approximate identity"*.

8.4. Ligocka's Nondegeneracy Condition

The nonvanishing of the determinant (8.8) will be an elementary consequence of the nonvanishing of another determinant. Because of the fundamental importance of this latter condition we make the following formal definition.

Definition. The domain D in $\mathbb{C}^n$ is said to satisfy **condition (B_k)** for some $k \geq 1$ if

(i) for each $a \in D$ one has $K_D(\cdot, a) \in C^k(\bar{D})$, and
(ii) for each $P \in \bar{D}$ there are points $a_0, \ldots a_n \in D$ with

$$K_D(P, a_0) \neq 0 \tag{8.16}$$

and

$$\det \begin{bmatrix} K_D(P, a_0) \cdots & K_D(P, a_n) \\ \frac{\partial K_D}{\partial z_1}(P, a_0) \cdots & \frac{\partial K_D}{\partial z_1}(P, a_n) \\ \vdots & \\ \frac{\partial K_D}{\partial z_n}(P, a_0) \cdots & \frac{\partial K_D}{\partial z_n}(P, a_n) \end{bmatrix} \neq 0. \tag{8.17}$$

The derivatives in (8.17) are taken with respect to the first variable in $K_D(\cdot, a_j)$. Notice that (i) implies that all entries in the determinant (8.17) are well defined even at points $P \in bD$; this is the crucial case for our purposes.

Condition (B_k) was introduced by E. Ligocka in 1979 [Lig 1]. A closely related condition was formulated earlier by S. Webster [Web].

We can now easily prove the main result of this section.

Theorem 8.7. *A strictly pseudoconvex domain D in $\mathbb{C}^n$ with boundary of class C^{2k+4} with $k \geq 1$ satisfies condition (B_k). Moreover, every smoothly bounded domain for which condition (R_k) holds satisfies (B_k).*

Proof. Part (i) of condition (B_k) is obvious by Theorem 8.1 and Theorem 8.2. To prove (ii), fix $P \in \bar{D}$ and suppose the determinant (8.17) were 0 for all $(a_0, a_1, \ldots, a_n) \in D^{n+1}$. By using Theorem 8.4 and the multilinearity and continuity of the determinant as a function of the columns, it would follow that for every collection of functions $g_0, g_1, \ldots, g_n \in A^\infty(D)$ one has

$$\det \begin{bmatrix} g_0(P) & \cdots\cdots g_n(P) \\ \frac{\partial g_0}{\partial z_1}(P) \cdots\cdots & \frac{\partial g_n}{\partial z_1}(P) \\ \vdots & \vdots \\ \frac{\partial g_0}{\partial z_n}(P) \cdots\cdots & \frac{\partial g_n}{\partial z_n}(P) \end{bmatrix} = 0.$$

But this is clearly impossible: take, for example, $g_0 \equiv 1$ and $g_j(z) = z_j$ for $1 \leq j \leq n$. Thus (8.17) must hold for some $(a_0, a_1, \ldots, a_n) \in D^{n+1}$. In particular, the first row contains a nonzero entry, and, after renumbering, we may assume that (8.16) holds as well. ∎

Condition (B_k) is not limited to domains with differentiable boundary. For example, one has the following result.

Lemma 8.8. *If the regions $D_i \subset\subset \mathbb{C}^{n_i}$, $i = 1, 2$, satisfy condition (B_k), so does $D = D_1 \times D_2$.*

The proof is left to the reader (use Theorem IV.4.7!).

We can now prove the existence of the special homogeneous Bergman kernel coordinates mentioned in §8.2.

Corollary 8.9. *Suppose $D \subset \mathbb{C}^n$ has C^k boundary and satisfies condition (B_k). Then for each $P \in bD$ there are a neighborhood Ω of P and points $a_0, a_1, \ldots, a_n \in D$, such that the map $u = (u_1, \ldots, u_n)$ defined by*

$$u_j(z) = \frac{K_D(z, a_j)}{K_D(z, a_0)}$$

has components in $A^k(\Omega \cap D)$ and satisfies

$$\det u'(P) \neq 0. \tag{8.18}$$

Moreover, u is a holomorphic coordinate system on $\Omega \cap D$ whose inverse u^{-1} has components in $A^k(u(\Omega \cap D))$.

Proof. Given $P \in bD$, choose $a_0, \ldots, a_n \in D$ according to condition (B_k), so that (8.16) and (8.17) hold. By continuity, there is $\Omega = \Omega(P)$, so that $K_D(z, a_0) \neq 0$ for $z \in \bar{\Omega} \cap \bar{D}$, and hence, by (i) in (B_k), $u_1, \ldots, u_n$ are well defined functions in $A^k(\Omega \cap D)$. To see (8.18) notice that

$$\frac{\partial u_j}{\partial z_l}(P) = K_D(P, a_0)^{-1} \frac{\partial K_D}{\partial z_l}(P, a_j) - \left(\frac{\partial K_D}{\partial z_l} K_D^{-2}\right)(P, a_0) K_D(P, a_j).$$

Hence, after multiplying the $(l + 1)$st row of the determinant in (8.17) by $K_D(P, a_0)^{-1}$ and then subtracting from it

$$\left(\frac{\partial K_D}{\partial z_l} K_D^{-2}\right)(P, a_0)$$

times the first row, for $1 \leq l \leq n$, the nonvanishing of (8.17) implies that

$$\det \begin{bmatrix} K_D(P, a_0) \cdots K_D(P, a_n) \\ 0 \\ \vdots \quad \left(\dfrac{\partial u_j}{\partial z_l}(P)\right)_{\substack{l=1,\ldots,n \\ j=1,\ldots,n}} \\ 0 \end{bmatrix} \neq 0,$$

and (8.18) follows. The remaining statement, with a perhaps smaller neighborhood Ω, is now a consequence of the inverse function theorem, as follows. Since u_j is holomorphic on $\Omega \cap D$, one has $(\partial u_j/\partial \bar{z}_l)(P) = 0$ for $i \leq j$, $l \leq n$, so that (8.18) implies that $\det J_{\mathbb{R}} u(P) \neq 0$, by Lemma I.2.1. Since bD is of class C^k near P, one can extend the functions u_j as C^k functions to a full neighborhood Ω of P (see Exercise E.III.1.15), so that the standard inverse function theorem from real calculus applies, showing that—after perhaps shrinking Ω—the map $u\colon \Omega \to u(\Omega)$ is a C^k diffeomorphism onto the open set $u(\Omega)$. Finally, since u is holomorphic on $U \cap D$, it follows that u^{-1} is holomorphic on $u(\Omega \cap D)$. ■

Remark. The conclusions of Corollary 8.9 remain true under much weaker regularity assumptions on bD, but we shall not pursue such technical generalizations here. The interested reader may consult Exercise E.8.5.

8.5. The Differentiability of Biholomorphic Maps at the Boundary

We now have available all the tools required for the proof of the following regularity result for biholomorphic maps.

Theorem 8.10. *Let D_1 and D_2 be bounded domains in $\mathbb{C}^n$ with boundary of class C^k which satisfy condition (B_k). Then every biholomorphic map $F\colon D_1 \to D_2$ is in $C^k(\bar{D}_1)$.*

The statement $F \in C^k(\bar{D}_1)$ means, of course, that every component of $F = (f_1, \ldots, f_n)$ is in $C^k(\bar{D}_1)$. It follows that F has a C^k extension to a neighborhood of $\bar{D}_1$ (see Exercise E.III.1.15). The differentiability assumptions on bD can be weakened—see the Remark at the end of the preceding section.

By combining Theorem 8.10 with Theorem 8.7 one immediately obtains the following more precise version of Fefferman's Mapping Theorem.

Theorem 8.11. *Suppose D_1 and D_2 are bounded strictly pseudoconvex domains in $\mathbb{C}^n$ with boundary of class C^{2k+4}, where $k \geq 1$. Then every biholomorphic map $F\colon D_1 \to D_2$ is in $C^k(\bar{D}_1)$.*

Fefferman's original proof [Fef] dealt with the case $k = \infty$. The version given here was proved by E. Ligocka in 1980 [Lig 2].

Remark 8.12. By applying Theorem 8.10 or Theorem 8.11 to F and F^{-1}, it easily follows that the biholomorphic map F extends to a C^k diffeomorphism of the closures $\overline{D_1}$ and $\overline{D_2}$.

PROOF OF THEOREM 8.10. The proof will involve three elementary steps.

Step 1. *There are constants c_1, c_2 such that $0 < c_1 \leq |\det F'(z)| \leq c_2 < \infty$ for all $z \in D_1$.*

Proof. Arguing by contradiction, suppose there is a sequence $\{p_\nu\} \subset D_1$ such that $\det F'(p_\nu) \to 0$ as $\nu \to \infty$. Since $\det F' \neq 0$ on D_1, we may assume—after passing to a subsequence—that $\lim p_\nu = P \in bD_1$. By the transformation formula for the Bergman kernel (Theorem IV.4.9.) we have

$$K_{D_1}(p_\nu, a) = \det F'(p_\nu) K_{D_2}(F(p_\nu), F(a)) \overline{\det F'(a)} \tag{8.19}$$

for every $a \in D_1$. Since condition (B_k) implies that $K_{D_1}(\cdot, a)$ and $K_{D_2}(\cdot, F(a))$ are, in particular, continuous on $\overline{D_1}$ and $\overline{D_2}$, respectively, (8.19) implies that $K_{D_1}(P, a) = \lim_{\nu\to\infty} K_{D_1}(p_\nu, a) = 0$. Since $a \in D_1$ is arbitrary, this contradicts (8.16) in condition (B_k) for D_1. The same argument applied to F^{-1} shows that $|\det(F^{-1})'(w)| \geq c > 0$ for all $w \in D_2$, which implies $|\det F'(z)| \leq 1/c$ for all $z \in D_1$.

Step 2. *F extends continuously to* $\overline{D_1}$.

Proof. It is enough to show that all partial derivatives $\partial f_j/\partial z_l$, $1 \leq j, l \leq n$, of the components $f_1, \ldots, f_n$ of F are bounded on D_1.

Arguing again by contradiction, assume that there is a sequence $\{p_\nu\} \subset D_1$ such that

$$\max_{1\leq j,l\leq n} \left| \frac{\partial f_j}{\partial z_l}(p_\nu) \right| \to \infty \qquad \text{for } \nu \to \infty. \tag{8.20}$$

Passing to a subsequence, we may assume that $p_\nu \to P \in bD_1$ and $q_\nu = F(p_\nu) \to Q \in bD_2$ as $\nu \to \infty$. Let $b_0, \ldots, b_n \in D_2$ be points associated to $Q \in bD_2$ according to condition (B_k) for D_2, so that $K_{D_2}(Q, b_0) \neq 0$ and that, by Corollary 8.9, $v = (v_1, \ldots, v_n)$ defined by

$$v_j(w) = \frac{K_{D_2}(w, b_j)}{K_{D_2}(w, b_0)}, \qquad 1 \leq j \leq n, \tag{8.21}$$

satisfies $\det v'(Q) \neq 0$, and v gives a coordinate system in $A^k(\Omega_2 \cap D_2)$ for some neighborhood Ω_2 of Q. Set $a_j = F^{-1}(b_j)$ for $j = 0, 1, \ldots, n$. Then (8.19) with $a = a_0$ and Step 1 imply that $K_{D_1}(P, a_0) \neq 0$. So there is a neighborhood Ω_1 of P, such that

$$u_j(z) = \frac{K_{D_1}(z, a_j)}{K_{D_1}(z, a_0)} \tag{8.22}$$

defines a function in $A^k(\Omega_1 \cap D_1)$ for $j = 1, \ldots, n$. Choose ν_0, so that $p_\nu \in \Omega_1 \cap D_1$ and $F(p_\nu) \in \Omega_2 \cap D_2$ for $\nu \geq \nu_0$. The transformation formula (8.19) for the Bergman kernel and the definitions (8.21) and (8.22) imply

$$u_j(p_\nu) = v_j(F(p_\nu))\lambda_j \qquad \text{for } i \leq j \leq n \text{ and } \nu \geq \nu_0, \tag{8.23}$$

where

$$\lambda_j = \overline{\left(\frac{\det F'(a_j)}{\det F'(a_0)}\right)}.$$

By introducing

$$\Lambda = \begin{bmatrix} \lambda_1 & 0 & \cdots & 0 \\ 0 & \lambda_2 & & \vdots \\ \vdots & & \ddots & 0 \\ 0 & & \cdots & \lambda_n \end{bmatrix}$$

and writing u and v as column vectors, (8.23) can be stated as

$$u(p_\nu) = \Lambda\, v(F(p_\nu)). \tag{8.24}$$

By differentiating (8.24) and by the chain rule one obtains

$$u'(p_\nu) = \Lambda\, v'(F(p_\nu))F'(p_\nu), \tag{8.25}$$

and hence

$$F'(p_\nu) = v'(F(p_\nu))^{-1}\Lambda^{-1}u'(p_\nu). \tag{8.26}$$

Since $\det v'(Q) \neq 0$, (8.26) implies that the coefficients of $F'(p_\nu)$ remain bounded as $\nu \to \infty$; but this contradicts the initial assumption (8.20), and the proof is complete.

Step 3. *F is in $C^k(\overline{D_1})$.*

Proof. Once we know, by Step 2, that F extends continuously to $\overline{D_1}$, the arguments given in the proof of Step 2 can be applied in a full neighborhood of an arbitrary point $P \in bD_1$, as follows. Set $Q = F(P)$ and define v and u as in Step 2 (cf. (8.21) and (8.22)). Since $|\det F'| \geq c_1 > 0$ by Step 1 and $\det v'(Q) \neq 0$, (8.25) implies that $\det u'(P) \neq 0$ as well. We can therefore choose the neighborhoods Ω_1 of P and Ω_2 of Q so that u and v are C^k coordinates on $\Omega_1 \cap \bar{D}_1$ and $\Omega_2 \cap \bar{D}_2$, respectively, and so that $F(\Omega_1 \cap \bar{D}_1) \subset \Omega_2 \cap \bar{D}_2$. Equation (8.24) then holds for all $z \in \Omega_1 \cap \bar{D}_1$, i.e., one has

$$F(z) = v^{-1}(\Lambda^{-1}u(z))$$

on $\Omega_1 \cap \bar{D}_1$, showing that $F \in C^k(\Omega_1 \cap \bar{D}_1)$. ∎

8.6. Complete Reinhardt Domains

Now that the reader has seen the far-reaching implications of the regularity condition (R), it might be useful to verify this condition for some class of domains other than the strictly pseudoconvex ones. This we now do for complete Reinhardt domains with smooth boundary. In contrast to the strictly pseudoconvex case discussed in §7, the techniques used here are quite elementary. Moreover, it is of particular interest that no pseudoconvexity assumptions are required.

We first show that condition (R) is a consequence of a special growth estimate for the Bergman kernel K_D.

Theorem 8.12. *Let $D \subset\subset \mathbb{C}^n$ have C^∞ boundary. Given $k \in \mathbb{N}$, suppose there are an integer s_k and a constant c_k such that*

$$|K_D(\cdot, a)|_{k,D} \leq c_k \delta_D^{-s_k}(a) \tag{8.27}$$

for all $a \in D$. Then D satisfies condition (R_k).

Remark. The estimate (8.27) for some s_k is in fact equivalent to condition (R_k).

In order to prove Theorem 8.12 we will need the following technical generalization of Lemma 8.5.

Lemma 8.13. *For each positive integer s there are an integer N_s and a linear operator*

$$T^{(s)}: C^{N_s}(\bar{D}) \to C^s(\bar{D})$$

with the following properties:

(i) $T^{(s)}$ *is bounded, i.e.,* $|T^{(s)}f|_s \lesssim |f|_{N_s}$;
(ii) $\mathbf{P}_D \circ T^{(s)} = \mathbf{P}_D$ *on* $C^{N_s}(\bar{D})$;
(iii) $D^\alpha(T^{(s)}f)(z) = 0$ *for* $z \in bD$, $|\alpha| \leq s$, *and* $f \in C^{N_s}(\bar{D})$.

The proof of Lemma 8.5 gives Lemma 8.13 in case $s = 2$, with $N_s = 6$, by setting $T^{(2)}f = f - q$. The general case is proved by the same method. The details are left to the reader.

Proof of Theorem 8.12. We will prove that the Bergman projection $\mathbf{P}_D$ is bounded from $C^{m_k}(\bar{D})$ to $C^k(\bar{D})$, where m_k is the integer N_{s_k} associated to $s = s_k$ according to Lemma 8.13. For $f \in C^{m_k}(\bar{D})$ we have

$$\mathbf{P}_D f(z) = \mathbf{P}_D(T^{(s_k)}f)(z) = \int_D T^{(s_k)}f(\zeta) K_D(z, \zeta)\, dV(\zeta). \tag{8.28}$$

By (8.27) we may differentiate under the integral sign in (8.28) up to k times, obtaining

$$|D^\alpha \mathbf{P}_D f(z)| \lesssim \int_D |T^{(s_k)}f(\zeta)| \delta_D^{-s_k}(\zeta)\, dV(\zeta) \tag{8.29}$$

for any multi-index $\alpha \in \mathbb{N}^n$ with $|\alpha| \leq k$. Now notice that Lemma 8.13(iii) implies

$$|T^{(s_k)}f(\zeta)| \lesssim |T^{(s_k)}f|_{s_k} \delta_D^{s_k}(\zeta),$$

so (8.29) implies

$$|D^\alpha \mathbf{P}_D f(z)| \lesssim |T^{(s_k)}f|_{s_k} \lesssim |f|_{N_{s_k}}$$

for all $z \in D$, and the desired conclusion follows. ∎

Next we consider a special property of the Bergman kernel K_D of a complete Reinhardt domain D which will allow us to prove the estimate (8.27), and hence establish condition (R) for such domains.

For λ and $w \in \mathbb{C}^n$, we set $\lambda w = (\lambda_1 w_1, \ldots, \lambda_n w_n)$.

Lemma 8.14. *Let D be a complete Reinhardt domain with center* 0. *Then*

$$K_D(\lambda\zeta, z) = K_D(\zeta, \bar{\lambda}z) \tag{8.30}$$

for all $\lambda, \zeta, z \in \mathbb{C}^n$ with $\zeta, z, \lambda\zeta, \bar{\lambda}z \in D$.

Proof. The formula (8.30) is an obvious consequence of the representation

$$K_D(\zeta, z) = \sum_{\nu \in \mathbb{N}^n} c_\nu \zeta^\nu \bar{z}^\nu \qquad \text{on } D \times D, \tag{8.31}$$

with suitable constants c_ν, valid for the Bergman kernel of a complete Reinhardt domain D. In order to see (8.31) notice that the set $\mathcal{M} = \{z^\nu : \nu \in \mathbb{N}^n\}$ of holomorphic monomials is orthogonal in $L^2(D)$ (just integrate in n-fold polar coordinates), and recall from Corollary II.1.7 that the Taylor series of any $f \in \mathcal{O}(D)$ converges to f on D. This latter fact implies that $\mathcal{M}$ is complete in $\mathscr{H}^2(D)$ (Exercise!), and hence

$$\{\varphi_\nu := \|z^\nu\|_{\mathscr{H}^2}^{-1} z^\nu : \nu \in \mathbb{N}^n\}$$

is an orthonormal basis for $\mathscr{H}^2$. The representation (8.31) then follows from Theorem IV.4.3, with $c_\nu = \|z^\nu\|_{\mathscr{H}^2(D)}^{-2}$. ∎

Theorem 8.15. *Let $D \subset\subset \mathbb{C}^n$ be a smoothly bounded complete Reinhardt domain. Then the estimate* (8.27) *holds for each $k \in \mathbb{N}$, and hence D satisfies condition* (R).

Proof. The idea is quite simple: given $\zeta, z \in D$, we use (8.30) to shift ζ towards the interior of D to a point ζ' by an amount which is a small, but fixed multiple of $\delta_D(z)$, while moving z a corresponding amount toward the boundary. This allows to fit inside D a polydisc with center ζ' and multiradius estimated by $\delta_D(z)$, and to obtain the required estimates by applying the Cauchy estimates to K_D on that polydisc.

Let us first set up the necessary geometric details. Since D has a smooth boundary, there is $\varepsilon > 0$ such that for each $\zeta \in \bar{D}$ with $\delta_D(\zeta) \le 3\varepsilon$ there is a unique point $P_\zeta \in bD$ which minimizes the distance between ζ and bD, and which depends smoothly on ζ. It follows that for $\zeta \in D$

$$\mathbf{n}(\zeta) = (P_\zeta - \zeta)/|P_\zeta - \zeta|$$

is the unit outer normal to bD at P_ζ, and that $\mathbf{n}(\zeta)$ extends smoothly to $\bar{D}$. Notice that $\mathbf{n}_j(\zeta) = 0$ if $\zeta_j = 0$. In fact, if $\zeta_j = 0$, and $P_\zeta = (p_1, \ldots, p_n)$, then $P'(\theta) = (p_1, \ldots, e^{i\theta} p_j, \ldots, p_n)$ lies in bD (here we use that D is a Reinhardt domain!) and $|P'(\theta) - \zeta| = |P_\zeta - \zeta| = \delta_D(\zeta)$ for all $0 \le \theta \le 2\pi$. The uniqueness of the minimizing point then implies that $p_j = 0$, so that $\mathbf{n}_j(\zeta) = 0$ as well.

Because $\mathbf{n}(\zeta)$ depends smoothly on ζ, we can therefore find a constant $A \geq 1$ so that

(8.32) $$|\mathbf{n}_j(\zeta)| \leq A|\zeta_j| \qquad \text{for } 1 \leq j \leq n.$$

We also choose M with $|\zeta| \leq M$ for all $\zeta \in D$.

Claim. *Fix $z \in D$ and set*

$$\tau = \min\left\{\varepsilon, \frac{1}{2A}, \frac{\delta_D(z)}{4AM\sqrt{n}}\right\}.$$

If $\zeta \in D$ and $\delta_D(\zeta) < \varepsilon$, then $\lambda = (\lambda_1, \ldots, \lambda_n) \in \mathbb{C}^n$ defined by $\lambda\zeta = \zeta - \tau\mathbf{n}(\zeta)$ satisfies

(8.33) $$|\lambda_j| \geq \tfrac{1}{2} \qquad \text{for } 1 \leq j \leq n,$$

(8.34) $$P(\lambda\zeta, \tau/\sqrt{n}) \subset D,$$

(8.35) $$\bar{\lambda}^{-1}z := (\bar{\lambda}_1^{-1}z_1, \ldots, \bar{\lambda}_n^{-1}z_n) \in D,$$

(8.36) $$\delta_D(\bar{\lambda}_1^{-1}z) \geq \delta_D(z)/2.$$

Proof of the Claim. Since $\lambda_j\zeta_j = \zeta_j - \tau\mathbf{n}_j(\zeta)$ and $|\tau\mathbf{n}_j(\zeta)| \leq \tau A|\zeta_j|$ by (8.32), we have $|1 - \lambda_j| \leq \tau A \leq \frac{1}{2}$, which implies (8.33). Since $\tau \leq \varepsilon$, it is clear that (8.34) holds as well. Next we estimate

$$|\bar{\lambda}_j^{-1}z_j - z_j| = \left|\frac{1 - \bar{\lambda}_j}{\bar{\lambda}_j}\right||z_j| \leq 2\tau A\, M \leq \delta_D(z)/2\sqrt{n}.$$

Hence $|\bar{\lambda}^{-1}z - z| \leq \delta_D(z)/2$, and this implies (8.35) and (8.36).

We are now ready to estimate $D_\zeta^\alpha K_D(\zeta, z)$ for an arbitrary multi-index $\alpha \in \mathbb{N}^n$. We will need the estimate

(8.37) $$\|K_D(\cdot, a)\|_{L^2(D)} \lesssim \delta_D^{-n}(a) \qquad \text{for } a \in D$$

(see IV(4.2)). Fix $z \in D$ and assume first that $\delta_D(\zeta) < \varepsilon$. Choose $\tau > 0$ and $\lambda \in \mathbb{C}^n$ as in the Claim. By Lemma 8.14 we have $K_D(\zeta, z) = K_D(\lambda\zeta, \bar{\lambda}^{-1}z)$, so that, with $\zeta' = \lambda\zeta$,

(8.38) $$D_\zeta^\alpha K_D(\zeta, z) = \lambda^\alpha D_{\zeta'}^\alpha K_D(\zeta', \bar{\lambda}^{-1}z).$$

By applying the Cauchy estimate (1.20) in Theorem I.1.6 to the polydisc $P = P(\zeta', \tau/\sqrt{n})$ and by (8.37) we obtain

(8.39) $$\begin{aligned} D_{\zeta'}^\alpha K_D(\zeta', \bar{\lambda}^{-1}z) &\lesssim \frac{1}{\tau^{|\alpha|+2n}}\|K_D(\cdot, \bar{\lambda}^{-1}z)\|_{L^1(P)} \\ &\lesssim \frac{1}{\tau^{|\alpha|+n}}\|K_D(\cdot, \bar{\lambda}^{-1}z)\|_{L^2(P)} \\ &\lesssim \frac{1}{\tau^{|\alpha|+n}}\delta_D^{-n}(\bar{\lambda}^{-1}z). \end{aligned}$$

Since $\tau = \tau(z) \gtrsim \delta_D(z)$ when z is close to bD, (8.38), (8.39), and (8.36) imply the required estimate

$$|D_\zeta^\alpha K_D(\zeta, z)| \lesssim \delta_D^{-|\alpha|-2n}(z)$$

in case $\delta_D(\zeta) < \varepsilon$. Finally, since $\Omega = \{\zeta \in D: \delta_D(\zeta) \geq \varepsilon\}$ is compact in D, Corollary I.1.7 and (8.37) imply

$$\sup_{\zeta \in \Omega} |D_\zeta^\alpha K_D(\zeta, z)| \lesssim \|K_D(\cdot, z)\|_{L^2(D)} \lesssim \delta_D^{-n}(z).$$

The proof of Theorem 8.15 is complete. ■

Exercises

E.8.1. Let $D \subset\subset \mathbb{C}^n$ have C^∞ boundary. Show that if $\mathbf{P}_D f \in C^\infty(\bar{D})$ for every $f \in C^\infty(\bar{D})$, then D satisfies condition (R). (This follows by using the Closed Graph Theorem in Fréchet spaces.)

E.8.2. Let $B = B(a, r)$ be a ball in $\mathbb{C}^n$. Show that if $f \in A(B)$, then

$$f(a) = (\text{vol } bB)^{-1} \int_{bB} f\, dS.$$

(This mean value property holds more generally for harmonic functions in $\mathbb{R}^n$.)

E.8.3. Prove Lemma 8.13.

E.8.4. Prove Lemma 8.8.

E.8.5. A domain $D \subset\subset \mathbb{R}^n$ is said to satisfy "*minimal regularity conditions*" if its boundary bD can be described locally as the graph of a Lipschitz function in suitable coordinates for $\mathbb{R}^n$. In particular, every domain with C^1 boundary satisfies minimal regularity conditions.

Show that if $D \subset\subset \mathbb{C}^n$ satisfies minimal regularity conditions and condition (B_k), then the conclusions of Corollary 8.9 remain true. (Hint: The main difficulty is a technical generalization of the Inverse Function Theorem; this is obtained, for example, by an analysis of the proof of that theorem based on the Fixed Point Theorem for contractions in complete metric spaces. Alternatively, one may use the Whitney Extension Theorem to extend $f \in C^k(\bar{D})$ to a C^k function on a neighborhood of $\bar{D}$. See [Lig 1] and [SteE 3], Chapter VI, for more details.)

E.8.6. Show that if D_1 and D_2 are bounded domains in $\mathbb{C}^n$ with C^k boundary which satisfy condition (B_k), then every biholomorphic map $F: D_1 \to D_2$ can be extended to a C^k diffeomorphism $\hat{F}: W_1 \to W_2$ of open neighborhood W_1 and W_2 of $\bar{D}_1$ and $\bar{D}_2$, respectively.

E.8.7. Suppose $D \subset\subset \mathbb{C}^n$ has C^∞ boundary and satisfies condition (R_k). Show that the estimate (8.27) in Theorem 8.12 holds for some integer s_k.

E.8.8. Let D be a complete Reinhardt domain D with center 0. Show that the (orthogonal) set $\mathcal{M} = \{z^\nu: \nu \in \mathbb{N}^n\}$ of monomials is dense in $\mathcal{H}^2(D)$ (and hence is a basis).

E.8.9. Let D_1 and D_2 be domains in $\mathbb{C}^n$ with *connected* boundary of class C^1. Show that if $n > 1$ and if there is a CR-diffeomorphism $F: bD_1 \to bD_2$ (i.e., a C^1 diffeomorphism whose components are CR-functions), then D_1 and D_2 are biholomorphic.

§9. The Reflection Principle

It is well known that a biholomorphic map $F: D_1 \to D_2$ between two domains in the complex plane extends holomorphically across the boundary bD_1 if both domains have real analytic boundaries. The proof involves two parts: first one shows that F extends continuously to the closure $\bar{D}_1$—by Carathéodory's Theorem this holds under much weaker restrictions on the boundaries—and then one applies the Schwarz reflection principle.

In the preceding paragraph we have generalized the first step to several variables, proving not only continuous, but even *differentiable* extension of biholomorphic maps to the boundary under appropriate hypotheses. We will now consider a corresponding generalization of the reflection principle. The main result is local, and its rather elementary proof could have been presented in Chapter II; but in contrast to the one-variable case, the a priori knowledge of a C^1 extension seems to be unavoidable, and therefore the application of the reflection principle in several variables does require results of the type discussed in §8.

9.1. Reflection on Real Analytic Arcs in the Plane

We shall first review the classical reflection principle in the complex plane, emphasizing those features which will help to understand the generalization to higher dimensions given later on.

The "refection" $z \mapsto \bar{z}$ on the real axis in $\mathbb{C}$ is generalized to arbitrary real analytic arcs as follows. Recall that a subset $\gamma \subset \mathbb{C}$ is a regular **real analytic arc at the point** $P \in \gamma$ if γ has a regular real analytic parametrization in some neighborhood Ω of P, i.e., if there is a real analytic map $\varphi: I \to \Omega$, where $I \subset \mathbb{R} \subset \mathbb{C}$ is an interval containing 0, such that $\varphi(0) = P$, $\varphi'(x) \neq 0$ for $x \in I$, $\varphi: I \to \varphi(I)$ is one-to-one and $\Omega \cap \gamma = \varphi(I)$. If this property holds at every point $P \in \gamma$ we say that γ is a regular real analytic arc. We will only consider *regular* arcs (i.e., the parametrizations φ satisfy $\varphi' \neq 0$), and hence we will omit "regular" in the following. By replacing x with z in the power series expansion of φ at 0 we extend φ to a holomorphic function φ on a disc Δ with center 0. Since $\varphi'(0) \neq 0$, we may assume that $\varphi: \Delta \to \varphi(\Delta)$ is biholomorphic; moreover, we choose $\Omega = \varphi(\Delta)$. We then define the **reflection** σ_γ **on the arc** γ in Ω by

$$\sigma_\gamma(z) = \varphi(\overline{\varphi^{-1}(z)}) \qquad \text{for } z \in \Omega. \tag{9.1}$$

It follows that $\sigma_\gamma\colon \Omega \to \Omega$ is an involution (i.e., $\sigma_\gamma \circ \sigma_\gamma \doteq$ identity) whose complex conjugate $\overline{\sigma_\gamma}$ is holomorphic, and $\sigma_\gamma(z) = z$ for $z \in \gamma$. By the Identity Theorem applied to $\overline{\sigma_\gamma}$ these properties define σ_γ uniquely. In particular, the definition (9.1) is independent of the parametrization φ, and if γ is real analytic at every point $P \in \gamma$, then the reflections defined locally by the above procedure can be glued together to a reflection σ_γ on a suitable neighborhood Ω of γ. Also notice that (9.1) implies that σ_γ maps points on one side of γ to the other side (this statement makes sense for points z sufficiently close to γ).

Let us summarize the above discussion.

Theorem 9.1. *If $\gamma \subset \mathbb{C}$ is a connected real analytic arc, then there are a connected neighborhood Ω of γ and a diffeomorphism $\sigma_\gamma\colon \Omega \to \Omega$ with the following properties:*

(i) *$\overline{\sigma_\gamma}$ is holomorphic on Ω.*
(ii) *$\sigma_\gamma(z) = z$ for $z \in \gamma$.*

The properties (i) *and* (ii) *determine σ_γ uniquely in the following sense: if $\sigma^{(j)}\colon \Omega_j \to \Omega_j$, $j = 1, 2$, are two maps which satisfy* (i) *and* (ii), *then $\sigma^{(1)} \equiv \sigma^{(2)}$ on the connected component of $\Omega_1 \cap \Omega_2$ which contains γ.*

For our purposes it will be convenient to consider real analytic arcs which are given in implicit form. By standard analysis techniques, γ is a real analytic arc at $P \in \gamma$ if and only if there are an open neighborhood Ω of P and a real analytic function $r\colon \Omega \to \mathbb{R}$ such that $dr \neq 0$ on Ω and $\Omega \cap \gamma = \{z \in \Omega\colon r(z) = 0\}$. The reflection σ_γ on γ is obtained from the defining function r as follows. For simplicity, assume $P = 0$. By replacing $x = (z + \bar{z})/2$ and $y = (z - \bar{z})/2i$ in the power series expansion of $r(x, y)$ we write $r(z)$ in the form $r(z, \bar{z})$, where $r(z, w)$ is holomorphic on a bidisc $\Delta \times \Delta$ in $\mathbb{C}^2$ centered at 0 with $\partial r/\partial w(0, 0) \neq 0$. By the Implicit Function Theorem I.2.4, if $\Delta' \subset \Delta$ are sufficiently small, the equation $r(z, w) = 0$, $(z, w) \in \Delta' \times \Delta$, has a unique solution $w = t(z)$ for each $z \in \Delta'$; moreover, $t(z)$ is holomorphic in z. We may assume $\Delta' \subset \Omega$. Since for $z \in \Delta' \cap \gamma$ the point $w = \bar{z}$ also solves $r(z, w) = 0$, it follows that $t(z) = \bar{z}$ for all $z \in \Delta' \cap \gamma$. Hence, by the uniqueness statement in Theorem 9.1, $\sigma_\gamma(z) = \overline{t(z)}$ for $z \in \Delta'$! We have thus shown the following result.

Lemma 9.2. *If γ is defined near $P \in \gamma$ by $\{z\colon r(z, \bar{z}) = 0\}$ as above, then the reflection σ_γ is the (locally) unique solution of $r(z, \bar{\sigma}_\gamma(z)) = 0$ in a neighborhood of $(P, \bar{P})$.*

Example. Suppose γ is a circle with center a and radius R, i.e., $\gamma = \{z \in \mathbb{C}\colon r(z, \bar{z}) = 0\}$, where $r(z, \bar{z}) = (z - a)(\bar{z} - \bar{a}) - R^2$. Then $t(z) = \bar{a} + R^2/(z - a)$, and the reflection σ_γ is given by $\sigma_\gamma(z) = a + R^2/(\bar{z} - \bar{a})$ for $z \in \mathbb{C} - a$.

It is now easy to describe the process of analytic continuation by reflection.

Theorem 9.3. *Let D be open in $\mathbb{C}$ and $P \in bD$. Suppose there is an open disc centered at P such that $\gamma = \Delta \cap bD$ is a real analytic arc which bounds D from one side. If $f \in \mathcal{O}(D)$ extends continuously to $D \cup \gamma$ and $f(\gamma) \subset \Gamma$, where Γ is some other real analytic arc in $\mathbb{C}$, then f has a holomorphic extension $\hat{f}$ to a neighborhood of P. If Δ is sufficiently small, the extension $\hat{f}$ is given by*

$$\hat{f}(z) = \begin{cases} f(z) & \text{for } z \in \Delta \cap \bar{D} \\ \sigma_\Gamma(f(\sigma_\gamma(z))) & \text{for } z \in \Delta - \bar{D}. \end{cases} \tag{9.2}$$

Proof. We leave it to the reader to verify that $\hat{f}$ in (9.2) is defined for $z \in \Delta$ provided Δ is sufficiently small, and that $\hat{f}$ is holomorphic on $\Delta - \gamma$. The hypothesis on f implies that $\hat{f}$ extends continuously from $\Delta - \bar{D}$ to γ, and since $\sigma_\Gamma(f(\sigma_\gamma(z))) = f(z)$ for $z \in \gamma$, (9.2) shows that $\hat{f}$ is continuous on Δ. It then follows by a standard result that $\hat{f}$ is holomorphic on Δ (see Exercise E.9.2.). ■

Remark 9.4. An analysis of the construction of $\hat{f}$ reveals the following: in order to find the holomorphic extension of f across P it is enough to find a holomorphic function $t(z)$ on $\Delta \cap D$ which extends continuously to $\Delta \cap \bar{D}$, so that $t(z) = \overline{f(z)}$ for $z \in \Delta \cap bD$. The extension $\hat{f}$ is then defined by $\hat{f}(z) = \overline{t(\sigma_\gamma(z))}$ for $z \in \Delta - \bar{D}$. To find such a function t, one uses the hypothesis $\rho(f(z), \overline{f(z)}) = 0$ for $z \in \Delta \cap bD$, where ρ is a real analytic defining function for the arc Γ. $t(z)$ is then the (locally) unique solution of the equation $\rho(f(z), t(z)) = 0$ in a neighborhood of $(f(P), \overline{f(P)})$. In several variables we will carry out analogs of these two steps separately in §9.3 and §9.4, as it is not possible to directly use a formula like (9.2) to define the holomorphic extension.

9.2. Real Analytic Hypersurfaces in $\mathbb{C}^n$

Passing to several variables, we first introduce the higher dimensional analog of a real analytic arc. A set $S \subset \mathbb{C}^n$ is called a **real analytic hypersurface** if for each $P \in S$ there are a neighborhood U of P and a real analytic function $r: U \to \mathbb{R}$ with $dr \neq 0$ on U, such that

$$S \cap U = \{z \in U: r(z) = 0\}. \tag{9.3}$$

We call such a function r a **defining function for S in U**. In analogy to the situation in $\mathbb{C}^1$ we write (9.3) in the form

$$S \cap U = \{z \in U: r(z, \bar{z}) = 0\}, \tag{9.4}$$

where the function $r(z, w)$ is holomorphic in a neighborhood of $(P, \bar{P})$ in $\mathbb{C}^{2n}$.

An open set D in $\mathbb{C}^n$ is said to have **real analytic boundary at $P \in bD$** if there is a neighborhood U of P such that $S = U \cap bD$ is a real analytic hypersurface so that $D \cap U$ lies on one side of S. Without loss of generality we choose U and the defining function r so that

$$D \cap U = \{z \in U: r(z, \bar{z}) < 0\}.$$

D is said to have **real analytic boundary** if the above holds at every point $P \in bD$.

We now want to study reflections on a real analytic hypersurface S. First, observe that even in the special case that S is a linear real hyperplane in $\mathbb{C}^n$, the natural (geometric) reflection on S is *not* the conjugate of a holomorphic map if $n \geq 2$! In fact, suppose $S = \{(z', z_n) \in \mathbb{C}^n: \operatorname{Im} z_n = 0\}$, where $z' \in \mathbb{C}^{n-1}$; then the reflection σ_S on S is given by $\sigma_S(z', z_n) = (z', \bar{z}_n)$, so $\overline{\sigma_S}(z) = (\bar{z}', z_n)$ is holomorphic only in z_n, but not in the tangential coordinates $z_1, \ldots, z_{n-1}$. This situation persists no matter how one tries to define σ_S. If $n \geq 2$, there is no holomorphic map σ from a neighborhood of 0 into itself which satisfies $\sigma(z) = \bar{z}$ for $z \in S$. Thus we see that for a general real analytic hypersurface S, the best one can hope for is a "reflection" σ_S whose complex conjugate is holomorphic in the "normal direction". This can indeed be done quite easily—at least locally. The remarkable fact is that this limited information suffices—under suitable additional technical hypothesis—to obtain the holomorphic continuation by reflection across S of certain holomorphic maps defined on one side of S.

Suppose that the real analytic hypersurface S is described near $P \in S$ as in (9.4). After a holomorphic change of coordinates we may assume that $P = 0$ and that $(\partial r/\partial w_n)(z, w) \neq 0$ on a polydisc $P(0, \varepsilon) \times P(0, \varepsilon)$ in $\mathbb{C}^{2n}$ for some $\varepsilon > 0$. It follows that for $c' \in P'(0, \varepsilon) \subset \mathbb{C}^{n-1}$ the set

$$\gamma_{c'} = \{(c', z_n): |z_n| < \varepsilon \text{ and } r((c', z_n), (\bar{c}', \bar{z}_n)) = 0\}$$

is a real analytic arc in the plane $\Pi_{c'} = \{(c', z_n): z_n \in \mathbb{C}\}$. Denote by $\sigma_{c'}$ the reflection on the arc $\gamma_{c'}$ in the plane $\Pi_{c'}$, and define the **reflection σ_S on the hypersurface** S by

$$\sigma_S(z', z_n) := (z', \sigma_{z'}(z_n)). \tag{9.5}$$

Lemma 9.5. *For ε sufficiently small the reflection σ_S is defined and real analytic on the polydisc $P(0, \varepsilon)$. Moreover, $\sigma_S(z) = z$ for $z \in S$, and $\overline{\sigma_S}(z', z_n)$ is holomorphic in z_n for $|z_n| < \varepsilon$.*

PROOF. By Lemma 9.2, $\bar{\sigma}_{z'}(z_n) = w_n(z', z_n, \bar{z}')$, where $w_n(z', z_n, w')$ is the unique holomorphic solution of the equation $r(z', z_n, w', w_n) = 0$. The desired conclusions are then obvious. ∎

Remark 9.6. In contrast to the situation in $\mathbb{C}^1$, the reflection σ_S defined above is not determined uniquely by intrinsic properties of S when $n \geq 2$. Definition (9.5) is valid only locally and it depends on the choice of coordinates.

We now use the reflection σ_S to prove an extension theorem which will be needed later on.

Lemma 9.7. *Suppose σ_S is defined on $P(0, \varepsilon)$ and let φ be real analytic in a neighborhood of S. Then, after perhaps shrinking ε, there is a real analytic* *function* $\hat{\varphi}$ *on* $P(0, \varepsilon)$ *which is holomorphic in* z_n *and which satisfies* $\hat{\varphi}(z) = \varphi(z)$ *for* $z \in P(0, \varepsilon) \cap S$.

PROOF. We may write $\varphi = \varphi((z', z_n), (\bar{z}', \bar{z}_n))$, where $\varphi(z, w)$ is holomorphic on $P(0, \varepsilon) \times P(0, \varepsilon)$ (shrink ε, if necessary). The function $\hat{\varphi}(z) := \varphi(z, \overline{\sigma_S}(z))$ is then an extension of $\varphi|_S$ with the required properties. ■

Lemma 9.7 generalizes the well known one variable result that a real analytic function defined on a real analytic arc $\gamma \subset \mathbb{C}$ has a holomorphic extension to a neighborhood of γ.

9.3. Analytic Continuation by Reflection

The following result, which is motivated by the discussion of the one variable case in Remark 9.4., shows how the reflection on a real analytic hypersurface, even though conjugate holomorphic in only one variable, can nevertheless be used to obtain holomorphic continuation in all variables.

We assume that $r(z, \bar{z})$ is a real analytic defining function for S in the polydisc $P(0, \varepsilon) \subset \mathbb{C}^n$ with $\partial r/\partial \bar{z}_n \neq 0$ and set $D = \{z \in P(0, \varepsilon): r(z, \bar{z}) < 0\}$.

Proposition 9.8. *Let* $F: D \to \mathbb{C}^n$ *be a holomorphic map which extends continuously to* $\bar{D} \cap P(0, \varepsilon) = \{z \in P(0, \varepsilon): r(z, \bar{z}) \leq 0\}$. *Suppose there is a continuous map* $T: \bar{D} \cap P(0, \varepsilon) \to \mathbb{C}^n$ *such that*

(9.6) $\quad T(z', z_n)$ *is holomorphic in* z_n *for* $(z', z_n) \in D$,

and

(9.7) $$T(z) = \bar{F}(z) \qquad \text{for } z \in S \cap P(0, \varepsilon).$$

Then F *extends holomorphically to a neighborhood of* 0.

PROOF. Recalling the discussion of the one variable case it is clear how one should define the extension $\hat{F}$ of F. For $z \in P(0, \varepsilon)$—shrink ε, if necessary—we set

$$\hat{F}(z) = \begin{cases} F(z) & \text{if } r(z, \bar{z}) \leq 0 \\ \overline{T(\sigma_S(z))} & \text{if } r(z, \bar{z}) > 0. \end{cases}$$

Since by (9.7) one has $\overline{T(\sigma_S(z))} = \overline{T(z)} = F(z)$ for $z \in S$, it follows that $\hat{F}$ is continuous on $P(0, \varepsilon)$. Furthermore, by (9.6) and Lemma 9.5, if z' is fixed, $\hat{F}(z', z_n)$ is holomorphic in z_n if $r((z', z_n), (\bar{z}', \bar{z}_n)) \neq 0$. Therefore, as in the proof of the one variable reflection principle, $\hat{F}(z', \cdot)$ is holomorphic for $|z_n| < \varepsilon$. We now show that $\hat{F}$ is holomorphic at 0 in *all* variables! Fix $a = (0, a_n)$ and $\delta > 0$ with $P(a, \delta) \subset \{z \in P(0, \varepsilon): r(z, \bar{z}) < 0\}$. Then $\hat{F}$ is holomorphic on $P(a, \delta)$ in all variables. Each component of $\hat{F}$ thus satisfies the hypotheses of Lemma 9.9 below, and the desired conclusion follows. ■

Lemma 9.9. *Suppose f is a continuous function on the polydisc $P(0, (\varepsilon', \varepsilon_n))$ in $\mathbb{C}^n$ which satisfies*

(9.8) *for each fixed $z' \in P'(0, \varepsilon')$, $f(z', z_n)$ is holomorphic in z_n for $|z_n| < \varepsilon_n$;*

(9.9) *there are a_n and $\delta > 0$ such that f is holomorphic on $P'(0, \varepsilon') \times \{z_n: |a_n - z_n| < \delta\}$.*

Then f is holomorphic on $P(0, (\varepsilon', \varepsilon_n))$.

PROOF. By Corollary I.1.5 and by (9.8) it is enough to show that f is separately holomorphic on $P(0, (\varepsilon', \varepsilon_n))$ in each of the other variables $z_1, \ldots, z_{n-1}$. This will follow from Morera's Theorem if we show for $1 \leq j \leq n - 1$ and $(z', z_n) \in P(0, (\varepsilon', \varepsilon_n))$ that

$$(9.10) \qquad I_\gamma^{(j)}(z', z_n) := \oint_\gamma f(z_1, \ldots, z_{j-1}, \zeta, z_{j+1}, \ldots, z_n)\, d\zeta = 0$$

for every closed curve $\gamma \subset \{\zeta \in \mathbb{C}: |\zeta| < \varepsilon'\}$.

To prove (9.10) we first show that $I_\gamma^{(j)}(z', \cdot)$ *is holomorphic in z_n on the disc* $\Delta_n = \{|z_n| < \varepsilon_n\}$. In fact, let γ_n be any closed curve in Δ_n. Since f is continuous in all variables, we may interchange the order of integration to obtain

$$(9.11) \qquad \oint_{\gamma_n} I_\gamma^{(j)}(z', z_n)\, dz_n = \oint_\gamma \left[\oint_{\gamma_n} f(z_1, \ldots, z_{j-1}, \zeta, \ldots, z_n)\, dz_n \right] d\zeta.$$

Because of (9.8), the inner integral on the right in (9.11) is 0 by Cauchy's Theorem; hence the integral on the left in (9.11) is 0, and the statement above follows by Morera's Theorem.

To complete the proof, notice that because of (9.9) and Cauchy's Theorem the integral in (9.10) is zero for z_n with $|z_n - a_n| < \delta$. Since we just saw that $I_\gamma^{(j)}(z', z_n)$ is holomorphic in z_n, the Identity Theorem implies that $I_\gamma^{(j)}(z', z_n) \equiv 0$ for $|z_n| < \varepsilon_n$, and we are done. ■

9.4. A Sufficient Condition for Analytic Continuation

In order to apply Proposition 9.8, one is left with the task of finding the map T which satisfies (9.6) and (9.7). As in case $n = 1$ (see Remark 9.4) we shall try to find T as the solution of a certain system of equations. Consider the setting in Proposition 9.8. Without loss of generality let $F(0) = 0$. Moreover, let us assume that $F(S \cap P(0, \varepsilon))$ is contained in a real analytic hypersurface $S^\#$ defined near $0 = F(0)$ by the real equation $\rho(u) = \rho(u_1, \ldots, u_n, \bar{u}_1, \ldots, \bar{u}_n) = 0$, where $\rho(u, v)$ is holomorphic in a neighborhood of 0 in $\mathbb{C}^{2n}$, and $\partial\rho/\partial v_n(0) \neq 0$. Thus

$$(9.12) \qquad \rho(F(z), \overline{F(z)}) = 0 \qquad \text{for } z \in S \cap P(0, \varepsilon),$$

and (9.7) implies that $T(z)$ must satisfy

$$\rho(F(z), T(z)) = 0 \qquad \text{for } z \in S. \tag{9.13}$$

Clearly (9.13) is not sufficient to determine $T(z)$ if $n > 1$. We will therefore generate additional equations which must be satisfied by $T(z)$ by differentiating (9.12) along S. For this purpose we must assume that F extends not only continuously, but as a C^1 map from D to $\bar{D} \cap P(0, \varepsilon)$, and some other technical conditions will be required to ensure that the resulting system of equations is analytically independent.

We now proceed with the details. Without loss of generality we choose the coordinates so that $\partial r/\partial z_j(0) = \partial\rho/\partial u_j(0) = 0$ for $1 \le j \le n-1$ and $\partial r/\partial z_n(0) = \partial\rho/\partial u_n(0) = 1$. Define the (1, 0) vector fields L_j, $1 \le j \le n-1$, by

$$L_j = \frac{\partial r}{\partial z_n}(z, \bar{z})\frac{\partial}{\partial z_j} - \frac{\partial r}{\partial z_j}(z, \bar{z})\frac{\partial}{\partial z_n}. \tag{9.14}$$

Clearly $L_1, \ldots, L_{n-1}$ are tangential to S, and $\bar{L}_1, \ldots, \bar{L}_{n-1}$ are tangential Cauchy–Riemann operators. Thus $L_j\bar{F} = \overline{(\bar{L}_j F)} = 0$ on D, and—by continuity—along $S \cap P(0, \varepsilon) \subset bD$ as well. Application of L_j to (9.12) gives

$$\sum_{\nu=1}^{n} \frac{\partial\rho}{\partial u_\nu}(F(z), \overline{F(z)})L_j F_\nu(z) = 0 \qquad \text{for } z \in S \cap P(0, \varepsilon). \tag{9.15}$$

For $u, v \in \mathbb{C}^n$ and $w = (w_{11}, \ldots, w_{j\nu}, \ldots, w_{n-1,n}) \in \mathbb{C}^{n(n-1)}$ we set

$$H_j(u, v, w) = \sum_{\nu=1}^{n} \frac{\partial\rho}{\partial u_\nu}(u, v)w_{j\nu} \qquad \text{for } 1 \le j \le n-1 \tag{9.16}$$

and

$$H_n(u, v, w) = \rho(u, v). \tag{9.17}$$

Also, let $w^{(0)} = (L_1 F_1(0), \ldots, L_j F_\nu(0), \ldots, L_{n-1}(F_n(0))$.

Lemma 9.10. *Suppose*

$$\det\left(\frac{\partial H_j}{\partial v_k}(0, 0, w^{(0)})\right) \ne 0. \tag{9.18}$$

Then there are $\varepsilon > 0$ and a continuous map T on $\bar{D} \cap P(0, \varepsilon)$ which satisfies the hypotheses (9.6) *and* (9.7) *in Proposition* 9.8.

Proof. Application of the Implicit Function Theorem I.2.4 to the holomorphic system of equations $H_j(u, v, w) = 0$, $1 \le j \le n$ gives δ', $\delta > 0$ such that if $(u, w) \in P(0, \delta') \times P(w^{(0)}, \delta')$, then there is a unique solution $v = G(u, w) \in P(0, \delta)$ of this system. Moreover, $G(u, w)$ is holomorphic in u and w. The uniqueness of solution together with (9.12), (9.15), and (9.16) implies that

$$G(F(z), (\ldots, L_j F_\nu(z), \ldots)) = \bar{F}(z) \qquad \text{for all } z \in S \cap P(0, \varepsilon), \tag{9.19}$$

provided ε is sufficiently small. It remains to be seen that the left side of (9.19) has a continuous extension T to $\bar{D} \cap P(0, \varepsilon)$ which is holomorphic in z_n. Obviously $F(z)$ and the partial derivatives $\partial F_\nu/\partial z_k$ have such an extension, and since by (9.14) the coefficients of L_j are real analytic functions, it follows from Lemma 9.7 that $L_j F_\nu|_S$ also has such an extension $\widehat{L_j F_\nu}$ to $\bar{D} \cap P(0, \varepsilon)$ if ε is sufficiently small. Since G is holomorphic, it follows that

$$T(z) := G(F(z), (\ldots, \widehat{L_j F_\nu}(z), \ldots))$$

has all the required properties. ∎

Corollary 9.11. *If* (9.18) *holds, then* F *has a holomorphic extension across* 0.

Let us analyze the condition (9.18) in more detail. From (9.16) and (9.17) one obtains

$$(\partial H_j/\partial v_k)(0, 0, w^{(0)}) = \sum_{\nu=1}^{n} \frac{\partial^2 \rho}{\partial u_\nu \partial \bar{u}_k}(0) L_j F_\nu(0) \qquad \text{if } j < n,$$

and

$$\partial H_n/\partial v_k(0, 0, w^{(0)}) = \partial \rho/\partial \bar{u}_k(0) = \delta_{kn},$$

where we again view ρ as a function $\rho(u) = \rho(u, \bar{u})$ of u by substituting $v = \bar{u}$. Therefore,

$$\left[\frac{\partial H_j}{\partial v_k}(0, 0, w^{(0)})\right]_{j,k=1,\ldots n}$$

$$(9.20) \qquad = \begin{bmatrix} \left[\sum_{\nu=1}^{n} \frac{\partial^2 \rho}{\partial u_\nu \partial \bar{u}_k}(0)(L_j F_\nu)(0)\right]_{j,k=1,\ldots n-1} & \begin{matrix} * \\ \vdots \\ * \end{matrix} \\ 0 \quad 0 \ldots\ldots\ldots\ldots\ldots\ldots 0 & 1 \end{bmatrix}.$$

Lemma 9.12. *Suppose the Levi form of the hypersurface* $S^\#$ *is nondegenerate at* 0 *and that the map* F *satisfies* $\det F'(0) \neq 0$. *Then the hypothesis* (9.18) *in Lemma* 9.10 *is satisfied.*

Proof. The first hypothesis means that the Levi form

$$(t, t') \mapsto L_0(\rho; t, t') = \sum_{\nu,k=1}^{n} \frac{\partial^2 \rho}{\partial u_\nu \partial \bar{u}_k}(0) t_\nu \bar{t}'_k$$

at 0 of the defining function ρ for $S^\#$ restricts to a nondegenerate Hermitian form on the complex tangent space $T_0^{\mathbb{C}} S^\#$ of $S^\#$ at 0. Since L_j is tangential to S and $F(S) \subset S^\#$, it follows that $V_j := (L_j F_1(0), \ldots, L_j F_n(0)) = (\partial F_1/\partial z_j(0), \ldots, \partial F_n/\partial z_j(0))$ lies in $T_0^{\mathbb{C}} S^\#$ for $j = 1, \ldots, n-1$. Moreover, the fact that $F'(0)$ is nonsingular implies that $\{V_1, \ldots, V_{n-1}\}$ is linearly independent and hence a basis for $T_0^{\mathbb{C}} S^\#$. The nondegeneracy hypothesis therefore implies that

$$\det\left[\sum_{\nu=1}^{n} \frac{\partial^2 \rho}{\partial u_\nu \partial \bar{u}_k}(0)(L_j F_\nu)(0)\right]_{\substack{j=1,\ldots,n-1 \\ k=1,\ldots,n-1}} \neq 0.$$

Hence the matrix on the left in (9.20) also has nonzero determinant, i.e., (9.18) holds. ∎

9.5. The Lewy–Pinčuk Reflection Principle

By combining the results obtained in Sections 9.3 and 9.4, one obtains the following several variables version of the reflection principle.

Theorem 9.13. *Let $F: D \to \mathbb{C}^n$ be a holomorphic map on the open set $D \subset \mathbb{C}^n$. Suppose D has real analytic boundary at the point $P \in bD$ and that there is a neighborhood U of P such that*

(i) *F extends as a C^1 map to $U \cap \bar{D}$ with* $\det F'(P) \neq 0$, *and*
(ii) *$F(U \cap bD)$ is contained in a real analytic hypersurface whose Levi form is nondegenerate at $F(P)$. Then F extends holomorphically to a neighborhood of P.*

Remark 9.14. Theorem 9.13 is not true without any restrictions on the real analytic hypersurface which contains $F(U \cap bD)$ in (ii). Consider, for example, $D = \{(z_1, z_2) \in \mathbb{C}^2, \operatorname{Im} z_2 > 0\}$. Then $bD = \{z \in \mathbb{C}^2 \colon \operatorname{Im} z_2 = 0\}$ is real analytic. Let h be holomorphic function on the upper half-plane which has a C^1 extension to a neighborhood of 0 but no holomorphic extension across 0 (such functions do exist!), and define $F(z_1, z_2) = (z_1 + h(z_2), z_2)$ for $(z_1, z_2) \in D$. Then F maps D biholomorphically onto D, F extends as a C^1 map to $U \cap \bar{D}$ for some neighborhood U of 0, and $\det F'(0) = 1$, so the hypothesis (i) in Theorem 9.13 holds. Moreover $F(U \cap bD) \subset bD$, but clearly the map F has no holomorphic extension to 0. Notice that the Levi form of bD is identically zero at every point of bD. This implies that no matter how often one differentiates Equation (9.12) along complex tangential directions to bD, the resulting equations will not be analytically independent of (9.12). On the other hand, if the Levi form of $S^{\#} \supset F(U \cap bD)$ degenerates only to "finite order", it may be possible to save the reflection principle. The reader may consult [Der] and the references therein for such generalizations.

We conclude with the main application of Theorem 9.13.

Theorem 9.15. *Let D_1 and D_2 be bounded strictly pseudoconvex domains in $\mathbb{C}^n$ with real analytic boundary and let $F: D_1 \to D_2$ be biholomorphic. Then F has a holomorphic extension to a neighborhood of $\bar{D}_1$.*

PROOF. By Fefferman's Mapping Theorem, F extends as a C^∞ map to $\bar{D}_1$, and we saw in Step 1 of the Proof of Theorem 8.13 that $|\det F'(z)| \geq c > 0$ on D_1, hence $\det F'(z) \neq 0$ for $z \in bD_1$ as well.[1] Clearly $F(bD_1) \subset bD_2$, and the strict

[1] Alternatively, apply Fefferman's Theorem to F^{-1} and use that $F'(z) \cdot [(F^{-1})'(F(z))]$ is the identity matrix for all $z \in D_1$, hence also for $z \in \bar{D}_1$.

pseudoconvexity of D_2 implies the nondegeneracy of the Levi form of bD_2. Hence the hypotheses of Theorem 9.13 are satisfied at every boundary point P, and it follows that F has a holomorphic extension F_P to some ball B_P centered at P for each $P \in bD$. We may assume that if $P, Q \in bD$ and $B_P \cap B_Q \neq \emptyset$, then $D \cap (B_P \cap B_Q) \neq \emptyset$. Since $F_P = F_Q = F$ on $D \cap B_P \cap B_Q$, the Identity Theorem implies $F_P \equiv F_Q$ on $B_P \cap B_Q$. Therefore there is a unique holomorphic function $\hat{F}$ on $\Omega = (\bigcup_{P \in bD} B_P) \cup D$ such that $\hat{F} = F$ on D and $\hat{F} = F_P$ on B_P for all $P \in B_P$. ■

Exercises

E.9.1. Show that if σ_Γ and σ_γ are the reflections on two real analytic arcs in the plane and if f is holomorphic, then $\sigma_\Gamma \circ f \circ \sigma_\gamma$ is holomorphic wherever defined.

E.9.2. Suppose γ is a closed, regular differentiable arc in the disc Δ. Let f be continuous on Δ and holomorphic on $\Delta - \gamma$. Show that f is holomorphic on Δ.

E.9.3. Generalize E.9.2 to $\mathbb{C}^n$, replacing the arc γ by a closed C^1 submanifold of dimension $2n - 1$.

E.9.4. Let $\gamma \subset \mathbb{C}$ be a real analytic arc and suppose $\varphi: \gamma \to \mathbb{C}$ is real analytic. Show that φ has an extension to a holomorphic function on an open neighborhood Ω of γ.

E.9.5. Let $S \subset \mathbb{C}^n$ be a real analytic hypersurface with $0 \in S$. Show that if $n > 1$ there is no holomorphic map $T: U \to \mathbb{C}^n$ defined on some neighborhood U of 0 such that $T(z) = \bar{z}$ for $z \in U \cap S$.

E.9.6. Suppose $D \subset\subset \mathbb{C}^n$ and $\Omega \cap bD$ is real analytic for some open set Ω. Show that if $f \in \mathcal{O}(D)$ extends continuously to $\Omega \cap \bar{D}$ and if $f(\Omega \cap bD)$ is contained in a real analytic arc in $\mathbb{C}$, then f extends holomorphically to $\Omega \cap \bar{D}$.

E.9.7. Let D_1 and D_2 be bounded strictly pseudoconvex domains in $\mathbb{C}^n$ with C^2 boundary. Suppose $F: D_1 \to D_2$ is a proper holomorphic map which extends as a C^2 map to $\bar{D}_1$. Prove that $\det F'(z) \neq 0$ for $z \in bD_1$. (Hint: If r is a C^2 strictly plurisubharmonic defining function, use the Hopf Lemma (see E.II.4.10) to show that $\varphi = r \circ F$ can be extended to a C^2 defining function for D_1. Use the proof of Proposition II.2.14 to construct from φ a strictly plurisubharmonic defining function $\psi = e^{A\varphi} - 1$, and compare the Levi forms of ψ and r.)

Notes for Chapter VII

Part A

Here we give references and some additional results directly related to the topics discussed in this chapter.

§1. The construction of the Cauchy kernel given here is due to N. Kerzman and E.M. Stein [KeSt], except for the technical simplifications due to the use of the explicit integral operator $\mathbf{T}_1^{V,D^{\#}}$ instead of Hörmander's L^2 solution operator for $\bar{\partial}$. This makes it elementary to keep track of differentiability in both ζ and z. Theorem 1.4 had been obtained first for the kernel of Henkin and Ramirez (§3). In particular, part (iv) was already proved by G.M. Henkin [Hen 1], while (iii) (in strengthened form) was proved, independently, by P. Ahern and R. Schneider [AhSc] and D.H. Phong and E.M. Stein [PhSt]. Regularity properties of the Cauchy kernel on the Hardy spaces $H^p(bD)$ were studied by E.L. Stout [Sto]. Kerzman and Stein obtain (iii) (and much more) as a consequence of earlier results of G.B. Folland and E.M. Stein [FoSt].

§2. The proof of Theorem 2.1 given here follows Henkin's proof [Hen 1]. Henkin proved the crucial Lemma 2.3 and mentions the Mergelyan type approximation result only in a footnote. By the example of Diederich and Fornaess [DiFo 2] the approximation theorem does not hold in general for smoothly bounded pseudoconvex domains. So far, attempts to find a good general version for weakly pseudoconvex domains have been unsatisfactory. E. Bedford and J.E. Fornaess [BeFo] obtain the result for pseudoconvex domains with real analytic boundary, but their method works only in $\mathbb{C}^2$. Other generalizations can be found in [BeRa] and in the references given there.

§3. Compared to Henkin's paper [Hen 1], the construction of E. Ramirez [Ram] is technically much more demanding; also, it is limited to the case of C^∞ boundaries. The presentation here involves ideas of both authors, simplified further by the use of the explicit integral solution operator $\mathbf{T}_1^{V,V_0}$, which gives the required optimal differentiability properties by elementary means. In particular, we follow Ramirez in constructing the function g so that $\operatorname{Re} g(\zeta, z) > 0$ on $bD \times \bar{D}$ for $\zeta \neq z$ (cf. Proposition 3.1). This feature is of interest in the context of peaking functions (cf. Corollary 3.2). The identification of the principal part of $\Omega_0(W^{\mathrm{HR}})$ involving the geometric information given in Lemma 3.9 is in [KeSt].

§4. The history and principal references are given in §4.1. The idea of the proof presented here is due to P. Ahern and R. Schneider [AhSc], who obtain the corresponding result for functions in $A^k(D)$ as well. Modulo Fatou's Theorem (see [SteE 2]), the arguments presented here can easily be adapted to functions in $H^\infty(D)$. Quite recently J.E. Fornaess and N. Øvrelid [FoØv] solved Gleason's problem for pseudoconvex domains in $\mathbb{C}^2$ with real analytic boundary; unfortunately, it is not at all clear how to extend their methods to dimension ≥ 3.

§5. L^p estimates for $\bar{\partial} u = f$ were obtained first by N. Kerzman [Ker 1] in case f is a (0, 1)-form, and in the general case by N. Øvrelid [Øvr 1], except for the classical case $p = 2$, due to J.J. Kohn [Koh 1], and the case $p = \infty$ already discussed in Chapter V. The construction of the parametrix ${}_{\bar{\partial}}\mathbf{T}_q$ presented here appears implicitly in [LiRa 2], who modified the construction from [Ran 6]. *Optimal L^p estimates* have been investigated by S. Krantz [Kra 1].

The nonisotropic nature of the estimates (see E.V.3.2) and the relationship with analysis on the Heisenberg group has been discovered by E.M. Stein [SteE 1] and developed extensively by him and his collaborators (see [FoSt], [GrSt], and [PhSt]). R.M. Range proved Hölder estimates for solutions of $\bar{\partial}$ on some special *weakly* pseudoconvex domains [Ran 4]. Generalizations of Range's results have recently been obtained by J. Bruna and J. del Castillo [BrCa], but so far no satisfactory general result of this type is known in the weakly pseudoconvex case. In particular, it is widely believed that Hölder estimates for $\bar{\partial}$ should hold on pseudoconvex domains with real analytic boundary, but even in case of a *convex* domain with real analytic boundary, where an explicit integral solution operator is well known (see Chapter IV, §3.4), such estimates are known only in $\mathbb{C}^2$ [Ran 4]. Most recently, J.E. Fornaess [For 3] obtained L^∞ estimates for solutions of $\bar{\partial}$ on a class of weakly pseudoconvex domains in $\mathbb{C}^2$, which includes the Kohn–Nirenberg example (see E.II.2.6) which, in particular, is *not* locally biholomorphic to a *convex* domain.

§6. The proof of the approximation theorem for $A(D)$ based on the solution of a Cousin I problem with bounds is due to N. Kerzman [Ker] and I. Lieb [Lie 1]. Theorem 6.3 was proved by Grauert and Lieb [GrLi]. Approximation in L^p norm, $p < \infty$, and pointwise bounded approximation for $f \in H^\infty(D)$ are discussed in [Ker] and in [CoRa]. For other related work, see the discussion of §2 above.

§7. The Bergman projection $\mathbf{P}_D$ is intimately related to the solution $\mathbf{S}^\perp f$ of $\bar{\partial}u = f$ which is orthogonal (in the L^2 sense) to ker $\bar{\partial}$: $L^2(D) \to L^2_{0,1}(D)$, i.e., to the space $\mathcal{O}L^2(D)$: clearly one has $\mathbf{P}_D u = u - \mathbf{S}^\perp \bar{\partial} u$ for $u \in \operatorname{dom} \bar{\partial}$. By results of Hörmander, $\mathbf{S}^\perp$ exists on arbitrary pseudoconvex domains ([Hör 1]). Regularity at the boundary for $\mathbf{P}_D$ is thus a consequence of deep results for $\mathbf{S}^\perp$ obtained by J.J. Kohn in the theory of the $\bar{\partial}$-Neumann problem (see [Koh 1, 3] and [FoKo]). An alternate approach for strictly pseudoconvex domains, based on Fefferman's asymptotic expansion of the Bergman kernel [Fef], was developed in [AhSc] and [PhSt]. The more elementary methods presented in §7, which give more precise results than those which follow from Kohn's work, are based on ideas of N. Kerzman and E. Stein [KeSt] and E. Ligocka [Lig 2]. In 1977 Kerzman and Stein investigated symmetries of the Henkin–Ramirez kernel and used them to obtain a representation of the abstract Szegö kernel in terms of explicit kernels. Ligocka then adapted these methods to the Bergman kernel. Ligocka lectured on her results at a meeting in Oberwolfach in 1980, but published them only much later. In the meantime, part of Ligocka's results were included as a special case in [LiRa 2]. Lemma 7.4, which is crucial, is already in [KeSt]. A calculus for a class of admissible kernels similar to the one considered here was developed by C. Fefferman in 1974 [Fef]. Techniques of integration by parts similar to those used here have been used by many others in related contexts (see, for example, [AhSc], [PhSt], and [LiRa 1]). A more general class of admissible kernels involving singularities of mixed type is introduced in [LiRa 2]. The admissible kernels in §7.5 are a subclass of the *pure* admissible kernels in that paper.

§8. (See also the introduction to §8.) One of the earliest results on continuous extension to the boundary of biholomorphic maps in more than one variable seems to be due to W. Rothstein ("Zur Theorie der analytischen Abbildungen im Raum zweier komplexer Veränderlichen," Dissertation, Univ. of Münster, 1935), who considered special analytic polyhedra in $\mathbb{C}^2$. No progress was made on this important question until 1973, when, independently, G.M. Henkin [Hen 4], S. Pinčuk [Pin 1], and N. Vormoor [Vor] proved the continuous extension to the boundary (and even Hölder continuity) of biholomorphic maps between bounded strictly pseudoconvex domains. These proofs made use of estimates for the Carathéodory metric which, in turn, were obtained by means of the L^∞ estimates for solutions of $\bar{\partial}$ which had become available just a few years earlier. R.M. Range [Ran 2] then combined the techniques of Henkin and Vormoor with the estimates for $\bar{\partial}$ of Range and Siu [RaSi] in order to obtain the corresponding result for domains with *piecewise* strictly pseudoconvex boundary. This result is of interest since the continuous extension to the boundary already fails for biholomorphic maps between bounded pseudoconvex domains if the boundary is only piecewise differentiable (see B. Fridman [Fri] for the relevant example). No such counterexample is known for bounded domains with smooth boundary, pseudoconvex or not! On the other hand, it is easy to find counterexamples for unbounded domains. We leave it to the reader to suitably modify the example in Remark 9.14. Various generalizations to weakly pseudoconvex domains were obtained thereafter (see, for example, [Ran 3] and [DiFo 4]), but, together with the earlier results mentioned above, they became mostly obsolete once S. Bell and E. Ligocka [BeLi] introduced new methods which not only yielded a simple proof of Fefferman's Mapping Theorem, but which were applicable in much more general situations.

Most of §8 is based on [BeLi]. Bell's Density Lemma was published first in [Bel 1]. The results in §8.6 on complete Reinhardt domains are due to S. Bell and H. Boas [BeBo].

In 1980 S. Bell [Bel 2] extended the new methods further, obtaining a new simple proof of Fefferman's Mapping Theorem in the following general form: *A biholomorphic map $F: D_1 \to D_2$ between bounded pseudoconvex domains in $\mathbb{C}^n$ with C^∞ boundary extends smoothly to the boundary as soon as at least one of the domains satisfies Condition* (R). The proof uses a regularity theorem of J.J. Kohn for solutions of $\bar{\partial}$ on smoothly bounded pseudoconvex domains [Koh 2] which is beyond the scope of this book. The mapping theorem may very well hold for arbitrary bounded domains with C^∞ boundary, even though it is known that Condition (R) does not hold in this generality (see [Bar]). In 1981 S. Bell discovered how to apply the new methods to *proper* holomorphic maps [Bel 3]. This area has been investigated vigorously since then by Bell and many others, and we refer the reader to the recent survey article by E. Bedford [Bed] for a thorough discussion and for additional references.

§9. The main results here are due to H. Lewy [Lew] and S.I. Pinčuk [Pin 2]. As mentioned in Remark 9.14, it is possible to extend the reflection principle to certain weakly pseudoconvex cases. Most recently, M.S. Baouendi, H. Jacobowitz, and F. Treves [BJT] have introduced some far-reaching new ideas into this area, proving the real analyticity of CR-diffeomorphisms between real analytic manifolds under very general hypotheses which are satisfied in a wide range of applications.

Part B

Here we briefly mention several other significant topics related to complex analysis on strictly pseudoconvex domains and/or integral representations. In some instances partial results are known for special classes of weakly pseudoconvex domains. This list is by no means intended to give a complete picture, but merely to encourage the reader to explore some other areas of contemporary research which are close to the topics discussed in this book.

1. The $\bar{\partial}$-Neumann Problem. This was briefly mentioned above in the Notes to §7. It involves existence and regularity theory in L^2 Sobolev spaces for the boundary value problem for the complex Laplacian $\square$ defined with respect to some Hermitian metric, and it leads to information about the (unique) solution to $\bar{\partial}$ orthogonal to ker $\bar{\partial}$. The main results are due to J.J. Kohn [Koh 1, 2, 3]. In order to apply them to pseudoconvex domains with real analytic boundary, some additional geometric information due to K. Diederich and J.E. Fornaess [DiFo 3] is required. This seems to be the area in which the theory for weakly pseudoconvex domains has been most successful, though the results are by no means complete. Kohn's results are the principal tool for establishing condition (R) on weakly pseudoconvex domains, i.e., the crucial hypothesis in the study of boundary regularity of proper holomorphic maps. For a quick introduction to the $\bar{\partial}$-Neumann problem the reader should consult the survey article by Kohn [Koh 4].

Quite recently I. Lieb and R.M. Range [LiRa 2, 3] have studied the $\bar{\partial}$-Neumann problem on strictly pseudoconvex domains by means of explicit integral representations of the type considered in this book, thus obtaining in particular a satisfactory theory of C^k estimates for solutions of $\bar{\partial}$. The interested reader is referred to [LiRa 3] for background and additional references.

2. Characterization of zero sets of functions in the Nevanlinna class. The classical Blaschke condition for zero sets of functions in the Nevanlinna class on the unit disc in $\mathbb{C}$ has a natural generalization to several variables. G.M. Henkin [Hen 5] and H. Skoda [Sko] proved that on strictly pseudoconvex domains the Blaschke condition still characterizes the zero sets of functions in the Nevanlinna class, even though—except for the case of one variable—the Blaschke condition for an analytic set A is not sufficient for A to be the zero set of a bounded holomorphic function, or even of a function in any of the Hardy spaces $H^p(bD)$, $p > 0$. A proof of the Henkin–Skoda Theorem for the ball is in [Rud 3]. The interested reader should also consult N. Varopoulos [Var] and the references given there.

3. Estimates for $\bar\partial$ on domains with non-smooth boundary. There is a vast literature, mainly due to G.M. Henkin and his collaborators, which deals with generalizations of the Bergman–Weil formula to obtain integral solution operators for $\bar\partial$ on analytic polyhedra which satisfy L^∞ estimates. G.M. Henkin and J. Leiterer have also obtained the Hölder 1/2-estimate for $\bar\partial$ on strictly pseudoconvex domains with not necessarily differentiable boundary. The reader is referred to [HeLe] and [AiYu] for details and references.

4. Peak points and zero sets. The question of whether a point $p \in bD$ is a peak point for $A(D)$ (or for some other algebra $\mathscr{A} \subset A(D)$), i.e., whether there is $h \in \mathscr{A}$ with $h(p) = 1$ and $|h(z)| < 1$ for $z \in \bar{D} - \{p\}$, is intimately related to the local complex geometry of bD. Notice that by Theorem VI.1.13, if D is strictly pseudoconvex, then every $p \in bD$ is a peak point for $\mathcal{O}(\bar{D})$, but in case $D = P(0, 1)$, no point in $bP - b_0P$ is a peak point for $A(P)$ (use E.I.1.11). The general situation is still understood very little. E. Bedford and J.E. Fornaess [BeFo] proved that if $D \subset\subset \mathbb{C}^2$ is pseudoconvex with real analytic boundary, then every $p \in bD$ is a peak point for $A(D)$, but no such result is known in higher dimensions. Notice that under the above hypothesis there are points $p \in bD$ which are not a peak point for $\mathcal{O}(\bar{D})$ (see [KoNi]) or even for $A^1(D)$ (see [For 2]). On strictly pseudoconvex domains attention has focused on characterizations of *peak sets* and *zero sets* in bD for $A(D)$ and $A^\infty(D)$. The reader should consult the recent articles [FoHe] and [ChCh] and the references therein.

5. Kernels with weights. Recently B. Berndtsson and M. Anderson have introduced weight factors in the integral representation formula of Henkin and Ramirez [BeAn]. These methods allow for a new degree of freedom in the kernels and should have interesting applications.

6. Harmonic analysis. There are deep connections between several complex variables on one side, and harmonic analysis in Euclidean spaces and in more abstract settings on the other side. In particular, the osculation of the boundary of a strictly pseudoconvex domain by the Heisenberg group has been an extremely fruitful idea. The reader should consult, for example, [FoSt] and Chapter 8 in [Kra 2].

7. Invariant metrics. The classical Poincaré metric on the unit disc can be extended in a number of different ways to higher dimensions. The Bergman metric (see E.IV.4.2–4.4) is just one of these possibilities. These *invariant metrics* have many interesting geometric applications, and their behavior near the boundary on strictly pseudoconvex domains is fairly well understood. A standard reference for the basics is S. Kobayashi [Kob]. References to more recent results may be found in [GrKr], [DFH], and in Chapter 10.2 of [Kra 2].

Appendix A

We give a self-contained proof of the "Morse Lemma" needed in Chapter II, §2.10. The main step is the following special case of Sard's Theorem (see also [Nar 4]).

Lemma 1. *Let $D \subset \mathbb{R}^n$ be open and suppose $F: D \to \mathbb{R}^n$ is of class C^1. Let $A \subset D$ be the set of critical points of F, i.e., the set of those points $a \in D$, for which* $\det J_{\mathbb{R}} F(a) = 0$. *Then $F(A)$ has measure 0 in $\mathbb{R}^n$.*

PROOF. It is clearly enough to show that $F(Q \cap A)$ has measure 0 for any compact cube $Q \subset D$. For all $x, y \in Q$ we have

$$F(x) - F(y) = dF_y(x - y) + r(x, y),$$

where $r(x, y) = o(|x - y|)$ uniformly on $Q \times Q$; hence there is a function $\lambda: \mathbb{R}^+ \to \mathbb{R}^+$ with $\lambda(t) \to 0$ as $t \to 0$, such that

$$|r(x, y)| \leq \lambda(|x - y|)|x - y|.$$

Suppose $Q_\varepsilon \subset Q$ is a compact cube with side $\varepsilon > 0$ which contains a point $a \in A$. By hypothesis, $V_a = dF_a(\mathbb{R}^n)$ has dimension $< n$. After an orthogonal change of coordinates in the image space $\mathbb{R}^n$ we may assume that $V_a \subset \{u_n = 0\}$. Thus, if ε is sufficiently small and $x \in Q_\varepsilon$, then $F(x) \subset \{u \in \mathbb{R}^n: |u_n - f_n(a)| \leq \lambda(\sqrt{n}\varepsilon)\sqrt{n}\varepsilon\}$. Also, by the Mean Value Theorem, there is $M < \infty$, such that $F(x)$ lies in a cube of side $M\varepsilon$ with center $F(a)$ (with sides parallel to the u-coordinate axis). Since the measure of $F(Q_\varepsilon)$ in $\mathbb{R}^n$ is invariant under orthogonal coordinate changes, it follows from the above that the measure of $F(Q_\varepsilon)$ is $O(\varepsilon^n \lambda(\sqrt{n}\varepsilon))$. Let l be the length of a side of Q. Divide Q into $N = (l/\varepsilon)^n$ cubes $Q_1, \ldots, Q_N$ of side ε. It then follows that

$$\text{measure } F(Q \cap A) \leq \sum_{i: Q_i \cap A \neq \varnothing} \text{measure } F(Q_i \cap A) \lesssim N\varepsilon^n \lambda(\sqrt{n}\varepsilon)$$
$$\lesssim \lambda(\sqrt{n}\varepsilon).$$

Now let $\varepsilon \to 0$; since $\lambda(\sqrt{n}\varepsilon) \to 0$, measure $F(Q \cap A) = 0$. ■

Lemma 2. *Suppose $D \subset \mathbb{R}^n$ is open and $f: D \to \mathbb{R}$ is of class C^2. If $a \in D$ is a nondegenerate critical point of f (i.e., $\det[\partial^2 f/\partial x_j \partial x_k(a)] \neq 0$), then there is a neighborhood U of a, such that a is the only critical point of f in U.*

PROOF. After an orthogonal change of coordinates we may assume that $(\partial^2 f/\partial x_j \partial x_k(a))$ is diagonal with nonzero entries $\lambda_1, \ldots, \lambda_n$ in the diagonal. In a neighborhood U of a one then has (notice that $df_a \equiv 0$)

$$f(x) = f(a) + \sum_{j=1}^{n} \lambda_j (x_j - a_j)^2 + r(x),$$

where $r(x) = o(|x - a|^2)$ as $|x - a| \to 0$. It follows that

$$|\text{grad } f(x)| \geq 2\left(\min_{1 \leq j \leq n} |\lambda_j|\right)|x - a| - o(|x - a|),$$

which implies $|\text{grad } f(x)| \neq 0$ for $x \neq a$ if x is in a sufficiently small neighborhood U of a. ■

We can now prove Lemma II.2.22.

Lemma 3. *Let $D \subset \mathbb{R}^n$ be open and suppose $g \in C^2(D)$ is real valued. Then there is a set $E \subset \mathbb{R}^n$ of measure 0, such that for all $u \in \mathbb{R}^n - E$ the set A_u of critical points of $g_u: D \to \mathbb{R}$, defined by $g_u(x) = g(x) - (u, x)$, is discrete in D.*

PROOF. Define $F: D \to \mathbb{R}^n$ by

$$F(x) = \left(\frac{\partial g}{\partial x_1}(x), \ldots, \frac{\partial g}{\partial x_n}(x)\right).$$

Then F is of class C^1. By Lemma 1, there is $E \subset \mathbb{R}^n$ of measure 0 such that $F^{-1}(E)$ contains all critical points of F. Fix $u \in \mathbb{R}^n - E$; then $L_u = \{x \in D: F(x) = u\}$ contains no critical point of F. This means that $(\partial^2 g/\partial x_j \partial x_k(x))$ is nonsingular for $x \in L_u$, and since this matrix does not change if g is replaced by g_u, and since $a \in D$ is a critical point of g_u if and only if $a \in L_u$, we see that all critical points of g_u are nondegenerate. The desired conclusion for g_u then follows by Lemma 2. ■

Appendix B

The following result is a generalization of the classical Young inequality for convolution integrals.

Theorem. *Let (X, μ) and (Y, ν) be two measure spaces, and suppose that K is a measurable function on $X \times Y$ (with respect to product measure), which satisfies*

(i) $$\int_X |K(x, y)|^s \, d\mu(x) \le M^s \qquad \text{for almost all } y \in Y$$

and

(ii) $$\int_Y |K(x, y)|^s \, d\nu(y) \le M^s \qquad \text{for almost all } x \in X$$

for some $M < \infty$ and $s \ge 1$. Then the linear operator $f \mapsto \mathbf{T}f$ defined ν-a.e. by

$$\mathbf{T}f(y) = \int_X K(x, y) f(x) \, d\mu(x)$$

is bounded from $L^p(X)$ to $L^q(Y)$ with norm $\le M$ for all $1 \le p, q \le \infty$ with

(iii) $$\frac{1}{q} = \frac{1}{p} + \frac{1}{s} - 1,$$

with the usual conventions in case q or p are ∞.

PROOF. We will prove the theorem in case $q < \infty$ and $1 < p, s < \infty$; the remaining cases are simpler and are left to the reader. Suppose $f \in L^p(X)$. For ν-almost all y we write

$$\mathbf{T}f(y) = \int_X (K^s f^p)^{1/q} K^{1-s/q} f^{1-p/q}\, d\mu(x)$$

and apply Hölder's inequality (with three factors) to the integral on the right. Notice that $(K^s(\cdot, y)f^p)^{1/q} \in L^q$ for ν-almost all y, $[K(\cdot, y)]^{s(1/s-1/q)} \in L^{p/p-1}$ (by (i) and (iii)), $f^{p(1/p-1/q)} \in L^{s/s-1}$ (by (iii) and $f \in L^p$), and that by (iii) one has

$$\frac{1}{q} + \frac{p-1}{p} + \frac{s-1}{s} = 1.$$

We thus obtain

$$|\mathbf{T}f(y)| \le \left(\int_X |K(\cdot, y)|^s |f|^p\right)^{1/q} \left(\int_X |K(\cdot, y)|^s\right)^{(p-1)/p} \left(\int_X |f|^p\right)^{(s-1)/s}.$$

Therefore

$$\int_Y |\mathbf{T}f|^q\, d\nu(y) \le \int_Y \left(\int_X |K(x, y)|^s |f(x)|^p\, d\mu(x)\right) d\nu(y) \cdot M^{sq(p-1)/p} \| f \|_{L^p}^{pq(s-1)/s},$$

where we have used (i). By Fubini–Tonelli we may interchange the order of integration in the double integral, and because of (ii) it then follows that

$$\| \mathbf{T}f \|_{L^q}^q \le M^s \| f \|_{L^p}^p M^{sq(p-1)/p} \| f \|_{L^p}^{pq(s-1)/s} = M^q \| f \|_{L^p}^q,$$

where we have again used (iii) in the last equation. ∎

Appendix C

The result below holds in more general settings, but to keep matters simple we only state the version which is needed in Chapter VII, §7.

Theorem. *Let $D \subset\subset \mathbb{R}^n$ be open and suppose $K(x, y)$ is measurable on $D \times D$ (with respect to Lebesgue measure) and that there is $C < \infty$ such that*

$$\int_D |K(x, y)|\, dV(x) \leq C \qquad \text{for all } y \in D, \tag{1}$$

and

$$\int_D |K(x, y)|\, dV(y) \leq C \qquad \text{for all } x \in D. \tag{2}$$

Then the linear operator $\mathbf{K}: L^p(D) \to L^p(D)$ defined by

$$\mathbf{K}f(y) = \int_D K(x, y) f(x)\, dV(x)$$

is compact for all p with $1 \leq p < \infty$.

For the proof we will need the following

Lemma. *Let $\{K_\nu, \nu = 1, 2, \ldots\}$ be a sequence of measurable functions on $D \times D$ such that*

$$|K_\nu(x, y)| \leq M < \infty \qquad \text{for } x, y \in D \text{ and } \nu = 1, 2, \ldots, \tag{3}$$

and

$$\lim_{\nu \to \infty} K_\nu(x, y) = 0 \qquad \text{for } x, y \in D. \tag{4}$$

If $\|\mathbf{K}_\nu\|_p$ *denotes the operator norm of the corresponding operator* $\mathbf{K}_\nu$: $L^p(D) \to L^p(D)$, *then* $\lim_{\nu\to\infty} \|K_\nu\|_p = 0$ *for* $1 \le p < \infty$.

PROOF OF THE LEMMA. Fix $p < \infty$ and let q be the conjugate exponent. If $f \in L^p(D)$, Hölder's inequality applied to $K_\nu f = K_\nu^{1/q}(K_\nu^{1/p} f)$ implies

$$|\mathbf{K}_\nu f(y)| \le \left[\int_D |K_\nu(x, y)|\, dV(x)\right]^{1/q} \left[\int_D |K_\nu(x, y)|\,|f(x)|^p\, dV(x)\right]^{1/p}.$$

Hence

$$\|\mathbf{K}_\nu f\|_{L^p}^p \le I(K_\nu) \int_D \int_D |K_\nu|\,|f(x)|^p\, dV(x)\, dV(y), \tag{5}$$

where

$$I(K_\nu) = \int_D \left[\int_D |K_\nu|\, dV(x)\right]^{p/q} dV(y). \tag{6}$$

By integrating first in y, (3) and (5) imply

$$\|\mathbf{K}_\nu f\|_{L^p}^p \lesssim I(K_\nu)\, \|f\|_{L^p}^p. \tag{7}$$

Finally, (3) and (4), together with repeated applications of Lebesgue's Convergence Theorem, imply that $\lim_{\nu\to\infty} I(K_\nu) = 0$; (7) then implies the desired conclusion. ∎

We now come to the proof of the theorem. We first assume that $K(x, y)$ is *bounded* on $D \times D$, say be $C < \infty$. We then choose a sequence $\{K_\nu(x, y), \nu = 1, 2, \ldots\}$ of simple functions, each being a finite linear combination of characteristic functions of product sets in $D \times D$, such that $|K_\nu(x, y)| \le 2C$ for $\nu = 1, 2, \ldots$, and $K_\nu \to K$ pointwise almost everywhere on $D \times D$. This implies that the corresponding operator $\mathbf{K}_\nu$: $L^p(D) \to L^p(D)$ has finite dimensional range, and hence is compact for $\nu = 1, 2, \ldots$. By applying the Lemma to the sequence $\{K - K_\nu\}$, it follows that $\|\mathbf{K} - \mathbf{K}_\nu\|_p \to 0$ as $\nu \to \infty$ if $p < \infty$. Since the subspace of the Banach space of bounded operators on $L^p(D)$ consisting of all *compact* operators is closed in the operator norm, it follows that $\mathbf{K}$: $L^p(D) \to L^p(D)$ is compact.

For the general case, define

$$K^{(j)}(x, y) = \begin{cases} K(x, y) & \text{if } |K(x, y)| \le j \\ 0 & \text{otherwise} \end{cases}$$

for $j = 1, 2, \ldots$. Then $K^{(j)}$ is bounded, and hence $\mathbf{K}^{(j)}$ is compact on $L^p(D)$, by the first part of the proof. We now apply the arguments in the proof of the Lemma to $K - K^{(j)}$ instead of K_ν. By integrating first in y in (5) and by (2) it follows that

$$\|(\mathbf{K} - \mathbf{K}^{(j)}) f\|_{L^p}^p \lesssim \|f\|_{L^p}^p I(K - K^{(j)}),$$

where $I(K - K^{(j)})$ is defined in (6). If $g_j(y) = \int_D |K(x, y) - K^{(j)}(x, y)|\, dV(x)$, then $g_j \to 0$ pointwise and $|g_j(y)| \leq 2C$ for $y \in D$ by (1). By Lebesgue's Dominated Convergence Theorem we again obtain

$$I(K - K^{(j)}) = \int_D [g_j(y)]^{p/q} \, dV(y) \to 0$$

as $j \to 0$, i.e., we have proved that $\mathbf{K}^{(j)} \to \mathbf{K}$ in the operator norm on $L^p(D)$. Since $\mathbf{K}^{(j)}$ is compact, so is $\mathbf{K}$. ■

Bibliography

[AhSc] Ahern, P., and Schneider, R.:
Holomorphic Lipschitz functions in pseudoconvex domains. *Amer. J. of Math.* **101**(1978), 543–565.

[Ahl] Ahlfors, L.V.:
Complex Analysis, 3rd ed., McGraw-Hill Book Co., New York, 1979.

[AiYu] Aizenberg, L.A., and Yuzhakov, A.P.:
Integral Representations and Residues in Multidimensional Complex Analysis. Transl. Math. Monographs **58**, *Amer. Math. Soc.*, Providence, R.I., 1983.

[Ale] Aleksandrov, A.B.:
The existence of inner functions in the ball. *Mat. Sb.* **118**(1982), 147–163. Engl. transl.: *Math. USSR Sb.* **46**(1983), 143–159.

[AnFr] Andreotti, A., and Fraenkel, T.:
The Lefschetz theorem on hyperplane sections. *Ann. of Math.* **69**(1959), 713–717.

[AnVe] Andreotti, A., and Vesentini, E.:
Carleman estimates for the Laplace–Beltrami equations on complex manifolds. *Publ. Math. Inst. Hautes Études Scient.* **25**(1965), 81–130.

[BJT] Baouendi, M.S., Jacobowitz, H., and Treves, F.:
On the analyticity of CR mappings. *Ann. of Math.* **122** (1985), 365–400.

[Bar] Barrett, D.:
Irregularity of the Bergman projection on a smooth bounded domain in $\mathbb{C}^2$. *Ann. of Math.* **119**(1984), 431–436.

[BeRa] Beatrous, F., and Range, R.M.:
On holomorphic approximation in weakly pseudoconvex domains. *Pacific J. Math.* **89**(1980), 249–255.

[Bed] Bedford, E.:
Proper holomorphic mappings. *Bull. Amer. Math. Soc.*, (New Ser.) **10**(1984), 157–175.

[BeFo] Bedford, E., and Fornaess, J.E.:
A construction of peak functions on weakly pseudoconvex domains. *Ann. of Math.* **107**(1978), 555–568.

[Beh] Behnke, H.:
Généralisations du théorème de Runge pour des fonctions multiformes de variables complexes. Coll. fonct. plus. variables, Bruxelles 1953, 81–96.

[BeSt] Behnke, H., and Stein, K.:
Konvergente Folgen von Regularitätsbereichen und die Meromorphiekonvexität. *Math. Ann.* **116**(1983), 204–216.

[BeTh] Behnke, H., and Thullen, P.:
Theorie der Funktionen mehrerer komplexer Veränderlichen. 2nd ed., Springer-Verlag, Berlin–Heidelberg, 1970.

[Bel] Bell, S.R.:
1. Non-vanishing of the Bergman kernel function at boundary points of certain domains in $\mathbb{C}^n$. *Math. Ann.* **244**(1979), 69–74.
2. Biholomorphic mappings and the $\bar{\partial}$-problem. *Ann. of Math.* **114**(1981), 103–113.
3. Proper holomorphic mappings and the Bergman projection. *Duke Math. J.* **48**(1981), 167–175.

[BeBo] Bell, S.R., and Boas, H.P.:
Regularity of the Bergman Projection in weakly pseudoconvex domains. *Math. Ann.* **257**(1981), 23–30.

[BeLi] Bell, S.R., and Ligocka, E.:
A simplification and extension of Fefferman's Theorem on biholomorphic mappings. *Invent. Math.* **57**(1980), 283–289.

[Ber] Bergman, S.:
1. Über eine in gewissen Bereichen gültige Integraldarstellung der Funktionen zweier komplexer Variablen. *Math. Zeit.* **39**(1934), 76–94; 605–608.
2. The Kernel Function and Conformal Mapping. Math. Surveys **5**, *Amer. Math. Soc.*, Providence, R.I., 1950.

[BeAn] Berndtsson, B., and Anderson, M.:
Henkin–Ramirez formulas with weight factors. *Ann. Inst. Fourier* (Grenoble) **32**(1983), 91–110.

[Blo] Bloom, T.:
C^∞ peak functions for pseudoconvex domains of strict type. *Duke Math. J.* **45**(1978), 133–147.

[Boc] Bochner, S.:
Analytic and meromorphic continuation by means of Green's formula. *Ann. of Math.* **44**(1943), 652–673.

[Bre] Bremermann, H.J.:
1. Über die Äquivalenz der pseudokonvexen Gebiete und der Holomorphiegebiete im Raum von n komplexen Veränderlichen. *Math. Ann.* **128**(1954), 63–91.
2. Complex Convexity. *Trans. Amer. Math. Soc.* **82**(1956), 17–51.

[BrCa] Bruna, J., and del Castillo, J.:
Hölder and L^p-estimates for the $\bar{\partial}$-equation in some convex domains with real analytic boundary. *Math. Ann.* **269**(1984), 527–539.

[BuSh] Burns, D., and Shnider, S.:
Pseudoconformal geometry of hypersurfaces in $\mathbb{C}^{n+1}$. *Proc. Nat. Acad. Sci. U.S.A.* **72**(1975), 2433–2436.

[Car] Cartan, H.:
Collected Works. R. Remmert and J.-P. Serre, eds. Springer-Verlag, New York, 1979.

[Car 1] Cartan, H.:
Séminaire, École Norm. Sup., Paris, 1951/52.

[CaTh] Cartan, H., and Thullen, P.:
Zur Theorie der Singularitäten der Funktionen mehrerer komplexen Veränderlichen. *Math. Ann.* **106**(1932), 617–647.

[Cat] Catlin, D.:
Boundary behavior of holomorphic functions on weakly pseudoconvex domains. Dissertation, Princeton Univ., 1978.

[ChCh] Chaumat, J., and Chollet, A.M.:
Ensembles de zéros et d'interpolation à la frontière de domaines strictement pseudoconvexes. Arkiv för matematik 1986 (in press).

[Che] Chern, S.S.:
Complex Manifolds without Potential Theory. 2nd ed., Springer-Verlag, New York, 1979.

[ChMo] Chern, S.S., and Moser, J.K.:
Real hypersurfaces in complex manifolds. *Acta Math.* **133**(1974), 219–271.

[CoRa] Cole, B., and Range, R.M.:
A-measures on complex manifolds and some applications. *J. Funct. Anal.* **11**(1972), 393–400.

[Der] Derridj, M.:
Le principe de reflexion en des points de faible pseudo-convexité, pour des applications holomorphes propres. *Invent. Math.* **79**(1985). 197–215.

[Die] Diederich, K.:
Das Randverhalten der Bergmanschen Kernfunktion und Metrik in streng pseudokonvexen Gebieten. *Math. Ann.* **187**(1970), 9–36.

[DiFo] Diederich, K., and Fornaess, J.E.:
1. Pseudoconvex domains: Bounded strictly plurisubharmonic exhaustion functions. *Invent. Math.* **39**(1977), 129–141.
2. Pseudoconvex domains: An example with nontrivial Nebenhülle. *Math. Ann.* **225**(1977), 275–292.
3. Pseudoconvex domains with real analytic boundary. *Ann. of Math.* **107**(1978), 371–384.
4. Proper holomorphic maps onto pseudoconvex domains with real analytic boundary. *Ann. of Math.* **110**(1979), 575–592.

[DFH] Diederich, K., Fornaess, J.E., and Herbort, G.:
Boundary behavior of the Bergman metric. *Proc. Symp. Pure Math.* **41**, 59–67; *Amer. Math. Soc.*, Providence, R.I., 1984.

[DoGr] Docquier, F., and Grauert, H.:
Levisches Problem und Rungescher Satz für Teilgebiete Steinscher Mannigfaltigkeiten. *Math. Ann.* **140**(1960), 94–123.

[Dol] Dolbeault, P.:
Formes différentielles et cohomologie sur une variété analytique complexe. I: *Ann. of Math.* **64**(1956), 83–130; II: *Ann. of Math.* **65**(1957), 282–330.

[Ehr] Ehrenpreis, L.:
A new proof and an extension of Hartogs' theorem. *Bull. Amer. Math. Soc.* **67**(1961), 507–509.

[Fab] Faber, G.:
Über die zusammengehörigen Konvergenzradien von Potenzreihen mehrerer Veränderlicher. *Math. Ann.* **61**(1905). 289–324.

[Fef] Fefferman, C.:
The Bergman kernel and biholomorphic mappings of pseudoconvex domains. *Invent. Math.* **26**(1974), 1–65.

[FoKo] Folland, G.B., and Kohn, J.J.:
The Neumann Problem for the Cauchy–Riemann Complex. *Ann. of Math. Studies* **75**, Princeton Univ. Press, Princeton, N.J., 1972.

[FoSt] Folland, G.B., and Stein, E.M.:
Estimates for the $\bar{\partial}_b$-complex and analysis on the Heisenberg group. *Comm. Pure Appl. Math.* **27**(1974), 429–522.

[For] Fornaess, J.E.:
1. An increasing sequence of Stein manifolds whose limit is not Stein. *Math. Ann.* **223**(1976), 275–277.
2. Peak points on weakly pseudoconvex domains. *Math. Ann.* **227**(1977), 173–175.
3. Sup-norm estimates for $\bar{\partial}$ in $\mathbb{C}^2$. *Ann. of Math.* **123**(1986) (in press).

[FoHe] Fornaess, J.E., and Henriksen. B.S.:
Peak sets for $A^k(D)$. *Proc. Symp. Pure Math.* **41**, 69–75. *Amer. Math. Soc.*, Providence, R.I., 1984.

[FoØv] Fornaess, J.E., and Øvrelid, N.:
Finitely generated ideals in $A(\Omega)$. *Ann. Inst. Fourier* (Grenoble) **33**(1983), 77–85.

[Fri] Fridman, B.L.:
One example of the boundary behavior of biholomorphic transformations. *Proc. Amer. Math. Soc.* **89**(1983), 226–228.

[Gam] Gamelin, T.W.:
1. Uniform Algebras. Prentice-Hall, Englewood Cliffs, N.J., 1969.
2. Polynomial approximation on thin sets. Symposium on Sev. Compl. Var., Park City, 1970, *Springer Lect. Notes* **184**(1971), 50–78.

[God] Godement, R.:
Théorie des Faisceaux. Hermann, Paris, 1958.

[Gra] Grauert, H.:
On Levi's problem and the embedding of real-analytic manifolds. *Ann. of Math.* **68**(1958), 460–472.

[GrFr] Grauert, H., and Fritzsche, K.:
Several Complex Variables. Springer-Verlag, New York, 1976.

[GrLi] Grauert, H., and Lieb. I.:
Das Ramirezsche Integral und die Lösung der Gleichung $\bar{\partial}f = \alpha$ im Bereich der beschränkten Formen. Proc. Conf. Complex Analysis, Rice University, 1969. *Rice Univ. Studies* **56**(1970), 29–50.

[GrRe] Grauert, H., and Remmert, R.:
1. Theorie der Steinschen Räume. Springer-Verlag, Heidelberg, 1977. Theory of Stein Spaces. Transl. by A. Huckleberry, Springer-Verlag, New York, 1979.
2. Coherent Analytic Sheaves. Springer-Verlag, New York, 1984.

[GrKr] Greene, R.E., and Krantz, S.G.:
Stability of the Carathéodory and Kobayashi metrics and applications to biholomorphic mappings. *Proc. Symp. Pure Math.* **41**, 77–93, *Amer. Math. Soc.*, Providence, R.I., 1984.

[GrSt] Greiner, P.C., and Stein, E.M.:
Estimates for the $\bar{\partial}$-Neumann Problem. Princeton Univ. Press, Princeton, N.J., 1977.

[Gun] Gunning, R.:
Lectures on Complex Analytic Varieties: The Local Parametrization Theorem. Princeton Univ. Press, Princeton, N.J., 1970.

[GuRo] Gunning, R., and Rossi, H.:
Analytic Functions of Several Complex Variables. Prentice-Hall, Englewood Cliffs, N.J., 1965.

[Har] Hartogs, F.:
1. Einige Folgerungen aus der Cauchyschen Integralformel bei Funktionen mehrerer Veränderlichen. *Münch. Ber.* **36**(1906), 223–242.
2. Zur Theorie der analytischen Funktionen mehrerer unabhängiger Ve änderlichen, insbesondere über die Darstellung derselben durch Reihen, welche nach Potenzen einer Veränderlichen fortschreiten. *Math. Ann.* **62**(1906), 1–88.

[HaLa] Harvey, R., and Lawson, B.:
On boundaries of complex analytic varieties. I. *Ann. of Math.* **102**(1975), 233–290.

[HaPo] Harvey, R., and Polking, J.:
Fundamental solutions in complex analysis. I: The Cauchy–Riemann operator. *Duke Math. J.* **46**(1979), 253–300.

[Hef] Hefer, H.:
Zur Funktionentheorie mehrerer Veränderlichen. Über eine Zerlegung analytischer Funktionen und die Weilsche Integraldarstellung. *Math. Ann.* **122**(1950), 276–278.

[Hen] Henkin, G.M.:
1. Integral representations of functions holomorphic in strictly pseudoconvex domains and some applications. *Mat. Sb.* **78**(1969), 611–632; Engl. Transl.: *Math. USSR. Sb.* **7**(1969), 597–616.
2. Integral representations of functions in strictly pseudoconvex domains and applications to the $\bar{\partial}$-problem. *Math. Sb.* **82**(1970), 300–308; Engl. Transl.: *Math. USSR Sb.* **11**(1970), 273–281.
3. Approximation of functions in pseudoconvex domains and the theorem of Z.L. Leibenson. *Bull. Acad. Polon. Sci. Ser. Sci. Math. Astr. Phys.* **19**(1971), 37–42.
4. An analytic polyhedron is not holomorphically equivalent to a strictly pseudoconvex domain. *Dokl. Akad. Nauk SSSR* **210**(1973), 1026–1029; Engl. transl.: *Soviet Math. Dokl.* **14**(1973), 858–862.
5. Solutions with bounds for the equations of H. Lewy and Poincaré-Lelong. Construction of functions of Nevanlinna class with given zeroes

in a strongly pseudoconvex domain. *Dokl. Akad. Nauk SSSR* **224**(1975), 771–774; Engl. transl.: *Soviet Math. Dokl.* **16**(1975), 1310–1314.

[HeLe] Henkin, G.M., and Leiterer, J.:
Theory of Functions on Complex Manifolds. Birkhäuser, Boston, Mass., 1984.

[HeRo] Henkin, G.M., and Romanov, A.V.:
Exact Hölder estimates for the solutions of the $\bar{\partial}$-equation. *Izv. Akad. Nauk SSSR* **35**(1971), 1171–1183; Engl. Transl.: *Math. USSR Izvestija* **5**(1971), 1180–1192.

[Hor] Hortmann, M.:
Über die Lösbarkeit der $\bar{\partial}$-Gleichung mit Hilfe von L^p, C^k und D'-stetigen Integraloperatoren. *Math. Ann.* **223**(1976), 139–156.

[Hör] Hörmander, L.:
1. L^2 estimates and existence theorems for the $\bar{\partial}$-operator. *Acta Math.* **113**(1965), 89–152.
2. An Introduction to Complex Analysis in Several Variables. 2nd ed., North Holland Publishing Co., Amsterdam, 1973.

[Hua] Hua, L.K.:
Harmonic Analysis of Functions of Several Complex Variables in the Classical Domains. Science Press, Peking, 1958; *Amer. Math. Soc. Transl., Math. Monograph* **6**(1963).

[KaKa] Kaup, L., and Kaup, B.:
Holomorphic Functions of Several Variables. Walter de Gruyter, Berlin-New York, 1983.

[Ker] Kerzman, N.:
1. Hölder and L^p estimates for solutions of $\bar{\partial}u = f$ in strongly pseudoconvex domains. *Comm. Pure Appl. Math.* **24**(1971), 301–379.
2. The Bergman kernel function. Differentiability at the boundary. *Math. Ann.* **195**(1972), 149–158.

[KeNa] Kerzman, N., and Nagel, A.:
Finitely generated ideals in certain function algebras. *J. Funct. Anal.* **7**(1971), 212–215.

[KeRo] Kerzman, N., and Rosay, J.P.:
Fonctions plurisousharmoniques d'exhaustion bornées et domaines taut. *Math. Ann.* **257**(1981), 171–184.

[KeSt] Kerzman, N., and Stein, E.M.:
The Szegö kernel in terms of Cauchy–Fantappiè kernels. *Duke Math. J.* **45**(1978), 197–224.

[Kob] Kobayashi, S.:
Hyperbolic Manifolds and Holomorphic Mappings. M. Dekker, New York, 1970.

[Koh] Kohn, J.J.:
1. Harmonic integrals on strongly pseudoconvex manifolds. I: *Ann. of Math.* **78**(1963), 112–148; II: *Ann. of Math.* **79**(1964), 450–472.
2. Global regularity for $\bar{\partial}$ on weakly pseudoconvex manifolds. *Trans. Amer. Math. Soc.* **181**(1973), 273–292.
3. Subellipticity of the $\bar{\partial}$-Neumann problem on pseudoconvex domains: sufficient conditions. *Acta Math.* **142**(1979), 79–122.

4. A survey of the $\bar{\partial}$-Neumann Problem. *Proc. Symp. Pure Math.* **41**, 137–145, *Amer. Math. Soc.*, Providence, R. I., 1984.

[KoNi] Kohn, J.J., and Nirenberg, L.:
A pseudoconvex domain not admitting a holomorphic support function. *Math. Ann.* **201**(1973), 265–268.

[Kop] Koppelman, W.:
1. The Cauchy integral for functions of several complex variables. *Bull. Amer. Math. Soc.* **73**(1967), 373–377.
2. The Cauchy integral for differential forms. *Bull. Amer. Math. Soc.* **73**(1967), 554–556.

[Kra] Krantz, S.G.:
1. Optimal Lipschitz and L^p regularity for the equation $\bar{\partial}u = f$ on strongly pseudoconvex domains. *Math. Ann.* **219**(1976), 233–260.
2. Function Theory of Several Complex Variables. John Wiley & Sons, New York, 1982.

[Lel] Lelong, P.:
1. Les fonctions plurisousharmoniques. *Ann. Éc. Norm. Sup.* **62**(1945), 301–338.
2. La convexité et les fonctions analytiques de plusieurs variables complexes. *J. Math. Pures Appl.* **31**(1952), 191–219.
3. Domaines convexes par rapport aux fonctions plurisousharmoniques. *Jour. d'Analyse* **2**(1952), 178–208.

[Ler] Leray, J.:
1. Fonction de variable complexes: sa répresentation comme somme de puissances négatives de fonctions linéaires. *Rend. Accad. Naz. Lincei, ser. 8*, **20**(1956), 589–590.
2. Le calcul différentiel et intégral sur une variété analytique complexe: Problème de Cauchy III. *Bull. Soc. Math. France* **87**(1959), 81–180.

[Lev] Levi, E.E.:
1. Studii sui punti singolari essenziali delle funzioni analitiche di due o più variabili complesse. *Ann. Mat. pura appl.* **17**(1910), 61–87.
2. Sulle ipersuperficie dello spazio a 4 dimensioni che possono essere frontiera del campo di esistenza di una funzione analitica di due variabili complesse. *Ann. Mat. pura appl.* **18**(1911), 69–79.

[Lew] Lewy, H.:
On the boundary behavior of holomorphic mappings. *Acad. Naz. Lincei* **35**(1977), 1–8.

[Lie] Lieb, I.:
1. Ein Approximationssatz auf streng pseudokonvexen Gebieten. *Math. Ann.* **184**(1969), 56–60.
2. Die Cauchy–Riemannschen Differentialgleichungen auf streng pseudokonvexen Gebieten. *Math. Ann.* **190**(1970), 6–44.

[LiRa] Lieb, I., and Range, R.M.:
1. Lösungsoperatoren für den Cauchy–Riemann-Komplex mit C^k-Abschätzungen. *Math. Ann.* **253**(1980), 145–164.
2. On integral representations and a priori Lipschitz estimates for the canonical solution of the $\bar{\partial}$-equation. *Math. Ann.* **265**(1983), 221–251.
3. Integral representations and estimates in the theory of the $\bar{\partial}$-Neumann problem. *Ann. of Math.* **123**(1986), 265–301.

[Lig] Ligocka, E.:
1. Some remarks on extension of biholomorphic mappings. Proc. 7th Conf. on Analytic Functions, Kozubnik, Poland, 1979. *Springer Lect. Notes in Math.* **798**(1980) 350–363.
2. The Hölder continuity of the Bergman projection and proper holomorphic mappings. *Studia math.* **80**(1984), 89–107.

[Løw] Løw, E.:
A construction of inner functions on the unit ball of $\mathbb{C}^p$. *Invent. Math.* **67**(1982), 294–298.

[Mal] Malgrange, B.:
Lectures on the Theory of Functions of Several Complex Variables. Tata Institute, Bombay 1958.

[Mar] Martinelli, E.:
1. Alcuni teoremi integrali per le funzioni analitiche di più variabili complesse. *Mem. della R. Accad. d'Italia* **9**(1938), 269–283.
2. Sopra una dimostrazione di R. Fueter per un teorema di Hartogs. *Comm. Math. Helv.* **15**(1942/43), 340–349.

[Mil] Milnor, J.:
Morse Theory. *Ann. Math. Studies* **51**, Princeton Univ. Press, Princeton, N.J., 1963.

[MoKo] Morrow, J., and Kodaira, K.:
Complex Manifolds. Holt, Rinehart and Winston, Inc., New York, 1971.

[Nar] Narasimhan, R.:
1. The Levi problem for complex spaces II. *Math Ann.* **146**(1962), 195–216.
2. Introduction to the Theory of Analytic Spaces, *Lect. Notes* **25**, Springer-Verlag, New York, 1966.
3. Several Complex Variables. Univ. of Chicago Press, Chicago, 1971.
4. Analysis on Real and Complex Manifolds. 2nd ed., Masson & Cie, Paris, 1973.

[Nor] Norguet, F.:
1. Sur les domains d'holomorphie des fonctions uniformes de plusieurs variable complexes (passage du local au global). *Bull. Soc. Math. France* **82**(1954), 137–159.
2. Problèmes sur les formes differentielles et les courants. *Ann. Inst. Fourier* **11**(1960), 1–88.

[Oka] Oka, K.:
Collected Papers. Transl. by R. Narasimhan, with comments by H. Cartan, R. Remmert, ed. Springer-Verlag, New York, 1984.

[Osg] Osgood, W. F.:
Lehrbuch der Funktiontheorie, Vol. II, 2nd ed., Teubner Verlag, Leipzig 1929. Reprinted by Chelsea Publ. Co., New York, 1965.

[Øvr] Øvrelid, N.:
1. Integral representation formulas and L^p-estimates for the $\bar{\partial}$-equation. *Math. Scand.* **29**(1971), 137–160.
2. Generators of the maximal ideals of $A(D)$. *Pacific J. Math.* **39**(1971), 219–223.

[PhSt] Phong, D.H., and Stein, E.M.:
Estimates for the Bergman and Szegö projections on strongly pseudoconvex domains. *Duke Math. J.* **44**(1977), 695–704.

[Pin] Pinčuk, S.I.:
1. On proper holomorphic mappings of strictly pseudoconvex domains. *Sib. Math. J.* **15**(1975), 644–649.
2. On the analytic continuation of holomorphic mappings. *Mat. Sbornik* **98**(1975), 416–435; Engl. Transl.: *Math. USSR Sbornik* **27**(1975), 375–392.

[Pol] Poljakov, P.L.:
1. The Cauchy–Weil formula for differential forms. *Mat. Sb.* **85**(1971), 388–402; Engl. Transl.: *Math. USSR Sb.* **14**(1971), 383–398.
2. Banach cohomology on piecewise strictly pseudoconvex domains. *Mat. Sb.* **88**(1972), 238–255; Engl. Transl.: *Math. USSR Sb.* **17**(1972), 237–256.

[Ram] Ramirez, E.:
Ein Divisionsproblem und Randintegraldarstellungen in der komplexen Analysis. *Math. Ann.* **184**(1970), 172–187.

[Ran] Range, R.M.:
1. Approximation to bounded holomorphic functions on strictly pseudoconvex domains. *Pacific J. Math.* **41**(1972), 203–212.
2. On the topological extension to the boundary of biholomorphic maps in $\mathbb{C}^n$. *Trans. Amer. Math. Soc.* **216**(1976), 203–216.
3. The Carathéodory metric and holomorphic maps on a class of weakly pseudoconvex domains. *Pacific J. Math.* **78**(1978), 173–188.
4. On Hölder estimates for $\bar{\partial}u = f$ on weakly pseudoconvex domains. Proc. Int. Conf. Cortona 1976–1977. Scuola Norm. Sup. Pisa 1978, 247–267.
5. A remark on bounded strictly plurisubharmonic exhaustion functions. *Proc. Amer. Math. Soc.* **81**(1981), 220–222.
6. An elementary integral solution operator for the Cauchy–Riemann equations on pseudoconvex domains in $\mathbb{C}^n$. *Trans. Amer. Math. Soc.* **274**(1982), 809–816.

[RaSi] Range, R.M., and Siu, Y.T.:
Uniform estimates for the $\bar{\partial}$-equation on domains with piecewise smooth strictly pseudoconvex boundaries. *Math. Ann.* **206**(1973), 325–354.

[Rei] Reinhardt, K.:
Über Abbildungen durch analytische Funktionen zweier Veränderlichen. *Math. Ann.* **83**(1921), 211–255.

[ReSt] Remmert, R., and Stein, K.:
1. Über die wesentlichen Singularitäten analytischer Mengen. *Math. Ann.* **126**(1953), 263–306.
2. Eigentliche holomorphe Abbildungen. *Math. Z.* **73**(1960), 159–189.

[Rha] de Rham, G.:
Variétés Différentiables. Hermann, Paris, 1960.

[Ros] Rosay, J.P.:
Injective holomorphic mappings. *Amer. Math. Monthly* **89**(1982), 587–588.

[Rss] Rossi, H.:
The local maximum modulus principle. *Ann. of Math.* **72**(1960), 1–11.

[Rud] Rudin, W.:
1. Real and Complex Analysis. McGraw-Hill Book Co., New York, 1966.
2. Functional Analysis. McGraw-Hill Book Co., New York, 1973.

3. Function Theory in the Unit Ball of $\mathbb{C}^n$. Springer-Verlag, New York, 1980.
4. New Constructions of Functions Holomorphic in the Unit Ball of $\mathbb{C}^n$. CBMS Regional Conference Series, Amer. Math. Soc., Providence, R.I., 1986. (To appear.)

[Ser] Serre, J.P.:
1. Quelques problèmes globaux relatifs aux variétés de Stein. Coll. Plus. Var., Bruxelles, 1953, 57–68.
2. Une propriété topologique des domaines de Runge. *Proc. Amer. Math. Soc.* **6**(1955), 133–134.

[Sib] Sibony, N.:
Un exemple de domain pseudoconvexe regulier ou l'équation $\bar{\partial}u = f$ n'admet pas de solution bornée pour f bornée. *Invent. Math.* **62**(1980), 235–242.

[Siu] Siu, Y.T.:
The $\bar{\partial}$-problem with uniform bounds on derivatives. *Math. Ann.* **207**(1974), 163–176.

[Sko] Skoda, H.:
Valeurs au bord pour les solutions de l'opérateur d″, et caractérisation des zéros des fonctions de la classe de Nevanlinna. *Bull. Soc. Math. France* **104**(1976), 225–299.

[Som] Sommer, F.:
Über Integralformeln in der Funktionentheorie mehrerer komplexer Veränderlichen. *Math. Ann.* **125**(1952), 172–182.

[Spa] Spanier, E.H.:
Algebraic Topology. McGraw-Hill, New York, 1966.

[SteE] Stein, E.M.:
1. Boundary values of holomorphic functions. *Bull. Amer. Math. Soc.* **76**(1970), 1292–1296.
2. Boundary Behavior of Holomorphic Functions of Several Complex Variables. Princeton Univ. Press, Princeton, N.J., 1972.
3. Singular Integrals and Differentiability Properties of Functions. Princeton Univ. Press, Princeton, N.J., 1970.

[SteK] Stein, K.:
1. Topologische Bedingungen für die Existenz analytischer Funktionen komplexer Veränderlichen zu vorgegebenen Nullstellenflächen. *Math. Ann.* **117**(1941), 727–757.
2. Analytische Funktionen mehrerer komplexer Veränderlichen zu vorgegebenen Periodizitätsmoduln und das zweite Cousinsche Problem. *Math. Ann.* **123**(1951), 201–222.
3. Überlagerungen holomorph-vollständiger komplexer Räume. *Archiv Math.* **7**(1956), 354–361.

[Sto] Stout, E.L.:
H^p functions on strictly pseudoconvex domains. *Amer. J. Math.* **98**(1976), 821–852.

[Tan] Tanaka, N.:
On generalized graded Lie algebras and geometric structures. *J. Math. Soc. Japan* **19**(1967), 215–254.

[Var] Varopoulos, N.Th.:
Zeroes of H^p functions in several complex variables. *Pacific J. Math.* **88**(1980), 189–246.

[Vor] Vormoor, N.:
Topologische Fortsetzung biholomorpher Funktionen auf dem Rande bei beschränkten streng pseudokonvexen Gebieten im $\mathbb{C}^n$ mit C^∞-Rand. *Math. Ann.* **204**(1973), 239–261.

[War] Warner, F.W.:
Foundations of Differentiable Manifolds and Lie Groups. Springer-Verlag, New York, 1983.

[Web] Webster, S.:
Biholomorphic mappings and the Bergman kernel off the diagonal. *Invent. Math.* **51**(1979), 155–169.

[Wei] Weil, A.:
L'intégral de Cauchy et les functions de plusieurs variables. *Math. Ann.* **111**(1935), 178–182.

[Wel] Wells, R. O.:
Differential Analysis on Complex Manifolds. 2nd ed., Springer-Verlag, New York, 1980.

[Wer] Wermer, J.:
1. The hull of a curve in $\mathbb{C}^n$. *Ann. of Math.* **68**(1958), 550–561.
2. An example concerning polynomial convexity. *Math. Ann.* **139**(1959), 147–150. Addendum: *Math. Ann.* **140**(1960), 322–323.

[Wu] Wu, H.:
Function Theory on Noncompact Kähler Manifolds. In: Complex Differential Geometry, DMV-Seminar 3. Birkhäuser, Basel, 1983.

Glossary of Symbols and Notations

General

(a, b)	standard Hermitian product of $a, b \in \mathbb{C}^n$, 2
$\|a\|$	Euclidean norm of $a \in \mathbb{C}^n$, 2
$\langle v_P, w_P \rangle_P$	inner product between tangent vectors at P, 131
$\langle \varphi, v \rangle_P$	action of 1-form φ on tangent vector v (also $\varphi(v)$), 168
$\langle \varphi, \psi \rangle_P$	inner product between forms at P, 132, 133
$(\varphi, \psi)_M$	integral inner product over M of forms φ and ψ, 134
$\\|\varphi\\|_M$	$= (\varphi, \varphi)_M^{1/2}$, 134
$A \lesssim B$	$A \leq cB$ for some constant c, 157
ε_B^A	sign of permutation, 135
$\tau(D)$	image of $D \subset \mathbb{C}^n$ in absolute space, 3
$D^\alpha, D^{\bar\beta}, D^{\alpha\bar\beta}$	partial differentiation operators, 5
$J_{\mathbb{R}}(F)(a)$	real Jacobian matrix of the map F, 19
$F'(a)$	complex Jacobian matrix, or derivative, of the holomorphic map F, 19
$dF_a, dF(a)$	differential of the map F at a, 19, 107
$\Omega \subset\subset D$	Ω is relatively compact in D, 2
bA	topological boundary of A, 2
$\operatorname{dist}(A, B)$	Euclidean distance between sets A and B, 2
$\delta_D(z)$	Euclidean distance from $z \in D$ to bD, 2
$\delta_D^{(r)}(z)$	distance from z to bD with respect to polydisc $P(0, r)$, 74
$\delta_{D,u}(z)$	distance from z to bD in u-direction, 94, 95
σ_γ	reflection on the real analytic curve γ, 340

$(R), (R_k)$	regularity conditions for the Bergman projection, 325, 326
(B_k)	regularity condition for the Bergman projection, 331

Special Sets

$B(a, r)$	ball of radius r and center a, 2
$P(a, r)$	polydisc of multiradius $r = (r_1, \ldots, r_n)$, 3
$b_0 P$	distinguished boundary of the polydisc P, 8
$H(r)$	Hartogs domain, 3
D_c	$\{z: r(z) < c\}$, where r is the given defining function for D, 223, 274
$Z(f, U)$	zero set of f in U, 28
Δ_{bD}	diagonal in $bD \times bD$, 295
$S, \partial S$	analytic disc and its boundary, 93
$(\Gamma, \hat{\Gamma})$	standard Hartogs frame, 49
$(\Gamma^*, \hat{\Gamma}^*)$	general Hartogs figure, 49
$\hat{K}^c, \hat{K}_{\mathscr{L}(\mathbb{R}^n)}$	linearly convex hull of K, 67
$\hat{K}_{\mathcal{O}(D)}$	holomorphically convex hull of K in D, 68
$\hat{L}_{\mathcal{O}(K)}$	holomorphically convex hull of L in K, 77
$\hat{K}_{\mathscr{M}}$	hull of K with respect to monomials, 79
$\hat{K}_{\mathscr{P}}$	polynomially convex hull of K, 218
$\hat{K}_{PS(D)}$	plurisubharmonic hull of K in D, 93
$H_d^r(M)$	de Rham cohomology group of M, 129
$H_{\bar\partial}^q, H_{\bar\partial}^{p,q}, H_{\bar\partial,c}^{p,q}$	$\bar\partial$-cohomology groups, 129, 130
$\Gamma(Y, \mathscr{S})$	sections of the sheaf $\mathscr{L}$ over Y, 253
$H^q(\mathscr{V}, \mathscr{S})$	Čech cohomology group, 255
$H^q(X, \mathscr{S})$	sheaf cohomology group of X, 255

Spaces of Functions, etc., and Norms

$C(D), C^k(D)$	continuous, and k times continuously differentiable functions on D, 4
$C_0(D)$	continuous functions with compact support in D
$C^k(\bar{D})$	functions in $C^k(D)$ with k-th order derivatives in $C(\bar{D})$, 4
$\lvert f\rvert_{k,D}, \lvert f\rvert_k, \lvert f\rvert_D$	C^k norms over D, 5
$\Lambda_\alpha(D)$	Lipschitz functions of order α, 156
$\lvert f\rvert_{\alpha,D}$	Lip-α norm over D, 156
$C^{k,\infty}(M \times N)$	C^k functions on $M \times N$ which are C^∞ in $y \in N$, 172
$\mathcal{O}(D)$	holomorphic functions on D, 5
$\mathscr{M}(D)$	meromorphic functions on D, 233
$\mathcal{O}(K)$	holomorphic functions on neighborhoods of the compact set K, 5
$\mathcal{O}_a, {}_n\mathcal{O}_a$	germs of holomorphic functions at a $(\in \mathbb{C}^n)$, 231

$\mathcal{M}_a$, ${}_n\mathcal{M}_a$	germs of meromorphic functions, 232
$\mathcal{O}^*(U)$, $\mathcal{M}^*(U)$, $\mathcal{C}^*(U)$	invertible holomorphic, meromorphic, and continuous functions on U, 237
$A(P)$	polydisc algebra, 10
$A(D)$	$C(\bar{D}) \cap \mathcal{O}(D)$, 17
$A^k(D)$	$C^k(\bar{D}) \cap \mathcal{O}(D)$, 165
$\mathcal{O}L^p(D)$	holomorphic L^p functions on D, 9
$\mathcal{H}^2(D)$	holomorphic L^2 functions on D (also $\mathcal{O}L^2(D)$), 9, 179
$PS(D)$	plurisubharmonic functions on D, 88
$\mathcal{P}$	holomorphic polynomials on $\mathbb{C}^n$, 218
$\mathcal{P}(K)$	uniform closure of $\mathcal{P}$ in $C(K)$, 218
$\mathcal{D}(D)$	divisors on D: in $\mathbb{C}^1$, 246; in $\mathbb{C}^n$, 247
$T_P(bD)$	tangent space to $bD \subset \mathbb{C}^n$ at P, 52
$T_P^{\mathbb{C}}(bD)$	complex tangent space to bD, 53
$T_P(M)$	tangent space of M at P, 107
$\mathbb{C}T_PM$	complex valued tangent vectors, 123
$T_P^{1,0}M$, $T_P^{0,1}M$	tangent vectors of type (1, 0) and (0, 1), 126
$T_P^{1,0}(bD)$, $T_P^{0,1}(bD)$	tangent vectors to bD at P of type (1, 0) and (0, 1), 164
T_P^*M	1-forms at P, 108
$\mathbb{C}T_P^*M$	complex valued 1-forms at P, 123
$\Lambda^r T_P^*M$	r-forms at P, 109
$C_r^l(M)$	r-forms on M of class C^l, 110, 123
$\Omega^r(D)$	holomorphic r-forms on D, 228
$\mathcal{G}_P(M)$	Grassman algebra of forms at P, 109
$\mathcal{G}^l(M)$	Grassman algebra of forms of class C^l on M, 123
$\Lambda^{p,q}(T_P^*M)$, $\Lambda_P^{p,q}(M)$	forms of type (p, q) at P, 126, 127
$C_{p,q}^k(M)$	forms of type (p, q) of class C^k, 127
$C_{p,q}^{k,\infty}(bD \times D)$	forms of type (p, q) with coefficients in $C^{k,\infty}(bD \times D)$, 172
$L_{p,q}^2(M)$	square integrable forms of type (p, q), 134
$L_{p,q}^s(D)$	(p, q)-forms with coefficients in $L^s(D)$, 134

Special Functions and Forms

$L_P(r; \cdot)$	Levi form of the function r at p, 56
$F^{(r)}(\zeta, \cdot)$	Levi polynomial of the function r at ζ, 60
$F^{\#}(\zeta, \cdot)$	modification of $F^{(r)}(\zeta, \cdot)$, 193
$\Phi(\zeta, z)$	smooth globalization of the Levi polynomial, 193
$\hat{\Phi}(\zeta, z)$	$= \Phi(\zeta, z) - r(\zeta)$, 295
$K_D(\zeta, z)$	Bergman kernel of D, 179
β	$= \|\zeta - z\|^2$, 146
$\Gamma \; (= \Gamma^{(n)})$	Newtonian solution kernel for $\square$ on functions in $\mathbb{C}^n$, 146

Γ_q	Newtonian solution kernel for $\square$ on $(0, q)$-forms, 150
$B = \partial\beta/\beta$	generating form for the Bochner–Martinelli–Koppelman kernel, 169
$C^{(r)}$	canonical generating form for a convex domain defined by r, 171
C_D	Cauchy kernel for the convex domain D, 172
L_D	$= \sum (P_j/\Phi)\, d\zeta_j$, generating form for the strictly pseudoconvex domain D, 193
W^{HR}	generating form for the Henkin–Ramirez kernel, 286
dz^J	$= dz_{j_1} \wedge \ldots \wedge dz_{j_r}$ for $J = (j_1, \ldots, j_r)$, 127
$d\bar{z}^J$	$= \overline{dz^J}$, 127
dV	volume form (on $\mathbb{C}^n$), 133
dS	surface element on the boundary of a domain, 134
K_q	Bochner–Martinelli–Koppelman kernel for $(0, q)$-forms, 148 (for $q = 0$); 154 ($q \geq 0$)
$\hat{W}$	homotopy form associated to the generating form W, 173
$\Omega_q(W), \Omega_q(\hat{W})$	Cauchy–Fantappiè form of order q generated by W, resp. $\hat{W}$, 169 (for $q = 0$); 173 ($q \geq 0$)
$G_D(\zeta, z)$	principal part of the Bergman kernel of D, 309
$\mathscr{E}_j(\zeta, z)$	smooth function which is $O(\|\zeta - z\|^j)$, 316
$\mathscr{A}, \mathscr{A}_\lambda$	admissible kernels, 316, 317

Operators

d	exterior derivative, 111
$\partial f_a, \partial f(a)$	$\mathbb{C}$-linear, or (1, 0)-part, of the differential of f at a, 7, 127
$\bar{\partial}$	Cauchy–Riemann operator: on functions, 7, 127; on forms, 128
ϑ	formal adjoint of $\bar{\partial}$ in $\mathbb{C}^n$, 138
$\square$	complex Laplacian, 140
Δ	Laplace operator, 82, 139 (sometimes also the open unit disc in $\mathbb{C}$)
$d_{\zeta,\lambda}$	exterior derivative on $\mathbb{C}^n \times [0, 1]$, 173
$\bar{\partial}_{\zeta,\lambda}$	$\bar{\partial}_\zeta + d_\lambda$ on $\mathbb{C}^n \times [0, 1]$, 173
F^*	pullback on differential forms under the map F, 112
$*$	Hodge operator defined by Riemannian structure, 135
$\mathbf{K}^M$	Bochner–Martinelli transform on the hypersurface M, 160
$\mathbf{P}_D$	Bergman projection on D, 182

$\mathbf{G}_D$	explicit principal part of $\mathbf{P}_D$, 309, 313, 314
$\mathbf{T}_q^W$	parametrix for $\bar\partial$ associated to the generating form W, 174
$\mathbf{E}_q$	extension operator for $\bar\partial$-cohomology, 194
$\mathbf{T}_q^{V,V_0}$	integral solution operator for $\bar\partial$ in neighborhoods of a Stein compactum, 200
$\mathbf{S}_q^{(D)}$	integral solution operator for $\bar\partial$ on the strictly pseudoconvex domain D, 201
$\mathbf{S}_q^H$, $\mathbf{S}_q^{\perp}$	solution operator for $\bar\partial$ giving the solution orthogonal to ker $\bar\partial$, 302, 351

Index

Graduate Texts in Mathematics

continued from page ii

48 Sachs/Wu. General Relativity for Mathematicians.
49 Gruenberg/Weir. Linear Geometry. 2nd ed.
50 Edwards. Fermat's Last Theorem.
51 Klingenberg. A Course in Differential Geometry.
52 Hartshorne. Algebraic Geometry.
53 Manin. A Course in Mathematical Logic.
54 Graver/Watkins. Combinatorics with Emphasis on the Theory of Graphs.
55 Brown/Pearcy. Introduction to Operator Theory I: Elements of Functional Analysis.
56 Massey. Algebraic Topology: An Introduction.
57 Crowell/Fox. Introduction to Knot Theory.
58 Koblitz. p-adic Numbers, p-adic Analysis, and Zeta-Functions. 2nd ed.
59 Lang. Cyclotomic Fields.
60 Arnold. Mathematical Methods in Classical Mechanics.
61 Whitehead. Elements of Homotopy Theory.
62 Kargapolov/Merzljakov. Fundamentals of the Theory of Groups.
63 Bollabás. Graph Theory.
64 Edwards. Fourier Series. Vol. I. 2nd ed.
65 Wells. Differential Analysis on Complex Manifolds. 2nd ed.
66 Waterhouse. Introduction to Affine Group Schemes.
67 Serre. Local Fields.
68 Weidmann. Linear Operators in Hilbert Spaces.
69 Lang. Cyclotomic Fields II.
70 Massey. Singular Homology Theory.
71 Farkas/Kra. Riemann Surfaces.
72 Stillwell. Classical Topology and Combinatorial Group Theory.
73 Hungerford. Algebra.
74 Davenport. Multiplicative Number Theory. 2nd ed.
75 Hochschild. Basic Theory of Algebraic Groups and Lie Algebras.
76 Iitaka. Algebraic Geometry.
77 Hecke. Lectures on the Theory of Algebraic Numbers.
78 Burris/Sankappanavar. A Course in Universal Algebra.
79 Walters. An Introduction to Ergodic Theory.
80 Robinson. A Course in the Theory of Groups.
81 Forster. Lectures on Riemann Surfaces.
82 Bott/Tu. Differential Forms in Algebraic Topology.
83 Washington. Introduction to Cyclotomic Fields.
84 Ireland/Rosen. A Classical Introduction to Modern Number Theory.
85 Edwards. Fourier Series: Vol. II. 2nd ed.
86 Van Lint. Introduction to Coding Theory.
87 Brown. Cohomology of Groups.
88 Pierce. Associative Algebras.
89 Lang. Introduction to Algebraic and Abelian Functions. 2nd ed.
90 Brøndsted. An Introduction to Convex Polytopes.
91 Beardon. On the Geometry of Discrete Groups.
92 Diestel. Sequences and Series in Banach Spaces.

93 DUBROVIN/FOMENKO/NOVIKOV. Modern Geometry — Methods and Applications Vol. I.
94 WARNER. Foundations of Differentiable Manifolds and Lie Groups.
95 SHIRYAYEV. Probability, Statistics, and Random Processes.
96 CONWAY. A Course in Functional Analysis.
97 KOBLITZ. Introduction in Elliptic Curves and Modular Forms.
98 BRÖCKER/tom DIECK. Representations of Compact Lie Groups.
99 GROVE/BENSON. Finite Reflection Groups. 2nd ed.
100 BERG/CHRISTENSEN/RESSEL. Harmonic Analysis on Semigroups: Theory of Positive Definite and Related Functions.
101 EDWARDS. Galois Theory.
102 VARADARAJAN. Lie Groups, Lie Algebras and Their Representations.
103 LANG. Complex Analysis. 2nd. ed.
104 DUBROVIN/FOMENKO/NOVIKOV. Modern Geometry — Methods and Applications Vol. II.
105 LANG. $SL_2(\mathbf{R})$.
106 SILVERMAN. The Arithmetic of Elliptic Curves.
107 OLVER. Applications of Lie Groups to Differential Equations.
108 RANGE. Holomorphic Functions and Integral Representations in Several Complex Variables.
109 LEHTO. Univalent Functions and Teichmüller Spaces.